Culture *That* Counts

A Decade of Depth

with the Journal of Mathematics & Culture

Culture That Counts

A Decade of Depth
with the Journal of Mathematics & Culture

Tod Shockey, *Editor*

White Plum Publishing
Galena, Ohio
United States of America
©2019

Culture That Counts - A Decade of Depth with the Journal of Mathematics & Culture

Published by ***White Plum Publishing***,
an imprint of *Skye's The Limit Publishing & Public Relations, LLC*

Books published by Skye's The Limit Publishing & Public Relations, LLC may be available at special discounts for bulk purchases in the United State by corporation, institutions, and other organizations. For more information, please contact the Marketing Department at Skye's The Limit Publishing & Public Relations LLC, P.O. Box 133, Galena, Ohio 43021, (fax) 740-548-4929; or via e-mail at talk2stl@gmail.com

Unless otherwise noted, all images were supplied by the contributors.

Skye's The Limit Publishing & Public Relations
P.O. Box 133, Galena, Ohio 43021
(ph) 740-913-1439
(fax) 740-548-4929
stlpublishing.com

Published in the United States of America, Ohio, Galena.

ISBN-13: 978-1-939044-50-1

Website: CultureThatCounts.blogspot.com

Dedication

There are so many "giants" whose shoulders we have stood upon to make the Journal of Mathematics and Culture happen. In particular, we owe our gratitude to Professor Ubiratan D'Ambrosio. His pioneering work has led to a global engagement in Ethnomathematics. Ubi has brought forth the importance of justice, equity, caring and love for all of us to embrace in our scholarship.

Other giants we want to dedicate these volumes to include: Marcia Ascher, Carlos Aceves, Alan Bishop, Beatrice D'Ambrosio, Tobias Dantzig, Sandy Dawson, Irene Duranczyk, Ron Eglash, Hyman Gabai, Paulus Gerdes, Claudia Giamatti, Gloria Gilmer, George Gheverghese Joseph, Victor Katz, Morris Kline, Lawrence Lesser, Jerry Lipka, Clo Mingo, Swapna Mukhopadhyay, Nirmala Naresh, Otto Neugebauer, Sal Restivo, Patrick Scott, Lawrence Shirely, Susan Staats, Jerome Turner, Claudia Zaslavsky.

This is dedicated to the wonderful individuals that serve and have served on our editorial board.

Finally, for the love and dedication of Rick Silverman, Daniel Orey, Milton Rosa, Veroncia Albanase, Mariel Gavarrete Villaverde, Jim Barta, Iman Chahine and Katherine Shockey.

Love and Peace to each of you.

Acknowledgements

From our discussions all those years ago about a journal that could support our colleagues in Ethnomathematics, to this publication, we have so many to acknowledge. Each member of our editorial board and every author have contributed to our success. The Journal of Mathematics and Culture has been a journey, and we have been so very privileged with all of our wonderful encounters.

Thank you Professor Lawrence Lesser, you offered up a brilliant title for these volumes.

Table of Contents

Volume 1

Volume 2 ◆ Community

Volume 3 ◆ Teaching

Volume 4 ◆ Teacher Prep

Volume 1

Culture *That* Counts

Volume 1

Volume 1

Ten Years Later: A Look Back on the History of the Journal of Mathematics and Culture

Tod Shockey
University of Toledo USA
tod.shockey@utoledo.edu

Fredrick "Rick" Silverman
Northern Colorado University USA
flsilver@gmail.com

Abstract

Ten years and sixty manuscripts later, the *Journal of Mathematics and Culture* JMC celebrates the milestone of a tenth anniversary. To that end, a coding scheme was developed around the emerging themes from publications. This paper brings together a selection of these themes as a strategy to look back at the Journal's history. Beyond the reference section, we include a bibliography of the corpus of published papers. The bibliography is a living document, in that it will be published with abstracts with each future edition of the Journal. While the authors are acknowledged as the "founding" editors for the *Journal of Mathematics and Culture*, the JMC is the accomplishment of many. We conclude with acknowledgements of the individuals that have made this tenth anniversary issue a reality.

Journal of Mathematics and Culture

The journal's contents examine the intersections between mathematics and culture in both western and non-western societies, and among both math professionals (such as university mathematicians, mathematics educators, and cryptologists) and non-professionals (such as carpenters, indigenous healers, and hair stylists). We define culture broadly, to include all scales: ethnic groups, nations, labor communities, religious traditions, professions, and so on. Particular aspects of culture examined might include broad social dynamics such as race and gender, or micro-practices such as dancing or computer programming. Mathematical practices include symbolic systems, spatial designs, practical construction techniques, calculation methods, measurement in time and space, specific ways of reasoning and inferring, ordering, classifying, and other cognitive and material activities which can be translated to formal mathematical representation. Of particular interest are educational studies which take the classroom setting into account, such as pedagogical applications of ethnomathematics.

Keywords:

- Ethnomathematics

Introduction

We don't know if anyone planning for a journal such as this one sitting in the hotel lobby all those years ago would have believed that the JMC would celebrate 10 years, but here we are, indeed! The first decade is history. This has been a wonderful journey connecting ethnomathematics colleagues from around the globe. Today the Journal is truly international, accepting manuscripts in Arabic, English, and Portuguese.

We have been very fortunate with our editorial board, representing many countries and many educational institutions. Through the years members of the board have reviewed manuscripts from six of the world's seven continents. In the year 1977, Professor Ubiratan D'Ambrosio at the Annual Meeting of the Association for the Advancement of Science (AAAS) brought the term "Ethnomathematics" to the audience of his colleagues (Mesquita, Restivo, and D'Ambrosio, 2011). D'Ambrosio's 1985 article elaborated the meaning of this term. In the years since Professor D'Ambrosio brought this term, really, this concept, to light, we have seen shifts in attention to the scholarship identified with Ethnomathematics. The important work of mathematics (through a western view) of cultures continues to be an emphasis. We acknowledge tension exists amongst scholars about the "western view."

> *Writing about the topic of cultural mathematics for readers with backgrounds primarily in Western mathematics brings one to a dilemma: On one hand, using Western terminology and notation to describe mathematics of non-Western cultures is inherently inaccurate because people in such cultures would not think of the mathematical content in the same way as it is perceived in Western culture. On the other hand, if the goal is for people of Western backgrounds to understand how cultural activities can be understood as mathematics, then one must speak to readers in familiar mathematical terms. (Gilsdorf, 2012, p. xii).*

Looking Back

A quick count reveals that JMC has published nearly 60 manuscripts in our history. Included in this count is the focus issue of the International Congress of Ethnomathematics hosted by Dr. Lawrence Shirley. JMC published the abstracts of the ICEm hosted by our late colleague Dr. Paulus Gerdes. Unknowingly, the Journal has "previewed" work that has led to the publications of books, Gilsdorf, 2012; Katsap and Silverman 2016; and Vandendriessche, 2015. A unique characteristic of our editorial board is the opportunity they have to mark the last question on the review form: "If you are not recommending "Accept," would you be willing to work with the author(s) to get this manuscript into publishable form?" Checking "yes" on this space, our editorial board members have supported and mentored many of our colleagues. This academic altruism has led to a number of publications in the JMC that may otherwise have never been published nor certainly even read. We have been pleased to support junior colleagues, graduate students, and our senior colleagues, all who have chosen the Journal as an outlet for their important work.

NOTES

Anniversary Edition Motivations

A motivation for this anniversary issue was forward thinking, an opportunity for our authors to reconsider their papers and reflect back. We encouraged the reflection to be inclusive of pedagogical considerations. We are told that ethnomathematics is part of the Brazilian national curriculum. In the United States we are aware of ethnomathematics courses occurring in higher education. We felt it was important for a moment of reflection to consider connecting our scholarship to our classrooms.

A second motivation for this edition is for us as a community to continue to look forward. Of late, the pioneering work of Ethnomodelling (Bassenezi, 2001; Rosa & Orey, 2013) has established a firm agenda for the Ethnomathematics community. New questions arise and expanding views of the scholarship are emerging as pioneering work from Brazil and the United States becomes available to our global community.

We acknowledge and dedicate this anniversary edition to those that have allowed us to stand on their shoulders to see further: Ubiratan D'Ambrosio, Marcia Ascher, Beatriz D'Ambrosio, Sandy Dawson, Paulus Gerdes, Rex Matang, and Claudia Zaslavsky.

As the *Journal of Mathematics and Culture* continues we look forward to expanding our role to be inclusive and the representative of many global voices. In the words of our dearly departed friend and colleague Beatriz D'Ambrosio (personal communication, 2015):

> *The Journal of Mathematics and Culture has become an important outlet for work in ethnomathematics, mathematics for social justice, and political dimensions of mathematics teaching and learning. While still a relatively young journal, the high quality of the articles that have appeared to date are an indication that this journal will soon be ranked in the top tier in the country. The co-editors are responsible for recruiting authors of the highest quality...This initiative has made a huge impact on the field and opened a new outlet for researchers striving to participate in the knowledge production in mathematics education... This has been an enormous service to the field, while also being a significant contribution to the production of knowledge in mathematics education.*

Emerging Themes

An emerging ethnomathematical theme through the years has been "calendar." Bishop (1991) introduced six cultural activities that have served as a theoretical framework for many scholars working in Ethnomathematics, (see Figure 1).

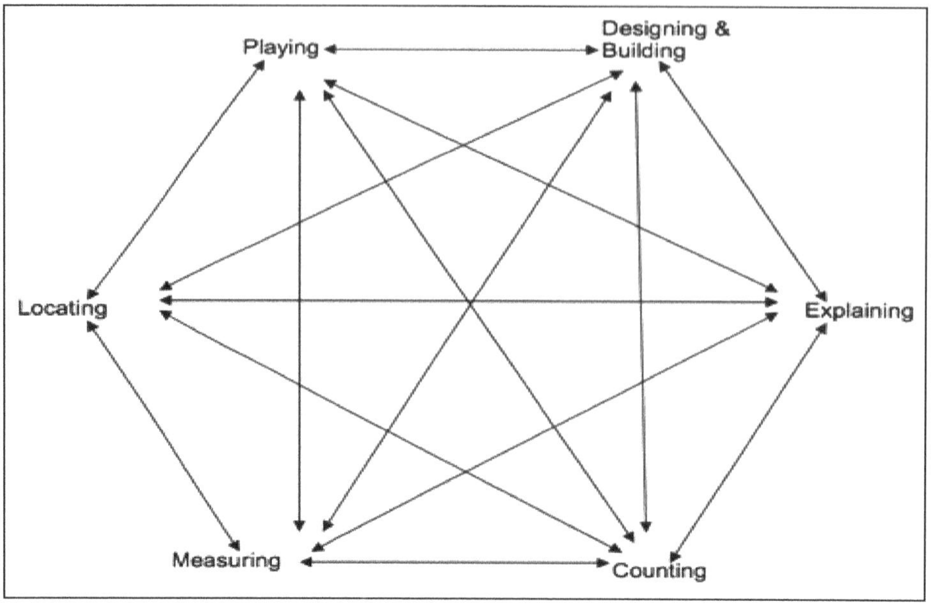

Figure 1. Bishop's (1991) Six, Adapted by Shockey, Mitchell and Barta

These universal, every day, naturally occurring activities are interrelated, which is clear from Bishop's (1991) work. When Shockey, Mitchell, and Barta considered Bishop's (1991) framework, they highlighted that these activities are not mutually exclusive. In the ten years of JMC, authors have brought the concept of calendar (Bjarnadottir, 2010; Sharp, 2015) to the forefront. To that end, we are reconsidering the rendering of Figure 1 to "grow" from a hexagonal representation to a heptagonal representation inclusive of calendar, (see Figure 2). We must consider whether calendrical activities fall sufficiently within the domain of naturally occurring measuring behavior, or whether at their root they are more sophisticated than the existing behaviors within Bishop's (1991) work.

NOTES

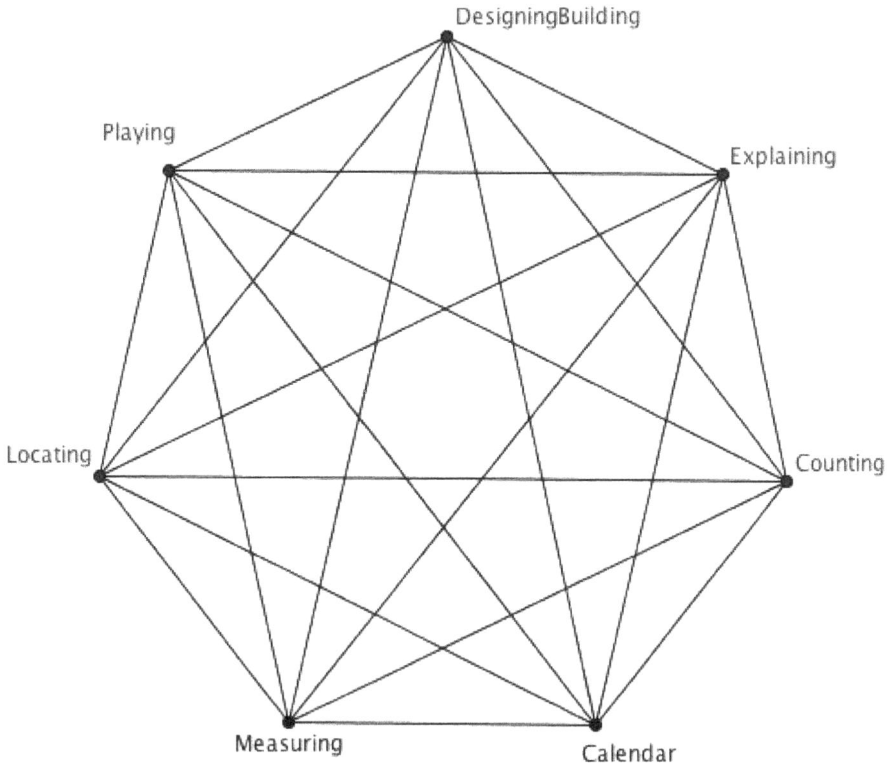

Figure 2.

We are suggesting that it may be worth considering Calendar as a stand alone activity as many activities have a time of year associated with them.

Fossa (2006) stated that: "Ethnomathematics is generally concerned with the resolution of practical problems" (p. 32). Practical problems or connections to everyday life have emerged as a theme through our history. Staats (2006) discussed her research on Ethnomathematics case studies for her students to "develop subjective, values based motivation to study mathematics" (p. 39). Palhares (2006) focused our attention on his findings from an after school program where "students worked on projects that connected mathematics to their lives" (p. 12). These connections explored through an ethnomathematical lens are found in the African drumming work of Sharp and Stevens (2007).

As many of our authors have emphasized connections, which is a point we attempt to illustrate in Figures 1 and 2 that each of the categories is connected to the others. In the teaching of a course in ethnomathematics, many of us have worked with our students to bring out this interconnectivity. Another component of connections has to do with linking research to the classroom through motivations of being culturally responsive (Bonner, (2010); Barta & Shockey (2006); Diez-Palomar, et. al (2007); Ealey & Henzel (2012); Engblom-Bradley (2006); Harding-DeKam (2007); Katsap & Silverman (2008); Massarwe, et. al (2010)).

Another theme that has become prevalent is the "voices" we "hear" as readers of work that is occurring globally. Owens (2012) brings the voices of New Guinea to our readership, we hear Native American and First Nations voices in the works of Barta and Shockey (2006), Galindo et. al (2009), Engblom-Bradley (2006), and Ealey and Henzel (2012). Chahine and Kinuthia (2013) bring a Zulu voice to our pages, Adams (2012) brings the voice of weaving, Goetzfridt (2012) brings us the voices of the Caroline Islands of Micronesia, and Naresh (2012) shares the voices of India bus conductors.

The stories and engagement that JMC has been privileged to be a part of are very rich. We hope that such stories and research continue to be shared through the *Journal of Mathematics and Culture* for many more years.

Acknowledgements

We acknowledge all of our important reviewers from our origins through today. This journey would not have been possible without each of them.

References

Bassenezi, R. C. (2002). Ensino-Aprendiqagem com modelagem matematica [Teaching and learning with mathematical modelling] Sao Paulo: Editora Contexto.

Bishop, A. J. (1991). *Mathematical enculturation.* Dordrecht, The Netherlands: Kluwer Academic Publishers

D'Ambrosio, U. (1985). Ethnomathematics and its place in the history and pedagogy of mathematics. *For the Learning of Mathematics, 5*(1), 44-48.

Gilsdorf, T. E. (2012). *Introduction to cultural mathematics*: with case studies in the Otomies and Incas. Hoboken, NJ: John Wiley & Sons Inc.

Mesquita, M., Restivo, & S., D'Ambrosio, U. (2011). *Asphalt children and city streets: A life, a city, and a case study of history, culture, and ethnomathematics in São Paulo*. Rotterdam, Holland: Sense Publishers.

Rosa, M., & Orey, D. (2013). Ethnomodelling as a methodology for ethnomathematics. In G. A. Stillman, K. Gabriele, W. Blum & J. P. Brown (Eds.), *Teaching mathematical modelling*: connecting to research and practice (pp. 77-88). Dordrecht, The Netherlands: Springer.

NOTES

Adam, N. (2012). "Weaving mathematics and culture: Mutual interrogation as a methodological approach." Journal of Mathematics and Culture **6**(1): 88-103.

This paper describes an adaptation of a methodology called 'mutual interrogation' to a study on Malay weaving. The interrogation was focussed on the weaving technique, framework construction and pattern formation. Mutual interrogation is a process of implementing a critical dialogue between two systems of knowledge. Using this approach, a three-cycle dialogue between several weavers and mathematicians was implemented, with the researcher playing the role of a mediator. The interactions between the concepts of the weavers and the conventions of the mathematicians have uncovered several interesting perspectives, which consequently dispute several critics of ethnomathematical research.

Adams, T. and G. Harrell (2010). "A study of estimation by professionals at work." Journal of Mathematics and Culture **5**(1): 1-15.

Estimation is a learned skill that supports learning and applications in measurement as well as other quantitative contexts. Estimation is an important skill to learn in that it provides a foundation for measurement – it helps one determine the most appropriate tools for measuring specific items. Estimation is equally important as it helps one determine whether or not an actual measurement obtained is reasonable for a given context. It can be beneficial for learners to explore the application of estimation so that they have context and motivation for learning to estimate. The authors engaged with selected professionals during the course of their work to obtain information about when they use estimation, how they use estimation, how they determine if estimation is helpful and when they decide not to use estimation. The authors share estimation activities of the professionals, which result in "estimation declarations" and suggestions for the teaching of estimation.

Albanese, V. and F. Perales (2015). "Enculturation with ethnomathematical microprojects: From culture to mathematics." Journal of Mathematics and Culture **9**(1): 1-11.

Ethnomathematics is a research program that focuses on the relationships between mathematics and culture. What impact does this vision have on mathematics education? We approach this question from two points of view that require: 1) some substantial methodological changes and 2) some changes on how to approach the content. We wish to focus here on teachers training in primary education in Argentina. We present a way to integrate methodology and content: the Microproject and we report an experience with Argentinian pre-service teachers.

Barta, J. and T. Shockey (2006). "The mathematical ways of an Aboriginal people: The Northern Ute." Journal of Mathematics and Culture **1**(1): 79-89.

Bjarnadottir, K. (2010). "Ethnomathematics at the margins of Europe: A Pagan calendar." Journal of Mathematics and Culture **5**(1): 21-42.

In 930, at the close of the settlement period in Iceland, a week-based calendar was adopted. Observations of the solar cycle soon revealed errors of the calendar, which were cleverly amended. In the 12th century, the week-based misseri calendar was adjusted to the Roman calendar used by the Christian Church. It remained in common use for secular purposes until the 19th century, and detailed guides to it were written. Special occasions related to it are still celebrated.

Bolanos, J. and M. Oliveras (2012). "Etnomatematicas y pintaderas canarias." Journal of Mathematics and Culture **6**(1): 104-118.

The ancient culture of the island of Great Canary relegates an archaeological pieces of high symbolic value for the islander population, known like "pintaderas canarias". His designs, entirely geometric, lead us to the search of a method that allows us to interpret them from a mathematical point of view for his later introduction in the education as a real cultural resource.

Bonner, E. (2010). "Promoting culturally responsive teaching through action research in mathematics methods course." Journal of Mathematics and Culture **5**(1): 16-33.

Culturally responsive pedagogy is one way that teacher educators and teachers can address the unique sociocultural and cognitive needs of learners from diverse backgrounds while shifting away from traditional school practices and mathematics teaching methods. This article describes a teacher educator's inquiry project that aimed to encourage and enhance culturally responsive thoughts, discussions, understandings, and actions among pre-service mathematics teachers in a mathematics methods course for elementary school teachers. Results show that in constructing meaningful in and out of class experiences that focus explicitly on culture, pre-service teacher understandings of culturally responsive mathematics teaching are enhanced and deepened.

Castagnolo, A. (2012). "La etnomatematica subyacente en los textiles." Journal of Mathematics and Culture **6**(1): 119-134.

Chahine, I. and W. Kinuthia (2013). "Juxtaposing form, function, and social symbolism: An ethnomathematical analysis of indigenous technologies in the Zulu culture." Journal of Mathematics and Culture 7(1): 1-30.

The purpose of this paper is to highlight the role of ethnomathematics as a model for translating and interpreting mathematical structures inherent in existing indigenous technologies used by South African indigenous people as means of survival and adaptation in a variety of environments. The present study focuses on the riches of two indigenous technologies as media for transmitting specialized mathematical knowledge produced and disseminated by the Zulu culture, namely beadwork and basketry. We argue that prospective work that investigates other indigenous technologies with underlying ethnomathematical ideas can be highly enlightening and reflect major breakthroughs in the field of indigenous mathematical knowledges.

D'Ambrosio, U. (2006). "The program ethnomathematics: A theoretical basis of the dynamics of intra-cultural encounters." Journal of Mathematics and Culture **1**(1): 1-7.

D'Ambrosio, U. (2012). "An ethnomathematics view of space occupation and urban culture." Journal of Mathematics and Culture **6**(1): 1-19.

Diez-Palomar, J., et al. (2007). ""Math is Everywhere": Connecting mathematics to students' lives." Journal of Mathematics and Culture **1**(2): 20-36.

This article describes an experience of teaching and learning mathematics in an after- school program based in a southwestern city, in which the students worked on projects that connected mathematics to their lives. In the project that we describe, the students discussed and solved problems about their school, their neighborhood, and immigration from a critical perspective. The pedagogical approach of our "Math Club" is grounded in socio-cultural approaches, from a critical point of view. This experience highlights the connections between mathematics and everyday life that are brought out when emphasizing culture, language, and dialogue among mathematics learners. Implications are made for drawing on student agency through mathematics learning.

Duranczyk, I. and J. Higbee (2012). "Constructs of integrated multicultural instructional design for undergraduate mathematics thinking courses for nonmathematics majors." Journal of Mathematics and Culture **6**(1): 148-177.

This paper introduces a new inclusive theoretical model, integrated multicultural instructional design (IMID), and discusses its potential for application within undergraduate mathematical thinking courses for nonmathematics majors. The guiding principles of IMID build upon the constructs of multicultural education, ethnomathematics, and universal instructional design (UID), a model for inclusion of students with disabilities in higher education. The paper addresses how IMID is compatible with research and practices promoted by international and national mathematics organizations by maximizing equity, access, and success for all learners. The international goal of increased mathematical literacy necessary for all to achieve effective global citizenship and participation is addressed.

Ealey, K. and C. Henzel (2012). "From "What Ought" to "What Is" infusion of First Nations, Metis and Inuit perspectives in mathematics: First steps on a journey in high school mathematics." Journal of Mathematics and Culture **6**(1): 178-190.

Academic and educational success for all students has become an international focus. In Alberta, Canada, there is concerted effort to increase high school completion rates, specifically within First Nations, Métis and Inuit (FNMI) populations. To this end, FNMI content and perspectives are being infused into all curricular areas to ensure the educational engagement of FNMI students and others. This paper traces the journey of the contextualized infusion of FNMI content and perspectives into the resources produced to support the Common Curriculum Framework for Grades 10—12 Mathematics, Western and Northern Canadian Protocol, January 2008

Elena, C. (2012). "Elementos geometricos del pueblo originario Mapuche." Journal of Mathematics and Culture **6**(1): 135-147.

Engblom-Bradley, C. (2006). "Learning the Yup'ik way of navigation: Studying time, position, and direction." Journal of Mathematics and Culture **1**(1): 90-126.

This paper is about the use of mathematics in Yup'ik navigation strategies, as practiced by Fred George of Akiachak, Alaska. Fred George travels by snow machine over snow covered frozen lakes and tundra in the Yukon-Kuskokwim Delta. In day light, he uses the position of the sun and time of day to determine his direction. On clear nights, he uses the position of the Big Dipper and time of night to determine his direction. In addition, he observes the frozen grass, isolated trees, and/or snow waves to reinforce his direction. The sun, Big Dipper, frozen grass, isolated trees, and snow waves function as natural compasses for Fred George.

Originally mentored by his father when he was a boy, Fred has continued to develop his navigational skills on the tundra for over 60 years to hunt and fish for his family of eight children and many grandchildren. He passionately wants to pass his navigational skills on to the young people in Akiachak. He knows young people are no longer being mentored by their families to navigate. Yet many young people drive snow machines, and many become lost on the tundra.

To help Fred George with this mission, this paper was written for the teachers in Akiachak and other native villages. The paper contains descriptions and explanations of the mathematics in navigation practices used by Fred George, plus appropriate classroom activities, which facilitate student understanding of natural compasses, available on the tundra. This valuable knowledge will help young people find their direction when they venture out on a snow machine over frozen lakes and snow covered tundra. In addition, the connection of mathematics to navigation and subsistence supports the cultural knowledge of Yup'ik students, and will contribute to their successful transition into adulthood throughout West Central Alaska.

Volume 1

Fossa, J. (2006). "Ethnomathematics and cooperativism." Journal of Mathematics and Culture **1**(1): 32-38.

Ethnomathematics is generally concerned with the resolution of practical problems and thus it is natural to investigate how this new science can be combined with cooperativism in order to foster sustained economic development. It is maintained that one way of developing this kind of investigation would include some business mathematics, informal logic and practical ethics. A project, implemented in a rural area of the Brazilian Northeast, is described in terms of both research into the community's ethnomathematical knowledge as well as pedagogical situations in which this knowledge and academic knowledge are taken into account.

Francois, K. (2012). "Ethnomathematics in a European context: Towards an enriched meaning of ethnomathematics." Journal of Mathematics and Culture **6**(1): 191-208.

This paper considers the field of enquiry called ethnomathematics and its role within mathematics education. I elaborate on the shifted meaning of 'ethnomathematics'. This "enriched meaning" impacts on the philosophy of mathematics education. Currently, the concept is no longer reserved for the so called 'nonliterate' people, but also includes diverse mathematical practices even within Western European classrooms. Consequently, mathematics teachers are challenged to handle people's cultural diversity occurring within every classroom setting. Ethnomathematics has clearly gained a prominent role, within Western European curricula, becoming meaningful in the exploration of various aspects of mathematical literacy. I discuss this enriched meaning of ethnomathematics as an alternative, implicit philosophy of school mathematical practices.

Furato, L. (2010). "Public good or private commodity? Mathematics education in Japan and implications for the U.S." Journal of Mathematics and Culture **5**(1): 34-64.

Given the flow of global policy and educational reform, a number of countries are approaching junctures on issues of equal access and opportunity. This study begins with the major role supplemental mathematics education plays in Japan. Research was conducted as a visiting scholar at the University of Tōkyō, and supported by classroom experience. The contextual background then leads to a discussion on sector relationships in the U.S., where a culture has emerged of relying on privatized education such as academic tutoring and test preparation. Although educators in both countries question the impact of growing sector disparities, the reality is that they have just begun to address issues of equity and accessibility to ensure education is provided as a public good.

Furato, L. (2013). "Bridging policy and practice with ethnomathematics." Journal of Mathematics and Culture **7**(1): 31-57.

In an effort to bridge policy and practice in diverse populations, research was conducted at U.S. higher educational institutions (Harvard, UCLA, University of Hawai'i) in the field of ethnomathematics. Ethnomathematics addresses first, how cultural values affect teaching, learning, and curriculum; and second, how mathematics education affects schooling process dynamics. A macroperspective provides a foundation for exploring the policy framework that will be bridged by ethnomathematics to explore practices at the school level (equity and quality education, building successful partnerships, mentoring in diverse populations). This has been accomplished primarily through the Hōkūle'a, a voyaging canoe internationally renowned for rekindling the Pacific tradition of celestial navigation to locations such as Tahiti, Rarotonga, Micronesia, Japan, and the U.S., and is preparing for a worldwide voyage in 2013 of which the author will be participating in. It is a vehicle to explore real-world mathematics applications in global communities, and represents resourcefulness, inventiveness, wisdom grounded in the past, and hope for the future.

Galindo, E., et al. (2009). "Compassion: A hearts-on paradigm for training Native American students into a STEM university environment." Journal of Mathematics and Culture **4**(1): 1-22.

While many studies report dismal American Indian/Alaska Native dropout rates, this paper provides insights from American Indian/Alaska Native teachers, Shoshone-Bannock Tribal members, and University staff on a successful education paradigm for Shoshone-Bannock students. Our students are finding success in projects that join traditional knowledge and respect of the Shoshone-Bannock people with scientific guidance from a local university. Our goal is to strengthen the links between science and math programs at the Shoshone-Bannock High School and Idaho State University, and to use this relationship to draw students from secondary schools into the university setting. The newly established Portneuf River Ecosystem Project (PREP) is but one example of a research focus for the component of this partnership. In 1998, two Shoshone-Bannock students were interested in higher education. In the years (2003, 2004), ten students applied to institutions of higher education. A question was asked by an Elder from the Shoshone-Bannock Indian Reservation located in Southern Idaho, "Of all the things you are teaching in science and math, are you teaching the most important subject of all...compassion for all that live on and with Mother Earth?"

Gavarrete, M. and M. Oliveras (2012). "Matematicas, culturas y formacion de profesores en Costa Rica." Journal of Mathematics and Culture **6**(1): 209-222.

This document relates to the role that should have the Ethnomathematics in teacher training in Costa Rica. Starting with a brief overview of some aspects multicultural social de Costa Rica, we discussed the importance of introducing in the training of teachers of mathematics knowledge implicit in the living cultures of this country. We show some conclusions of the consultation which took place with a group of specialist about Ethnomathematics and teacher training of Costa Rican as part of the Final Master Work presented at the Department of Mathematics Education at the University of Granada, Spain, in September 2009.

Gilsdorf, T. (2009). "Mathematics of the Hnahnu: The Otomies." Journal of Mathematics and Culture **4**(1): 84-105.

This article is about the mathematics of the cultural group known as the Otomies, of present day central Mexico. In particular, I discuss the Otomi number system and a comparison of that system with Aztec counting, Otomi art and decoration, mathematical symbols that appear in some Mesoamerican codices, and the Otomi calendar.

Goetzfridt, N. (2012). "Pacific ethnomathematics: The richness of environment and practice." Journal of Mathematics and Culture **6**(1): 223-252.

Contrasting geographic features that are evident between atolls and islands in Micronesia and in large Pacific land masses, particularly those of Papua New Guinea, have historically compelled indigenous peoples to apply distinctive linear systems related to distance, place, and ancestral origins. In the Caroline Islands of Micronesia – particularly on the islands of Puluwat and Satawal – traditional land finding techniques have applied linear approaches to the rising and setting of stars as they relate to specific sea routes between home islands and target islands. While linear concepts used in the culturally nuanced, earth grounded sand drawings in the Melanesian nation of Vanuatu are unique in their own right, linear based mnemonic systems for the recollection of ancestral origins and prestige granting cultural knowledge at sea in the Caroline Island and on the ground in Melanesia have interesting similarities. A greater academic concentration on the rich ethnomathematics of the Pacific could yield a trove of revelations in comparative cultural practices.

Harding-DeKam, J. (2007). "Foundations in ethnomathematics for prospective elementary teachers." Journal of Mathematics and Culture **1**(2): 1-19.

Ethnomathematics is a term that has been coined to elucidate that everyone uses and can learn mathematics. This manuscript demonstrates how a college Mathematics Methods Course at a Doctoral Intensive University in the Rocky Mountain Region can implement the idea of using the culturally diverse background of students as a foundation to teach children learning second languages and children of diverse cultures. This includes one year of quantitative and qualitative research about prospective elementary teachers who were instructed on how to teach children learning second languages during their Mathematics Methods Course. These prospective teachers then applied these methods while completing their student teaching.

Hare, A. (2010). "International service-learning for preservice teachers: Strengthening mathematics literacy in West Africa." Journal of Mathematics and Culture **5**(1): 65-79.

This article describes the benefits of service-learning for undergraduate preservice teachers, specifically through the example of an international partnership between education students at a liberal arts college in central Pennsylvania and primary school teachers of children with disabilities in the country of Burkina Faso, West Africa. Four undergraduate college students studying mathematics, education, statistics, and art wrote and illustrated a children's counting book that is culturally appropriate in a French West African context. Through this project, the student team learned firsthand about issues of language, culture, mathematics, literacy, and special education. The partnership, which has since expanded in scope and involvement to school library development and robotics, is rich in connections to diverse learning environments and provides meaningful resources and motivation for preservice teachers on multiple levels.

Herron, J. and J. Barta (2009). "Culturally relevant word problems in second grade: What are the effects?" Journal of Mathematics and Culture **4**(1): 23-49.

In this study researchers investigated the effects of the use culturally relevant word problems on second graders' mathematics achievement. The participating elementary schools were in one of three groups: control, word problems, or culturally relevant word problems. A pre- and post-test design measured the effectiveness of the interventions. The two intervention schools taught problem solving for 15 minutes twice a week. The culturally relevant word problem group used modified problems based upon a home survey while the word problem group used textbook problems. Findings from this study suggest further research investigating the use of culturally relevant problem solving instruction is warranted.

Hirsch-Dubin, F. (2009). "Maguey elders, Mayan mathematics, and the weaving of resistance in Maguey bag production." Journal of Mathematics and Culture **4**(1): 63-83.

This article documents maguey bag production by two Mayan elders and presents an emergent analysis of the Mayan ethnomathematics involved in the procedure. Maguey bags are created using the vigesimal system, which provides continuity to a process transmitted orally and by example since the time of the ancient Maya. This method, handed down by Mayan elders to their grandsons, offers a glimpse into the role of mathematics in strengthening the identity and political resistance in autonomous rural communities in Chiapas, Mexico.

Hoyles, C. (2010). "Creating an inclusive culture in mathematics through subject-specific teacher professional development: A case study from Englan." Journal of Mathematics and Culture **5**(1): 43-61.

Katsap, A. and F. Silverman (2008). "A case study of the role of ethnomathematics among teacher education students from highly diverse cultural backgrounds." Journal of Mathematics and Culture **3**(1): 66-102.

Mathematics prospective teachers (MPT), Jewish and Bedouin alike, who came together in the Ethnomathematics Program (EP) in the Kaye College of Education (Israel), explored, identified, and learned naturally occurring mathematics in the cultures of their own people. They developed projects based upon their explorations and shared them in class. The MPT participated in collaborative discussions and formed intercultural understanding within a purposefully conducive classroom atmosphere. In particular, they shared with one another classroom activities for their classmates, plans for transferring their accomplishments to school children, ethnomathematics activities derived directly from cultural experiences, as well as mathematics content and pedagogy. This case study investigated and documented the intentional integration of two people's cultures, traditions, and their ethnomathematics, on the one hand, with mathematics content, history, and pedagogy, on the other. The MPT learned ethnomathematics in the cultures of their own people and came to perceive mathematical concepts as a mix of two significances: one meaning stemming from their abstract courses in academic mathematics, a focus preeminent in western culture, and the second, stemming from comprehension of practical math concepts rooted for both Jews and Arabs in their cultures of the East. The following perception arose during class debates, discussions, and sharing of projects: that there is some common religious knowledge (e. g., that of the number seven in passages in the Torah and in the Koran) and common ritual (e. g., Jewish and Muslim religious women must cover their heads) in both lines of the descendants of Abraham. However, the big take-away point is that just an exposure to mathematics rooted in socio-cultural settings of their own people was not sufficient to have altered the standpoints of the class members toward the way in which mathematical concepts should be seen. Inquiry into situations where mathematical practice actually occurs was necessary.

Leonard, M. and C. Shakiban (2010). "The Incan abacus: A curious counting device." Journal of Mathematics and Culture **5**(1): 81-106.

Little is known about the ancient calculating device of the Incans la yupana. It was a tablet upon which stones, grains, or beans were placed and manipulated to perform calculations. This paper critically examines the little information available on la yupana and proposes a new interpretation, taking into consideration both mathematical and cultural factors. The conclusion reached is that la yupana is an abacus that directly correlates with the quipus, performs addition, subtraction, multiplication, and division in base ten, can be used to decompose numbers, and reflects linguistic principles of the Aymara language. A better understanding of the Incan abacus could lead to further insights about the civilization.

Lesser, L. (2006). "Book of numbers: Exploring Jewish mathematics and culture at a Jewish high school." Journal of Mathematics and Culture **1**(1): 8-31.

Leyton, R. (2012). "Artisticas geometricas." Journal of Mathematics and Culture **6**(1): 333-343.

Massarwe, K., et al. (2010). "An ethnomathematics exercise in analyzing and constructing ornaments in a geometry class." Journal of Mathematics and Culture **5**(1): 1-20.

This paper presents two case studies that examine an approach to teaching geometry through an ethnomathematics exercise in analysis and construction of culturally meaningful ornaments. The exercise was given to students from the Arab sector high schools in Israel. The studies indicated that the students perceived the practice of constructing geometrical ornaments and discovery of their mathematical properties as a meaningful and enjoyable learning experience. This experience inspired emotions, lively discourse, and learning motivation. It arose into a geometrical and socio-cultural inquiry, reflecting the students' thirst for practical use of the acquired mathematical knowledge and their awareness of cultural identity.

Massarwe, K., et al. (2012). "Ethnomathematics and multi-cultural education: Analysis and construction of geometric ornaments." Journal of Mathematics and Culture **6**(1): 344-360.

This paper reports our study which was conducted in Israel in the framework of the ongoing project "Joymetry - Learning Geometry and Culture through Joyful Activity of Analysis and Construction of Ornaments". The project aims to raise students' and teachers' awareness of the interplay between geometry and culture as a lever for geometry and multi- cultural education, through the engagement of secondary school students, and teachers in ethnomathematical activities rooted in their own and other cultures. We present the core of the project, a new course "Issues in Ethnomathematics: Teaching Geometry in Socio-Cultural Context" that we have delivered to pre-service and in-service teachers at the Technion annually, since 2007/08 academic year. The course is given and is directed towards developing competence of teaching multi-cultural groups. It culminates in a workshop, where each student teaches analysis and construction of ornaments from a specific culture to a diverse group of Jewish and Arab school pupils. Our study indicated that the course helped the students to significantly enhance their content and pedagogical content knowledge in geometry, and to understand the importance of ethnomathematical learning activities related to the pupils' own and other cultures.

Meaney, T., et al. (2008). "The role of language in ethnomathematics." Journal of Mathematics and Culture **3**(1): 52-65.

Drawing on experiences from Mäori immersion schools in New Zealand, this paper examines the impact of language when ethnomathematical practices are discussed. The inclusion of ethnomathematical perspectives into the mathematics education of indigenous students is often described as being beneficial. In New Zealand, it is also perceived as supporting the revitalisation of the Mäori language, te reo Mäori. This paper will argue that cultural practices including ethnomathematical ones cannot be separated from the language in which they were developed. Changing the language or the linguistic register in which the practices are discussed will have an impact on how the practices are perceived by students. This could result in a loss in the fundamental values that would normally accompany the practices. Without proper consideration of this issue many of the benefits aligned with using these practices may be nullified.

Miarka, R. (2012). "The role of mathematics within ethomathematics descriptions." Journal of Mathematics and Culture **6**(1): 296-307.

This paper aims to discuss the description of a practice within ethnomathematics from a contemporaneous point of view, without separating describer from a practice described. For that, instead of looking for mathematics in a practice, we defend the idea of describing a practice mathematically. Besides that, we point out what would be the ethnomathematics reasercher role within this vision, and a methodological path based on Alangui's Mutual Interrogation methodology.

Naresh, N. (2012). "Bus conductors' use of mental computation in everyday settings - is it their ethnomathematics?" Journal of Mathematics and Culture **6**(1): 308-332.

Although over the past 15 years, mathematics education research has begun to explore the nature of the mathematics used in different workplaces, the research field of workplace mathematics is still in its infancy. Guided by the desire to add to the mathematics education research in India, the general aim of this study is to develop a better understanding of the mathematics used in everyday situations. To this end, I focused on the workplace mathematics of bus conductors in Chennai, India. More specifically, the purpose of this study is to observe, understand, analyze, and describe the mental mathematical practices of the bus conductors in their workplace and examine what this knowledge can add to the study of everyday mathematics.

Noblitt, B. and B. Richter (2013). "Using Vedic mathematics to make sense of the finger algorithm." Journal of Mathematics and Culture **7**(1): 58-73.

Vedic mathematics, based on sutras from ancient Hindu texts, was rediscovered in the early twentieth century by Bharati Krishna Tirthaji (Stemn & Collins, 2001). Vedic mathematics deals mainly with mentally carrying out tedious arithmetical operations. While studying a particular Vedic algorithm and its connections with standard arithmetic operations, the authors discovered connections to a nonstandard multiplication algorithm, the finger algorithm, found in a mathematics textbook for preservice elementary teachers. These connections provided justification for why the finger algorithm works. This paper discusses a justification of the Vedic algorithm and how it led to justification of the finger algorithm.

Orey, D. and M. Rosa (2006). "Ethnomathematics: Cultural assertions and challenges towards pedagogical actions." Journal of Mathematics and Culture **1**(1): 57-78.

In this article, the authors discuss ethnomathematical concepts that were found useful in the interpretation of assertions and challenges faced by the Ethnomathematics Program for teacher professional development, as described by D'Ambrosio. The authors share the student's teaching-learning process, and for mathematics education in our ongoing work in Brazil and the United States. The discussion begins with a brief examination of where ethnomathematics stands today in respect to the process of teaching and learning of mathematics.

Owens, K. (2012). "Papua New Guinea indigenous knowledges about mathematical concepts." Journal of Mathematics and Culture **6**(1): 20-50.

This paper presents some aspects of Indigenous mathematics in Papua New Guinea in order to establish its value for Indigenous people and their education. It also points out the value of research into Indigenous mathematics to strengthen mathematics education in general. It provides richer perspectives and understandings of mathematical relationships and language. The paper complements previous work undertaken on counting systems based on work by Glen Lean (Glen Lean Ethnomathematics Centre University of Goroka, 2003; Owens, 2000, 2001) and focusses on recent research in the space and geometry and measurement aspects of Papua New Guinea Indigenous mathematics.

Owens, K. (2012). "Policy and practices: Indigenous voices in education." Journal of Mathematics and Culture **6**(1): 51-75.

This paper presents some policies and practices that occur in Papua New Guinea and Australia. It is argued that school systems must recognise cultural mathematics and teachers need to establish dialogue with communities, especially Elders, in order to provide a socially just education for Indigenous students. Several recent policy recommendations are provided as examples but there is still much more to be done to provide adequately for Indigenous mathematics education. One aspect needing development is the making of links between culture and school mathematics in order to maintain and develop a richer mathematics. Several examples are provided.

Padron, O. (2012). "Una experiencia de capacitacion en etnomatematica, en docentes indigenas Venezolanos." Journal of Mathematics and Culture **6**(1): 286-295.

En este documento se abordan detalles referidos a la capacitación en Etnomatemática que se concretó con la construcción de Proyectos pensados como materiales formativos y derivados del contexto donde los participantes del curso ejecutan su acción docente. En los Proyectos elaborados se exploró la Matemática usada fuera de la escuela en función de la construcción y uso de objetos ancestrales de su cultura, encontrándose que allí subyacen abundantes contenidos matemáticos que sirven de insumo para la concreción de actividades escolares sustentadas en tales contenidos. Se concluye que es importante que estos docentes no solo reconozcan la diversidad cultural correspondiente a su comunidad sino que valoren, analicen y promuevan la construcción de Proyectos que emerjan de las prácticas matemáticas desarrolladas por los diferentes grupos culturales, tomando en cuenta aspectos tales como los sistemas simbólicos involucrados, los métodos y sistemas de conteo, cálculo y medición, las formas especificas de razonamiento e inferencia y otros referentes que pueden traducirse a representaciones de la Matemática formal.

Palhares, P. and F. Sousa (2015). "A ethnomatematica na comunidade piscatoria de camara de lobos em Portugal." Journal of Mathematics and Culture **9**(1): 12-29.

The mathematics education research community has been discussing the issue of the connections between mathematics and culture. This is the main concern underneath this work, as it tries to reveal everyday life mathematics of a fishing community in Câmara de Lobos, Madeira, Portugal. The study followed a qualitative approach based on an ethnographic model. In order to collect data the researcher remained during long periods of time in the field, using participant observation, interviews, photographic and video registers and document collection. Data analysis was of an interpretative nature. We focus here on the mathematical notions of ratio and fraction and how they are used within the community, especially by one caulker living there.

Pinto, C. (2012). "Apprendiendo geometria en ambientes interculturales: El caso de escolares crillos y tejedores warao (Venezuela)." Journal of Mathematics and Culture **6**(1): 266-285.

Ramirez, O., et al. (2015). "The quinceanera event: Pre-service teachers implementing a culturally relevant math activity in a Hispanic community." Journal of Mathematics and Culture **9**(1): 52-80.

Preparing pre-service teachers (PSTs) to work with culturally relevant mathematics and family math learning events (FMLEs) in mutual settings are important goals. This paper promotes these important teacher preparation goals as implemented in an after school program by a university math content faculty along with other FMLE support faculty. To illustrate, this paper focuses on a culturally relevant mathematics activity based on the popular social event known as the Quinceañera (the 15th birthday party for a young lady) presented at FMLEs by PSTs and which explores the customs of the Hispanic culture coupled with linkages to mathematical connections.

Rauff, J. (2009). "Native American dice games and discrete probability." Journal of Mathematics and Culture **4**(1): 50-62.

Dice games were a common feature of Native American culture at the turn of the century. This essay reports on my examination of 40 Native American dice games described by James Culin (1992) in his two-volume 1907 work on Native American games from the point of view of discrete probability theory. I provide some initial answers to the question of whether or not the point values assigned to the outcomes of these games reflect the a priori probabilities of the outcomes.

Rocche, D. (2012). "Conocimientos y saberes matematics en la cultura Maya." Journal of Mathematics and Culture **6**(1): 361-367.

Rojas, I. (2012). "Conocimientos matematicos de la mesoamerica precolumbina." Journal of Mathematics and Culture **6**(1): 253-265.

Rosa, M. and D. Orey (2013). "Culturally relevant pedagogy as an ethnomathematical approach." Journal of Mathematics and Culture **7**(1): 74-97.

The implementation of a culturally relevant pedagogy in the mathematics curriculum helps to develop students' intellectual, social, emotional, and political learning by using their own cultural referents to impart their previous knowledge, skills, and attitudes. A culturally relevant pedagogy provides ways for students to maintain their identity while succeeding academically. In this context, there is a need to examine the embeddedness of mathematics in culture by drawing from an ethnomathematical perspective that takes on the cultural nature of knowledge production into the mathematics curriculum. Ethnomathematics and culturally relevant pedagogy-based approaches to mathematics curriculum are intended to make school mathematics relevant and meaningful as well to promote the overall quality of students' educational experience.

Sharp, J. (2015). "Relationships between Hopi calendar and measurement concepts." Journal of Mathematics and Culture **9**(1): 30-51.

There is much to be relearned about mathematics by seeing it from a viewpoint of a particular society. This paper describes measurement concepts by illustrating how the ancient Hopi used them to actively mathematize their environment, building a calendar for celebrations and festivals necessary for their societal mission. The core mathematical concepts are: size, standard unit, zero-origin, tiling, continuousness, and part-whole. The Hopi calendar is presented in terms of the importance and development of the moon, as a counting unit. Ultimately, a 13th moon helps explain the Hopi calendar and further clarifies the mathematical concepts.

Sharp, J. and A. Stevens (2007). "Culturally-relevant algebra teaching: The case of African drumming." Journal of Mathematics and Culture **1**(2): 37-57.

Aside from being its own field of study, algebra has long been a servant to upper division courses and arguably to productive citizenship. Algebra plays a key role in courses like chemistry and trigonometry, which are the kinds of upper level courses required of high-tech, usually high-paying, occupations such as engineering. Keeping learners out of upper division courses also means limiting their potential for active citizenship in society and economic success (Povey, 2003). Many (see Burton, 2003) argue more philosophically that access to and knowledge of mathematics leads to an empowered populace, making it a key to social justice. Unfortunately, algebra can appear to be a tool of injustice when it protects upper division mathematics courses from potential students. Intentions are understandable since success in those courses requires algebraic thinking. Under-prepared students often fail. But, the net result has been to preclude students from those courses rather than to ensure students are prepared for those courses. Robert Moses (Moses & Cobb, 2001) suggests that algebra must move beyond its role as a pre-requisite hurdle. He refers to algebra as a civil right, suggesting that all learners have the right to know algebra. Viewing algebra in this manner elevates its justice status, allowing all students greater opportunity to economic success.

If such equity is a desirable goal, the mathematics education community should develop teaching strategies that resonate with all learners. For algebra teachers to provide teaching and learning opportunities that promote access for more students, we need insightful pedagogical knowledge and deep algebra knowledge. Practicing culturally relevant pedagogy is an approach that may require teachers to develop these kinds of knowledge. By posing problems that are relevant (Hart, 2003) and that allow students to respect, appreciate and/or celebrate other cultures (Boyer, 1990) teachers' pedagogical knowledge may become stronger. Driscoll (1999) suggests that a useful strategy to support efforts to deepen algebra knowledge is to provide teachers with experiences learning algebra in new ways and in unexpected settings. Together these kinds of knowledge impact teachers practice. In this paper, we describe an atypical context, African/ Afro- Cuban drumming that has the potential to connect teachers' knowledge about pedagogy and algebra by stimulating algebra learning.

Staats, S. (2006). "The case for rich contexts in ethnomathematics lessons." Journal of Mathematics and Culture **1**(1): 39-56.

This article outlines a "socially contextualized" approach for integrating ethnomathematics into undergraduate classes. In this approach, students engage issues associated with an application or case study, that are significant socially and anthropologically in tandem with the mathematics lesson. The ethnographic backgrounds of well-known ethnomathematics case studies, Tschokwe lusona and Vanuatuan nitu sand drawings, are used to develop examples of these anthropologically significant issues; a case study on the mathematics of theories of race is outlined as well. By engaging social issues associated with ethnomathematics case studies, students can develop subjective, values-based motivation to study mathematics.

Sternstein, M. (2008). "Mathematics and the Dan culture." Journal of Mathematics and Culture **3**(1): 1-13.

The author was a Fulbright Professor of Mathematics in Monrovia, Liberia, when he had the opportunity to travel upcountry and talk with informants from the villages of Dayala, Graie, Ziah, Zontuo, Zuolay, and Zwatuo, and from the towns of Karnplay, Tappita, and Towehtown. From market women and zo (powerful female medicine makers) to village elders and chiefs, the Dan tribal members showed tremendous pride in their heritage and an openness to discuss most aspects of their culture including how mathematical concepts are expressed in their language and embedded in their activities. It is unclear how this pride and openness have been affected by the recent civil war in Liberia in which half the Dan population were displaced and over ten percent were killed.

Tanase, M. and T. Lucey (2015). "Interdisciplinary connections: Teaching mathematics for social justice and financial literacy." Journal of Mathematics and Culture **9**(1): 81-118.

The authors argue for a broad mathematical literacy, and explore mathematical practices which concern personal finance and social justice issues. They base their claims on a review of literature that looks at connections among these three areas. Postulating the intersections among these areas, the authors claim that by understanding the connections among mathematics, financial literacy, and social justice, mathematics teachers may revitalize mathematics teaching and learning, such that it develops students' broad views of social issues, and it realizes their capacity for initiating social change.

Troutman, J. and L. McCoy (2008). "Re-membering mathematics." Journal of Mathematics and Culture **3**(1): 14-51.

This study examined the effect of math history lessons (which focused on the accomplishments of Latin American, African, African-American, and Middle Eastern individuals) on the attitudes of fifteen students enrolled in an Algebra 2 class at a predominantly African-American high school. Qualitative data taken during interviews with the participants showed some important subtle changes in the way they viewed their success in math, the usefulness of math, and the role of race in determining mathematical success. The effect of the lessons was analyzed according to three pedagogical theories: situated cognition, role model development, and culturally relevant pedagogy.

Vandendriessche, E. (2015). "Thomas Storer's heart-sequence: A formal approach to string figure-making." Journal of Mathematics and Culture **9**(1): 119-159.

String figure-making is a procedural activity carried out in many societies of oral tradition. It consists in producing geometrical forms using a string knotted into a loop. In 1988, an American mathematician, Thomas Storer (1938-2006), published a long article in which he developed several formal approaches of string figure-making. One of these is the Heart-sequence. Passing the string around a finger forms a "loop". The point is then to focus, during the process, on the movements of the loops without taking into account the way the fingers operate to make these loops, and to convert these movements into a mathematical formula. This mathematical approach to string figures allows a new reading of string figure procedures which makes it possible to classify them and is a promising way to carry out a comparative study of string figure corpora. This paper concentrates on string figure algorithms leading to "double-sided lozenge" final figures. Although various different methods in forming such a "double-sided lozenge" can be found in geographically and culturally distant societies, only two classes of heart-sequences will be identified.

Yow, J., et al. (2013). "Lessons learned from a community math project: Ethnomathematical games & opportunities for teacher leadership." Journal of Mathematics and Culture 7(1): 98-139.

Minority Access to Revolutionary Instructional Extensions (MATRIX) is a two-part supplemental elementary mathematics curriculum based on six games coupled with a focus on parental involvement and advocacy. One curricular goal incorporates student's culture into its design to show mathematics comes from many cultures and is an evolving discipline in which students can be active participants. The article data comes from a larger pilot study conducted in a rural African American community. This article speaks specifically about the ethnomathematical games in the curriculum in addition to discussing the "lessons learned" by the researchers for the opportunity for teachers to be leaders in incorporating community cultures into classroom practice.

The Ethnomathematics Program and a Culture of Peace

Ubiratan D'Ambrosio
Universidade Anhanguera Brasil
ubi@usp.br

Introductory Note

The *Journal of Mathematics and Culture*, founded 10 years ago under the sponsorship of the North American Study Group on Ethnomathematics (NASGEm), was a great achievement in the area. I was privileged to publish a paper on *The Program Ethnomathematics: A Theoretical Basis of the Dynamics of Intra–Cultural Encounters* in the inaugural issue of the journal and now I have the opportunity of submitting a paper on my own ideas about the intimate relation of Mathematics and Culture, which is the supporting basis for the Ethnomathematics Program. This paper deals primarily with Science, Technology and Mathematics and also with other forms of knowledge with the major goal of building-up a culture of peace.

From the Origins through Pre-history to the Future

What can we say about the beginnings of this movement? *How did it all begin?* This is the most intriguing and an unsolved question: *Can it ever be solved?* I doubt it. Answers have been given with resource to creating myths, religions and sciences, supported by explanations and theories given by philosophers and scientists, all of them questionable in the broadest sense and many are refutable.

Across cultures and time, there have been a repetition of arguments, which can be considered in two main lines of proposals about human origins. One is that it is the result of a supernatural design of gods, of deities, or of a single deity or of a single god (how did they appear?). Another is that it is the result of fortuitous encounters or changes of matter and energy (how these elements originate?).

In both cases, we deal with the *how* as an act of faith and of believing. In the sequel of reflections about the origins, there is the equal challenging concept of life, as a self-sustaining process. *What is the origin of life?* Particularly, how the homo species were originated? Again, this is a hard dispute, which is polarized in two main lines: creationism and evolutionism, with many variants. For the sake of space, I will not proceed with this discussion.

I side with paleontological and archaeological evidences that from the order of primates evolved, about 5 million years ago, the biped genus *Australopithecus,* which spread through the continent and, about 3 million years ago, evolved as the species *homo*, characterized mainly for an erectus position and increased neurons in the brain. These hominids spread throughout Earth and differentiated into *Homo Sapiens.*

The active research area of paleoanthropology is uncertain about when and were to locate early groups of *Homo Sapiens*. Probably about 300,000 years ago in many places of the Planet. There is evidence that these creatures looked for caves as shelter, learned about fire and developed stone tools. Later they left caves and socially organized as families, clans and tribes and villages, aggregating individuals with some form of affinity (parenthood, physical characteristics, language, believes, abilities and other kinds of identity).

A movie, *Quest for Fire*, from 1981, directed by Jean-Jacques Annaud, shows, in a lively and captivating way, the struggle to control fire. The cinematography was admirably successful in combining a travel in space and in time, in the Paleolithic Era. For a discussion of these early stages of humanity (D'Ambrosio & Almeida, to appear) I proceed with accepting the evolution of knowledge as a process.

Every other known animal species has developed strategies for coping and for dealing with the natural environment and with other individuals of the same species, with the objective of the *survival* of the individuals and of the species. Individuals, from birth to death, relate to nature according to principles of physiology, among individuals according to principles of socio-biology and, organized as groups and herds, relate with nature according to principles of ecology.

Thus, they instinctively try to satisfy the pulsion for survival. These three elements, the individual, the other(s) and nature and the three relations between them, that is individual↔other(s)/society↔nature, constitute the essence of life. I metaphorically represent life as the *primordial triangle* (figure 1).

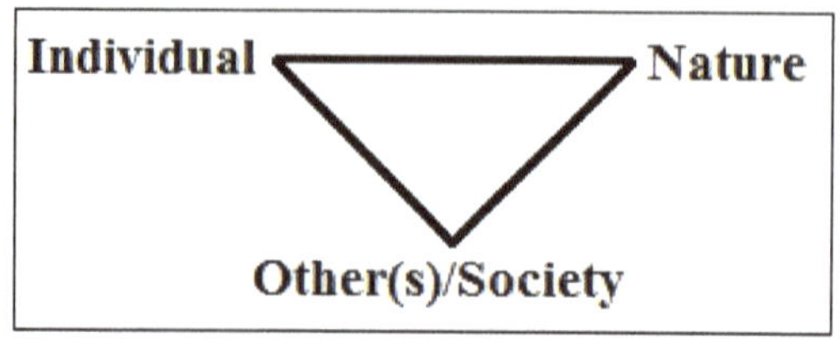

Figure 1. The Primordial Triangle

This is a mathematical metaphor. A triangle is a geometrical figure formed by six associated elements: 3 vertices and 3 sides. Suppressing any one the six, it is not a triangle anymore. All six elements are threatened by the Modern Civilization.

The species *Homo Sapiens* have a differential. Individuals developed unique strategies to keep within the primordial triangle. To intermediate the relationship between the individual and nature, humans created instruments, which gave new possibilities of acquiring nourishment and of clothing to protect the body. The relation of an individual with another individual of the same species makes use of another interaction between the developments of a common language, communication strategies based on complex articulation of sounds, unique to *Homo Sapiens*. The most important consequence

NOTES

was the discovery of the other, of *thee*, as a differentiated individual and, consequently, of the self, of *myself*[1].

The acquisition of will is a decisive step in the evolution of humans. The recognition of the self and of the *thee* affects how humans develop socially. As a consequence of will, the organization of herds or clans or tribes, focus on objectives and on distribution of tasks. Thus, it leads to transcending one's own space and time and projecting oneself onto the other, which is a step towards transcending one's own existence.

This results in probing beyond the span of individual life, asking *what before* birth and *what after* death. This search outside the limits of anything, both in time and in space is called transcendence. Thus, the human species developed, beyond the pulsion of survival, common to all animal species, the pulsion of transcendence. Both pulsions are of a symbiotic relation and, together, they constitute the essence of being human.

The recognition of the self, and of the *thee* and the search for a *common thee*, for a group of individuals with traits in common, such as sharing a common language, lead to generate emotions, like/dislike, prefer/reject, and to create myths and symbols, traditions and norms, wisdom and knowledge, and to organize themselves as hierarchies, castes and social classes. All these categories, developed by the kind characterize what is called *culture*.

These culturally identify a group as it comes to develop deeper relations with nature. According to the needs and interests of the group and will of individuals, some shared by the groups. These new relations can often violate spontaneous principles of ecology, essential for maintaining the primordial triangle. Thus, as agriculture and breeding arose and became dominant in the relation to the individual↔other(s)/society↔nature[2].

These intermediacies continue to affect the primordial triangle. For example, agriculture and breeding demand more water and energy and soil treatment, violating principles of ecology. Results are seen in changes in soil composition, in flora and fauna and, consequently, in meteorological regimes. In addition, the ever increasing demand on previously untapped resources, such as minerals and fossil fuels is drastically affecting the triangle. This demand has now reached a critical situation, as observed by the eminent mathematician Mikhail L. Gromov (Raussen & Skau, 2010):

> *Earth will run out of the basic resources, and we cannot predict what will happen after that. We will run out of water, air, soil, and rare metals, not to mention oil. Everything will essentially come to an end within fifty years. What will happen after that? I am scared. It may be okay if we find solutions, but if we don't then everything may come to an end very quickly. Mathematics may help to solve the problem, but if we are not successful, there will not be any mathematics left, I am afraid!*

We are faced with the relations individual↔other(s)/society↔nature and how humans developed, since early times, strategies to deal with these relations, as discussed above. Obviously, these strategies profoundly affected the behavior of individuals (conflicts with self), the harmony

[1] How did our species reach these incredibly diverse and beautifully unique strategies? Both evolutionism and creationism recognize specificities in human intelligence. Even more challenging is how beliefs, traditions, religions, arts and sciences, enchained in the manifestations of humankind throughout history. Curiously, the enchantment is very similar to what happens in child development. Essentially, this is the result of the search for explanations, for understanding, which go together with the search for predictions. One explains in order to anticipate, thus building up systems of explanations (beliefs) and of behavior (norms, precepts). These are, until nowadays, the common grounds on which both religions and sciences are justified.

[2] Interesting that this appears in the Genesis as Adam and Eve were expelled from Paradise.

of societies (conflicts with thee), and the equilibrium of natural resources (conflicts with nature).

Humanity needs *peace* in the broadest sense of the word. To avoid the increasingly more complex conflicts is a road to peace in all its dimensions:

- individual peace (eliminating conflicts with the self),
- social peace (eliminating conflicts with Thee),
- and, environmental peace (eliminating conflicts with nature).

Consequently, we may obtain *military peace*, eliminating wars in all senses. The continuation of civilization depends on obtaining peace in all dimensions. Yet, we need a true system of knowledge that offers the possibility of peace, relying on respectful, harmonious and constructive relations among humans and between humans and nature, as a response to the *ethics of diversity*:

- respect for the other (even if different);
- solidarity with the other (even if different);
- cooperation with the other (even if different).

The practice of ethics of diversity is the only hope we have for achieving a just relation among humans and between humans and nature, and peace as a consequence. If education is to contribute to achieving a just social order, we contend that educational practices should be grounded in ethics of diversity.

It is a mistake to claim, as many mathematicians do, that this refers to other forms of knowledge, and that mathematicians and mathematics educators have little to do with this situation. We are as responsible as every individual human being is, in facing the serious threats to the future of civilization.

As Gromov said, "Mathematics may help to solve the problem". However, we ask "what kind of Mathematics?" I believe we have to look into some form of ethics in mathematics, that is, an *ethical mathematics*. It is generally agreed that ethics is a cultural trait. It is undeniable that mathematics is a cultural endeavor. This *Journal of Mathematics and Culture* has been conceived to emphasize Mathematics as a cultural endeavor. What kind of ethics do we recognize in mathematics? There is an ethics of rigor, which permeates most lines of Philosophy of Mathematics, and also a code of behavior, impregnated of corporate concerns (AMS, 2005).

However, there are only a few references to cultural differences in consideration to and the elaboration of strategies of mathematical nature, such as the often diverse strategies humans have develop for observing, comparing, classifying, ordering, measuring, quantifying and inferring. As a manifestation of *abstract thinking,* every culturally identified group has generated specific strategies of a mathematical nature. The recognition of these abstract modes of thinking in every cultural group is the quintessence of the ethics of diversity.

NOTES

The Ethnomathematics Program

The Ethnomathematics Program began as a research program in the History and Philosophy of Mathematics with the objectives of understanding how strategies of observing, comparing, classifying, ordering, measuring, quantifying and inferring were developed by humans in different natural and socio-cultural environments, in response to the pulsions for survival and transcendence.

The research relied on the history of mathematics and in the history of science and technology. But, the necessity of advancing into the analysis of language, myths and symbols and of other cultural traits was recognized early on in the development of the program. The importance of human relations and socio-cultural institutions, such as education and academies, belief systems and religions, diverse practices of bartering, commerce, money and economics, all of them were essential in building-up civilizations.

These strategies are synthesized in the ways, modes and styles, arts and techniques, developed to cope, to explain, to learn and to understand, to know reality as a whole, that is how the natural, social, cultural, mythical and imaginary environment combine.

An etymological exercise lead me to identify some Greek roots meaning these concepts: *techné* ≈ *tics*: arts and techniques, modes and styles, *mathemá*: to cope, to explain, to learn and to understand, to know, and *ethno*: reality as a whole, the natural, social, cultural, mythical and imaginary environments. This etymological exercise lead me to construct the concept of *tics* of *mathema* in distinct *ethnos*, and, rearranging, *ethno+mathema+tics*, hence Ethnomathematics. This is a broad theoretical concept, indeed a proposal of a general theory of knowledge.

So, it is that the actual concept of ethno+mathema+tics is primeval in nature since it recognizes different perceptions of space and time and the emergence of specific ways for observing, comparing, classifying, ordering, measuring, quantifying and inferring as strategies to satisfy the pulsions of survival and transcendence. This means the emergence of abstract thinking, in every corner of the planet.

Hence, I have developed the Ethnomathematics Program to contribute to the restoration of cultural dignity, and to offer the intellectual tools for the exercise of planetary citizenship. This Program enhances creativity, reinforces cultural self-respect, and offers a broad view of humanity that provides for a theoretical framework that establishes the foundation for organizing diverse practices and systems of explanations, which have been developed by our species in order to survive and to transcend throughout our complicated evolution.

The Program I developed seeks to add towards building a civilization that rejects inequity, arrogance, and bigotry; where education gives special attention to the redemption of people that have been for a very long time subordinated and that seeks to constitute excluded sectors of societies. Real goals of *empowerment*, a phrase commonly used in education, can be achieved by the implications of the Program Ethnomathematics in the curriculum (D'Ambrosio & D'Ambrosio, 2013).

The concept of Ethnomathematics, as proposed by many authors forms a mathematics that has a distinct Ethnos, or Ethno+Mathematics. Mathematics, as it is understood in the Modern World, is mistaken as a category of knowledge developed as only the elaboration of the *tics* and *mathemá* and forms of *ethnos* of the peoples of the Mediterranean region, which acquired specific style of narrative and criteria of truth in Ancient Greece.

Ethnomathematics is certainly, but limited to, or rooted in the perceptions of the world of the Mediterranean perspective, which include their relation to nature, myths, history, and traditions. These perceptions as synthesized imply the concepts of space and time, which form the base of every *tics* of *mathemá* thinking in western civilization, but differ enormously in other diverse regions of the planet.

For example, calendars, forms of territorial demarcation, shelter and nourishment of the people originating in Mediterranean cultures are absolutely meaningless for the Inuit or Amazonian populations. The same can be said for people and traditions of people of Mayan and Andean descent and the Aboriginal populations in the Pacific.

Each one is strange to the others. Obviously, each one developed their own *tics* of *mathema* in their *ethnos*, which have nothing to do, conceptually, with mathematics developed by the Greeks. Some coincidences, mainly in dealing with quantities and forms, may occur, but are fortuitous.

This is why, the question whether ethnomathematics is considered as true research or practice is recurrent. The school practice of ethnomathematics arises from the research, and this is the reason for calling ethnomathematics as a research program. I have come to guide my investigation to three basic questions:

1. How do *ad hoc* practices and solution of problems develop into methods?
2. How do *methods* develop into theories?
3. How do *theories* develop into invention?

As discussed in the first section of this paper, the main objective for the Program Ethnomathematics is based on the interconnectedness, as it were, of these relations as given by ***practices ⇒ methods ⇒ theories ⇒ invention***.

This is fundamental to the evolution of our species. The work of many ethnomathematicians with different cultural environments, describes the diverse practices and theoretical ideas of other cultures, mainly those dealing with strategies for observing, comparing, classifying, ordering, measuring, quantifying, and inferring. This has come to form an essential component of the Ethnomathematics Program.

It is important here to note that the many cultural groups studied by researchers seeking to apply aspects of the Ethnomathematics Program most certainly include indigenous populations, but also cultures of labor and artisan groups, communities in urban environments and in the periphery, farm communities, and all types of professional groups. All these groups have specific strategies of a mathematical nature, that is, they have their own strategies for observing, comparing, classifying, ordering, measuring, quantifying and inferring.

We must devote special attention to how these ideas, which I believe are essential for peace, and how they are incorporated into distinct pedagogical practices. This requires us to explore programs that develop a disposition of teachers and leaders to conceptualize goals for education and consider the well-being of all humanity as the main concern of education, which needs to rethink teacher preparation (D'Ambrosio & Kastberg, 2012).

NOTES

As a Conclusion

Most pungent is the appeal of the Manifesto of Albert Einstein and Bertrand Russell (1955) seems most important here:

> *There lies before us, if we choose, continual progress in happiness, knowledge, and wisdom. Shall we, instead, choose death, because we cannot forget our quarrels? We appeal as human beings to human beings: Remember your humanity, and forget the rest. If you can do so, the way lies open to a new Paradise; if you cannot, there lies before you the risk of universal death*[3].

The Pugwash Movement or *Pugwash Conferences on Science and World Affairs*, which were awarded a Nobel Price for Peace in 1995, was founded to give sequence to the Manifesto of Einstein and Russell, in the small village of Pugwash, in new Scotia, Canada[4].

The Pugwash Movement adopted the motto *Thinking in a New Way.* Indeed, in all sectors of societies, everywhere in the World, we need new thinking, a new institutional order, mainly in economy, in styles of governance, in cities, urban and rural structures, in demographic demarcation.

It is obvious that the capitalistic system has reached a point of exhaustion; the war-like structures it has created increasingly reveal inefficacy to keep its economic, political and social order, with ethical and moral values to be ignored. These are problems that must be solved for the survival of our civilization.

We need to develop a culture of peace associated with mathematics and mathematics education. This requires new thinking. Repeating, once again, what Gromov said, "Mathematics may help to solve the problem". I, as well repeat this by asking "what kind of mathematics?".

The strategies of a mathematical nature are a very powerful instrument to propose new thinking and new institutional order, as they were determinant in every moment and in every region of the planet in the evolution of the human species.

The kinds of mathematics we need are those complex strategies of a mathematical nature impregnated with the ethics of diversity, which are the basic pillars of the Ethnomathematics Program.

Volume 1

[3] https://pugwash.org/1955/07/09/statement-manifesto/

[4] The Russell-Einstein Manifesto. *Scientists in the Quest for Peace: A History of the Pugwash Conferences.* Ed. Joseph Rotblat, Cambridge: The MIT Press, 1972; pp. 137-140.

Volume 1

References

AMS (2005). *Policy statements on ethical guidelines.* American Mathematical Society, 2005. Retrieved from http://www.ams.org/secretary/ethics.html.

D'Ambrosio, U.; & Almeida, M. C. (to appear). Ethnomathematics and the emergence of mathematics: the nature and development of mathematics. In Barmby, P.; Adams, J.; & Mesoudi, A. (Eds.), *Cross-Disciplinary Perspectives on Cognition, Learning and Culture*. Routledge.

D'Ambrosio, U.; & D'Ambrosio, B. S. (2013).The role of ethnomathematics in curricular leadership in mathematics education. *Journal of Mathematics Education at Teachers College*, 4, Spring–Summer, 19-25.

D'Ambrosio, B. S.; & Kastberg, S. (2012). Giving reason to prospective mathematics teachers. *For the Learning of Mathematics* 32(3), 22-27.

Raussen, M.; & Skau, C. (2010). Interview to Mikhail Gromov. *Notices of the AMS, 57*(3), 391-409.

What do Math and Language have in Common? Lessons from Foreign Language Learning

Molly Tun
St. Olaf College
tun1@stolaf.edu
1520 St. Olaf Avenue
Northfield, MN 55057

Abstract

In this reflection on my previous JMC publication "The Incan Abacus: a Curious Counting Device" and current research trajectory I focus on math and language. Because math and language are both semiotic systems involving memorization, manipulation, and interpretation, they can be taught with similar pedagogical approaches, including the task-based approach. To illustrate this, I analyze the ethnomathematics seminars I have designed around the use of a multilingual mathematics medium, the quipu, in diverse and multicultural settings in order to advocate for greater inclusivity and diversity in mathematics.

Keywords:

- Mathematics
- Language
- Quipu

Introduction

Unbeknownst to me, my own research trajectory ultimately has its origin in my freshman year of college when my advisor presented me with a question (which would become very common in future years): *What will you do with a double major in Math and Spanish? What do those fields have in common, if anything?* Like most teenagers my age I wasn't too fazed by this predicament and continued studying and taking classes that interested me, mostly for the simple pleasure of learning. Throughout the years, however, I found that my seemingly disjoint combination of studies wasn't so unheard of; whether by chance or design I came to know numerous students and faculty that study or had studied math in combination with other languages. Now, as faculty, my colleagues send their language-learning math majors to me for help with advising.

What had started as a very practical question (*what do math and language have in common?*) soon became a theoretical inquiry. As an undergraduate I was faced with the logistics of bringing together a committee composed of faculty members from my different majors in order to present a senior honors paper. More so for the benefit of my audience, I pondered a way to make connections between classes and combine ideas from both fields of thought as to not bore anyone in the presentation. I decided that ethnomathematics (see Tun, 2014a) was the solution to this question that had accompanied me all these years as it provided the language and theoretical background necessary to analyze math practices contextually; my first exposure to ethnomathematics literature and ideas made me consider how numbers could be socially-constructed (see Urton, Gary, 1997), how math practices could be situated (see D'Ambrosio, 1986), and how the commonly-promulgated history of mathematics could be greatly limited and selective (see Selin, 2001). Using this starting point, I wrote a paper about the Inca abacus and it's possible connection to indigenous language; it was precisely this reflection that materialized in the article I published in JMC a few years later: "The Incan Abacus: A Curious Counting Device" (Leonard & Shakiban, 2011). In this current reflection, in commemoration of the 10th Anniversary edition of JMC, I will further elaborate on the connections between math and language by focusing on pedagogy (informed by second-language acquisition techniques) and the specific teaching materials (based on the Inca quipus) I have designed to teach math and culture in multilingual and multicultural contexts.

Math and Language

Although the traditionally quiet, note-taking math classes seem at odds with the loud, noisy, cacophony of voices of a second-language classroom, this opposition is actually more a matter of pedagogy than anything else. While language-learning classrooms have traditionally privileged oral input and production, and math classrooms have generally placed emphasis on the written symbol, both are forms of "foreign language," in so far as they are semiotic systems that are new, and in a way, "foreign" to the student. As in any foreign context with unknown expressions, body language, and vocabulary, a math student that does not master the content, understand the operant rules and norms, or gets lost in the flow of formulas, variables, and calculations,

NOTES

could consider his/her experience in a math classroom very similar to that of "learning a foreign language."

As semiotic systems, both math and language learning entail a process of a) memorization, b) manipulation, and c) interpretation. Learning to communicate in a new semiotic system is a skill that begins with memorizing the given "chunks" of information of that system, whether in the form of letters (á, ñ, ll, m…), morphemes (suffixes, roots…), and words (*amigo, fiesta, adios…*) of a Spanish travel dictionary, or in the form of symbols (∞, , e, x…), concepts (radians, *sin(x)*, |x|…), and formulas ($a^2+b^2=c^2$, $y=mx+b$, $A=1/2bh$….) of the reference page of a math textbook. Once a basic collection of memorized building blocks is established, students are introduced to rules and operant norms of manipulating, combining, and calculating such segments. In learning a foreign language, for example, one learns how to apply formulas for conjugating verbs ("go yo, drop o, add opp" to form the Spanish subjunctive tense), string words together to make basic sentences (subject + verb + complement), and manipulate punctuation (Let's eat grandma ≠ Let's eat, grandma). In a similar fashion, math students learn how to perform operations (3+5, 9/3, 2^4), solve equations and isolate variables (2x+1=5 à 2x=4 à x=2), and use logic to prove certain arguments (such as " is irrational"); the mathematician Michael Hutchings effectively articulates this connection between math and language when constructing an argument: "there is a certain vocabulary and grammar that underlies all mathematical proofs… by 'grammar', I mean that there are certain common-sense principles of logic, or proof techniques, which you can use to start with statements which you know and deduce statements which you didn't know before" (Hutchings, n.d., p. 1). This linguistic and mathematical process of constructing and manipulating a shared symbolic code via grammatical rules is only mastered when one is able to accurately and effectively interpret the results, whether this is an answer you calculate after solving a complicated word problem or the understanding you draw from a command or text in a foreign language (or correspondingly, the command you give or text you write in another language). Ultimately, both math and language acquisition can be measured by one's ability to a) code, b) construct, and c) communicate meaning in a "foreign" but shared semiotic system.

While mathematics and languages have many *structural* similarities, it would be unproductive to minimize their *contextual* differences; both mathematic and linguistic traditions are framed and influenced by their users or their distinct communities of practice. In language studies these groups are known as "speech communities" and can be defined as the "speakers of the same language that negotiate and establish the extent of common meaning in the linguistic tokens with which they operate" (from Brumfit, 1984, p. 28). Linguistic speech communities tend to be regionally/geographically and culturally delineated while mathematics communities of practice might be harder to define or visualize; this stems back to the emphasis a speech community places on interpersonal relationships and oral negotiation while a mathematics community of practice is connected in a more discreet way through written code and negotiation. This is not to say that oral and written negotiation of meaning is strictly applicable to linguistic and mathematics communities of practice respectively (as both kinds of communities engage in both kinds of negotiation), but could explain why both fields (math and language) *have been perceived as* fundamentally different. Students would corroborate that the communities established by Math majors and those of Spanish majors seem very different, at least on a stereotypical level. Regardless of contextual differences between the math and language communities of practice, the structural similarities of math and language learning allow for similar pedagogical approaches in the teaching of math and language, as proposed below.

Math and Language Pedagogy in Multicultural Contexts

Since the 1990s pedagogy across disciplines has focused on student-centered learning, specifically, a move from teaching *about* the operative norms of a field to facilitating a *practice of* those skills and ways of thinking. In language classrooms this can be seen as students and instructors are continually talking less *about* Spanish grammar and conjugation and dedicating more time for students to *practice using* and *producing* the language. Likewise, I'm familiar with more math instructors that are "flipping the classroom" so that students watch the lectures *about* course content at home in order to *practice* and *discuss* math topics and problems in class. This pedagogical shift is more prevalent or visible in disciplines or classrooms in which the instructor considers a main course objective to be the development of a skill, as opposed to simply the mastery of content.

To the extent that math and language studies are related to skill-development in conjunction with content-mastery,[1] it could be mutually productive to compare the evolution of their pedagogies and draw connections between the strategies and approaches used in the teaching of each. Although before my time, my own Spanish professors would recall the audio-lingual method of the 1950s and 60s when they would repeat the instructor's perfectly-articulated questions and responses in drill-based fashion; this teacher-dominant approach fueled by American politics and the Army Method of language learning, tape-recorder vendors, and behavioral psychology (Barker, 2001) led to student language production with highly accurate pronunciation in a limited range of expressions, but little ability or creativity to truly engage in communication in the target language (see Mueller, 1971). This method might be likened to the arithmetic drills common in an elementary school setting in which students memorize a series of teacher prompts (4x3) and student response (12) within a limited range (such as the memorization of multiplication tables up to 10x10 or 12x12).

Subsequent decades were witness to the development and emphasis of the communicative method (still prevalent in schools and universities today) which no longer taught language as a means to communication but rather communication via language (Brumfit, 1984, p. 41); this approach transformed the classroom from an atlas model with solely teacher-student interactions to a student-centered model that privileged pair-work and group activities. Such pedagogy might be envisioned in the math classroom as one that does not teach math for the sake of problem-solving but rather as one focused on problem-solving via mathematics; one such pedagogical shift in a calculus classroom, for example, could be facilitated by focusing on the conceptual

[1] Here, the idea of "content-mastery" refers to an advanced knowledge of the underlying structure of math or language, independent of the ability to produce math or language. In the context of a Spanish classroom this might entail knowledge of the rules of conjugation, grammar, vocabulary, or syntax, independent from the ability to write, read, speak, or comprehend the language in a variety of contexts. This is unrelated to the content-based pedagogical approach that teaches language via content (which could be considered a specific application of a task-based approach that provides students with tasks centered on culturally relevant content and material).

NOTES

problems (finding maximum/minimum, areas/volumes, etc.) and then identifying and learning the skills necessary to solve the problem at hand as opposed to the traditional framework in which skills (derivatives, integrals, etc.) are the organizing principle of the content and the applications are presented as an afterthought.

The communicative approach served as a natural segue to one of the most prominent second-language acquisition pedagogies of today: the task-based approach (or similarly, task-design in mathematics education). In a way, this pedagogy is an extension of the previous in that tasks are completed via communication via language; in this type of classroom "the primary focus is on meaning, and learners comprehend, manipulate, produce, and interact in the target language to achieve an objective" (de la Fuente, Peris, & Baulenas, 2007, p. xvi). Providing students with concrete and relevant tasks which can later be evaluated as completed (or not) and presented to the classroom as tangible evidence of language acquisition and skill-development could evoke higher levels of motivation and drive as opposed to a more general, repetitive objective of "communicate in Spanish with your classmate." Likewise, current research in math task-design advocates for the creation of interest-dense tasks as connected to the perceived needs of students (Ainley & Margolinas, 2015, p. 133; Bikner-Ahsbahs & JanBen, 2013).

One challenge I envision of the application of this task-based pedagogy is the creation of relevant and interest-dense tasks inclusive of a wide range of learners in multicultural and multilingual classrooms and contexts. Tasks are inherently culturally-framed; while it may be easier to find inclusive tasks for a language classroom (such as "plan a party," "organize a trip," "design a home") that allow for multiple answers and for students to shape the outcome of the task in an infinite number of ways, tasks in mathematics textbooks and classrooms commonly have only one right answer associated with a very confined and culturally-driven context (for example "optimize the area of a swimming pool," "find the time it takes coffee to cool down to room-temperature," "calculate total revenue"). In general, word problems in math textbooks in the U.S. tend to be Western-culture specific and not necessarily reflective of the diversity of learners or cultural math practices that place value on other distinct kinds of tasks, problems, and numeric ways of distributing and organizing.

Furthermore, in mathematics, tasks are not only culturally but linguistically framed; while tasks in a second-language classroom are designed to place all students into a new linguistic framework, that of the target language, mathematics tasks are framed by the specific speech community of the dominant culture from which all students are introduced to a new mathematical "language." The speech community privileged in a math classroom might be that of the larger community or determined by the student population and instructor profile; a student of mine actually complained about a math class she took at the University of Minnesota in which the students and professor comprised a unique dominant culture in which much of the classroom discussion of math problems was carried out in Chinese! In addition to the cultural nature of math tasks, the speech-community dynamic of a mathematics classroom can influence a student's ability to successfully complete these tasks assigned, granting an advantage to members of the dominant speech community and challenges to those coming from other lingual backgrounds. Oftentimes one's native language has an influence on the numeric base system one is comfortable using (Li, 2007, p. 25) or the way in which one develops numeric reasoning. These differences in logic can lead to differences in the social sphere as well. Richard Day (1982) has shown how "language attitudes and perceptions develop early, so that speakers from majority and minority groups already have strong in- and out-group perceptions, on the basis of language, by the age of three" (cited in Brumfit, 1984, p. 26); this results in the creation of "in- and out-groups" when completing culturally and linguistically framed tasks in a mathematics classroom.

While impossible to dissociate culture and language from tasks in the math classroom, we should consider ways to accommodate diverse speech communities as well as promote a more inclusive community of practice in mathematics. This might be done by envisioning and incorporating multilingual and multicultural tasks into the mathematics curriculum—a goal actively pursued by researchers and instructors working from an ethnomathematics perspective (Gavarrete Villaverde, 2013). Such an inclusionary task-design takes into consideration distinct learning styles and backgrounds, allows for diverse forms of knowledge production, provides for alternative forms of evaluation, promotes collaboration between diverse lingual and learning groups, and may even bring awareness to underrepresented topics in mathematics.

Teaching with Quipus: A Multilingual Approach

One set of multilingual and multicultural tasks that could be incorporated into mathematics education of diverse contexts and objectives are those I have designed in relation to Inca counting and calculating practices. Since the publication of "The Incan Abacus: A Curious Counting Device" in JMC I have continued researching Andean mathematics (Tun & Díaz, 2015) and numeric devices including the Inca abacus/yupana (Tun, 2014c), the taptana (Tun & Luis Octavio, 2014), and the Inca knot records/quipus (Tun, 2014b). These three Andean counting and recording devices (yupana, taptana, and quipu) are multilingual and multicultural in nature in so far as they are not explicitly connected to any one language (as opposed to a math textbook or calculator articulated in English). In this way, these devices can easily be used by distinct cultural and speech communities. In the case of the Inca Empire, for example, the quipu was a numeric record system whose general semiotic structure of strings and knots (see Fig. 1) allowed for various ethnic groups of diverse languages to partake in this tradition when submitting tribute to a larger, more heterogeneous political entity (Tun, 2015, p. 70).

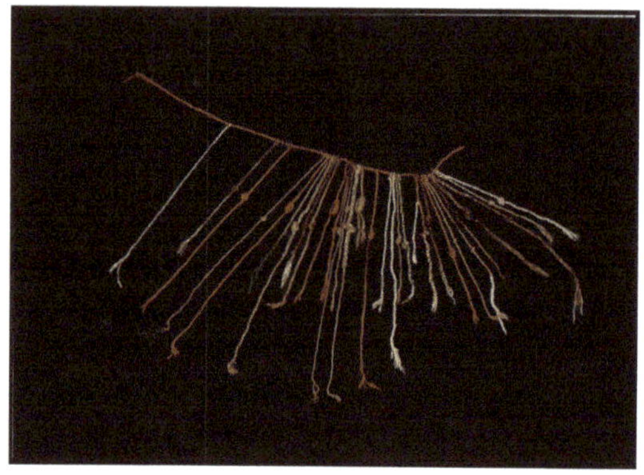

Figure 1. Quipu. Image Courtesy of the Rare Book Collection at the Wilson Library, University of North Carolina (Chapel Hill).

My time spent teaching both Math and Spanish at both the University of Minnesota and the University of St. Thomas afforded me the time, opportunity, and intellectual locality from which I could apply this research in practical ways through the development of ethnomathematics teaching and seminar materials. I have introduced the quipu in diverse contexts including a) the Gran Bretaña elementary school in Lima, Peru, b) the Girls Experiencing Math in the Summer (GEMS) high-school math camp hosted by the University of St. Thomas, and c) the

NOTES

undergraduate history course Early Latin America-1825 at the University of Minnesota. Each of these three settings exemplifies how the quipu can be taught in different cultural and lingual contexts and adapted to the particular needs of each learning community, as outlined below, respectively:

Autochthonous Application. In this context, math practices endemic to a particular location or community are promoted. This might entail the reintroduction of traditional methods forgotten or eradicated over time or the further expansion and dissemination of current local practices and patrimony. Many ethnomathematics publications discuss this dynamic in relation to the incorporation of indigenous practices in corresponding local school curriculums, as seen in the efforts to incorporate the use of the Inca abacus into various school settings in Peru (see Chirinos Rivera, Andrés, 2010).

Academic Numeracy. In this context, different cultural math traditions are prefaced, discussed, and interpreted in connection with values associated with the Westernized discipline of academic mathematics. Through this lens, the mathematical traditions and ideas presented are those that contribute to the advancement of the Western narrative of mathematics or that could be translated into Western terms in order to develop math skills and logic typical of academic mathematics in a Western context. An example of this tendency can be found in the article "The Inka Numeration System in the Yupana and Khipu" (Rojas-Gamarra & Stepanova, 2015) which identifies itself as the first study that formalizes an interpretation of the Inca abacus in terms of algebraic expressions (54).

Anthropological Exploration. In this context, diverse math traditions and practices are studied from an ethnographic viewpoint in order to highlight concepts related to history, culture, social organization, economics, etc. Given the colonial origins of anthropology, the discussion of "other" math practices has generally been one about non-Western cultures as witnessed by Western, academic practice. This dynamic is particularly prominent in anthropological descriptions of math practices, as found in some of the nascent publications of ethnomathematics, including the work of Otto Friedrich Raum (1938) and Claudia Zaslavsky (1973) in describing and analyzing mathematical practices in Africa.

The specific task design of the quipu classes and seminars I organized varied according to the objectives and dynamic of the contexts listed above; in general, however, each session included a cultural and mathematical discussion of the quipu, a hands-on introduction to coding a quipu, and a final interpretation and evaluation of the group's results. This activity design brings students through the stages, discussed previously, of learning and mastering a new linguistic or numeric semiotic system: memorization, manipulation, and interpretation. In general, students were asked to use the coding features of a quipu (mainly color, cord placement, knot position, and knot quantity) to represent a given quantity of goods, exchange this recorded information with other classmates, encode the subsequent results, and finally interpret and report the quantities and outcomes of the simulation.

Figure 2: Images of students making and reading quipu at the Gran Bretaña elementary school in Lima, Peru.

This quipu simulation is representative of the way the quipu has been used to encode, exchange, and recall numeric information pertinent to the organization and distribution of goods, services, and people in the Andes for centuries. It was used an on imperial scale during the Inca empire with evidence suggesting these records were moved through bureaucratic and geographic hierarchies by summing together local quantities in order to evaluate global supplies, and/or conversely dividing global or ideal quantities in order to determine local contributions. Gary Urton and Carrie Brezine's article "Information Control in the Palace of Puruchuco: An Accounting Hierarchy in a Khipu Archive

NOTES

from Coastal Peru"(2007) is an excellent resource on the cultural and numeric basis for this information exchange. The quipu (or versions of it) have been used in other settings all throughout Andean time and space, including those used in more recent times in the context of hacienda accounting (Mackey, 1970). The colonial chronicles are also rich sources of quipu description and images (see Parssinen & Kiviharju, 2011). The depth of cultural and historical background provided students in each context was dependent on the students' previous knowledge and future goals.

The memorization stage of the exercise requires students to learn (or memorize) the operating parts and components of a numeric quipu. The basic structure is composed of a main cord from which pendant cords are hung, each cord indicative of a distinct category. Cord colors and position can indicate category, whether through associating a certain good with a certain color (for example, bars of gold with a yellow cord) or associating position with a certain progression (for example the fourth cord indicating the fourth month of a year). Numbers operate in the decimal system and are encoded (as a string of single knots or as long knots with a certain number of turns) in a vertical fashion with the highest unit values towards the main cord and spaces in between unit values. To help students understand the underlying structure and numeric function of the quipu I oftentimes present them with a drawing of the quipu that Locke used (1923) to determine that the knots operate in base-10 and that the values of the pendant cords can be added together and recorded on the summation cord (134+366+250+55=805, see Fig. 3). Marcia and Robert Ascher's *Mathematics of the Incas: Code of the Quipu* (1981) provides a comprehensive introduction to these structural features as well as instructions of how to spin your own cords and exercises to practice these concepts. The number of features introduced and encoded in the activity is highly variable.

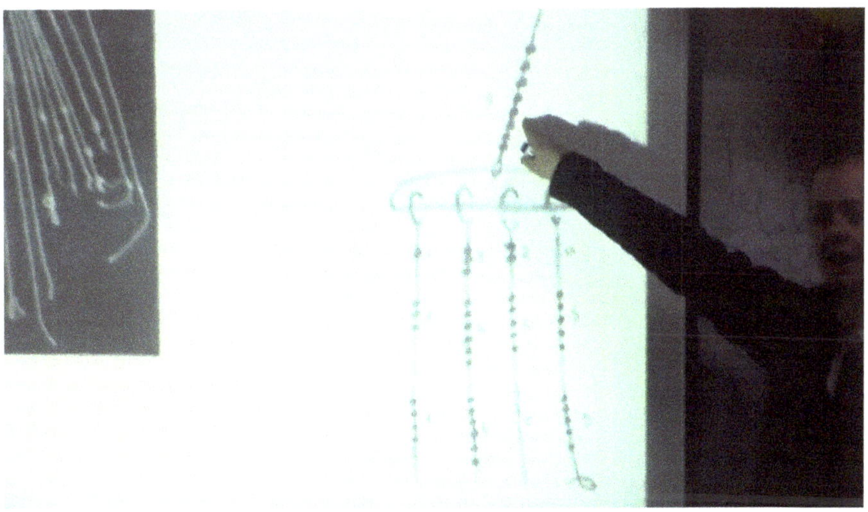

Figure 3: Exploring the structural and numeric elements of the quipu with students.

The dynamic of the manipulation stage of the activity is dependent on the specific prompt given or task assigned (and its designed outcome), but generally includes an encoding, sharing, and reporting of data. In the context of the elementary school in Peru, students worked in groups to make a quipu with different color pendants representing the different quantities and categories provided; the groups had to come to a consensus as to the colors they would use to encode each item. In the activity prepared for the high school math camp, teams competed in a relay race in which student #1 prepared the quipu structure by attaching pendant cords to the main cord and indicating the starting point, student #2 knotted the numeric quantities provided into each cord, student #3 totaled local quantities and knotted the regional totals, and student #4 totaled regional quantities and knotted the global total; this activity relied on mental calculations but could also incorporate the use of the Inca abacus when calculating sums. In the case of the undergraduate history course, students used colonial records and attempted to transfer the data onto quipus.

This task lends itself quite easily to group-oriented interpretation and evaluation of results, as opposed to teacher-dominated feedback. The elementary school students exchanged their quipus with other groups to see if those groups encoded categories with the same colors and if the knots were readable and accurate; because this activity was perceived as a fun challenge for this group of young, energetic students, they were all eager and more than willing to check their classmates' work, give feedback, compare results, and even strategize about how to standardize methods in the future. These students enjoyed learning this "secret code" and were interested in the process as well as result. The high school math students also enjoyed the challenge and competition of the task; in their relay race student #5 actually had to read the knots of the final summations in front of the class to see which groups successfully completed the challenge and in the shortest amount of time. The undergraduate history students struggled through the calculations and encoding but came to a new appreciation of the colonial history they had been studying all semester; these students discussed the results in terms of empire, distribution, conquest, power relations, and colonialism in a hands-on and minds-on way that illustrated many of these pre-Colombian and colonial dynamics.

Results and Reflections

These quipu tasks were successful in teaching mathematics concepts in diverse settings when evaluated against their distinct objectives: autochthonous application, academic numeracy, and anthropological exploration. At the Gran Bretaña elementary school in Lima, Peru, the students were appreciative to learn more about a cultural symbol quite familiar to them and took ownership of the task; in a way, by learning how to use and communicate with quipus they re-appropriated this medium as an important part of their identity (I'm told that after the seminar the students even hung a calendar quipu on the blackboard from which they kept track of the days and months of the remaining school year). The advanced high school math students who attended the GEMS camp used the activity to practice the kinds of math games and calculations pertinent to their own academic formation and as an unintended consequence began thinking about other numeric semiotic codes present in their own cultural context (such as the Dewey decimal system and bar codes). In the case of the undergraduate history students at the University of Minnesota, it might have been the first time they worked together to solve group math problems, which heightened their awareness of the ways in which math shapes societies and can be a shared experience that influences history and its outcomes.

NOTES

Because this activity challenged the lingual norms of the classroom by introducing new codes and rules of operation, the in- and out-groups of the classrooms shifted, allowing for a greater inclusion of diverse students. The instructor of the 4th grade classroom, Miguel Ángel Díaz, was astonished to find students, ones who never normally contributed to math class and were otherwise considered poor math students, participating in this quipu activity; when presented with images of the Inca abacus one of these seemingly unengaged students was able to identify that the abacus structure (of 1, 2, 3, and 5 token spaces) was related to Fibonacci's sequence in which each term is the summation of the previous two terms (see Fig. 4). On the other hand, a few of the students with the highest grades in their normal math class could not figure out how to tie numbers onto the cords and became part of the new out-group unable to make quipus. In general, this hands-on, tangible encounter with diverse math coding practices led to the inclusion of a new set of students, and when paired with current task design in the math classroom could lead to a greater diversity of student participation, group dynamic, and evaluation.

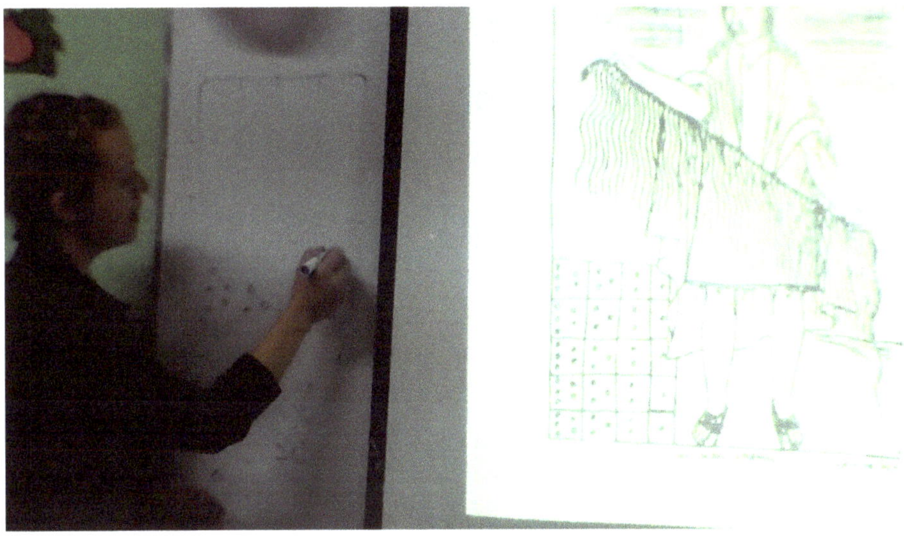

Figure 4: Introducing students to Guaman Poma de Ayala's drawing of the Inca abacus.

While ethnomathematics has addressed selected issues related to a universally-applied Western mathematics curriculum, or conversely, the benefits of developing local curricula that address local needs and objectives, this reflection presents a unique proposal, informed by math and language pedagogy, to design multilingual mathematics tasks that are more inclusive and applicable to numerous cultural and lingual communities than are dominant approaches. I'm not advocating for a universal call to teach quipus in all settings, but rather to consider mathematics tasks and mediums that might be more accessible to underrepresented cultural groups in a multicultural classroom or that connect to diverse speech communities in a less unilateral way than a language-specific textbook. Ethnomathematics has broadened the definition of mathematics to include a diverse range of human activity, and mathematics pedagogy, task design, and evaluation should be diversified in a similar way. Math and language are both semiotic systems used to cognize and transform reality and as such should be at the forefront of our own systems used to introduce students to that reality all around us.

References

Ainley, J., & Margolinas, C. (2015). Accounting for Student Perspectives in Task Design. In A. Watson & M. Ohtani (Eds.), *Task Design in Mathematics Education: an ICMI study 22*. New York: Springer.

Ascher, M., & Ascher, R. (1981). *Mathematics of the Incas: Code of the Quipu*. Mineola, New York: Dover Publications, Inc.

Barker, J. L. (2001). On the Mortality of Language Learning Methods. *Speech*, (November 8).

Bikner-Ahsbahs, A.-A., & JanBen, T. (2013). Emergent tasks--spontaneous design supporting in-depth learning. In C. Margolinas (Ed.), *Task Design in Mathematics Education* (pp. 153–162). Oxford.

Brumfit, C. (1984). *Communicative Methodology in Language Teaching: The Roles of Fluency and Accuracy*. Cambridge: Cambridge University Press.

Chirinos Rivera, Andrés. (2010). *Quipus del Tahuantinsuyo: Curacas, Incas, y su saber matemático en el siglo XVI*. Lima: Editorial Commentarios.

Claudia Zaslavsky. (1973). *Africa counts; number and pattern in African culture*. Boston, Prindle, Weber & Schmidt.

D'Ambrosio, U. (1986). Ethnomathematics and its Place in the History and Pedagogy of Mathematics. *For the Learning of Mathematics, 5*(1), 44–48.

Day, R. R. (1982). Children's attitudes towards language. *Ryan and Giles*, 116–132.

De la Fuente, M. J., Peris, E. M., & Baulenas, N. S. (2007). *Gente* (2nd ed.). Upper Saddle River: Pearson.

Gavarrete Villaverde, M. E. (2013). La Etnomatemática como campo de investigación y acción didáctica: su evolución y recursos para la formación de profesores desde la equidad. *Revista Latinoamericana de Etnomatemática, 6*(1), 127–149.

Hutchings, M. (n.d.). Introduction to Mathematical Arguments (background handout for courses requiring proofs). Retrieved from https://math.berkeley.edu/~hutching/teach/proofs.pdf

Leonard, M., & Shakiban, C. (2011). The Incan Abacus: A Curious Counting Device. *The Journal of Mathematics and Culture, 5*(2), 81–106.

Li, Y. (2007). Curriculum and Culture: An Exploratory Examination of Mathematics Curriculum Materials in Their System and Cultural Contexts. *The Mathematics Educator, 10*(1), 21–38.

Locke, L. L. (1923). *The Ancient Quipu or Peruvian Knot Record*. New York: The American Museum of Natural History.

Mackey, C. (1970). *Knot Records in Ancient and Modern Peru* (PhD Dissertation). Department of Anthropology, University of California, Berkeley.

NOTES

Mueller, T. H. (1971). Could the New Key Be a Wrong Key? *The French Review, 44*(6), 1085–1093.

Parssinen, M., & Kiviharju, J. (2011). *Textos andinos I* (Vol. 1). Madrid: Inst. Iberoamericano Finlandia.

Raum, Otto. (1938). *Arithmetic in Africa*. London: Evans Brothers.

Rojas-Gamarra, M., & Stepanova, M. (2015). Sistema de numeración Inka en la Yupana y el Khipu. *Revista Latinoamericana de Etnomatemática, 8*(3), 46–68.

Selin, H. (Ed.). (2001). *Mathematics Across Cultures: The History of Non-Western Mathematics*. Springer Science & Business Media.

Tun, M. (2014a). Ethnomathematics. In H. Selin (Ed.), *The Encyclopedia of the History of Science, Technology, and Medicine of Non-Western Cultures* (3rd ed.). Springer Reference.

Tun, M. (2014b). Quipu. In H. Selin (Ed.), *The Encyclopedia of the History of Science, Technology, and Medicine of Non-Western Cultures* (3rd ed.). Springer Reference.

Tun, M. (2014c). Yupana. In H. Selin (Ed.), *The Encyclopedia of the History of Science, Technology, and Medicine of Non-Western Cultures* (3rd ed.). Springer Reference.

Tun, M. (2015). *El quipu: escritura andina en las redes informáticas incaicas y coloniales* (PhD Dissertation). University of Minnesota, Minneapolis.

Tun, M., & Díaz, M. A. (2015). Recuperar la memoria histórica y las matemáticas andinas. *Revista Latinoamericana de Etnomatemática, 8*(1), 67–86.

Tun, M., & Luis Octavio, M. C. (2014). Taptana. In H. Selin (Ed.), *The Encyclopedia of the History of Science, Technology, and Medicine of Non-Western Cultures* (3rd ed.). Springer Reference.

Urton, G., & Brezine, C. (2007). Information Control in the Palace of Puruchuco: An Accounting Hierarchy in a Khipu Archive from Coastal Peru. In J. Pillsbury & J. Quilter (Eds.), *Variations in the Expression of Inka Power: A Symposium at Dumbarton Oaks, 18 and 19 October 1997* (pp. 357–384). Washington, D.C.: Harvard University Press.

Urton, Gary. (1997). *The Social Life of Numbers: A Quechua Ontology of Numbers and Philosophy of Arithmetic*. Austin: University of Texas Press.

Pacific Strata

Nicholas J. Goetzfridt, Ph.D.
University of Guam
ngoetzfr@gmail.com

Volume 1

Abstract

Mnemonic patterning schemes in the Pacific engender the acquisition of esoteric knowledges through multifaceted processes of doing, traditionally providing capacities to not only achieve target destinations (most dramatically exemplified by Central Caroline navigators reaching an island landfall in a vast and complex Pacific world) but to also claim clan allegiance and land possession facilitated by these schemes and communal epistemologies. Traditional narratives expand beyond these practical applications, challenging colonially induced notions of island isolation and dependency and opening potential instructional use of mathematically reflective cosmologies of which Pacific-wide heavenly strata provide but one example. As a long-time practicing librarian in the Pacific – and not as a mathematician – I have found over that time an appreciation for the wordless dimensions of indigenous practices within lives that exhibit the deep communal nature of environmental interpretations that have helped me better appreciate my own wordless dimensions of understanding the world around me.

Keywords:

- Lines
- Indigenous Cartography
- Oceania
- Epistemology
- Colonialism - Spatial Perspective
- Mnemonics.

It appears that I may have actually come to appreciate ethnomathematics – coincidentally as a librarian - from an unwritable - perhaps even an expressively unreachable - appreciation for the epistemological fractions of Pacific cultures that have complimented my own largely ingrained and evolutionary responses to environments into which I was born and through which I have imaginatively been propelled throughout my life. Perhaps the only discernable and justifiable distinctiveness of those environments is the contrast that exists in my own reflections of them now – namely the far Minnesota fields of snow and ice that I once traversed on skis and snowmobiles and which I now traverse only in memory and the Pacific 'fields' of an endless ocean that have surrounded me for so long. That in itself does not make me unique except in the context that it does for everyone – namely in our distinctive interpretations of reality based not only upon the unmapped dimensions that draw from our origins but also upon those nuances of "culture" that situate our sensitivities toward influences. My overall proclivity toward the linear dimensions of those 'fields' may actually and subsequently exhibit the extent to which my past and future experiences in ethnomathematical considerations have served and will serve to satisfy that wordless appreciation of cultures not my own. As such, I exist within these fields not so much as an academic sojourner planning future researches but rather as one speaking for now from the consequences of those nuances and influential contexts.

I recently published a paper (Goetzfridt 2016) that attempts to tease out a few examples of praxis and the rootedness of cultural lines that slowly became clearer to me during my decades-long life in the Pacific. They reflect communal values and perceptions in indigenous identities that are each distinctive in the context not only of environments that have had such powerful roles in these perceptions and within ethnomathematical applications that have functioned so effectively. But these consciousnesses are also distinctive within the many Pacific histories and characteristics of colonialism, acculturation and inevitable social and cultural changes. The modern cartographic line is distinct from traditional means of conveying a story or even the components of an esoteric language, reflecting a society's cultural context upon which surface lines of interpretation indeed represent a conflicting means to an end. But it is that means – its functionality in the context of "culture" and history – which makes for an inevitable clash – and if not a clash then a necessary search for appeasement on the side of the colonialist – between whether this cultural line and story of origin leading from one place to another can co-exist in a contemporary need for speed, directness, and arrival at a destination. The extent of this struggle (if it can be called a struggle) varies not in the context of my own outsider imaginings but rather in the context of reality – THAT reality upon which the lines have originated from in the first place.

The simple line, as a generic function of space construction and expression in designs or in cartographic renditions of geography, can often become the antithesis of an independent and sovereign basis for communal realities and legitimacies. Contemporary maps subsume the wayfarer's historical and culturally based forging and expression of that forging into a path that epitomizes an ownership of space and its containment through featureless, functional lines moving a traveler purely from Point A to Point B and that insist on a linear universality. This universality cuts spaces into territorial lines of control as the historical and cultural distinctions between wayfaring and transport are lost. However, in the context of indigenous, communal arts of both physical expression and action – of movement – the line possesses the power of sovereignty of thought, history, and ideas of space and, indeed, of movement. These lines form the patterns that identify the values and realities that are integral to a people's being, their understandings of the world, and the will to express them. In another sense the simple line can indeed be the antithesis of this freedom but at the same time, it is a powerful instrument for rendering space, distance, travel, and communal values in patterns as the foundations for cultural and societal preservation.

The problem with referring to this vast part of the planet as "the Pacific" or even worse, "the South Pacific" (there is also the "North Pacific" influenced and historically

NOTES

administered by the United States) is that it provides a convenience of reference for "us" in a slightly more abstract sense than was the case when French explorer Jules Dumont d'Urville lineally divided Oceania into three sections in 1831 within which were also interjected racial structures that made it much easier for "us" to classify peoples as either Micronesian, Polynesian, or Melanesian. One could then make wide ethnic, observational swaths across the Pacific after the first intrepid European explorers recorded lines of contact that became historically distinct in terms of the pseudo-universal impacts that imperialism brings in time – degrees being relative to the conditions upon which the reigns of such impacts influence communal consciousnesses and lines of thinking as a whole. The contrast and in many cases, the clash between European assumptions of individualism – the notion of discrete and unassailable personhood – and the communal nature of Pacific peoples both established and absorbed by their histories and their distinctive recollections on the basis of local political and social expediencies – can be extreme. The extent of this extremity may in turn have a direct, albeit winding road – or line – to the befuddlement colonial policy makers have sometimes expressed to charges of usurping indigenous methods of knowing that form values and veins of historical identity through the imposition of universal notions of, in our case, educational approaches to the teaching of mathematics. For many European policy makers, these have not been impositions but really derivatives from another communal nature that draws its foundations from the achievements and successes of European discoverers throughout the ages. One could actually believe that its communal nature is not the antithesis of the communal natures of the Pacific and is in fact much greater in terms of its human content and time and is bound together through a shared history and subsequent verifications. Is that not the essence of a communal environment? The problem, however, is that its comprehension rewards the success of individuals along with the subsequent, at best, neutrality of those unsuccessful in its mastery. There are consequences to be had in this communal regiment.

As I noted in my Journal of Mathematics & Culture paper (Goetzfridt 2012), the long history of colonial influences and the dramatic and elongated colonial interventions in the Pacific (generally "America" to the North, England to the South, and France to the southeast) have predisposed scholarship to those antecedents of scope engendered by respective colonial powers – albeit for the positive, integrative impact that Pacific focused conferences have had in the past couple decades. I noted that Per Hage's 1978 explication of mnemonic structures used by traditional navigators from Polowat atoll in the Central Caroline Islands is a particularly noteworthy, early example of this integrative, hegemonically freed scholarship, given the juxtaposition of this analysis with comparative elements that are reflective of epistemologies among the Iatmul people (Sepik River area, Papua New Guinea) in the mnemonic retrieval schemes used as important clan mythological naming referents (Goetzfridt 2012, 230). Hage outlines the "phonetically and semantically interacting" patterning of mnemonic name cycles that aid in the recollection of totemic names which are crucial for clan identification and communal and land standing or position. The geographic, cartographic schemata used in Polowat voyaging repertoires offer - in similar structures - linear links between destination points composed of real and imagined reference islands and marine creatures indigenous to the islands concerned. Mnemonic patterning in both systems engenders the acquisition of esoteric knowledges in which names in relation to "myths" in both socio-cultural instances are learned, remembered, and used (perhaps most importantly in the assertion of one's position or one's clan's position in either society) through the process of doing. While as both Hage and Diaz (2015) note, the resultant implications are essential not only for, in the case of Polowat navigators, achieving landfall, they are also culturally significant in terms of the recollection of "traditions", "myths" or stories derived from the practical utilization and, in the case of the Polowatese people, mnemonic linear structures that have both practical and deeper meanings that guide the social worlds of their practitioners. One might consider the Hage example to be indicative or at least suggestive of what a breakage of the continental imposition of the image of the isolated "island" (Ibid.) might convey in terms of the egalitarian nature of the academic and human imperative against imperialism.

The premise of Pacific personhood is, by its very nature, contradictory and perhaps even ironic. Complex responses and issues of acculturation within distinctively colonized Pacific societies have led to individualized, inter-island and inter-societal responses. Lieber considered a person of Oceania to actually be "a locus of shared biographies: personal histories of people's relationships with other people and with things. The relationship defines the person, not vice versa" (Leiber 1990, 72). Rather than think of indigenous Pacific people as individuals, Geertz (1973, 364-367) also considered them better referred to as "cosociates" although this has also historically facilitated the classification of native peoples into what Howard describes as a ranking of "peoples along a linear sequence from savagery to civilization" with race becoming "indexical of group character, including moral character" (Howard 1990, 264). And it may very well be that restrictive island land spaces and subsequent close associations with kinship links and thus linkages to one's environment have subsequently ensured, throughout time, the contribution of the most organic of elements to the construction of these Pacific personhoods. European mobility and the diasporic feigning of geographic links have led to its antithesis and to the individualism that subsequently makes contemporary opposition to the universality of mathematics so, well, mysterious to these individuals that have had a role in its propagation.

But as Tongan scholar Epeli Hau'ofa (1994) has so famously conceptualized, the ocean has always been a complex composite of routes of communication that, in contrast to the forever "belittling" attitudes, policies, and practices of colonial administrators, established and continues to sustain a "sea of islands" in a vast expanse that requires even traditional epics to embellish themselves in order to adequately reflect this vastness – a vastness that completely contradicts the idea of isolated islands that have justified colonial ideas of native dependency. And in the age of the internet and jet travel, communication and mutual support between islands and kinships only continue to grow on a global scale.

In an environmentalist sense of this Pacific time and place, distinctive as those times and places are, it also behooves us to be reminded of the role of an educational system's curriculum, for good or worse, as essentially being designed "to pursue goals created, negotiated, and accepted by society" (D'Ambrosio 2013). 'Environment' can be as nebulous and thus as undeveloped a concept as the concept of 'information' – particularly if it is encompassed in an island environment where this curriculum - negotiated elsewhere - is imposed and thus denuded of its evolutionary foundations and linked to the 'information' upon which this foundation is subsequently dependent. At one level, it is the same old dualistic conflict between a formerly righteous narrativity of colonial power and influence debunked through

NOTES

forces of time and a kind of metaphysical recapitulation of wrong and right, leading to influential reconciliation with the oppressed through global forces of change. At another level – and one could argue at a higher and more ingrained, indigenously cellular level – is a conflict of native "primordial connectedness of landedness" (Diaz 2015, 91) from the perspective of indigenous epistemologies and cartographies as they apply to traditional Micronesian navigational techniques. Colonial constructions of "the island" and the physical isolationism with which "the island" is imbued miss, in "related apprehensions of land" (Ibid. 90) made conventional, defendable, and simply assumptive by these continentalisms, the indigenous epistemologies clearly evident in native cartographies on land and at sea (Goetzfridt 2016).

In his examination of the aesthetics of this narrativity of an Oceanic place, Diaz (2015) focuses on the navigational techniques of *etak* and *pookof* – the former involving the segmentation of a voyage to conceptualize time and space in a scheme of moving islands and a stationary canoe (Goetzfridt 2012, 225-227) and the latter concerned with an environmental reading of sea creatives indigenous to islands in this scheme whose realization - in the face of life and death scenarios at sea – depends upon ritualistic narratives of ancestral and contemporary master navigators (Ibid., pp.230-234). Both involve the faith of vision in the destination island before a voyage even commences that is in turn dependent upon these narratives that, for example, lead to pilot whales awaiting the voyager to the east of Guam before the island can even be seen.

Besides the fact that these natural elements, these creatures of this indigenous space and traditional chants or "words of work" on land and in sea often lead to accomplished results (survival at sea not being the least) and in time in many societies, honorific titles, is the understanding of wayfinding and island reaching discernments – birds, phosphorescence, the creatures of *pookof* - that expand islands many miles outward and thus diminish the isolative nature of an "island" that, in its "political fiction" of isolation, is however "necessary . . . in the constructedness of ideas of continentalness. Products of continental thinking, islands can better be understood as instrumentalities in opposition to which continental thinking could imagine itself into being, and imagine this being to be epicenters as big, cosmopolitan, sophisticated, worldly things and people" (Diaz 2015, p. 101). And therein lie the contours and the contexts of forms upon which analytical frameworks are or are not grounded upon this cosmopolitan conceptualization that in political, conceptual, and historically driven acculturative presences, restrict the means by which a cartography of indigenous proportions and locations can take its place in the expression of thought, practice, and by extension, the justices of freedom and peace.

The problem of course with looking at mathematics as a culture free, universal reality of the world is that it requires distinctive peoples – and in this case, in the Pacific – to adapt. They must drop or at least seriously compromise whatever self-imposed and self-developed pretentions they have about their environment and their communal values connected to that environment and adapt to what this universal western mathematical and scientific world tells them to adapt to – that is, if they really want to understand reality for what it is. After all, if indigenous values related to an environment and to practices tied to coping or dealing or thriving in that environment are to be sacrificed for some assumed greater measure of the world, then it can only be done by imperialistic or colonial methods. Either that or we begin to understand these Pacific indigenous environments for what they are and separated from the demand of a universal interpretation of mathematical approach and of reality itself. One writer who has done considerable work on the diverse counting systems in Papua New Guinea and on how indigenous peoples of this country respond or do not respond to the teaching of western mathematics in the classroom, considered western mathematics to be, as his paper title in an issue of Race & Class reads, "The Secret Weapon of Cultural Imperialism" (Bishop 1990). He rejected the conveyance of Western mathematics to Papua New Guinea school children as being a "dehumanized, objectified, ideological world-view" as well as the use of this conveyance to control social and even physical environments. Here again we have the simple questions of what constitutes reality for what people and what determines the applicability of a mathematical approach bound within a specific indigenous heritage?

While one could paint a life and death-like scenario when thinking and revealing cultural and societal heritages and histories in terms of what has been lost and what is unrecoverable, one can also recognize that many things are not really permanent or ground into a consciousness of a people to such an extent that even the all-encompassing acculturating forces of America in Micronesia, for example, have had no impact on the way children count, draw lines, and judge distances. This is a way of thinking however that honors the past and it is a way of thinking that is responsible for the common thread of cultural expression that we often hear about, particularly as it relates to indigenous, noninstrumental navigators of the past (and the small numbers of navigators of the present), to family structure and communal values and practices, and so on. By thinking that many things are not really permanent, we can perhaps believe that there are values that nevertheless allow even traditions that are not practiced in the present to live on and to live on with pride. These values resurrect the navigators of the past, the experts of knot divination, and as we see very clearly in contemporary handicrafts, traditional ideas of patterns and the meanings of those patterns.

There was a time when Hawaiians understood cosmology to be a number of circular and domed heavens. Like people of the Tuamotu Archipelago, the Hawaiians had a division of several strata. Each Polynesian society had distinctive understandings of these strata and their hierarchical structures in relation to the heavenly bodies that inhabited each level. Part of the Hawaiian word for horizon – *kukulu-o-ka-lani* – *kukulu* – refers to a "'wall' or 'vertical structure' and was used to denote the four pillars that were the principle supports of the heavenly dome" (Akerblom 1968, 15). Hawaiians also had a concept of a celestial sphere composed of zones with each division defined by the "parallels of declination" that denoted the sun's path at the points of the summer and winter solstices as well as the celestial equator and as one writer described it, "a southerly parallel of declination whose angular distance from the equator was twice that of the other two parallels" (Ibid., 19).

According to anthropologist Ward Goodenough - one of the first anthropologists to study Chuukese culture and society after the end of World War II - the Chuukese people also had a sky world in their cosmology and called it *Kachaw* which was often mistaken by European scholars as a Carolinian name for the island of Kosrae. A German scholar from the 1920s wrote that this sky world was conceptualized as a "disc bounded by the horizon" and the "expanse of heaven . . . likened to a great, upturned bowl. The space between the world below and the zenith was divided into several layers or distinct heavens" (Bollig 1927, 3-6, – cited in Goodenough, 1986, 557). Goodenough described characteristics of individual layers, beginning with the lower level, "Soaring Heaven," as being composed of winds, air, and birds; and the "Singing Place," which lay

NOTES

next to a path the souls of the dead followed on their way to heaven, although the path itself turned these souls to irresponsible dance and song that distracted them away from their final heavenly abode. Two other lands above the "Singing Place" provided places of rest for souls traveling to find their final resting places. Goodenough also noted that "as objects," these places "may have been the evening and morning star." And as I noted (Goetzfridt 2012), Goodenough also described layers beyond these transitory levels that were occupied by the gods, including the center of the Under Vault or Under Brow level where the chief of the gods resided. From the Chuukese perspective, "the sky world and the sea world seemed to merge at the horizon" beyond which was a "vast, unified expanse" (Goodenough 1986, 558).

What Micronesian student or any student for that matter would not be fascinated by this description of an indigenous perception – and there are other descriptions less and more complicated – of the way that their ancestors looked toward the sky and at the sky and the importance these conceptualizations held in their lives? That, and the fact that Polynesians – Hawaiians for example – held somewhat similar views of the structures of the sky and the levels of which it was composed. These similarities are in themselves interesting but even more interesting is the question of why such distinct and distant cultures had such similarities. Could they be based on ancient communications between these peoples that are lost to time? Or is it simply a matter of acknowledging a fundamental Pacific mid-set if you will or capacity or a tendency to think of the distant horizons which create such fascinating conceptualizations? We can even go further with this if we wish. And like much of Pacific histories and cultures, the extent that we can go is really only limited by our imaginations. And by that I don't mean making things up but rather the imagination as it relates to how these histories and cultures are to be approached, expressed, and examined down one road or another road or another road or another road.

We are not in these cases harnessing an ethnomathematical practice of the moment but we are rather harnessing the potential of a child's imagination based upon the cultural premises of his or her own experiences in that culture. If we wanted to use a somewhat more straightforward ethnomathematical practice of Pacific heritage, we could point to the navigator's concept of *etak*. *Etak*, in its most straightforward form, is simply a method of dividing a journey between a home island and an out-of-sight island and thus being able to tell where one is on that journey. In a voyaging strategy with six *etak* segments between two islands, a reference island off to the side is aligned with specific stars in order to create individual segments. When one has gone through three segments, one is of course halfway to one's destination. Simple division and a heritage based concept of alignment are applicable and teachable here.

Given that the assignment for this paper is to reflect upon my own past in relation to my past scholarship, I would imagine that there may be at least some nuances of expectations of definitiveness in terms of what I might tell readers outside of the Pacific about the Pacific and particularly as it relates to the core missions of the Journal of Mathematics and Culture. Nuances and instances and examples I can give but a finality of expression I cannot. I am reminded in this sense of my own ignorance of place, space, and time when my initial adaptability to the Pacific as I first arrived from the metropolis of Milwaukee was bound by the safeguards not only of my expectations and the initial administrative protections I consciously gleaned from as a beginning Peace Corps Volunteer in Fiji but also by the defensive nature of perceptual classifications. And these classifications, although I could not name them at the time, essentially revolved around race, poverty, and practices that were not only uncommon to me but which constituted the wrongful aberrations of the one reality from which I had yet to be liberated. My story would have to be one of those psychological explanations of how one encounters and ultimately deals with grief and death. It is particularly the denial stage that protects one from ever having to encounter anger – anger ultimately not in the context of these practices which nevertheless were carried out for some time in a kind of tender tolerance for my belittling of them but rather that anger that rises up after the fact and which must be then reconciled into the realities upon which one has been lifted. And it was and it is in that lifting – upon my own personalized scales of effort that nevertheless could not have been possible save for the patience and compassion of "the other" that I so stubbornly held for some time at arm's length – that has made it possible for me to understand what little I do about the Pacific and its peoples today.

One can always say that when one changes it is a bad idea to revert one's self emotionally back to this ignorance, given that the reliving of it is really not only trying to re-realize the incorrigible man but in a sense to also assail the progress made and the realities seen. But perhaps this time travel led by discernable points of consciousness is needed to help insure the humility upon which the finality of ideas about other societies, places, dreams, and senses of love is as unreachable as the strata of horizons toward which we all aspire.

NOTES

References

Akerblom, K. (1968). *Astronomy and navigation in Polynesia and Micronesia: A survey*. Stockholm: Ethnogratiska Museet.

Bishop, A. J. (1990). Western mathematics: The secret weapon of cultural imperialism. Race & Class 32(3), 51-65.

Bolling, P. L. (1927). Die bewohner der Truk-Inseln. *Anthropos Ethnologische Bibliothek*, 3(1). Münster i. W.: Aschendorffsche Verlagsbuchhandlung.

D'Ambrosio, U. and B. S. D'Ambrosio. (2013). The role of ethnomathematics in curricular leadership in mathematics education. *Journal of Mathematics Education at Teachers College*, 4, 19-25.

Diaz, V. (2015). No island is an island. In S. N. Teves, A. Smith & M. H. Raheja (Eds.), *Native studies keywords* (pp. 90-108). Tucson, AZ: The University of Arizona Press.

Geertz, C. (1973). *The interpretation of cultures*. New York: Basic Books.

Goetzfridt, N. J. (2012). Pacific ethnomathematics: The richness of environment and practice. *Journal of Mathematics & Culture* 4, 223-252.

Goetzfridt, N. J. (2016). The sovereignty of the line: A case for the indigenous swirl. *Pacific Asia Inquiry* 6(1), 130-143.

Goodenough, W. H. (1986). Sky world and this world: The place of Kachaw in Micronesian cosmology. *American Anthropologist* 88(3), 551-568.

Hage, P. (1978). Speculations on Puluwatese mnemonic structure. *Oceania* 49(2), 81-95.

Hau'ofa, E. (1994). Our sea of islands. *The Contemporary Pacific* 6(1), 147-161.

Howard, A. (1990). Cultural paradigms, history, and the search for identity in Oceania. In J. Linnekin and L. Poyer (Eds.), *Cultural identity and ethnicity in the Pacific* (pp. 259-279). Honolulu, HI: University of Hawai'i Press.

Lieber, M. D. 1990. Lamarckian definitions of identity on Kapingamarangi and Pohnpei. In J. Linnekin and L. Poyer (Eds.), *Cultural identity and ethnicity in the Pacific* (pp. 71-101). Honolulu, HI: University of Hawai'i Press.

Ethnomodelling as a Creative Insubordination Approach in Mathematics Education

Volume 1

Milton Rosa, Ed.D.
Universidade Federal de Ouro Preto
milton@cead.ufop.br

Daniel Clark Orey, Ph.D.
Universidade Federal de Ouro Preto
oreydc@gmail.com

Abstract

The application of modelling methods usually makes sense to researchers and educators (global, outsiders) when they understand that these techniques exam mathematical patterns developed by the members of distinct cultural groups (local, insiders). What is less evident is how these professionals attempt to make sense of the underlying cultural frameworks within which diverse mathematical ideas, procedures, and practices are embedded. Currently, an important dilemma in mathematics education is its overwhelming bias against a local orientation in its research paradigm. Thus, a search for innovative methodologies such as ethnomodelling is necessary to record historical forms of mathematical ideas, procedures, and practices developed for use in diverse cultural contexts. It is important to highlight that ethnomodelling is not an attempt to replace globalized academic mathematics, yet, at the same time, it is necessary to acknowledge the existence of local mathematical knowledge in mathematics curriculum. In this context, the insubordination triggered by ethnomodelling is creative and evokes a disturbance that causes a review of rules and regulations in the mathematical modelling process. This process increases the potential for continual growth in the debate about the nature of mathematics as it relates to culture since it proposes a dialogue between local and global approaches to the construction of mathematical knowledge.

Keywords:

- Ethnomodelling
- Local Approach
- Global Approach
- Glocal Approach
- Creative Insubordination.

Introduction

The acknowledgement of the relation between culture and mathematics can be interpreted as a reaction to *cultural imperialism* that imposed its version of mathematical knowledge on colonized communities around the world with the expansion of the great navigations from the fifteenth century (D'Ambrosio, 1990). In order to keep up with modern Western developmental models, other cultures have been forced to adapt to these paradigms or perish. In this regard, mathematics perpetuates to some extent imperialist goals, thus, it is perceived as a secret weapon that maintains the imposition and domination of Western cultural values on local cultures (Bishop, 1990).

School mathematics has also been criticized because it helps to reinforce a Western/ Eurocentric approach in the prevailing curriculum as well as for helping to globalize particular kinds of mathematical technologies and ideologies (D'Ambrosio & D'Ambrosio, 2013) that supports the maintenance of cultural imperialism. However, the development of non-prescribed strategies to solve problems in diverse societal domains is an alternative method as well as an important tool for identifying innovative problem-solving techniques and mathematical ideas, procedures, and practices in ethnomodelling research.

The reaction to this cultural imperialism can be also related to the development of the concepts of *creative insubordination* (Crowson & Morris, 1982), *responsible subversion* (Hutchinson, 1990), and *positive deviation* (Zeitlin et al.; 1990). These concepts are equivalent as they "relate to the flexibility of rules and regulations in order to achieve the welfare of the members of distinct cultural groups" (Rosa & Orey, 2015a, p. 133). In this paper, we apply these three terms interchangeably because the amplitude of their concepts embraces innovative solutions in the pedagogical action of mathematical through ethnomodelling, which helps to confront the belief that persists in contemporary society that mathematics is a culturally neutral knowledge.

The historical and contemporary relations between culture and mathematics illustrate that mathematics is not culture-free. Accordingly, the culturally specific nature of mathematics should be acknowledged in order to describe mathematical ideas and procedures practiced among the members of distinct cultural groups such as tribal societies, labor groups, professional groups, social classes, and children of a certain age group (D'Ambrosio, 1985).

Consequently, it is important to search for alternative methodological approaches as Western mathematical practices are accepted worldwide in order to record historical forms of mathematical ideas, notions, procedures, and practices that occur in different cultural contexts. Hence, the members of distinct cultural groups apply innovative mathematical solutions to the challenges faced by society, which are identified and refined from the ideas, procedures, and practices they develop from generation to generation.

One alternative methodological approach is *ethnomodelling* that may be considered as the practical application of ethnomathematics that adds cultural perspectives to the modelling process (Rosa & Orey, 2012). The "application of ethnomathematical techniques and the tools of mathematical modelling

NOTES

allow us to see a different reality and give us insight into science done in a different way" (Orey, 2000, p. 250). The pedagogical approach that connects the cultural aspects of mathematics to mathematical modelling is *ethnomodelling* (Rosa & Orey, 2010).

As a creative insubordination process, ethnomodelling seeks to change the "outside existing paradigms and conflicts with prevailing values and norms" (Marzano et al., 2005, p. 113) since it stands for the development of mathematical ideas, procedures, and practices that have its roots within distinct cultures. In this context, ethnomodelling binds contemporary views of ethnomathematics and, simultaneously, recognizes the need for a culturally based view on the modelling concepts and processes. Studying the unique cultural differences in mathematics encourages the development of new perspectives on the scientific questioning methods.

Research on culturally bound modelling ideas may address the issue of mathematics education in non-Western societies by bringing the local cultural aspects into mathematical teaching and learning processes (Eglash, 1999). This approach reveals responsible aspects of subversion in the ethnomodelling process that are identified with an ongoing movement that challenges the *status quo* of academic mathematical knowledge by aiming at to alter the system in creative ways (Lyman et al., 2005) in order to better serve the needs of the students. Essentially, it involves looking at issues from perspectives outside of existing educational systems and pedagogical models.

Similarly, ethnomodelling can be considered as insubordinate and creative educational approaches because it disrupts the existing order in the academic mathematics, as it does not follow the linear modelling approach prevalent in schools. It develops the study of ideas, procedures, and mathematical practices that are found in distinct cultural contexts. Hence, ethnomodelling attempts to break bureaucratic rules of academic mathematics in order to recognize different techniques and value diverse modes of producing mathematical knowledge in diverse cultural groups (Rosa & Orey, 2015a).

This context allows ethnomodelling to challenge the prevailing traditional mathematical ways of thinking, thus, this positive deviance approach involves thoughts and/or actions that differ *from* the imposed norms and regulations (Dehler & Welsh, 1998). In the anthropological and sociological point of view, this act of responsible subversion examines how individuals solve problems in spite of or in opposition to the formal system or the commonly accepted rules (Hutchinson, 1990). Historically, mathematical knowledge takes different forms in different cultures as well as Western worldviews on the ideas of modelling begin to shift in order to acknowledge that this process is culturally bound.

Local (Emic), Global (Etic) and Glocal (Dialogical) Aspects of Ethnomodelling

When researchers investigate the members of distinct cultural groups, they may be able to find distinctive characteristics of mathematical ideas and procedures that we might label as ethnomodelling. However, an outsiders' (global, etic) understanding of objectivated cultural traits is always an interpretation that may misinterpret mathematical practices developed by the members of these groups.

This happens because members of specific cultural groups often have their own unique interpretation of their cultural traits, including the mathematical knowledge passed from one generation to another. This is the *local* (emic[1]) approach opposed to an outsider's interpretation of these cultural traits, which is the *global* (etic) approach. According to this context, there are three approaches (global, local, and glocal) to be considered in order to investigate and study the mathematical ideas, procedures, and practices developed by members of distinct cultural groups.

a) Global (Etic) Approach

Global or etic approaches can best be defined as an *external* or *outsider's* view on the beliefs, customs, scientific and mathematical knowledge developed by members of distinct cultural groups. The ongoing globalization process we are all part of, deals with cultural convergences that perceive mathematical knowledge in terms of increasing homogenization or sameness. It, either willingly or not, denies and ignores the diversity of cultures across social, cultural, political, economic, environmental, and geographic boundaries; often viewing western/academic forms of mathematics as correct, universal and monolithic and that can only emanate from the Euro/academic tradition. This process has inculcated an ethnocentric fear of, or at very least disinterest and disrespect for the many mathematical differences regarding symptoms of *cultural alterity*[2]. This approach is considered as *culturally universal* (Sue & Sue, 2003).

b) Local (Emic) Approach

Local or emic approaches can best be defined as the insiders' view about their own customs, beliefs, and scientific and mathematical knowledge traditions; and encompasses the skills, abilities, competencies, experiences, and insights developed by the members of distinct cultural groups as they use it in their daily lives and perhaps to maintain or improve their livelihood. Focusing on the local entails a more conceptual and dynamic interaction that developed and adapted continuously to gradually changing environments and passed down from generation to generation and closely interwoven with people's cultural values. It is also the *cultural and social capital*[3] of the members of these groups; that is it is their main asset used to invest in the struggle for survival, to solve problems faced daily, and to achieve control of their own lives. Local knowledge influences conceptions of the greater world, but it is important to understand how it changes in different times and places in relation to the broader political,

---

[1] The terms emic and etic were originally introduced by Pike (1954) who drew on an analogy with the linguistic terminology phon***etics*** and phon***emics***. Phon***etics*** are the general aspects of all possible vocal sounds production in languages. Phon***emics*** are sounds locally significant that are used in a particular language. Just like in the study of a language's sound system, it is possible to apply at least two approaches in the investigation of distinct cultures, which are the point of view of either the *insiders* or the *outsiders*. In this context, researchers can conduct investigations in mathematics education from the *emic* and *etic* approaches.

[2] Cultural alterity is the process by which societies and cultures may exclude particular group of people because of their otherness. Thus, this term has embedded in it a negative connotation because of practices like stereotyping that allow people to use social markers to construct identity (Wexler, 2004).

[3] Social capital is the sum of the resources, actual or virtual, that accrue to an individual or a group by virtue of possessing a durable network of more or less institutionalized relationships of mutual acquaintance and recognition. Cultural capital includes non-economic resources that enable social mobility such as knowledge, skills, and education. Both concepts remind us that social networks and culture have value (Bourdieu & Wacquant, 1992).

NOTES

social, economic, environmental, and cultural contexts (Yifeng, 2009). This approach is considered as *culturally specific* (Sue & Sue, 2003).

c) Glocal (Dialogical) Approach

A glocal approach can best be considered as the dialogical relation between globalization (etic) and localization (emic). This approach involves blending, mixing, and adapting two or more distinct knowledge systems or cultures and is known as *glocalization*[4] (Robertson, 1995). In order to be meaningful, glocalization must include at least one component that addresses the local culture, system of values, or knowledge systems (Khondker, 2004). Glocalization describes the relationship between the local and the global as interdependent and mutually constitutive approaches and "challenges notions of cultural imperialism because the term suggests a negotiation process that appears to start from the inside out, i.e., a process that begins with a high regard for the local" (Maynard, 2003, p. 57). The main objective of this positive deviance procedure that departs from established norms or rules is to modify these regulations by applying innovation, creativity, and adaptability (Walker, 2005).

Because glocalization arises from culture and through interactions, adapts to the changes in culture, the dialogical approach is an active force of the ethnomodelling process. Indeed, our notion of contextualization is not an exclusionary view; rather it is an attempt to harness positive aspects of globalization of mathematical knowledge in a sustainable way. In this regard, well-conceived contextualized mathematical practices do not promote any forms of hegemony apart from the emphasis on dialogical relationships between sometimes opposing, or contradictory, and complementary approaches.

It is important to highlight here that in our point of view, the glocal approach is a reaction to globalization, or a reinforcement of cultural identity at the local community level. This means that, in ethnomodeling processes, mathematical knowledge might consider the worldwide connections, but also with the specific conditions of the local knowledge. This insubordination aspect of mathematics education is creative because it involves adapting rules in order to change, challenge, or even subvert the regulation of the implementation processes of the norms.

Dialogical approaches (glocalization) of ethnomodelling help us become more mindful of forms of the hegemony prevalent in mathematics through the application of dialogical approaches in ethnomodeling. Thus, it is necessary to incorporate cultural-based forms of knowledge and the notion of dialogical approaches and the continuous changes that arise in the process of teaching and learning mathematics.

It is necessary to state here that we are not enforcing another form of dualism, which is globalization (etic) versus localization (emic). Indeed, our intention is to contest the narrow view of globalization that allow for new or alternative traditions and developments of mathematical ideas, procedures, and practices. This is important in order to demonstrate that global approaches are not necessarily exclusive constructs, rather they coexist with localization

Ethnomodelling as a Glocalization Process

The debate between local (emic) and global (etic) mathematical knowledge traditions has a long tradition in conducting investigations. Some researchers have made important distinctions between the culturally specific and the culturally universal (Headland,

[4] Glocalization is a term coined by Robertson (1992) that merges the concepts of globalization and localization.

Pike, & Harris, 1990). Unfortunate misunderstandings have arisen with the assumption that this distinction implies dichotomous approaches to cultures (Pike, 1967), and in this case, diverse mathematical thinking or traditions.

A *local or emic approach* seeks to understand a particular phenomenon from the point of view of the members of distinct cultural groups (Pelto & Pelto, 1978) since they are the sole judges of the validity of emic descriptions on their own mathematical terms and contexts. Emic approaches focuses on studying sociocultural aspects of mathematical phenomena from within a specific cultural context and understanding, as the people from within that culture understand it (Gudykunst, 1997).

A *global or etic approach* seeks to understand particular phenomena by means of analytical tools and concepts drawn from the outsiders' worldviews (Pelto & Pelto, 1978). The etic approach analyzes human behavior and knowledge with focus on universals because they can be compared across cultures using common definitions and metrics (Berry, 1969). It relies upon extrinsic concepts and categories that have meaning for researchers and educators who are the sole judges of the validity of etic accounts.

In general, local refers to taking the insiders1viewpoint whereas global means taking the outsider's viewpoint. In this context, in a global approach, mathematical phenomena is cross-culturally understood rather than comprehended by culturally specific meanings and contexts. While local and global are often thought to create a conflicting dichotomy, they are, in reality, complementary viewpoints (Rosa & Orey, 2015b). Thus, rather than posing a dilemma, the use of both approaches deepens our understanding regarding cultural understandings and comprehensions.

However, cultural bias, is an unfortunate consequence, and occurs when researchers come to assume that a local (emic) construct is actually a global (etic) construct. For example, this result wrongly imposes the culturally-universal mathematical knowledge to culturally-specific mathematical ideas, procedures, and practices emically developed by the members of distinct cultural groups. The study of cultures according to pre-established etic procedures impedes the discovery of cultural diversity whereas emic analysis truly broadens this view (Headland, Pike & Harris, 1990). In this context, an emic analysis focuses on the behaviors and the knowledge unique to cultures or in diverse ways in which etic activities are carried out in a specific cultural settings.

Moreover, it is often presupposed that this methodological dichotomy corresponds to the opposition between the development of mathematical knowledge of the members of distinct cultural groups and their external experiences. In this context, the etic approach is equated with the objective explanation of sociocultural and mathematical phenomena from external points of view, while emic approaches are identified with the sympathetic comprehension of subjective experiences from the internal point of view (Harris, 1980).

Local and global mathematical knowledge traditions should not be studied in isolation from one another but as mutually constitutive parts of a single expression. Ethnomodelling provides a lens through which it is possible to perceive both homogeneity and heterogeneity of mathematical ideas, procedures, and

NOTES

practices by allowing members of distinct cultural groups to increase their understanding of specific responses to global forces.

Glocalization takes place when local (emic) and the global (etic) mathematical knowledge encounter each other, and refers to the mixing of *cultural traits*[5] developed in distinct cultures that have created cultural patterns that where not previously known by outsiders. It is the interpenetration of the local with global mathematical knowledge that can result in unique outcomes in different cultural groups. It can be perceived as the local globalization or interchangeably as global localization (Robertson, 1995). Thus, the ethnomodelling process values the fusion of various cultural inputs into new and unique products regarding to mathematical practices.

The exclusive view of globalization cannot help us realize the disempowerment and the loss of mathematical traditions or the limitations of a hegemonic worldview. On the other hand, an extreme advocacy of localization cannot empower students to apply multiple referents to their mathematical creativity.

Therefore, this dialogical process (glocalization) provides us with opportunities to challenge both forms of hegemony, thereby, opening for multiple opportunities through the interplay of mathematical ideas and actions in context. It helps researchers and educators to examine the limitations and advantages of knowledge systems arising from different worldviews, thereby developing an honest and more contextualized version in their pedagogy (Rosa & Orey, 2015a).

[5] A cultural trait is a socially learned system of beliefs, values, traditions, symbols, and meanings that the members of a specific cultural group acquire throughout history. It identify and coalesce a cultural group because traits express the cohesiveness of the members of the group. It is a deposit of knowledge, experiences, actions, attitudes, hierarchies, religion, notions of time, roles, spatial relations, concepts of the universe, and artifacts developed by the members of distinct cultural groups in the course of generations through individual and group strivings (Samovar & Porter, 2000).

Ethnomodelling considers the techniques and strategies rooted in local (emic) and glocal (etic) mathematical knowledge systems. Throughout history, humanity has developed tools that enable members of distinct cultural groups to explain, understand, and comprehend the world around them. Hence, transcendence was responsible for the development of the tools, techniques, codes and communication skills, which helped humanity to expand the perception of the past, present, and future (D'Ambrosio, 2015). In addition, mathematical facts are organized as techniques and strategies that evolve representations of reality (ethnomodels) explanations systems on the phenomena that occur in everyday life.

Ethnomodels as Representations of Local, Global, and Glocal Knowledge

By using ethnomodels, humanity tries to understand the world by means of explanations that are organized as procedures, techniques, methods, and theories, as it aims to explain and consider daily realities and phenomena. These strategies are historically organized in every culture as knowledge systems (D'Ambrosio, 2015).

Researchers and educators, if not blinded by their prior worldview, should come out with an informed sense of the diversity of mathematical ideas, procedures, and practices that are modeled since they should be able to tell outsiders (global, etic) what matters to insiders (local, emic) and vice versa (Rosa & Orey, 2011).

Ethnomodelling tends to privilege the organization and presentation of mathematical ideas, notions, and procedures developed by members of distinct cultural groups by encouraging the elaboration of local, global, and glocal ethnomodels by the people who actually do it. Thus, the:

> *(...) elaboration of models that represent these systems are representations that help the members of these groups to understand and comprehend the world by using small units of information, named* ***ethnomodels****, which link their cultural heritage with the development of the mathematical practice. This approach helps the organization of the pedagogical action that occurs in classrooms through the use of the local aspects of these mathematical practices (Rosa and Orey, 2015a, p. 140).*

Local (emic) ethnomodels reflect the information and observations that have come to represent the target population's own vocabulary, scientific and mathematical knowledge, conceptual categories, language of expression, and cultural belief system. It deals with the diversity of mathematical knowledge and traditions that makes a difference from the *insiders'* point of view. These ethnomodels are grounded in what matters in the world of the people being modeled (Rosa & Orey, 2013).

These ethnomodels attempt to investigate and understand phenomena and their structural interrelationships through the eyes of the members of a specific cultural group. The primary goal of these ethnomodels is to develop a descriptive idiographic orientation that describes the effort to understand the meaning of contingent, unique, and often-subjective mathematical phenomena since it emphases the uniqueness of mathematical practices developed by the members of distinct cultural groups.

Global (etic) ethnomodels reflect the information collected in terms of the conceptual systems and categories of the researchers. It attempts to identify lawful relations and causal explanations that are valid across different cultures. These ethnomodels are built on the outsiders' view of the world of the people being modeled (Rosa & Orey, 2013).

The analysis of global ethnomodels is comparative since it examines many different cultures by using standardized methods and categories. Thus, etic knowledge is essential for cross-cultural comparisons that demand standardized units and categories, which are the essential components of ethnology.

Glocal forms of ethnomodeling can capture the melding processes of interpretation. While rooted within the discursive arena of local mathematical ideas, procedures, and practices, these ethnomodels also includes uses of global mathematical knowledge. Rather than presenting local and the global perspectives as opposite sides of a single sociocultural spectrum, glocalized ethnomodels recognize the ways in which both local and glocal mathematical knowledges act upon one another.

It highlights both the structural constraints and the individual and collective agencies by referring to real world endeavors that may (re)contextualize global mathematical phenomena with respect to local cultural traditions. Thus, it critically transcends the binary oppositions of the emic and the etic mathematical knowledges and acknowledges the *co-presence* of sameness and differences, and the intensified interpenetration of the local and the global (Giulianotti & Robertson 2007) in the process of the elaboration of glocal ethnomodels.

NOTES

Thus, in the elaboration of ethnomodels, if researchers wish to make statements about universal (global, etic) aspects of mathematical knowledge, these statements need to be phrased in abstract ways as well as based on attributes of mathematical patterns found across many cultures. Conversely, if researchers wish to highlight the meaning of these generalizations in specific (local, emic) ways, then it is necessary to refer to a more precisely specified mathematical knowledge.

In this context, it is important to acknowledge the interplay of globalization and localization and the fact that we are microcosmic reflections of how globalization works on a local community level (Mendis, 2007). Thus, glocal ethnomodels incorporate knowledge systems arising from local and global cultural practices through dialogical approaches.

An Ethnomodel of the Gable

Informants from a roofing contractor cultural group can easily describe the practices acquired for the construction of a roof gable, which is the most commonly used type of pitched roof construction. After choosing the type of tile such as red roofing tiles or shingles to begin the construction of the roof, it is necessary that roofing contractors calculate the slopes for the beams that form the triangles in the gable.

Gabled roofs often possess a ridge near or at the center and slopes in two directions. It is simple and common in design, economical to construct, and can be used on any type of structure, and in any type of climate. Roofing contractors use triangles because they are stable, rigid and do not move. The main objective of the roof is to provide protection from climate because they must be strong enough to withstand high winds, and shed moisture, and often snow and ice quickly.

For example, Rosa (2014) states that in the case of many Brazilian roofs, roofers calculate the slope of the roof by applying a ratio between the height and the length of the gable, which they express as a percentage. In this context, the percentage of the slope (trim) for the roof to the tiles is at least 30% so that rainwater can quickly drain. According to this kind of mathematical knowledge they acquired through observation and experimentation, for each meter (100 cm) that runs horizontally, there is a vertical rise of 30 cm.

Thus, if the length of the gable is L = 8 meters, roofing contractors mentally perform the percentage calculation by using a = 4 meters, which is half of that measure. Then, they multiply it by the percentage of the slope of the roof. For example, 30% of 4 meters corresponds to the height of 1.20 m. Figure 1 shows the scheme of a gable used in most of the roof constructions in Brazil.

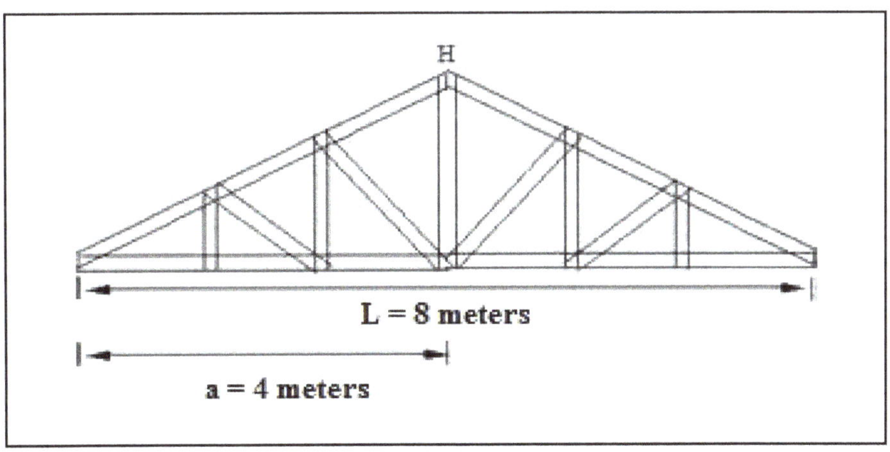

Figure 1: Scheme of a gable used in roof constructions in Brazil

This context allows researchers and investigators to describe this mathematical practice (emic, local) by using the Pythagorean Theorem (etic, global). However, it is important to understand the dialogical (glocal) relationship between these two approaches. For example, the informal calculation (emic knowledge) of the height (trim, flow) of the gable does not preclude the use of the Pythagorean Theorem (etic knowledge) by these professionals. This means that these professionals strive to compare, interpret and explain the type of mathematical knowledge they observe and that the members of this particular cultural group are experiencing.

The results of the study conducted by Rosa (2014) shows that the glocal observation of this mathematical practice tries to understand it from the perspective of the internal dynamics of the Brazilian roofers (local) while providing cross-cultural comparisons in order to comprehend it from the point of view of individuals from different cultural backgrounds (global). Thus, this approach is necessary to comprehend and explain this particular mathematical practice as a whole from a dialogical point of view.

Glocal Characteristics of Translation

Issues of cultural differences and translation strategies are examined in the context of ethnomodelling. Ethnomodelling investigators often use *translation* to describe the modelling process to translate emic knowledge systems (Rosa & Orey, 2013), which is a process that depends "on acts of translation between emic and etic perspectives" (Eglash et al., 2006, p. 347). Accordingly, ethnomodelling attempts to establish relations between emic mathematical ideas and procedures embedded in local mathematical practices (designs, patterns, and symmetry) to etic (global) conceptual frameworks (Rosa & Orey, 2011).

In this context, mathematical practices can be seen as arising from emic rather than etic origins. However, in some cases, the *translation* from emic to etic mathematical knowledge is direct and simple such as in examining diverse counting systems or calendars (Rosa & Orey, 2015b). In other cases, mathematical knowledge is *embedded* in processes such as iteration in beadwork and in Eulerian paths in sand drawings. As well, there is a need to be cautious in this process because it is easier to use numeric systems and counting procedures rather than to understand the embedded mathematical knowledge found and applied in architecture and crafts.

For example, results of a study conducted by Eglash at al. (2006) show that frequently local (emic) mathematical knowledge such as the application of the symmetry classifications from crystallography to local textile designs and patterns are merely analysed from an (global) etic approach. In this process, "translators need to understand the cultures towards which they are translating" (Séguinot, 1995, p. 56).

In this context, local mathematical knowledge as redefined within conceptual frameworks of glocalization are reinforced through translations that invite and introduce differences and allow or encourage members of distinct cultural groups to interact with each

NOTES

other. Thus, translation conceptualizes the relationship between the global knowledge and the local culture.

In mutual encounters of members of distinct cultural groups, the emerging *otherness* necessitates a translation, which is primarily concerned with giving it its due without subsuming it under pre-conceived notions. Such an attempt highlights the translatability, as an operational mode, that marks it from the underlying assumptions that guide both comparative and typologies of cultures (Iser, 1994).

In this regard, an important transfer takes place when two or more cultures meet and interact, as the language, scientific, and mathematical knowledge of one cultural group passes or is shared into the interpretative realm of another (Iser, 1994). In this process, the translation of mathematical ideas, procedures, and practices of the observed culture is understood and comprehended through glocal terms that are different in temporal and special frames and is transformed. This approach raises the issue of cultural and temporal situatedness and the effect it has on our lack of or ability to appreciate other cultures.

The dialogical approach (glocalization) of ethnomodelling explores the diverse forces shaping the products and processes of transcultural and translational mathematical phenomena. It also emphasizes the links and tensions between local (emic) and global (etic) approaches by looking for a more targeted study of mathematical ideas, procedures, and practices.

Ethnomodelling seeks to equip students with a sound critical, reflective, and methodological framework for analyzing the complex interactions within and across diverse communities of linguistic, cultural, historical, political, and social practices through the elaboration of ethnomodels. The corollary is a more complex relationship between local and global mathematical discourses as reflected in translation, which is a constant process of decolonization in its cultural reproduction open to cultural specifics inherent in different traditions.

An Ethnomodelling Curriculum

The concept of positive deviance is useful, offering researchers and educators a basis for decision making when expected actions collide with their perception of mathematics curriculum. It involves an intentional act of breaking the rules in order to serve the greater good of the students. Researchers and educators who are positive deviants must question and discuss opposing *status quo* of mathematical knowledge in order to enact meaningful changes into the mathematics curriculum.

One of the goals of ethnomodelling curriculum is to add cultural components to the modelling process. Thus, instead of being another research paradigm itself, ethnomodelling aims at encouraging the search for mathematical ideas, procedures, and mathematical practices that are culturally bound as well as to their examination and adoption into mathematics curriculum (Rosa & Orey, 2015c). Traditional mathematical modelling methodologies in school curricula do not fully take into account the implications of the cultural aspects of local systems.

Mathematical curriculum conceived in an ethnomodelling approach helps students to develop mathematical concepts and practices that originate in their own cultural traditions by linking them to formal academic mathematics. The understanding of conventional mathematics then feeds back and contributes to broader understandings of culturally based mathematical principles (Rosa, 2010).

Classrooms should not be isolated from the communities in which they are embedded, thus, they are part of a larger community with defined cultural practices. In this context, classrooms may be considered as learning environments that facilitate the application of pedagogical action developed through the application of an ethnomodelling curriculum, which allows for a broader analysis of the school context in which pedagogical actions transcend school environment since curricular practices embrace sociocultural contexts of the students.

An ethnomodeling-based curriculum can integrate mathematical ideas, procedures, and practices that originate in the cultural background of student's with the conventional, formal, and traditional mathematical knowledge. In such classroom environments, students build on what they know as well as on the experiences they have from their cultural environments. These experiences are used as part of understanding how mathematical ideas are developed and how they are built into systems, formulated, and applied in various ways within the culture.

The objectives for developing an ethnomodelling curriculum include: a) to assist students to become aware of how people mathematize and think mathematically in their own culture, b) to use this awareness to learn about formal mathematics, and c) to increase the ability to mathematize mathematical practices in distinct cultural contexts. This curriculum applies cultural experiences as vehicles to make mathematics learning more meaningful and, more importantly, to provide students with the insights of mathematical knowledge as embedded in their own sociocultural environments (Rosa & Orey, 2008).

This curriculum leads to the development of cultural activities that enable students to become aware of the mathematical potential found in their own communities and cultural traditions so that they are better able to understand the nature, development, and origins of academic mathematics (Rosa & Orey, 2010). This implies that an ethnomodelling curriculum is not just about the application of relevant connections in learning mathematics, but it is also about generating formal mathematics from cultural ideas.

The comprehension of the acts of creative insubordination in the process of learning mathematics generated from ethnomodeling processes enable the development of teaching strategies that help researchers and educators to apply methodological decisions related to teaching practices (Rosa & Orey, 2015a). This approach helps to improve the mathematics performance of students by modifying, adapting, and flexibilizing curricular practices.

Implications of an Ethnomodeling Curriculum

One of the primary issues regarding the mathematics curriculum is concerned to the position of researchers and educators in relation to the *etic* (global) and *emic* (local) approaches. In this regard, pedagogical work on mathematical content developed in classrooms may be based on the researchers and educators' own worldviews, which relates to culturally-universal (global, etic), culturally-specific (local, emic), or culturally-dialogical (glocal) approaches to mathematics education.

Both researchers and educators who operate from etic (global) positions have been taught to see mathematical ideas, procedures, and practices as occurring in the same way in every culture. They can learn, however, to

NOTES

base their beliefs in relation to Western versus non-Western mathematical traditions in which the members of distinct cultural groups construct, develop, acquire, accumulate, and diffuse the same kind of mathematical knowledge (Rosa & Orey, 2012).

Researchers and educators who take on an emic perspectives believe that many factors come into play when mathematical ideas, procedures, and practices are developed in regards to the cultural backgrounds of the members of distinct cultural groups. These factors include diverse sociocultural values, morals, and lifestyle. For example, different cultures have developed different ways of doing mathematics in order to understand and comprehend their own cultural, social, political, economic, and natural environments. It is necessary to highlight that students operate from an emic approach.

Since these professionals have come to believe that the cultural backgrounds and life experiences can greatly influence the overall development of mathematical knowledge of students, they then use of culturally specific strategies in the pedagogical work of teaching and learning mathematics. In this regard, they are come to see that current worldwide guidelines and standards for mathematical instruction are very much culturally bound (Rosa, 2010). According to this discussion, it is important that both researchers and educators understand that diverse experiences, lifestyles, cultural values, and overall worldviews influence the development of mathematical knowledge (Rosa & Orey, 2013).

Another issue to discuss here is to determine the belief systems related to researchers and educators in relation to cultural universality, especially that which focuses on similarities and minimization of cultural factors. As well their understanding of techniques and beliefs that focus on cultural differences. Characteristics of culturally dialogical approaches in ethnomodelling research provide conditions in regards to the development of intercultural competence, which is the "ability to communicate effectively in cross-cultural situations and to relate appropriately to a variety of cultural contexts" (Bennett & Bennett, 2004, p. 149).

Similarly, intercultural competence is the ability to "develop targeted knowledge, skills and attitudes that lead to visible behaviour and communication that are both effective and appropriate in intercultural interactions" (Deardorff, 2006). Figure 2 shows intercultural competence in the ethnomodelling process.

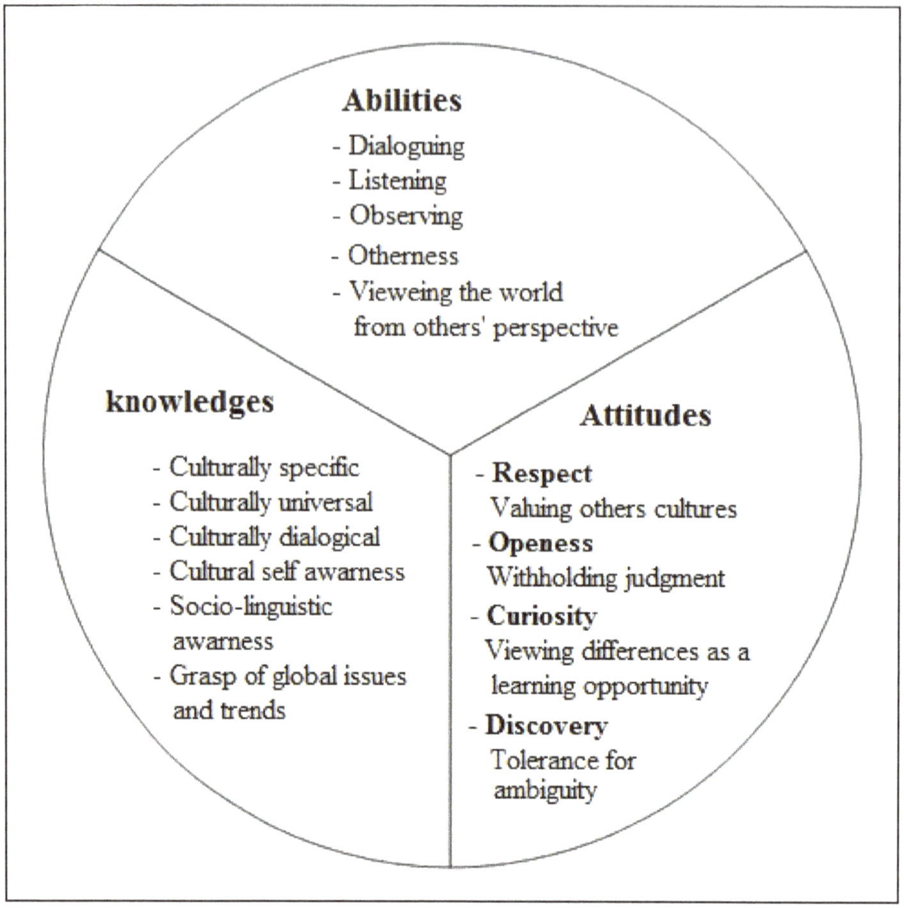

Figure 2: Intercultural competence in the ethnomodelling process
Source: Adapted from Deardorff (2006)

The question is, then, whether it is necessary to understand cultural specificity (local, emic) against the background of universal theories and methods (global, etic) that can be susceptible to cultural differences and to the demands of cultural contextualizations. However, results from culturally specific investigations encourage more cross-cultural research that supports the development of emic perspectives (D'Ambrosio, 1990; Eglash et al, 2006; Rosa & Orey, 2010).

This context strengthens the notion that mathematics cannot be conceived as a universal language because its principles, concepts, and foundations are not the same everywhere (Rosa & Orey, 2007). Conversely, it is naïve to state that the members of distinct cultural groups do not share universal mathematics ideas, thus, some mathematical activities are widely practiced across cultures.

For example, many of the everyday activities members often perform daily involve a substantial amount of mathematical

NOTES

application. In this regard, counting, measuring, designing, locating, explaining, and playing are six universal activities that are practiced by the members of any cultural group. These universal activities provide fundamental facets used to probe traditional daily living practices (Bishop, 1993). It is also important to highlight that these activities are intertwined with other aspects of the daily routine of these members.

On the other hand, there are cultural differences within these universal activities. Academic mathematics may look the same in many societies because there is a competitive social, economic, environmental, and political ethics that demand a competitive mathematics curriculum. In this regard, dominant cultures may have imposed their forms of mathematics on other societies since the beginning of the development of the mathematical knowledge.

In this regard, even though these activities are considered universal, it is important to recognize that they are merely universal to those individuals who share the same cultural characteristics and historical perspectives. On the other hand, it is equally naïve to believe that mathematical ideas and procedures do not reflect the cultural values and lifestyles of the members of cultural groups (Rosa, 2010).

By applying an innovative pedagogical approach to these opposing views, it may be helpful to understand the universality of mathematical ideas, procedures, and practices that could be relevant to researchers, educators, and the members of a specific cultural group (Rosa & Orey, 2015c). Hence, it is necessary that this approach takes into consideration the relationship between cultural norms, values, attitudes, and the manifestation of mathematical ideas, procedures, and practices developed in different knowledge fields as well as in the context of distinct cultures.

If researchers and educators become self-aware of themselves and their worldviews, cultural paradigm, and values, then they may be more open to apply aspects of ethnomathematics and modelling in their pedagogical practices through the application of ethnomodelling (Rosa, 2010). This may lead them to a clearer decisions in the application of local (emic), global (etic), and glocal (dialogical) approaches.

Final Considerations

The tragedy of the impending disappearance of local (emic) knowledge is most obvious to those who live it, but the implication for others can be detrimental as well when mathematical procedures and techniques, technologies, artifacts, and problem solving strategies are lost during the development of contemporary society. Defined in this manner, the usefulness of both emic and etic knowledge is evident and necessary.

The majority of mathematics researchers and educators have been enculturated to, more or less, one particular cultural worldview. They therefore need a means of distinguishing between the answers they derive as enculturated individuals and the answers they derive as observers. Defining emics and etics in epistemological terms provides a reliable means of making that distinction. From this perspective, the two definitions of culture, emic and etic, can be considered as two sides of the same coin. By utilizing the research provided by both approaches, we gain a more complete understanding of the culture(s) of interest.

Thus, one possible reason for many failing educational systems around the world could be that both policy makers and curriculum developers have ignored emic approaches in relation to school curricula, especially when it suggests the recognition of other epistemologies and of holistic and integrated natures of the mathematical knowledge developed by the members of other cultural traditions. It has been hypothesized that low attainment in mathematics could be due to lack of cultural consonance in the mathematics curriculum (Rosa, 2010).

When students come to school, they bring with them the values, norms, procedures, techniques, and concepts that they have acquired in their own sociocultural environment and some of these elements are mathematical in nature. However, mathematical concepts of the school curriculum are presented in a way that may not be related to their cultural background. Moreover, the inclusion of cultural aspects in the mathematics curriculum has long-term benefits for mathematics learners (Rosa & Orey, 2010).

Current mathematics curriculum lacks a dialogical approach in regards to the preparation of students for a living in a dynamic and diverse society. The lack of awareness of emic knowledge and the alienating effects of Western educational norms indicates a need for an ethnomodelling curriculum. The mathematical knowledge created based on an emic approach is a form of intellectual decolonization, and provides us with a major contribution to mathematics education and in the development of contemporary society. Ethnomodelling becomes the "joining of a new field of inquiry, which might be called communal transformation" (Block, 2010, p. vii).

In this context, an ethnomodeling-based curriculum comes to provide a theoretical basis for the teaching and learning process because it combines key elements of emic knowledge with an etic approach. The main goal of ethnomodelling is the acquisition of both emic and etic knowledge by applying dialogical approach.

Similarly, it is possible to define glocalization "by the social actor's fluid and critical engagement with, and reconstruction of, local and global phenomena" (Giulianotti & Robertson 2006, p. 173). When analyzing the diffusion of mathematical knowledge, it is necessary to look at the local and global reconstructions together. In this regard, cultural aspects contribute to recognizing mathematics as part of daily life, enhancing the ability to make meaningful connections, and deepening the understanding of mathematics (Rosa, 2010).

In this context, creative insubordination is necessary to serve students' learning (Ayers, 2001), which is the main concern of the educational system and, therefore, if rules and norms need to be bent to achieve this goal. Then, researchers and educators must address student's cognitive and pedagogical needs into the mathematics curriculum through ethnomodelling.

NOTES

References

Ayers, W. (2001). *To teach*: the journey of a teacher. New York, NY: Teachers College Press.

Bennett, J., & Bennett, M. (2004). Developing intercultural sensitivity: an integrative approach to global and domestic diversity. In D. Landis, J. Bennett, & M. Bennett (Eds.), *Handbook of intercultural training* (pp. 147–165). Thousand Oaks, CA: Sage.

Berry, J.W. (1969). On cross-cultural comparability. International Journal of Psychology, 4, 119-128.

Bishop, A. J. (1990). Western mathematics: the secret weapon of cultural imperialism. *Race & Class, 32*(2), 51-65.

Bishop, A. J. (1993). Influences from society. In A.J. Bishop, K. Hart, S. Lerman, & T. Nunes (Eds), *Significant influences on children's learning of mathematics* (pp. 3-26) Paris, France: UNESCO.

Block, P. (2010). Foreword. *Inviting everyone: Healing healthcare through positive deviance.* Bordentown, NJ: Plexus Press.

Bourdieu, P., & Wacquant, L. J. D. (1992). *An invitation to reflexive sociology*. Chicago, IL: University of Chicago Press.

Crowson, R. L. & Morris, V. C. (1982). The principal's role in organizational goal-attainment: discretionary management at the school site level. *Proceedings of the Annual Meeting of the American Educational Research Association* (pp. 19-23). New York, NY: AERA.

Dehler, G. E., Welsh R. W. (1998). Problematizing deviance in contemporary organizations: a critical perspective. Stamford, CT: JAI Press.

D'Ambrosio, U. (1985). Ethnomathematics and its place in the history and pedagogy of mathematics. *For the Learning of Mathematics: an International Journal of Mathematics Education, 5*(1), 44-48.

D'Ambrosio, U. (1990). *Etnomatemática* [Ethnomathematics]. São Paulo, Brazil: Editora Ática.

D'Ambrosio, U. (2015). Mathematical modelling as a strategy for building-up systems of knowledge in different cultural environments. In G. A. Stillman, W. Blum, & M. S. Biembengut (Eds.), *Mathematical modelling in education research and practice: cultural, social, and cognitive influences* (pp. 35-44). New York, NY: Springer.

D'Ambrosio, U., & D'Ambrosio, B. S. (2013). The role of ethnomathematics in curricular leadership in mathematics education. *Journal of Mathematics Education at Teachers College, 4*, 19-25.

Deardorff, D. K. (2006). The identification and assessment of intercultural competence as a student outcome of internationalization at institutions of higher education in the United States. *Journal of Studies in International Education, 10*, p. 241-266.

Eglash, R. (1999). *African fractals*: modern computing and indigenous design. New Brunswick, NJ: Rutgers University Press.

Eglash, R., Bennett, A., O'Donnell, C., Jennings, S., & Cintorino, M. (2006). Culturally situated designed tools: ethnocomputing from field site to classroom. *American Anthropologist, 108*(2), 347-362, 2006.

Giulianotti, R.; & Robertson, R. (2007). Recovering the social: globalization, football and transnationalism. *Global Networks, 7*(2), 144–186.

Gudykunst, W. B. (1997). Cultural variability in communication. *Communication Research, 24*(4), 327-348.

Harris, M. (1980). The epistemology of cultural materialism. In Harris, M. *Cultural materialism*: the struggle for a science of culture (pp. 29-45). New York, NY: Random House.

Headland, T. N.; Pike, K. L.; Marvin Harris (1990). *Emics and etics*: the insider/outsider debate. Newbury Park, CA: Sage Publications.

Hutchinson S. A. (1990). Responsible subversion: a study of rule-bending among nurses. *Scholarly Inquiry for Nursing Practice, 4*(1), p. 3-17

Iser, W. (1994). On translatability. *Surfaces, 4307*, 5-13.

Khondker (2004). Glocalization as globalization: evolution of a sociological concept. *Bangladesh e-Journal of Sociology, 1*(2), 12–20.

Levinas, E. (1999). *Alterity and transcendence*. Trans. Michael B. Smith. New York, NY: Columbia UP.

Lonner, W. L. & Berry, J. W. (1986). Field methods in cross-cultural research. Beverly Hills, CA: Sage publications, Inc.

Lyman, L. L., Ashby, D. E., & Tripses, J. S. (2005). *Leaders who dare: pushing the boundaries*. Lanham, MD: Rowman & Littlefield Education.

Marzano, R. J., Waters, T., & McNulty, B. A. (2005). *School leadership that works: from research to results*. Alexandria, VA: Association for Supervision and Curriculum Development.

Maynard, M. L. (2003). From global to local: how Gillette's sensor Excel accommodates to Japan. *Keio Communication Review, 25*, 57-75.

Mendis, P. (2007). *Glocalization*: the human side of globalization as if the Washington consensus mattered. Morrisville, NC: Lulu Press.

Orey, D. C. (2000). The ethnomathematics of the Sioux tipi and cone. In Selin, H. (Ed.). *Mathematics across culture: the history of non-western mathematics* (pp. 239-252). Dordrecht, The Netherlands: Kulwer Academic Publishers.

Pelto, P. J.; & Pelto, G. H. (1978) *Anthropological research: the structure of inquiry*. New York, NY: Cambridge University Press.

Pike, K. L. (1954). Emic and etic standpoints for the description of behavior. Glendale, IL: Summer Institute of Linguistics.

Pike, K. L. (1967). *Language in relation to a unified theory of the structure of human behaviour*. The Hague, Netherlands: Mouton.

Robertson, R. (1995) Glocalization: time-space and homogeneity- heterogeneity, M. Featherstone et al (Eds.), *Global modernities* (pp. 25-44), London, England: Sage.

Rosa, M. (2010). A mixed-methods study to understand the perceptions of high school leader about English language learners (ELL): The case of mathematics. Unpublished doctorate dissertation. College of Education. California State University, Sacramento.

NOTES

Rosa, M. (2014). Explorando saberes e técnicas locais no contexto da etnomodelagem: destacando-se as abordagens êmica, ética e dialógica [Exploring local knowledge and techniques in the ethnomodeling context: highlighting emic, etic, and dialogical approaches]. Post-doctorate report. Faculdade de Educação. São Paulo. SP, Brazil: Universidade de São Paulo.

Rosa, M., & Orey, D. C. (2007). Cultural Assertions and Challenges towards Pedagogical Action of an Ethnomathematics Program. *For the Learning of Mathematics, 27*(1), 10-16.

Rosa, M.; Orey, D. C. (2008). Ethnomathematics and cultural representations: teaching in highly diverse contexts. *ACTA SCIENTIA, 10*(1), 27-46.

Rosa, M.; Orey, D. C. (2010). Ethnomodelling: a pedagogical action for uncovering ethnomathematical practices. *Journal of Mathematical Modelling and Application, 1*(3), 58-67.

Rosa, M.; Orey, D. C. (2011). Ethnomathematics: the cultural aspects of mathematics. *Revista Latinoamericana de Etnomatemática, 4*(2), 32-54.

Rosa, M.; Orey, D. C. (2012). O campo de pesquisa em etnomodelagem: as abordagens êmica, ética e dialética [The field of research in ethnomodelling: emic, etic, and dialectical approaches]. *Educação e Pesquisa, 38*(4), 865-879.

Rosa, M.; Orey, D. C. (2013). Ethnomodelling as a research lens on ethnomathematics and modelling. In: G. A. Stillman, & J. Brown. (Orgs.). *Teaching mathematical modelling:* Connecting to research and practice (pp. 117-127). Dordrecht, The Netherlands: Springer Science+Business Media Dordrecht.

Rosa, M.; & Orey, D. C. (2015a). Evidence of creative insubordination in the research of pedagogical action of ethnomathematics program. In: B. S. D'Ambrosio & C. E. Lopes (Orgs.), *Creative insubordination in Brazilian mathematics education research* (pp. 131-146). Raleigh, NC: Lulu Press.

Rosa, M.; & Orey, D. C. (2015b). Three approaches in the research field of ethnomodelling: emic (local), etic (global), and dialogical (glocal). *Revista Latinoamericana de Etnomatemática, 8*(2), 364-380.

Rosa, M.; & Orey, D. C. (2015c). A trivium curriculum for mathematics based on literacy, matheracy, and technoracy: an ethnomathematics perspective. *ZDM, 47*(4), 587-598.

Samovar, L. A.; & Porter, R. E. (2000). *Communication between cultures*. Beijing, China: Foreign Language Teaching and Research Press.

Séguinot, C. (1995). Translation and advertising: going global. In C. Shäffner, & K. Holmes (Eds.), *Cultural functions of translation* (pp. 55-69). Philadelphia, PA: Multilingual Matters.

Sue, D. W.; & Sue, D. (2003). Counseling the culturally diverse: Theory and practice. New York, NY: John Wiley & Sons.

Walker L. O. (2005). *Avant KC strategies for theory construction in nursing*. Boston, MA: Boston Prentice Hall.

Wexler, J. (2004). *Alterity:* Theories of Media. Chicago, IL: University of Chicago. Retrieved from http://csmt.uchicago.edu/glossary2004/alterity.htm.

Yifeng, S. (2009). Cultural translation in the context of glocalization. *Ariel, 40*(2-3), 89–110.

Zeitlin, M., Ghassemi, H., & Mansur, M. (1990). *Positive deviance in child nutrition: with emphasis on psychological and behavioural aspects and implications for development.* Tóquio, Japão: The United Nations University.

Evolutions and Contradictions: A Pedagogical Reflection

Emily P. Bonner
University of Texas at San Antonio
College of Education and Human Development
Department of Interdisciplinary Learning and Teaching
One UTSA Circle
San Antonio, TX 78249
P: (210) 458-5402
F: (210) 458-7281
emily.bonner@utsa.edu

Volume 1

Abstract

This manuscript is a reflection on my 2010 article in the *Journal of Mathematics and Culture* entitled, "Promoting Culturally Responsive Teaching through Action Research in a Mathematics Methods Course". I couch this reflection, which has a pedagogical focus, within the current educational climate, and connect it to where I am today in terms of teaching and scholarship. Inherent in this work, which is in a state of current evolution, are a number of personal evolutions and professional contradictions. I outline framing concepts related to scholarly and pedagogical evolution and discuss various ambiguities that emerge as I engage in scholarly work.

> *"The goal of education is to expand one's knowledge base and critical thinking skills, rather than protect our preexisting opinions. Of course we have a right to our opinions, but there will be no personal or intellectual growth for us if we are not willing to think critically about them."*
>
> *-Sensoy & DiAngelo, 2012*

Introduction

Since the publication of *Promoting Culturally Responsive Teaching through Action Research in a Mathematics Methods Course* in the Journal of Mathematics and Culture 6 years ago, little has changed in the field of mathematics education in relation to tangible results that indicate reduced disparities between students of color and their counterparts in public schools. Though African American and Hispanic students will make up the majority of public school students in the next ten years (U.S. Census Bureau, 2015), pernicious gaps in achievement and opportunity continue persist. In mathematics, a 20-30 point gap in achievement on NAEP remains between students of color and their White counterparts (NCES, 2013) at all levels. While "gap gazing" (Gutiérrez & Dixon-Román, 2011) is too simplistic when describing inequities in mathematics education, these differences in opportunity between groups is a symptom of systemic, underlying racism and classism in the public school system.

School segregation continues to worsen with increasing numbers of apartheid schools receiving inferior services and resources, and even more control has been taken away from teachers, particularly those who work with traditionally underserved populations. Moreover, these schools continue to be underfunded and overcrowded with strict zero tolerance policies and few teachers of color. Under 28% of bachelors degrees are awarded to Black and Hispanic students combined, whereas White students receive almost 75% of conferred degrees, and the number of Hispanic high-school dropouts is increasing yearly (National Center for Education Statistics [NCES], 2013).

For both Hispanic and Black students, the educational system has failed to maintain an effective pipeline leading to advanced degrees and eventual careers in engineering or mathematics. This is notable, as many of these students have access to an outdated system that is not preparing them for 21st century jobs (Darling-Hammond, 2010). This trend begins early in schooling and declines over time. At the 8th grade level, for example, 19% of Hispanic students and 13% of Black students are proficient in mathematics (NAEP, 2015). By 12th grade, only 12% of Hispanic and 7% of Black students are proficient in mathematics (NAEP, 2013), both far below the 26% average. According to the National Center for Education Statistics (2012), fewer than 10% of Hispanics graduate with a baccalaureate degree and Hispanics constitute less than 1% of degree-earning students in engineering or mathematics. Hispanic students comprise less than 8% of students earning a master's degree in <u>any</u> field, 6% of students earning a Ph.D. in <u>any</u> field, and 2% of students receiving a Ph.D. in engineering or mathematics. These trends, which are a continuation of patterns that have been evident for over 20 years, point to the importance of teacher education, and illustrate the key role that mathematics teacher educators can play in promoting equity in schools, especially with the national push in STEM education.

Though the overall picture seems bleak, my engagement in scholarship since I published *Promoting Culturally Responsive Teaching* has provided me opportunities to meet colleagues and read their work. I am convinced that although the U.S. has vast challenges related to educational attainment, particularly for students of color, there are pockets of greatness wherein mathematics

NOTES

educators are doing exceptional work that is impacting their own community in profound ways, particularly through teacher education programs that promote equitable practice. The Teachers Empowered to Advance Change in Mathematics (*TEACh Math*) group, for example, is a multi-institutional consortium of women who have investigated and produced research-based resources for mathematics teacher educators that promote equitable practice. This work specifically focuses on *developing* effective mathematics teachers who explicitly focus on cultural funds of knowledge (see Bartell et al., 2013 and Turner et al., 2012, for example). Other programs such as CULTURES (Irvine, 2003), a professional development model that focuses on culturally responsive teacher development, and Teach for Diversity (TFD, Ladson-Billings, 2001), a graduate program with many goals, including preparing teachers who provide access and equitable education, have been highlighted in the literature.

The work of others in this realm informs my own in ways that transform my teaching and scholarship while challenging me to engage in innovative ways of thinking. While the beliefs and foundations expressed in my 2010 article still hold true, I now view my work in teacher education as more complex and multifaceted, constantly changed by scholarship (my own and others'), and potentially transformative in my own community. I am entrenched now in a community of scholars who are working to transform mathematics education through equitable practice and opportunity. This paper is a reflective piece through which I intend to capture that sense of camaraderie, and the ways in which my practice has changed or evolved through this type of engagement. I frame this progression with results from my research, and discuss how these inform my current teaching and scholarship, which are intricately connected. Lastly, I will discuss professional contradictions related to teacher access to research that have emerged as a result of these reflections.

Framing Concepts

Looking back at my scholarship over the past 6 years, it is clear that several main themes have emerged that currently frame my work. I will focus how these inform my work in teaching, specifically in teaching pre-service elementary school teachers (PSTs) in my mathematics methods course, since this was the focus of the 2010 article. These frameworks or practices have evolved over time, and will continue to develop and change. Further, these are not meant to encapsulate my scholarship in a holistic way, but rather to describe some aspects of my scholarship today while critically reflecting on the ways that this work has progressed.

Symbiotic Relationship Between Research and Practice

As my work has evolved throughout my time as a tenure-track faculty member, the ways in which my research informs my practice (and vice versa) has become more formal and intentional. Looking back over the course of my first 7 years, I see that this strong integration of research and practice has always been foundational to my scholarship, but more loosely than it is currently. For example, in my 2010 JMC piece, I mention showing a "videotape of a culturally responsive mathematics teacher who [was] well known in our geographical area" – my dissertation was based on the work of this teacher. However, the ways that I incorporated that research into my methods class were somewhat superficial in that we simply talked *about* her. While this was a reasonable first step in exposing PSTs to culturally responsive teaching (CRT), my approach now is focused on outcomes – what can we learn from this teacher and how does it apply to your classroom? I encourage students to try out some of her methods, and conduct an informal study on whether these methods, or modified methods, worked for them and their students. This could lead to a larger discussion about equity and designing instruction that is relevant to your particular population (as opposed to a one-size-fits-all curriculum that is prevalent now).

Currently, my courses closely and more intentionally align with the research I conduct in mathematics education. For example, one line of research in which I am engaged deals with developing culturally responsive mathematics teachers. Collecting data in this area involves going out to classrooms of highly successful teachers of traditionally underserved populations and documenting cornerstones of effective practice. I incorporate these into my course activities and assignments, and into my own teaching so that I continue to model an evolving, effective teacher. Working with students informs my research agenda, pushing me to explore particular aspects of teacher education that may benefit the larger community. This is evidenced by the numerous pieces I have published on teaching (Bonner, 2009; Kalinec-Craig & Bonner, 2015, for example) and with teachers (Muniz, Carrier, & Bonner, 2010; Meyer & Bonner, 2014, for example). Further, I see each course as a rich source of actionable data. I collect data from my students (obtained from class interactions, assessments, informal feedback, and observation) in an informal way that directly informs my teaching. After implementing changes based on these data, I reassess to determine effectiveness. I carry forth the goal of continuing to ground my research in practice, and my practice in scholarship. Courses should evolve as innovative, research-based teaching strategies are discovered.

Another connection to scholarship relates to my approach to teaching as student-centered with strong emphases on communication, knowledge and relationships (see figure 1). Each of these cornerstones, aided by a constant process of reflection and revision, work in conjunction with and inform one another, and shape the student experience. *Communication* includes communicating with students in culturally responsive ways. This includes communicating about content and knowledge, but also setting high expectations that build relationships and trust among the class and with me as a facilitator. *Knowledge* involves information related to the course content, but also knowledge about my students' lives, interests, and funds of knowledge that are important to my practice. This knowledge is obtained through communication, and fosters relationships in the class. *Relationships and trust* with and between students are foundational to my teaching, as I want prospective teachers to understand the importance of relationships and trust with their own students when they have their own classroom. These relationships are built through communication and shared knowledge in and out of the classroom. These relationships are squarely at the center of (and inform and affect) the cornerstones, and the cycles of reflection/revision and pedagogy/discipline are based on the student experience and feedback. There is also explicit and intentional focus on power structures in the classroom being fluid, and honoring students' funds of knowledge as an asset in instruction. In my 2010 piece, I tinkered with the idea of shifting power, but in many ways was simply designing a course that created an illusion of a true power shift. Looking back, I believe this was a necessary step in my continued work.

NOTES

This perspective has emerged from my own research, which focuses on deconstructing the practices of successful teachers and is evolving over time (Bonner, 2010; 2012; 2014) and directly informs my teaching. This graphic is constantly evolving as new research emerges, and through new experiences in teaching.

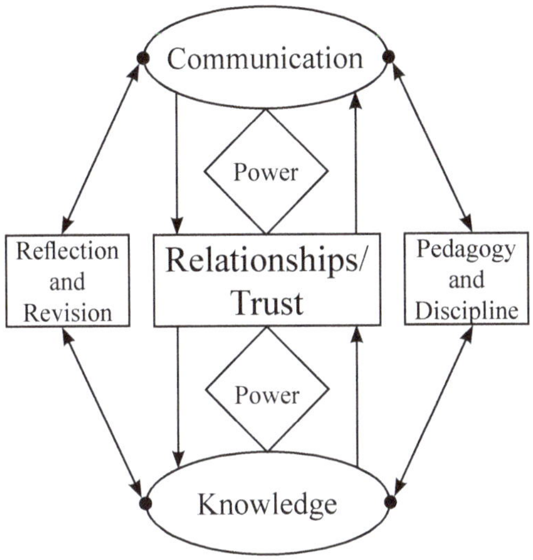

Figure 1: Bonner research-based approach to undergraduate teaching (Bonner, 2014)

Problem Solving as Equitable Teaching

In reading my 2010 piece it is clear to me that my thoughts about CRT and systemic inequality were evolving, but somewhat simplistic. As I critically reflect on this work, I perceive that in outlining activities that I incorporated into class in an effort to encourage a largely White population of PSTs, I may have communicated that CRT was an "extra" or special thing that should be paid attention to, but that did not need to be a foundational part of practice and, essentially, who you are as a person. In my current setting, I still utilize journaling, videos, and cultural explorations, but these are tools that facilitate discussions about deconstructing practice. For example, my entire course is centered on teaching mathematics through problem solving and orchestrating mathematical discussions around well-chosen, relevant tasks. If we do this work well, we are attending to issues of equity inherently in our teaching. These tasks are challenging, have multiple entry points, are relevant, require problem solving skills but no prescribed strategy, provide opportunities for written and oral communication, attend to multiple representations, and empower students to learn and do mathematics in culturally relevant ways.

Further, teaching my students through problem solving pushes their thinking in new ways, and forces them to get comfortable with uncomfortable situations. If students are always comfortable in my course, I am not effectively challenging their thinking. When students are forced to grapple with new ideas or difficult content through debate, discussion, and justification, new ideas are generated and transformative learning takes place. For example, in my methods course, we begin the first day with a non-traditional, open-ended mathematics task that is meant to stretch student thinking as it is difficult to solve using any traditional mathematical method (e.g. a traditional algorithm). This carefully selected task leads to a variety of unique student presentations, and lively discussions about particular solutions. Further, students are forced to justify their solution, since it is not a traditional representation. This activity immediately pushes students out of their comfort zone, and allows them to productively struggle with a task. We finish class by unpacking the lesson, discussing how and why this type of instruction is effective. I also apply this principle to myself, often trying new lessons that are slightly out of my comfort zone, but may provide a more effective experience for students. It is always through these experiments that I learn the most about teaching and learning.

Intentional and Authentic Assessment

My approach to assessment provides students with opportunities to think differently and deeply about material (that they may be uncomfortable with) while extending their thinking beyond the course. This creates a space where students are active participants not only in their own learning, but also in determining the degree to which they have met learning objectives. Further, I position myself as a support system for students rather than the "keeper of the knowledge". I learn from them as they learn from experiences in my course, and my assessments reflect this stance by providing many opportunities for feedback and reflection.

Literature related to effective and formative experiences for pre-service mathematics teachers, in addition to collegial relationships and collaborations, strongly guides my course development. Still rooted in promoting equitable practice, my materials draw from the work of the aforementioned *TEACh Math* Project. Projects and assignments are rooted in preparing teachers to examine and understand student conceptions of mathematics, while recognizing issues of equity in mathematics education, and capitalizing on student funds of knowledge in the classroom. In looking back at my 2010 article, I see that my goals are similar, but the intentionality and connectedness of the assignments is more salient, as development has been guided by research. For example, in 2010 I report on showing videos of culturally responsive teachers as a formative class activity. Currently, I still utilize videos, but focus on culturally responsive teaching as a foundation rather than a lens. So, rather than positioning one exemplar as an example of CRT, we discuss cultural aspects of any teacher, and deconstruct the ways in which teaching can empower or quiet underserved students. In 2010 I also discuss requiring students to attend a cultural event that would broaden their views on cultural differences. Again, this theme remains in my teaching, but permeates all aspects of the course. Students in my classes now do a School and Community Mathematics Project (adapted from the *TEACh Math* group), which requires that they explore the community around the school in which they have their field placement, document their exploration, and develop mathematical tasks that are relevant to students based on what they have learned. Students also participate in a case study, wherein they engage a student in task-based interviews that inform their perceptions and teaching.

One major goal, then, related to my teaching is to encourage teacher *noticing* that leads to *challenging* a pre-conceived notion about mathematics education, particular students, or other social justice issues, *acting* on this new information in ways that challenge the status quo, and finally *re-noticing* similar issues over time so that conceptions are constantly changing in order to promote equity. This cycle is outlined in figure 2, which a colleague and I published in 2015 (Kalinec-Craig & Bonner, 2015).

NOTES

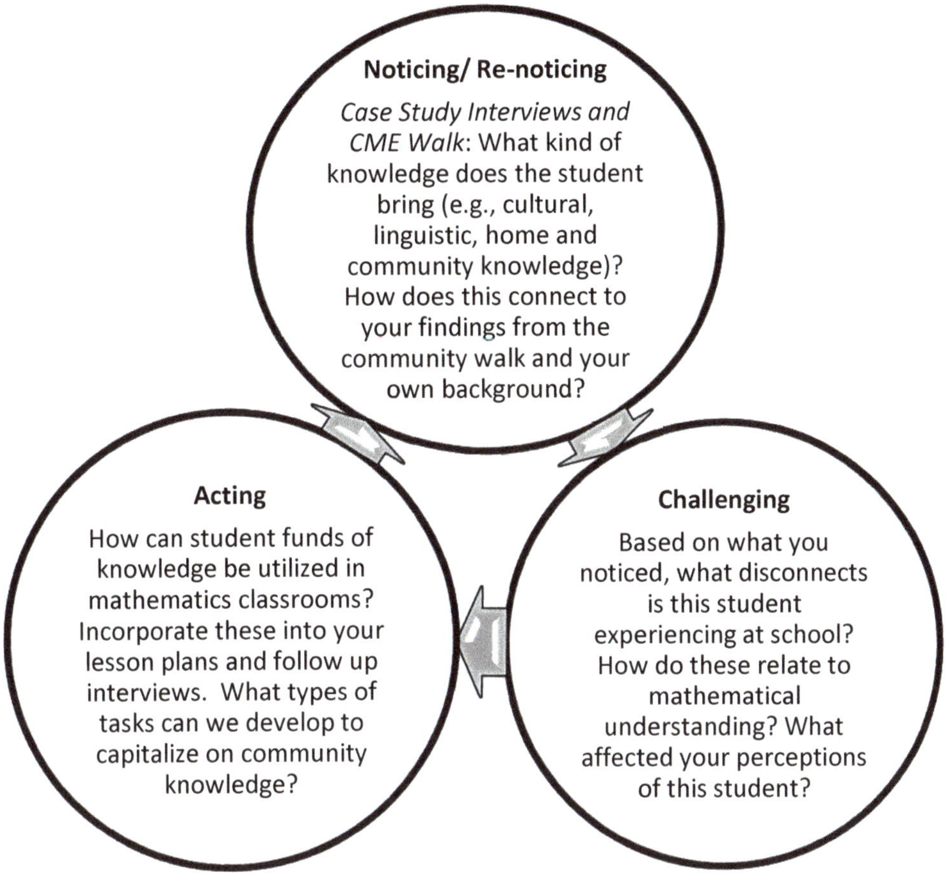

Figure 2. Cycle of noticing, challenging, acting, re-noticing (Kalinec-Craig & Bonner, 2015).

Attention to Systemic Challenges

In recent years, my research has expanded to include several funded professional development programs. Through these programs, I work with mathematics teachers in

high-need schools and districts to ultimately increase student access to mathematics through innovative teaching. This work informs my work with pre-service teachers, and has broadened my understandings of the political climate in which teachers are entrenched. Further, it has heightened my awareness of the highly charged deficit language around low-income students, and the ways in which this reinforces commonly held beliefs about students of color. This language, which is systemically accepted, centers on "achievement gaps" and "gap gazing" (Gutiérrez, 2008), and provides a very underdeveloped and limited understanding of inequities that exist in education. In working with teachers to understand equity work more broadly, colleagues and I developed an Equity-Focused Professional Development (EFPD, figure 3) model that outlines our approach. This is a younger framework; as such, we realize that it is not fully developed. We are consistently refining the framework to capture more aspects of our work, and to adjust as we find places in our work that are lacking or somehow reinforcing deficit thinking.

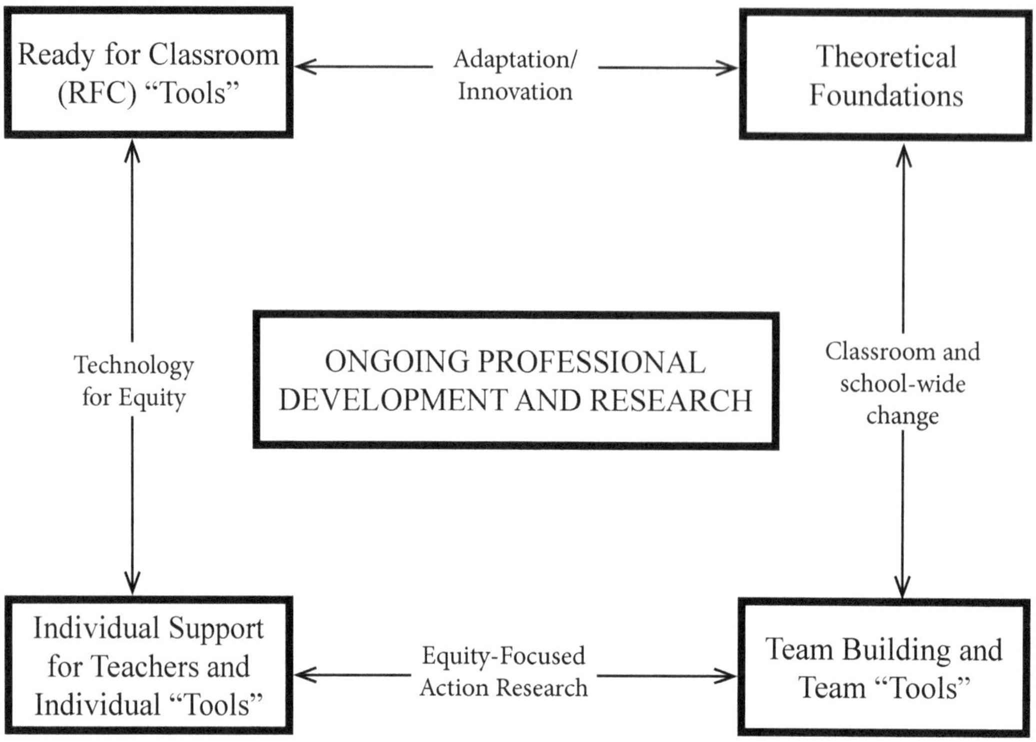

Figure 3. EFPD Framework

Ultimately, this work aims to aid teachers in "creative insubordination" (D'Ambrosio & Lopes, 2015) by providing creative ways to implement tasks that promote equity and access but fall within district and state mandates on practice.

Systemic challenges are not limited to K-12 classrooms. In fact, engaging in this reflective process has resulted in moments of cognitive dissonance for me in relation to the larger field of mathematics education, and as a result systemic issues in the field at large have come to the forefront. For example, we provide tools to mathematics teachers that allow them to engage in practice that provides their students with *access* to content. As mathematics teacher educators, we need to examine the ways in which we provide teacher *access* to our content, which is relevant research. In academia, publications in "top tier" journals are highly valued, as are presentations at conferences that are largely attended by other researchers. Publications in "practitioner" journals, for the most part, are not valued as heavily in the academic arena, though these pieces have the greatest potential to affect classroom practice.

It is counterintuitive that our goal as mathematics teacher educators is to impact student access and achievement in mathematics, but collectively we are replicating behaviors that we are fighting at the K-12 level. As an organization of scholars, we need to examine these types of

NOTES

disconnects, and the ways in which we can encourage a more symbiotic relationship between classroom teachers and researchers. Similarly, we need to examine our use of deficit language. When discussing students, we are conscious of this language, and more often mathematics educators find it unacceptable. I do not perceive this same standard, however, when we are talking about teachers. In all types of mathematics education research (including my own work), we discuss changing teachers rather than capitalizing on their funds of knowledge. As such, deficit language permeates teacher education research. Moreover, this type of language establishes stratifications in terms of status among teachers, teacher educators, and researchers.

These types of systemic issues are not hidden or implicit, but it are rarely discussed in mathematics teacher education communities. How can we, as mathematics teacher educators, engage in D'Ambrosio's notion of creative insubordination in our own settings? How can we provide access to teachers while maintaining our professional positions? Further, issues of power and status and the ways that we position ourselves in relation to each other and in relation to teachers are implicit in this discussion. How can we practice what we preach, so to speak, in ways that dismantle these barriers to more transformative experiences for K-12 students?

Future Directions

Engaging in this reflection allowed me to see the ways in which my work in culturally responsive mathematics education has progressed, but more importantly, it highlights the various and interconnected spaces where there is great need for growth and professional reflection. To summarize, moving forward the following questions will guide my thinking on these topics, particularly as they relate to my practices in teaching:

- How can we refine our EFPD model to most effectively (a) support classroom teachers in adopting effective practices, and ultimately (b) positively affecting traditionally underserved students' experiences in mathematics?
- What strengths and areas for improvement are students from our teacher education program carrying forth into the field? How can we better support them while they are in our program to develop the practices that will promote equitable practice?
- How can we more effectively support teachers to negotiate spaces where they must operate within district mandates that create barriers when implementing equitable practice?
- There seems to be a disconnect between research and practice. Can we, as a community of mathematics teacher educators, provide more accessible outlets for teachers where the conversation is not "top-down" but mutually informative and transformative?

Critical reflection is a necessary and ongoing process. It is my hope that these guiding questions will allow me to continue this process, both formally and informally, as I continue to explore issues of injustice in mathematics education.

References

Bartell, T. G., Foote, M. Q., Drake, C., Roth McDuffie, A., Turner, E. E., & Aguirre, J. M. (2013). Developing teachers of Black children: (Re)orienting thinking in an elementary mathematics methods course. In J. Leonard & D. B. Martin (Eds.), *The brilliance of Black children in mathematics: Beyond the numbers and toward a new discourse* (pp. 343 – 367). Charlotte, NC: Information Age.

Bonner, E. P. (2009). Achieving success with African American learners: A framework for culturally responsive mathematics teaching. *Childhood Education, 86*(1), 2-6.

Bonner, E. P. (2010). Promoting Culturally Responsive Teaching through Action Research in a Mathematics Methods Course. *Journal of Mathematics and Culture, 5*(2), 16-33.

Bonner, E. P. & Adams, T.L. (2012). Culturally responsive teaching in the context of mathematics: A grounded theory approach. *Journal of Mathematics Teacher Education, 15*(1), 25-38.

Bonner, E. P. (2014). Investigating practices of highly successful teachers of traditionally underserved students. *Educational Studies in Mathematics, 86*(3), 377-399.

D'Ambrosio, B. & Lopes, C. E. (2015). Ethics and solidarity in mathematics education: Acts of creative insubordination. In S. Mukhopadhyay & B. Greer (Eds.), *Proceedings of the Eighth International Mathematics Education and Society Conference v.2* (pp. 413-426).

Gutiérrez, R. (2008). A "gap-gazing" fetish in mathematics education? Problematizing research on the achievement gap. *Journal for Research in Mathematics Education, 39*(4), 357-364.

Gutiérrez, R., & Dixon-Román, E. D. (2011). Beyond gap gazing: How can thinking about education comprehensively help us (re)invision mathematics education? In B. Atweh, M. Graven, W. Secada, and P. Valero (Eds.), *Mapping equity and quality in mathematics education* (pp. 21-34). New York: Springer.

Irvine, J. J. (2003). *Educating teachers for diversity: Seeing with a cultural eye*. New York, NY: Teachers College Press.

Kalinec-Craig, C. & Bonner, E. P. (2015). Seeing the World with a New Set of Eyes: (Re)Examining Our Identities as White Mathematics Education Researchers of Equity and Social Justice. In N. M. Russell & F. Cobb (Eds.), *Interrogating whiteness and relinquishing power: White faculty's commitment to racial consciousness in STEM classrooms* (pp. 91-112). New York, NY: Peter Lang Publishing.

Ladson-Billings, G. (2001). *Crossing over to Canaan: The journey of new teachers in diverse classrooms*. San Francisco, CA: Jossey Bass.

Meyer, H. S.* & Bonner, E. P. (2014). Higher-Order Thinking in Algebra: Barriers and Opportunities. *Texas Mathematics Teacher, XLI*(1), 6-12.

Muniz, A., Carrier, P., & Bonner, E. P. (2010). Improving elementary school students' multiplication skills through small group intervention and recursive review. *Texas Mathematics Teacher, LVII*(2), 22-25.

NOTES

Turner, E. E., Drake, Roth McDuffie, A., Aguirre, J., Bartell, T. G. & Foote, M. Q. (2012). Promoting equity in mathematics teacher preparation: A framework for advancing teacher learning of children's multiple mathematics knowledge bases. *Journal of Mathematics Teacher Education*, 15, pp. 67-82.

U.S. Census Bureau. (2015). *State median family income by family size*. Retrieved from http://www.census.gov/topics/population.html.

U.S. Department of Education (2013). Institute of Education Sciences, National Center for Education Statistics. Retrieved from: http://www.nces.ed.gov/programs/digest/2013menu_tables.asp

Lessons Learned: Strengths-Based Approaches to Mathematics Education in the Pacific

Linda H.L. Furuto, Ph.D.
University of Hawai'i at Mānoa
1776 University Avenue
Everly 224
Honolulu, HI 96822
Email: lfuruto@hawaii.edu

Abstract

In this article, I examine through strengths-based approaches, current and future ethnomathematics research in the design and implementation of professional development for educators and teacher educators. The inclusion of mathematics indigenous to Hawai'i and the Pacific leads to synergistic efforts and collective impact in global indigenous communities. By drawing on our unique strengths, we strive to create spaces and support voices that have and will continue to emerge in this work. This is accomplished by honoring culturally responsive, place-based strategies such as learning within and outside of the classroom and reshaping the curriculum in a cultural context. The expertise that exists in our communities, schools, and families is the foundation in the production of knowledge for empowerment to achieve equitable and quality mathematics education.

Keywords:

- Mathematics Achievement
- Equity
- Ethnomathematics
- College, Career, and Community Readiness

Ma ka hana ka ʻike
In working one learns

The ʻōlelo noʻeau (proverb) above frames the content of this article (Pukui, 1993, p. 227). Through culturally responsive and place-based learning, we deepen our perspectives by doing mathematics that permeates our environments, values, and traditions.

Background

Several years ago, I published an article "Bridging Policy and Practice with Ethnomathematics" (2013) on research conducted at U.S. higher educational institutions (Harvard University, UCLA, University of Hawaiʻi) in the field of ethnomathematics. The discussion focused on first, how cultural values affect teaching, learning, and curriculum; and second, how mathematics education affects schooling process dynamics. Examining the Ethnomathematics Institute, a National Science Foundation-funded project, and the role of navigation in mathematics of the Pacific did specifically this. Through research and praxis, I highlighted the effectiveness of ethnomathematics as a bridge between policies introduced at the macroperspective level and hands-on learning experiences in the communities we are endeavoring to serve by building partnerships and mentoring in diverse populations. While these promising practices still hold true, new lessons have unfolded and been refined through strengths-based approaches with current and future implications.

Since 2013, collective impact in our networked communities and strategic partnerships (Kania & Kramer, 2011; Waitzer & Paul, 2011) has led to tremendous progress. Now in its eighth year, the Ethnomathematics and STEM Institute (EthnoSTEM) transitioned from a 2-week summer program to a yearlong professional development institute for P–20 STEM educators. Drawing on island wisdom to make global connections, EthnoSTEM is designed to utilize the local environment and community as a contextualized classroom. As this article will discuss, EthnoSTEM is guided by the mission and values of the Polynesian Voyaging Society and its traditional double-hulled voyaging canoe Hōkūleʻa. Hōkūleʻa, star of gladness, is internationally renowned for the role it has played in rekindling the Pacific Island tradition of non-instrument, wayfinding techniques based on celestial navigation (i.e., sun, moon, stars, winds). Over the past 40 years, Hōkūleʻa has sailed more than 150,000 nautical miles since the first trans-archipelagic, traditionally navigated voyage in 600 years—from Hawaiʻi to Tahiti—in 1976. The current 50,000 nautical mile, 2013–2017 Mālama Honua Worldwide Voyage is the ideal vehicle to explore real-world STEM applications through indigenous and 21st century learning. The author was invited to be part of the first international leg from Hawaiʻi to Tahiti as apprentice navigator and education specialist, and subsequent voyages to the Samoan Islands, Olohega (Swain's Island), Aotearoa (New Zealand), South Africa, Washington, D.C., and New York City.

In light of these developments, I will examine current and future ethnomathematics research in the design and implementation of professional development for educators and teacher educators. The inclusion of mathematics indigenous to Hawaiʻi and the Pacific leads to synergistic efforts and collective impact in global indigenous

NOTES

communities (Furuto, 2014). By drawing on our unique strengths, we strive to create spaces and support voices that have and will continue to emerge in this work. This is accomplished by honoring culturally responsive, place-based strategies such as learning within and outside of the classroom and reshaping the curriculum (Maton, Schellenbach, Leadbeater, & Solarz, 2003; Meyer, 2003). Place-based strategies recognize the importance of including students' cultural contexts in all aspects of learning (Ladson-Billings, 1995). The expertise that exists in our communities, schools, and families is the foundation in the production of knowledge for empowerment to achieve equitable and quality mathematics education (Fetterman, 2000; Hall, 1993).

From a body of research we deduce that strengths-based approaches are necessary to effectively teach and learn in diverse classrooms (PREL, 1995; Tuhiwai Smith, 1999). The literature highlights the need for pedagogy to align with assets of today's students, and "requires a strong commitment by participating researchers and practitioners to share their expertise with the people, while recognizing the communities directly involved" (Sohng, 1998, p. 187). According to Kanaʻiaupuni (2005), "Strengths-based research begins with the premise of creating social change. It means treating the subjects of study as actors within multi-layered contexts and employing the multiple strengths of individuals, families and communities to overcome or prevent difficulties…the purpose of strengths-based research and evaluation is to benefit the people involved in the study by giving them voice, insight, and political power" (p. 35). This is demonstrated in the words of EthnoSTEM teacher participant Kelsey Amos, "A foundational idea is that there is not just one math—the math we learned in school. Rather, all cultures develop mathematical practices and applied technologies, and these are worth paying attention to. Ethnomathematics is the act of recovering cultural dignity for all people by helping us strengthen our roots" (personal communication, December 13, 2015).

Pacific Ethnomathematics

Ethnomathematics is defined as the intersection of culture, historical traditions, and sociocultural roots, among others. It celebrates diverse systems and frameworks that have served many cultures well in the teaching and learning of mathematical activities (D'Ambrosio, 2001; Kyselka, 1987). Current efforts in this field orient educators to discover pathways for student engagement, and support multiple approaches to learning mathematics within formal and informal environments. A strong component of ethnomathematics is finding relevance in real-world applications through physical, environmental, spiritual, and cultural capacities (Gay, 2000; Kawakami, 2003; Ladson-Billings, 1995).

The Pacific region faces challenges and opportunities of diverse populations spread over expansive geographical areas, combined with a long history of local control of education and politicization of efforts to implement reforms and standards on a national scale (Ascher, 1998; Greer, Mukhodpadhyay, Powell, & Nelson-Barber, 2009; Weiss & Miller, 2006). The focus on local control provides a wealth of perspectives and approaches, and fertile soil in which to conduct mathematics education research. For example, the State of Hawaiʻi has been a national leader in place-based education and in connecting students' classroom learning to a broader cultural context through the field of ethnomathematics (Baybayan, Finney, Kilonsky, & Thompson, 1987; Chinn, 2011; Goetzfridt, 2008). It is through the lens of ethnomathematics that we explore *ma ka hana ka ʻike* (in working one learns).

A prominent example of *ma ka hana ka ʻike* is the 2013–2017 Mālama Honua Worldwide Voyage (WWV) of the double-hulled Pacific Island voyaging canoe Hōkūleʻa. By wayfinding without the guidance of modern instruments, the WWV blends indigenous traditions with innovative technology to map a course for the future. The mission of the voyage, *mālama honua* (to care for Island

Volume 1

Earth), invites people across the world to join us in exploring values and practices that can sustain our planet. Living on an island chain teaches us that our natural world is a gift with limits and that we must carefully steward this gift if we are to survive together. As we work to protect cultural and environmental resources for our children's futures, Pacific voyaging traditions and STEM tools allow us to venture abroad so that we may learn from and with others. Most recently, Hōkūle'a arrived in Cuba and is headed up the U.S. east coast, where the author will next join the crew in Washington, D.C. and New York City before the canoe sails through the Panama Canal back to Hawai'i in 2017.

Over the past several years, we have received declarations and pledges to care for Island Earth by French Polynesia President Gaston Flosse upon arrival in Tahiti, sailing with the United Nations Secretary General Ban Ki-moon in Samoa, discussing education and universal human values with His Holiness the 14th Dalai Lama, and conversing with Nobel Peace laureate Desmond Mpilo Tutu in South Africa. It was incredible being with these extraordinary leaders; however, the highlight is being with the children. Thousands of children have experienced the floating classroom Hōkūle'a, which is a model of island sustainability and exploration of STEM principles. According to His Excellency Ban Ki-moon in a handwritten message being carried on Hōkūle'a around the globe, "I am honored to be a part of the Worldwide Voyage. Its global mission inspires me. As you tour the globe, I will rally more leaders to our common cause of ushering in a more sustainable future, and a life of dignity for all" (B. Ki-moon, personal communication, August 31, 2014). Archbishop Desmond Mpilo Tutu added his support of STEM education while in South Africa, "We have just one planet home. This is an issue of whether we want to survive as a species or not, and we can do this through teaching the children science, technology, engineering, and mathematics" (D. Tutu, personal communication, November 14, 2015).

The WWV has impacted educational efforts abroad, as well as at home in the Pacific. The Polynesian Voyaging Society Education *Hui* (group) was formed to strategically align the missions, visions, and values of early childhood education through graduate institutions with the WWV. After participating in WWV in Aotearoa, University of Hawai'i System (UHS) President David Lassner and Polynesian Voyaging Society (PVS) Executive Director and Master Navigator Nainoa Thompson asked the author to be the UHS representative to the PVS Education Hui. In this capacity, we convened the UHS on both O'ahu Island and Hawai'i Island to strengthen current UHS initiatives in STEM, sustainability, and becoming a model indigenous-serving institution through the WWV. In addition, the UHS Board of Regents participated in a sail, and acknowledged the importance of *'ike Hawai'i* (knowledge of Hawai'i) in teaching, research, and service to our communities. According to UH Hilo Chancellor Donald Straney, "It is through this type of collaborative work, learning from each other, that the University of Hawai'i can fulfill our responsibilities to the indigenous people of Hawai'i" (D. Straney, personal communication, February 12, 2016).

As another direct result of PVS Education Hui collaborations, Nā Hopena A'o Cultural Pathways are one of the outcomes from the Hawai'i State Department of Education (HIDOE). Nā Hopena A'o Cultural Pathways

NOTES

are a statewide framework to develop the skills, behaviors and dispositions that are reminiscent of Hawai'i's unique context, and to honor the qualities and values of the indigenous language and culture of Hawai'i (HIDOE, 2015). According to the HIDOE Office of Hawaiian Education, "Nā Hopena A'o are a reflection of this special place, six outcomes to be strengthened in every student over the course of their K–12 learning journey—Belonging, Responsibility, Excellence, Aloha, Total Well-being, and Hawai'i. When taken together, these outcomes become the core BREATH that can be drawn on for strength and stability throughout school and beyond to succeed in college, careers and communities" (p. 1). The intra- and inter-organizational work we are engaged in is the foundation for current and future ethnomathematics research in the design of professional development for educators and teacher educators.

Professional Development for Educators and Teacher Educators

Now in its eighth year, the Ethnomathematics and STEM Institute (EthnoSTEM) is an effort to address issues of equitable and quality education through culturally responsive STEM pedagogy. From 2008–2012, the Ethnomathematics Institute was a 2-week summer program for undergraduate STEM majors when the author was an Associate Professor of Mathematics at the University of Hawai'i – West O'ahu. The past three years, 2013–2016, represent a shift to a yearlong institute focusing on P–20 STEM teacher education as the author is currently an Associate Professor of Mathematics Education at the University of Hawai'i at Mānoa. Research and data presented in this section are from the two completed years, 2013–2015, involving P–20 STEM educators. Our partners across Hawai'i and the Pacific include the University of Hawai'i System, Hawai'i State Department of Education, Hawai'i P–20 Partnerships for Education, Pacific American Foundation, Pacific Resources for Education and Learning, and Polynesian Voyaging Society. The WWV has served as a creative vehicle to develop alliances spanning P–20 education levels, public and private sectors, and invites new partners by leveraging social innovations (Bryk, Gomez, & Grunow, 2011).

The three main EthnoSTEM objectives are to: (1) explore promising practices in STEM courses in diverse, high needs populations through national and state standards (i.e., Mathematics Common Core State Standards, Next Generation Science Standards, Nā Hopena A'o Cultural Pathways); (2) strengthen campus-community partnerships and build a research consortium within Hawai'i and the Pacific; and (3) commit to high-quality professional learning that is relevant, contextualized, and sustainable.

EthnoSTEM encourages investigations and adaptations of traditional and 21st century knowledge within formal and informal environments. In addition to classroom learning, place-based learning occurs throughout the Hawaiian Islands from skies to seas. For example, through traditional wayfinding techniques with the Polynesian Voyaging Society, we integrate trigonometry and analytic geometry with celestial navigation utilizing the sky as a textbook for estimating latitude and position. At Waimea Valley, participants discover intersections of tropical ecology and measurement through traditional rock wall building using vectors without cement or modern implements. At the 400-year old Waikalua Loko I'a, participants help restore the fishpond and experiment with engineering skills involved in structural design and biodiversity of microclimates. Here, participants have an opportunity to investigate bioacoustics through sinusoidal curves and the impact of sound waves on fishpond walls. The culminating experience, Kalaupapa, Moloka'i, was once a leprosy community in isolation, and now serves as a place for education and contemplation. Our experiences are grounded in service learning, and we strive to always give back to Island

Earth. As an example of service learning, in the past we cleared the Baldwin Boys' Home to give the patients a place to rest and relax in their natural environment. In the process, we had active dialogues about concave down parabolas and vertices as we threw debris into the back of a truck.

The in-service and pre-service teacher participants are selected to create a diverse group spanning a broad range of experiences, subjects and grade levels, and geographical locations. Demographic information on the past two cohorts of EthnoSTEM participants representing 34 P–20 schools from across the Hawaiian Islands is provided in Table 1. One of the 40 participants was unable to complete the program.

NOTES

Table 1. Demographic Information on EthnoSTEM Participants.

Characteristics	No. of teachers
Grade level taught during study year (n=39):	
Elementary school (Grades K–5)	8
Middle school (Grades 6–8)	9
High school (Grades 9–12)	14
District resource teacher (K–12)	2
Undergraduate students	3
Post-secondary teachers	3
Island (n=39):	
O'ahu	30
Hawai'i	4
Kaua'i	2
Maui	3
Ethnic background (n=39):	
Asian	12
Caucasian	15
Native Hawaiian	9
Hispanic	2
Pacific Islander	2
Mixed	2
Number of years teaching (n=34)[a]:	
<1	1
1–4	11
5–10	5
11–15	6
>15	11

Characteristics	No. of teachers
School type (n=34)[a]:	
Public	29
Public charter	4
Private	1
Disciplines taught during study year (n=34)[ab]:	
Education	2
English	2
Math	21
Science	16
Technology	1
All subjects (elementary)	1
College preparation	1
Pre-service teachers—number of years until degree completion (n=8)[c]:	
<1 year	5
1 year	2
2 years	1
Subjects taught during the study year (n=34)[b]:	
Algebra, Art, Biology, Chemistry, College Preparation, Elementary Education, English, Graphic Technology, Health, Language Arts, Marine Science, Mathematics, Math for Elementary Teachers, Math Resource, Modeling Our World, Natural Resources, Physical Science, Physics, Pre-Algebra, Pre-Calculus, Robotics, Science, Science Literacy, Social Studies	

[a] *These questions were only asked of in-service teachers (n=34).*
[b] *The in-service teachers were asked to list all subjects taught during the program year.*
[c] *These questions were only asked of participants who identified themselves as students (n=8). Four of these participants were undergraduate pre-service teachers, four were full-time in-service teachers who were also enrolled in graduate school.*

Professional Development Structure

During the yearlong EthnoSTEM institute, participants meet four times face-to-face (with sessions lasting from one to seven days) and once online, for a total of 132 direct contact hours over 13 days. The extended intervention is designed to allow facilitators to build a supportive network and offer a variety of avenues for the participants to engage with the material. The professional development structure consists of a two-day intensive orientation focused on introducing and laying the foundation and building community, three one-day workshops centered around themes that expose the teachers to different ethnomathematics promising practices and strategies through field experiences (respectively called PD1, PD2, and PD3), and a one-week summer immersion *(see Figure 1)*.

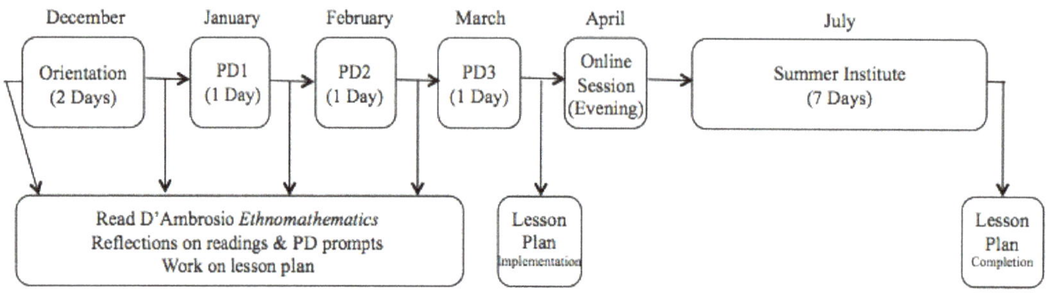

Figure 1 EthnoSTEM Professional Development Structure

Professional development workshops are linked through participant reflections on readings and individualized progress on the creation and implementation of lesson plans. As the EthnoSTEM culminating project, the final lesson plan is intended to reflect the participants' understandings of how to adapt concepts into their classrooms. In addition to having an overarching theme, each EthnoSTEM day focuses on a different Hawaiian value. These are aligned with the PVS and WWV values:

Aloha: To show kindness/compassion

ʻImi ʻike: To seek knowledge

Lokomaikaʻi: To share

Mahalo: To show gratitude/respect

Mālama: To care for

Naʻau pono: To be just

Olakino maikaʻi: To live healthily

Laulima: To work together

The 20 participants are divided into five *hui* (groups) of four participants each. Each group is mentored by a project team *hui* advisor. Two of the EthnoSTEM alumni have become *hui* advisors. After completing his first year as a participant, Kaipo Tam, Ke Kula Kaiapuni ʻo Ānuenue mathematics teacher, expressed the importance of the EthnoSTEM guiding values in his reflection (personal communication, July 20, 2014):

NOTES

The Ethnomathematics and STEM Institute has helped me personally and professionally. This institute has allowed me to see math and science in a way that our kids can learn that's meaningful, and that they can use in their everyday lives as we laulima. I love dancing hula, weaving lauhala, and I've always wondered how I could incorporate them into a math classroom. Through 'imi 'ike, I can show my kids that culture can be incorporated into math and love what our ancestors have done for centuries, and continue to be a 21st century student and citizen of this world. I'm so excited to go back to school in a couple of weeks and showcase what I've learned and mahalo the institute for everything it has done for us. We can make a big difference, not only in our classrooms but in the State of Hawai'i.

The EthnoSTEM website contain a curriculum database of lesson plans created and implemented by participants, including topics such as the "Fibonacci Sequence in Nature", "A Splash of STEM Through Aquaponics", "Cartesian Coordinates and Weaving", "Exploring Ocean Waves with the Sine Curve", and the "Hawaiian Star Compass and the Unit Circle" (see http://ethnomath.coe.hawaii.edu). EthnoSTEM has been featured in articles, press releases, television, and other media locally and nationally (Hawai'i Public Radio, 2010; Mathematical Association of America, 2011; Pacific Business News, 2007; PBS Hawai'i, 2016; University of Hawai'i Systemwide News, 2016).

Evaluation Results and Findings

The University of Hawai'i at Mānoa Curriculum Research & Development Group conducts project evaluation. The evaluation of EthnoSTEM utilizes a mixed-methods approach, and evaluators use instruments for collecting formative and summative information. The main methods of data collection are observations, surveys, focus group interviews, and reflections after every professional development component. The six evaluation questions are presented in Table 2. The analyses consist of descriptive statistics for quantitative data, and qualitative analysis using a grounded theory approach for constructed-response questions (Corbin & Strauss, 2008).

Volume 1

Table 2. EthnoSTEM Program Goals and Evaluation Questions.

Goals	Evaluation Questions
Increase content knowledge and pedagogy in culturally responsive STEM based on the Common Core State Standards (CCSS), Next Generation Science Standards (NGSS), and Nā Hopena Aʻo (HĀ) Cultural Pathways.	1. To what extent did the participants perceive that the professional development (PD) affected their knowledge of culturally responsive STEM pedagogies based on CCSS, NGSS, and HĀ? 2. Did the participants experience the project as valuable and relevant to their ongoing learning and teaching practices?
Explore connections to STEM courses through curriculum design, implementation, and assessment.	3. How did the participants perceive the process of their lesson plan development and implementation?
Foster teachers' understanding of the design and implementation of STEM units based on CCSS, NGSS, and HĀ that include a service-learning component.	4. To what extent did the participants perceive that their participation in the PD changed their teaching practice?
Establish campus-community partnerships in the State of Hawaiʻi and Pacific for sustainable classroom and community networks.	5. To what extent did the participants report the PD cultivated a supportive, sustained community? 6. To what extent were the PD's activities being conducted as expected?

Changes in Teaching Practices

Overall, the participants perceived the professional development experience to be valuable and relevant to their teaching practice (N = 39, M = 4.87, SD = 0.34). In addition, at the end of the yearlong program, the participants were most likely to agree they understood and could incorporate culturally responsive pedagogy, STEM content, CCSS, and NGSS into their teaching practice (see Table 3). Disaggregated, the four items on "understanding" had a reliability of 0.69 (Cronbach's alpha) and an overall mean of 4.18; the four items on "incorporation" had a reliability of 0.75 and an overall mean of 4.23.

NOTES

Table 3. Understanding and Incorporation of Pedagogy, Content, and Standards.

Prompt	Construct	N	M	SD
1. The professional development (PD) helped me to better my understanding of culturally responsive pedagogy.	Understand pedagogy	39	4.82	0.39
2. The PD helped me to understand how to incorporate culturally responsive pedagogy into my teaching practice.	Incorporate pedagogy	39	4.81	0.36
3. The PD helped me to better understand how to incorporate STEM content into my teaching practice.	Incorporate STEM content	39	4.46	0.70
4. The PD helped me to better my understanding of STEM content.	Understand STEM content	39	4.44	0.71
5. The PD helped me to better understand how to incorporate the math Common Core State Standards (CCSS) into my teaching practice.	Incorporate CCSS	39	3.95	0.89
6. The PD helped me to better my understanding of the math CCSS.	Understand CCSS	39	3.79	0.76
7. The PD helped me to better my understanding of the Next Generation Science Standards (NGSS).	Understand NGSS	39	3.68	0.96
8. The PD helped me to better understand how to incorporate the NGSS into my teaching practice.	Incorporate NGSS	39	3.68	1.01
9. The PD as a whole was valuable and relevant to my teaching practice.	Overall experience	39	4.87	0.34

When asked to give specific examples of how their teaching practices have changed, most participants indicated that the professional development encouraged them to modify curriculum materials to make them more relevant to their students through cultural activities and place-based connections. The teachers were also encouraged to do more hands-on activities, service learning, and project-based learning. One participant said the professional development validated his teaching philosophy, "Although I was encouraged to be culturally responsive in my teaching, I was not taught how to do it in my teacher preparation program. This PD showed me that we have resources in our own backyard that we can use to enhance our teaching."

The participants were separated into new (<5 years of teaching experience), intermediate (5-15 years of teaching experience), and experienced (>15 years of teaching experience) categories to assess the teachers' changes in practice. The new teachers tended to focus on making curriculum relevant to their students and building community in their classrooms. The experienced teachers were more likely to mention the professional development validated their teaching philosophy and motivated them to try new things such as service learning projects. For example, one experienced teacher participant commented, "Ethnomathematics focuses on valuing people and the knowledge they bring. It has opened my eyes to various types of culture not just Hawaiian, but math culture, art culture, science culture...the service learning projects that we did will always be ingrained in my mind because the people I was surrounded by made it easy to learn and also to LOVE learning. They showed me how we can give back and take care of the land and culture and feel good about it. It reeled me back into a culture that I know and am proud of."

Further Discussion

Due to its geographical position, Hawai'i often serves as a gateway between the mainland U.S. and Pacific. Through research and praxis, we actively seek opportunities to interact and collaborate, and hold a shared belief *ma ka hana ka 'ike* (in working one learns). The Mālama Honua Worldwide Voyage and Ethnomathematics and STEM Institute have strived to engage P–20 educators on significant themes to expand our network of strategic partners, including the University of Hawai'i System, Hawai'i State Department of Education, Hawai'i P–20 Partnerships for Education, Pacific American Foundation, Pacific Resources for Education and Learning, and Polynesian Voyaging Society. These efforts have led to successful outcomes both within and outside of the classroom, and offer insights into next steps. Through research-based design and implementation of effective indigenous STEM professional development, culturally responsive methods are being transferred directly into P–20 classrooms.

In the future, as outlined by the National Science and Technology Council's Federal STEM Education 5-Year Strategic Plan (2013), we will expand our approaches "to enable research and development at all levels on STEM instructional strategies, learning in informal environments, and ways to improve professional development of teachers and faculty" (p. 4). We will invite and involve larger communities to mālama honua collectively as we build indigenous-serving, strengths-based alliances in our voyages on land and sea. Through continued efforts, the common denominator of our work remains the same—the children remind us why we do the things we do, and help us envision beyond the horizon. As it says on the backbeam of the Hōkūle'a canoe, *kapu nā keiki* (hold sacred the children).

References

Ascher, M. (1998). *Ethnomathematics: A multicultural view of mathematical ideas*. New York, NY: Chapman & Hall/CRC.

Baybayan, C., Finney, B., Kilonsky, B., & Thompson, N. (1987). Voyage to Aotearoa. *The Journal of the Polynesian Society, 96*(2), 161–200.

Bryk, A., Gomez, L., & Grunow, A. (2011). *Getting ideas into action: Building networked improvement communities in education.* Retrieved April 27, 2016 from http://archive.carnegiefoundation.org/pdfs/elibrary/bryk-gomez_building-nics-education.pdf

Chinn, P. W. U. (2011). Developing a sense of place and an environmental ethic: A transformative role for Hawaiian/Indigenous science in teacher education? In J. Reyhner, W. S. Gilbert & L. Lockard (Eds.), *Honoring Our Heritage: Culturally Appropriate Approaches for Teaching Indigenous Students*, (pp. 75–95). Flagstaff, AZ: Northern Arizona University.

NOTES

Corbin, J., & Strauss, A. (2008). Basics of qualitative research: Techniques and procedures for developing grounded theory (3rd ed.). *Organizational Research Methods, 12*(3), 614–617.

D'Ambrosio, U. (2001). *Ethnomathematics link between traditions and modernity*. Rotterdam: Sense Publishers.

Fetterman, D. (2000). *Foundations of empowerment evaluation*. Thousand Oaks, CA: Sage Publications.

Furuto, L. (2014). Pacific ethnomathematics: Pedagogy and practices in mathematics education. *Teaching Mathematics and its Applications: International Journal of the IMA*. Oxford: Oxford University Press.

Furuto, L. (2013). Bridging policy and practice with ethnomathematics. *Journal of Mathematics and Culture, 7*(1), 31–57.

Gay, G. (2000). *Culturally responsive teaching: Theory, research, & practice*. New York, NY: Teachers College Press.

Goetzfridt, N. (2008). *Pacific ethnomathematics a bibliographic study*. Honolulu, HI: University of Hawaii Press.

Greer, B., Mukhodpadhyay, S., Powell, A., & Nelson-Barber, S. (2009). *Culturally responsive mathematics education*. New York, NY: Routledge Press.

Hall, S. (1993). Cultural identity and diaspora. In P. Williams & L. Chrisman (Eds.), *Colonial discourse and post-colonial theory: A reader.* London: Harvester Wheatsheaf.

Hawai'i Public Radio. (2010). *Studying the cultural nature of math*. Retrieved April 27, 2016, from https://www.hawaiipublicradio.org/news/podcasts/podcasts/kepa@hoakaleifoundation.org?page=13

Hawai'i State Department of Education. (2015). *Nā hopena a'o*. Retrieved April 27, 2016, from http://www.hawaiipublicschools.org/TeachingAndLearning/StudentLearning/HawaiianEducation/Pages/HA.aspx

Kana'iaupuni, S. (2005). Ka'akālai kū kanaka: A call for strengths-based approaches from a Native Hawaiian perspective. *Educational Researcher, 33*(9), 32–38.

Kania, J., & Kramer, M. (2011). *Collective impact*. Retrieved April 27, 2016 from http://www.ssir.org/articles/entry/collective_impact

Kawakami, A. (2003). Where I live, there are rainbows: Cultural identity and sense of place. *Amerasia Journal, 29*(2), 67–79.

Kyselka, W. (1987). *An ocean in mind*. Honolulu, HI: University of Hawaii Press.

Ladson-Billings, G. (1995). But that's just good teaching! The case for culturally relevant pedagogy. *Theory Into Practice, 34*(3), 159–165.

Mathematical Association of America Focus. (2011). *Ethnomathematics and mathematics professors*. Retrieved April 27, 2016, from http://www.maa.org/pubs/FOCUSApr-May11_Ethnomathematics.html

Maton, K., Schellenbach, C., Leadbeater, B., & Solarz, A. (2003). *Investing in children, youth, families and communities: Strengths-based research and policy.* Washington, D.C.: American Psychological Association.

Meyer, M. (2003). *Ho'oulu our time of becoming: Hawaiian epistemology and early writings*. Honolulu, HI: Native Books.

National Science and Technology Council. (2013). *Federal science, technology, engineering, and mathematics (STEM) education 5-year strategic plan.* Washington, D.C.: Office of Science and Technology Policy Executive Office of the President.

Pacific Business News. (2007). *Math students will use voyaging techniques.* Retrieved April 27, 2016, from http://www.bizjournals.com/pacific/stories/2007/12/17/daily36.html

Pacific Region Educational Laboratory. (1995). *Pacific standards for excellence in teaching, assessment and professional development.* Honolulu, HI; Pacific Mathematics and Science Regional Consortium.

PBS Hawaiʻi. (2016). *Long story short: Linda Furuto.* Retrieved April 27, 2016, from http://pbshawaii.org/lss/

Pukui, M. K. (1993). *ʻŌlelo noʻeau: Hawaiian proverbs and poetical sayings.* Honolulu, HI: Bishop Museum Press.

Sohng, S. (1998). Research as an empowerment strategy. In L.M. Gutierrez, R.J. Parsons, & E.O. Cox (Eds.), *Empowerment in social work practice: A source book.* (pp. 187–203). Pacific Grove, CA: Brooks/Cole.

Tuhiwai Smith, L. (1999). *Decolonizing methodologies: Research and indigenous peoples.* New York, NY: Zed Books.

University of Hawaiʻi Systemwide News. (2016). *UH maps way forward at second worldwide voyage meeting in Hilo.* Retrieved April 27, 2016, from http://www.hawaii.edu/news/2016/02/16/uh-maps-way-forward-at-second-worldwide-voyage-meeting-in-hilo/

Waitzer, J., & Paul, R. (2011). Scaling social impact, when everybody contributes, everybody wins. *Innovations, 6*(2), 143–155.

Weiss, I., & Miller, B. (2006). *Developing strategic leadership for district-wide improvement of mathematics education.* Lakewood, CO: National Council of Supervisors of Mathematics.

The Relationship Between Language, Culture and Ethnomathematics

Volume 1

Tony Trinick
University of Auckland
t.trinick@auckland.ac.nz

Tamsin Meaney
Bergen University College

Uenuku Fairhall
Te Kura Kaupapa Māori o Te Koutu

Abstract

In an earlier paper, we pointed to a number of unresolved issues pertaining to the introduction of ethnomathematics into classrooms and how Indigenous cultural practices and language might be adversely affected, particularly when attempts are being made to revitalize these. In this paper, we continue this discussion by describing theoretical and practical issues related to the connection between the language of instruction, culture and ethnomathematics. We illustrate these issues and propose a three step approach, cultural symmetry, for working with these issues by drawing on the ethnomathematical concept of location and direction pertaining to Māori, the Indigenous community of Aotearoa/New Zealand.

Keywords:

- Ethnomathematics
- Cultural Symmetry
- Māori Language Revitalisation
- Colonization

Introduction

In our earlier article in the *Journal of Mathematics and Culture* in 2008 (Meaney, Fairhall, Trinick, 2008), we raised concerns about the use of ethnomathematical activities as a means to improve mathematical understanding, without consideration of the value, in their own right, of the culture and language that the ethnomathematical activities were embedded in. In classrooms in Western societies, mathematics is often decontextualized and teaching about it in international languages, such as English and French, is rarely problematized. However, the consequence of this is that activities are often taught without regard for the original culture and languages in which the activities arose.

In many Western classrooms, this imperialistic approach to mathematics teaching and learning (Bishop, 1990) may not be considered problematic, but for marginalised Indigenous groups who are endeavouring to simultaneously save their Indigenous language and culture, these aspects need to be elevated. This is because as Fishman (1991) and others have argued, culture is in the language and that "take language away from the culture, and the culture loses its literature, its songs, its wisdom, ways of expressing kinships relations and so on" (Fishman, 1991, p. 72). Therefore, the potential loss of a language for discussing a cultural practice in educational situations has implications for the cultural knowledge connected to it. Therefore, we contend that revitalisation and maintenance of the language, including the teaching of mathematics, is insufficient unless cultural knowledge is also revitalised and maintained (Trinick, 2015).

In our earlier paper, we raised concerns related to the impact on the Indigenous language and culture of Aotearoa/New Zealand when *te reo Māori* was used as the medium instruction for the teaching of Western mathematics and/or ethnomathematics—in particular how this might affect the valuing or changing of those cultural activities. We acknowledge that all languages and cultures change over time. However, the issue is how to ensure that these changes do not result in undesirable consequences for the language and culture and, thus, the community of speakers. Our aim in this paper is to suggest an alternative, less detrimental, approach for using ethnomathematical approaches that supports both revitalisation and maintenance of an Indigenous culture and language, whilst also supporting students to gain Western mathematical understanding. To do this, we describe some of our recent theoretical and practical work around these issues, drawing on examples from the teaching and learning of location and direction.

Ethnomathematics is often cited as a potential solution to a range of different problems in mathematics education (Meaney & Lange, 2013). For example, "one of the goals of ethnomathematics is to contribute both to the understanding of culture and the understanding of mathematics, and mainly to lead to an appreciation of the connections between the two" (D'Ambrosio, 1999, p. 146). On the other hand, it has also been criticised for reinforcing colonial and hegemonic constraints about what counts as the valuable mathematics needed to gain entry to further study or high paying jobs (see for example, Vithal & Skovsmose, 1997; Jablonka & Gellert, 2010; Pais, 2011). If ethnomathematics is to be used constructively in educational situations so that it supports the

NOTES

decolonising of cultural knowledge, then there is a need, amongst other things, to recognise that cultural considerations are as important as mathematical ones. From a teaching and learning perspective, this means that when students engage in ethnomathematics, they should be aware that they are learning and responding to different forms of mathematics, as well as to cultural aspects of the activity. This should result in meta-discussions about what constitutes valuable mathematical knowledge and who gets to decide this, as these can lead to understandings about how mathematics education could be a contributor to colonisation or decolonisation.

Colonial Linguistic and Cultural Hegemony

At the time the first missionaries and settlers arrived in the 18th century, Māori had a robust system for educating their children to ensure the survival of their communities in Aotearoa (Riini & Riini, 1993). These included mathematical aspects connected to "the design (of canoes, buildings and tools); astronomical, meteorological and oceanographic information and analysis; geometries of pattern in weaving (rāranga), painting (kowhaiwhai) and carving (whakairo); and the logic of argument in oratory" (Barton & Fairhall, 1995, p. 1). After 1840 and the establishment of a British colony, European forms of government and schooling were instigated. The hegemonic function of the early schools was to provide a formalised context to assimilate Māori communities into European beliefs, attitudes and practices with the intent to "civilise" the Māori population (Simon, 1998). This goal of assimilation was maintained by successive governments and their agencies over the next 150 years, resulting in a range of educational policies, both overt (English-language-only schooling policy) and covert (English-language-only workplaces), to privilege English as the sole language of education (Trinick, 2015) and Western mathematics as the only mathematics worth teaching.

With the privileging of English and Western knowledge domains, such as mathematics, in both society and schooling, many Māori, came to believe that their language and knowledge was of limited use in this new environment. Apple considered (1979) that "to a large extent society as it exists ... is held together by implicit common sense rules and paradigms of thought by hegemony as well as overt power" (p. 96). The imposition of this hegemonic model; that is, where one language (English) and one form of ethnomathematics, Western mathematics, limited the use of other languages (Māori) and other forms of ethnomathematics, can be traced to ideologies and beliefs that informed the Eurocentric education of the time. English language hegemony was utilised as a means to build a nation state, and to take political power and control from Māori (May, 2012). Similarly, for Māori, "western mathematics has been one of the most powerful weapons in the imposition of western culture" (Bishop, 1990, p. 51). As in other situations of colonisation, the educational policies reflected strong assimilationist attitudes that linked proficiency in English to supposed best outcomes for Māori (de Bres, 2008).

While *te reo Māori* was excluded from schooling, it was still nurtured throughout the 19th and 20th centuries in largely Māori rural communities (May & Hill, 2005). However, throughout the period following World War II and into the 1960s, the urban migration of Māori considerably changed the demographics of Aotearoa/New Zealand. This shift to urban areas where *te reo Māori* was not strong, combined with other forces, impacted negatively on Māori language development and resulted in the younger generation of Māori rapidly losing their language and culture (King, 2007). By the 1970s, *te reo Māori* was an endangered language (Spolsky, 2005).

We can only speculate, but if schooling in Aotearoa/New Zealand had been allowed to continue to develop in the Indigenous language, from the colonial era into modern times, then an adapted form of Western mathematics would likely be present. This is because Māori understood that the knowledge associated with technology and trade could enhance their traditional ways of life (Spolsky, 2005). Māori, like all other cultural communities, do not stagnate in the development of cultural practices but adapt to fit changing circumstances. Although it is highly probable that any change would have caused angst in the community, survival including the need to adapt to changing economic times was paramount. In reality, Māori were denied the opportunity for over 150 years to incrementally develop subjects such as mathematics in their Indigenous language and based on their cultural perspectives.

Māori-Medium Schooling and Revitalisation Efforts

It was against the background of rapid and significant language loss that the Māori community initiated Māori-medium education, in particular many schools, such as *kura kaupapa Māori*, were founded. Te Koutu is one such school (Meaney, Trinick, Fairhall, 2012). However, to revitalise and reinvigorate an endangered language and culture primarily via schooling, as is the case in Aotearoa/New Zealand, is fraught with many tensions and issues. Initially, *kura kaupapa Māori* were developed outside the state system, not only to revitalise *te reo Māori*, but also as a resistance movement to the assimilationist nature of New Zealand European schooling (Penetito, 2010).

By the 1990s, *kura kaupapa Māori* had become government funded, which on one hand provided their schools with financial security, but on the other hand required *kura* to implement state-mandated curricula and assessment practices developed from Eurocentric perspectives (McMurchy-Pilkington & Trinick, 2008). On this basis, the Ministry of Education insisted that the structure of the inaugural Māori-medium mathematics curriculum "mirror" the hegemonic English-medium version (McMurchy-Pilkington & Trinick, 2008).

The requirement to essentially translate the English-medium version into Māori was met with some consternation and resentment (Trinick, 2015). Nevertheless, as McMurchy-Pilkington (2004) noted, Māori curriculum developers took advantage of opportunities to co-opt its development in order to elaborate the Māori language in a systemised way– thus enabling the teaching of mathematics in the medium of Māori to senior secondary school levels. This indirectly raised the status of the language within the Māori community as well as among educators. The development of a Māori-medium curriculum also encouraged the linking of mathematics to contemporary Māori contexts– albeit somewhat tempered by the requirements of the national curriculum and standardised national high stakes assessments.

Nevertheless at that time, few attempts were made to systematically incorporate traditional Māori mathematical practices into Māori-immersion mathematics classrooms. This was partly due to uncertainty about how it could be done without devaluing the cultural practices and in a manner that did not reinforce the hegemonic valuing of Western mathematics. There was some concern that using cultural activities in the teaching of Western mathematics would be akin to inviting the Trojan horse in because it had the potential to "'destroy the conceptual basis

NOTES

of Maori culture and language" (Barton & Fairhall, 1995, p. l). In the 1980s, attempts were made to connect Māori culture to mathematics by identifying symmetrical patterns found in traditional artefacts (Knight, 1984). However, by the 1990s, only focusing on patterns as exemplars of Western mathematics was disparaged as insufficient to count as ethnomathematics (Barton, 1993).

By thc late 2000s, the capacity to develop Māori-medium curriculum had expanded significantly because of lessons learnt from earlier developments and thus the "Ministry of Education appeared more accommodating of difference" (McMurchy-Pilkington, Trinick & Meaney, 2013, p. 357). This provided an opportunity for Māori-medium schools to be more proactive in reintroducing traditional Māori mathematical concepts and practices into the classroom, something which had been recognised as neglected by Māori themselves (Meaney, et al., 2012).

Cultural Symmetry

From our investigations of how to use mathematics lessons in the revitalisation of cultural practices, we have developed a three-step approach, which we call cultural symmetry (Trinick, Meaney, Fairhall, 2016). Our work in this area indicates that a set of principles is needed to ensure that the introduction of ethnomathematics-based lessons contributes to discussions with students about the cultural practice, the mathematics as well as how knowledge comes to be valued. In describing this three-step approach, we draw on earlier work we have done in regard to the spatial concepts of location and direction (see for example, Trinick, Meaney, Fairhall, 2015). This is an area of interest for Māori because traditionally they were renowned for their navigational expertise, being able to navigate using traditional techniques across vast distances in the Pacific Ocean (Best, 1922). Additionally, one of the authors had recorded information from elders in the 1990s that they considered important about location and direction for specific areas of Aotearoa/New Zealand (Trinick, 1999)—knowledge that perhaps is no longer readily available as it once was.

In research undertaken a few years ago at Te Koutu, the school where Uenuku is principal, it was clear that many of the students across all ages had difficulties using the environment to orientate themselves (Trinick, Meaney, Fairhall, 2015). We considered that just having theoretical knowledge about how to read a map was perhaps not adequate. We were also aware that the teachers at Te Koutu who had been very much inculcated into Western spatial orientation struggled with teaching Māori spatial orientation conventions and frameworks, which included an orientation to East/West rather than North/South. There was, therefore, a practical need to work on this issue.

Another consideration was that technology such as Global Positioning System (GPS) was changing what was valued as useful skills in work situations. Global Positioning System (GPS) locates specific points based on their latitude and longitude anywhere on Earth. It is now used almost exclusively to guide ships, airplanes, rescue and salvation crews, land surveys and so on, replacing the reliance of traditional compass knowledge. However, in school mathematics the incorporation of GPS into lessons of location and direction has been slow with few experimental studies being undertaken (see for example, Mooney & Laubach, 2002). Although incorporating the use of GPS might seem contradictory to revitalising traditional practices, one of the key aims of Te Koutu is to prepare students to live as Māori in a modern technological world. We considered that including discussions about the role of GPS would provide opportunities for students to evaluate how knowledge comes to be seen as valuable in school mathematics.

Step 1

The first step in cultural symmetry is for the cultural knowledge to be identified and then acknowledged as valuable. In regard to the spatial orientation concepts, it was considered important to have the students understand that location and direction are not just about finding one's way, but about how the features in the tribal landscape influenced cultural practices developed over a long period of time. In 2014, students in Year 8 were asked to create a map of their local tribal area from memory, adding significant cultural sites, place names, places of significance and a measurement scale of some sort. They were then told the story of Tuki and Ngāhuruhuru, two young men who were kidnapped from Northland, New Zealand, in 1793 and taken to Norfolk Island to teach the colony's prisoners about how to make rope from flax. While there, and amid questions about Aotearoa/New Zealand, Tuki drew a map of his homeland for Governor Philip Gidley King from memory (Binney, 2004). Specifically, the map provides an insight into how Tuki understood the world: the places important to him, his tribe, his enemies and included political and spiritual information (Binney, 2004). The area that Tuki knew best was disproportionate in size and area to the rest of Aotearoa/New Zealand, and it was also the most detailed (see Figure 1). However, it also shows that he had knowledge of distant places, such as Te Waipounamu (Pounammo on the map) which is about 1 240 km from northland (Tuki's area).

NOTES

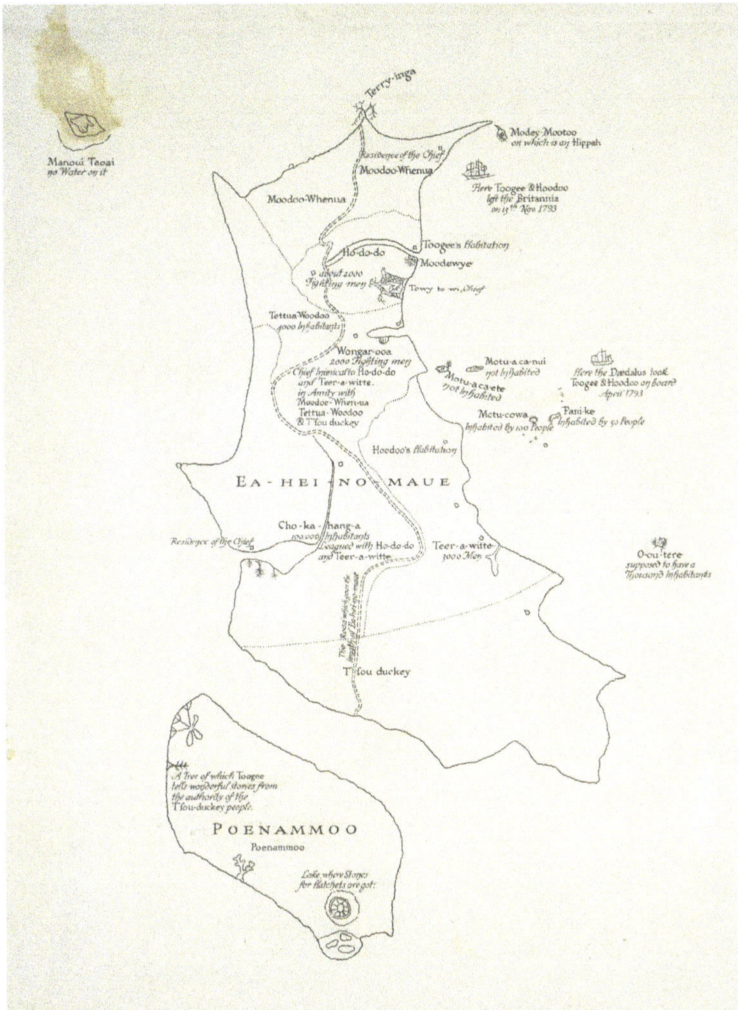

Figure 1: Tuki's Map. from: https://commons.wikimedia.org/wiki/File:New_Zealand-Map-by-Tuki-1798.jpg

The students then compared their maps with Tuki's map. Uenuku, in reflecting on the lesson, stated:

> *Most of the students' maps of Rotorua reflected Tuki's perceptions of scale etc. For example, the area where the students were familiar with and lived was out of scale with the rest of their map. Lake Rotorua varied between 5km and 35km. They surprised me in the depth of cultural detail they were able to add to their maps. They then appreciated the idiosyncratic nature of Tuki's map.*

This lesson highlighted cultural aspects of locating and directing by connecting the local area to larger geographical spaces. It also allowed for a discussion about the knowledge which is shown in Tuki's map and why he might have valued it so highly that he included it in his map. The comparison with the students' own maps enabled Tuki's map to be seen in the context of his experiences and what was considered important, for example who his friendly relatives were, and not judged for accuracy against its closeness to Western cartographical representations of New Zealand. In this way, the students came to see that models and maps of spatial environments are sociocultural tools as is the language that describes spatial orientation (Gauvin, 1991).

Step 2

The second step is to introduce the traditional ways for describing location and direction in *te reo Māori*. Māori used a variety of spatial frameworks abstracted from geomorphic or landmark-based systems to orientate themselves to the general direction of—east, west, north and south—and intermediary directions (Trinick, 1999). These references were derived from a mixture of phenomena, including the actions of the sun and wind, and geographical landforms. Figure 2 includes, amongst other things, some of the traditional wind names used to indicate directions.

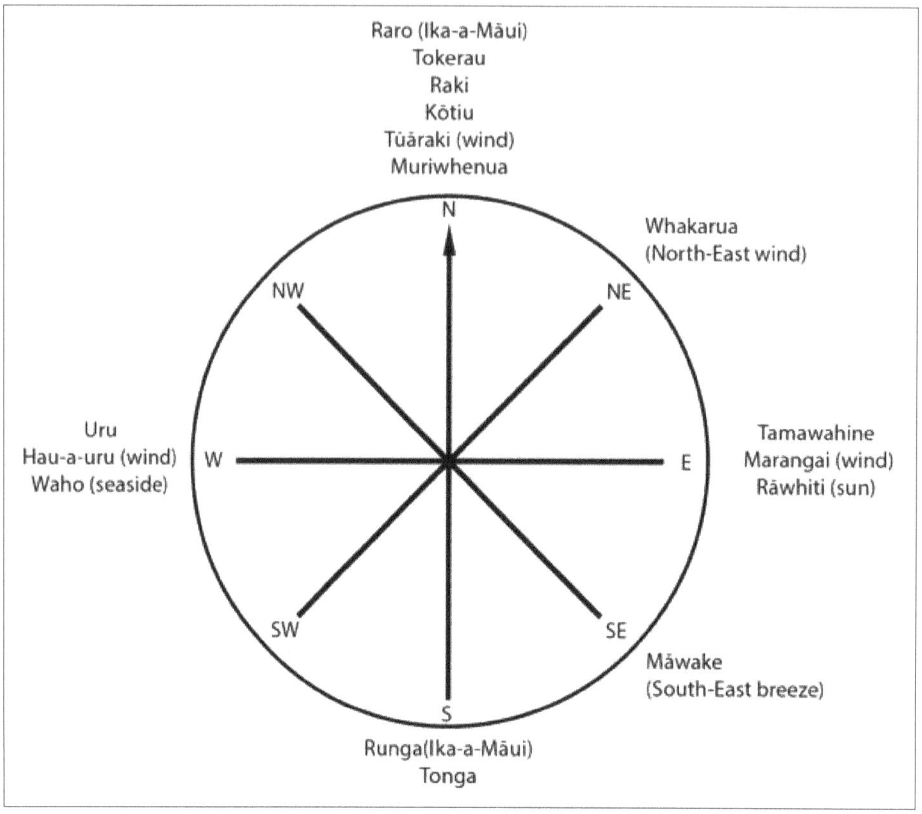

Figure 2: Māori geomorphic system for spatial orientation (Trinick, 1999)

This stage is important because the linguistic terms are linked to cultural frameworks and to mathematics. While there was minimal acknowledgement of traditional Māori mathematical knowledge in the first Māori-medium curriculum, the standardised directional terms had been based on the traditional orientation from East to West. However, in research (Meaney, et al., 2012), we found that those teaching or learning about location in *te reo Māori* had not grasped the importance of the east-west orientation. In this case, the link between language and culture had become disconnected and, therefore, could not act as a viaduct for learning. The hegemony of Western

NOTES

mathematics ways of valuing the North/South orientation made it challenging for both teachers and learners to consider other possibilities. In this case, explicit discussion about the east-west orientation was needed if language and cultural practices were to be revived in a meaningful way. This was done by having one of the authors, Tony, discuss with the teachers the background to why the directions were labelled as they were.

Step 3

The third and final step is to discuss the origins and frameworks that underpin location and direction in Western mathematics as well as traditional Māori practices and discuss advantages and disadvantages of both. A vehicle for contextualising the various ways of knowing was to use the story of a traditional explorer; Īhenga, (an ancestor of two of the authors) who after arriving on one of the first canoes to settle Aotearoa/New Zealand set off with his uncle Kahumatamomoe to explore a large area of the North Island of New Zealand.

Figure 3: A map of Īhenga's journey. (from: http://www.TeAra.govt.z/en/map/14143/kahumatamomoe-and-ihengas-journeys)

Quite clearly Īhenga did not have technology such as the compass, therefore it becomes essential for students to map the route by reading/listening to the traditional story. In so doing, Īhenga's *te reo Māori* descriptions of locations and directions have to be interpreted and connected to understanding of Western maps. Students can also find the GPS coordinates of major cultural sites/landmarks along the journey, either by going to the areas themselves and/or by using Google Maps. As it may not be practicable for students to visit all of the landmarks in the story because of the long distances, Google maps can be used to find the latitude and longitude of those outside of their visiting range. Students then calculate the direction from waypoint (cultural site) to waypoint. The students′ investigations can then be presented to peers in an oral/visual presentation which would combine the Western cultural tool of the map, with the oral language representation of the journey.

Although we have not yet tried this activity out with students using the distances in the Ihenga story, we consider that it would enable understanding of how Western mathematics, as one form of ethnomathematics, could add value to understanding traditional, cultural practices, rather than detracting value from those practices by purely focusing on Western mathematics knowledge. Comparing the advantages and disadvantages of the different sets of knowledge, traditional Māori, traditional Western and modern technological allows for a meta-discussion with students about why only some of this knowledge is required to be learnt by curricula. These sorts of discussions are vital if students are also to become aware of how hegemony operates to suppress non-Western forms of knowledge as valueless.

From our perspective, keeping in mind these three steps can contribute to cultural symmetry being achieved in mathematics classrooms. It is only over time and working with the teacher and students at Te Koutu that we have become aware that simply incorporating traditional activities into mathematics classrooms was unlikely to contribute to overcoming the impact of colonisation. For students to want to participate in the revitalising and maintaining of traditional practices, then like language learning, they need to see a purpose for it.

Developing activities in order to support the resurrection or revitalization of practices that are not necessarily evident in students' everyday lives raises the questions about who gains from this. Students may be resistant to participating in traditional practices if the goal of learning these practices are solely about maintaining knowledge and skills within the wider Indigenous community without any interest to the students themselves. Thus, finding ways to indicate the benefit of these knowledge and skills in the 21st century is essential. Otherwise the practices become exotic tokens, which do not need to be considered as valuable knowledge (Zaslavsky, 1973).

Conclusion

The acknowledged link between language and culture means that introducing Indigenous ethnomathematical practices into mathematics classrooms needs to be done with caution. The aim is not to marginalise and lower the value of the Indigenous knowledge by setting them up as "exotic tokens" and to elevate "proper Western mathematics", that is the mathematics of

NOTES

national assessments. While the goal is not to "freeze" language or cultural practices at a particular juncture, thus jeopardising language and culture revitalisation efforts, the community of users need to be aware of the implications of changes.

It is, therefore, important that if the traditional imperialistic role of Western mathematics is to be overcome, then students need to be involved in critical reflections on the processes by which practices and knowledge come to be valued. This is important for both Indigenous students and non-Indigenous students who may learn in their classrooms, perhaps unconsciously, that mathematics education, although not always considered enjoyable, is politically neutral, which of course is a fallacy. Introducing ethnomathematical practices into classroom practice can support this critical awareness but only if this becomes an aim for these lessons.

References

Apple, M. W. (1979). *Ideology and curriculum* London: Routledge & Kegan Paul.

Barton, B. (1993). Ethnomathematics and its place in the classroom. *Science and Mathematics Education Papers*, 1993, 46–67.

Barton, B., & Fairhall, U. (1995). *Is mathematics a Trojan horse? Mathematics in Maori education? Paper presented to the History and Pedagogy of Mathematics Conference,* Cairns, Australia. July, 1995.

Best, E. (1922). *The astronomical knowledge of the Maori: Genuine and empirical, including data concerning their systems of astrogeny, astrolatry, and natural astrology, with notes on certain other natural phenomena.* Wellington, NZ: Dominion Museum. Available from: https://www.knowledge-basket.co.nz/kete/taonga/contents/taonga/text/dm/dm3.html

Bishop, A. J. (1990). Western mathematics: The secret weapon of western imperialism. *Race & Class, 32*(2), 51–65.

Binney, J. (2004). Tuki's universe. *New Zealand Journal of History, 38*(2). 215-232.

D'Ambrosio, U. (1999). Literacy, mathematics and technology: A trivium for today. *Mathematical Thinking and Learning, 1*(2), 131–153.

de Bres, J. (2008). *Planning for tolerability: Promoting positive attitudes and behaviours towards the Māori language among non-Māori New Zealanders* (Unpublished PhD thesis). Wellington: Victoria University of Wellington, New Zealand.

Fishman, J. (1991). *Reversing language shift: Theoretical and empirical foundations of assistance to threatened languages*. Clevedon, England: Multilingual Matters.

Gauvain, M. (1993). The development of spatial thinking in everyday activity. *Developmental Review, 13*(1), 92-121.

Jablonka, E., & Gellert, U. (2010). Ideological roots and uncontrolled flowering of alternative curriculum conceptions. In U. Gellert, E. Jablonka, & C. Morgan (Eds.), *Proceedings of the Sixth International Mathematics Education and Society Conference, 20–25 March 2010*, (pp. 31–39). Berlin: Freie Universität Berlin. Retrieved from http://www.ewi-psy.fu berlin. de/en/v/mes6/research_papers/index.html

King, J. (2007). *Eke ki runga i te waka: The use of dominant metaphors by newly-fluent Māori speakers in historical perspective* (Unpublished PhD thesis). Christchurch: University of Canterbury, New Zealand.

Knight, G. H. (1984). The geometry of Maori art-rafter patterns. *NZ Mathematics Magazine, 21*(2), 36–41.

McMurchy-Pilkington, C. (2004). *Pāngarau Māori medium mathematics curriculum: Empowerment or new hegemonic accord?* Unpublished EdD thesis, University of Auckland, Auckland, NZ.

McMurchy-Pilkington, C., & Trinick, T. (2008). Potential & possibilities. In V. Carpenter, J. Jesson, P. Roberts & M. Stephenson (Eds.), *Ngā kaupapa here: Connections and contradictions in education* (pp. 133–144). Melbourne, Australia: Cengage Learning.

McMurchy-Pilkington, C., Trinick, T., & Meaney, T. (2013). Mathematics curriculum development and Indigenous language revitalisation: Contested spaces. *Mathematics Education Research Journal, 25*(3), 341–360.

May, S. (2005). Language rights: moving the debate forward. *Journal of Sociolinguistics, 9*(3), 319-347.

May, S. (2012). *Language and minority rights: Ethnicity, nationalism and politics of language* (2nd ed.). New York: Routledge.

May, S., & Hill, R. (2005). Māori medium education: Current issues and challenges. *International Journal of Bilingual Education and Bilingualism, 8*(5), 377–403.

Meaney, T., Fairhall, U. & Trinick, T. (2008). The role of language in ethnomathematics. *Journal of Mathematics and Culture, 3*(1). Available from: http://nasgem.rpi.edu/index.php?siteid=37&pageid=543

Meaney, T. & Lange, T. (2013). Learners in transition between contexts. In K. Clements, A. J. Bishop, C. Keitel, J. Kilpatrick, & F. Leung (Eds.), *Third international handbook of mathematics education* (pp. 169-202). New York: Springer.

Meaney, T., Trinick, T., & Fairhall, U. (2012). *Collaborating to meet languages challenges in Indigenous mathematics classrooms*. Dordrecht: Springer.

Mooney, M. A., & Laubach, T. A. (2002). Adventure engineering: A design centered, inquiry based approach to middle grade science and mathematics education. *Journal of Engineering Education, 91*(3), 309-318.

Pais, A. (2011). Criticism and contradictions of ethnomathematics. *Educational Studies in Mathematics, 76*(2), 209–230.

Penetito, W. (2010). *What's Māori about Māori education?* Wellington, NZ: Victoria University Press.

Riini, M., & Riini, S. (1993). Historical perspectives of Māori and mathematics. In M. Ohia (Ed.), *Pāngarau: Māori mathematics and education* (pp. 16–20). Wellington, New Zealand: Ministry of Māori Development.

Simon, J. (1998). *Ngā kura Māori: The native schools system 1867–1969*. Auckland, NZ: Auckland University Press.

NOTES

Spolsky, B. (2005). Māori lost and regained. In A. Bell, R. Harlow & D. Starks (Eds.), *Languages of New Zealand* (pp. 67–85). Wellington, New Zealand: Victoria University Press.

Trinick, T. (1999). *The relationships between Māori culture and Māori mathematical language* (Unpublished master's thesis). University of Auckland, New Zealand.

Trinick, T. (2015). *Te Reo Tātai: the development of a mathematics register for Māori-medium schooling* (Unpublished doctoral dissertation). Waikato University, Hamilton, New Zealand.

Trinick, T., Meaney, T., & Fairhall, U. (2015). Finding the way: Cultural revival through mathematics education. In S. Mukhopadhyay & B. Greer (Eds.) *Proceedings of the eighth international mathematics education and society conference* (pp. 164-168) Portland, OR: Portland State University. Available from: http://mescommunity.info/

Trinick, T., Fairhall, U. & Meaney, T. (2016). *Cultural and mathematical symmetry in Māori meeting houses (wharenui)*. Paper presented at Topic Study Group 35, International Congress of Mathematics Education 13, Hamburg, 24-31 July.

Vithal, R., & Skovsmose, O. (1997). The end of innocence: A critique of "ethnomathematics." *Educational Studies in Mathematics, 34*, 131–158.

Zaslavsky, C. (1973). *Africa Counts: Number and Pattern in African Culture*. Boston: Prindle, Weber and Schmidt.

Reflecting on "A Study of Estimation by Professionals at Work": Constants, Changes, Challenges, and Commitments

Thomasenia Lott Adams
University of Florida
tla@coe.ufl.edu

Gregory Harrell
Valdosta State University
gharrell@valdostastate.edu

Abstract

In reflecting on "A Study of Estimation by Professionals at Work", Adams and Harrell (2010), provide insight on constants, changes, challenges, and commitments. Examples of these points are mathematics content, focus on the intersection of mathematics and culture, development of new mathematics standards, and continuation of scholarship, respectively. The authors posit that the article remains relevant because it is grounded in a foundational concept in mathematics and the contexts used in the article are just a very few of the possibilities that teachers and students can explore to learn of the diverse ways mathematics is used in society.

Keywords:

- Estimation
- Geometry

Introduction

Along with number and geometry, measurement is a hallmark of early learning in mathematics as well as a strong presence in advanced mathematics, such as trigonometry. In early mathematics, measurement gives life to the application of number and number sense, especially when numerical estimates are used. Thus estimation, the act of approximating a measurement before a measurement is or can be taken, is a critical component of the process of measurement. Furthermore, geometry makes the application of measurement, as learners study space and the objects in space, necessary. Therefore, measurement may look different throughout the progression of mathematics, but we are very confident in the role that estimation plays in helping learners become proficient in measuring and using measuring tools. As we reflected on our article, "A Study of Estimation by Professionals at Work" (Adams & Harrell, 2010) to address the intersection of mathematics and culture in regards to professions, we considered the following questions since the article was published:

1. What has remained constant?
2. What has changed?
3. What are present challenges?
4. What is our on-going commitment?

The Constants

The primary aspect that has remained constant since we wrote the article is the mathematics content. We chose estimation as the mathematics content for the article because, first, as mathematics educators, estimation was of interest to us. We both enjoyed teaching geometry and working with the powerful relationship between geometry and measurement. However, we also knew that sometimes students struggle with geometry and measurement because for them it is a challenge to visualize and describe the objects of geometry and make hypotheses about the measurement of geometric objects, such as the length of line segments and the size of angles. So we began with a comprehensive description of what estimation is and the role it plays in learning and doing measurement. We still believe that estimation is critical to the experience of measurement, that estimation helps the learner to make sense of a potential measurement and provides a way for the learner to check the reasonableness of a confirmed measurement. We maintain our position that estimation should be an active, integral part of students' measurement activity. We agree with Van de Walle and colleagues (2013) that "measurement estimation [in particular] is the process of using mental and visual information to measure or make comparisons without using measuring instruments [and that] it is a practical skill used almost every day" (pg. 379). In addition, Van de Walle and colleagues (2013) suggest that estimation brings an element of fun to measurement as learners attempt to guess estimates that are on target with actual measurements. Thinking about our own experiences with estimation confirms that indeed, estimation allows flexibility in mathematics and often is the conduit for healthy competition as we have attempted to determine the best estimate in a measurement situation.

The second reason we wrote the article is that we had found in our mathematics teaching experiences that many estimation (as well as

NOTES

measurement) activities were very removed from real life experiences and not very inviting to learners. We found many estimation activities to be abstract and disconnected to possibilities for applying the act of estimating. We also found that many students were not engaged in mental mathematics beyond memorizing and repeating basic facts or memorizing and repeating definitions, and with this lack of connection to mathematics, students were tuning out the mathematics. We wanted to write an article that would demonstrate how to bring the world into the mathematics classroom and how to potentially take students out into the world to see mathematics, specifically, estimation in action. We still believe that this is a necessary means of engaging all learners.

The third, but not less important, reason we wrote the article is that we were and still are interested in scholarship that diversifies the face of mathematics. For instance, we wanted to demonstrate the applicability of mathematics across a variety of professions. We wanted to show that mathematics is important to the ways people work throughout society. While we were only able to incorporate four examples of professionals in our article, our larger study from which the article was drawn included many other professionals and ways they use estimation in the workplace. In our article, we discussed the ways that a patrolman, farmer, automotive technician and timber sales consultant apply skills of estimation to perform their job duties. We interviewed these professionals, and we observed them at work so we could see how they used estimation to perform their work duties. This experience gave us insight about the cultural influence of how people work and the different ways in which mathematics is used when people work.

However, we interviewed and observed 13 other professionals. They are as follows: (1) tow truck operator, (2) land surveyor, (3) seamstress, (4) nurse, (5-6) basketball referees, (7) post office staff, (8) butcher, (9) cosmetologist, (10) carpenter, (11) cook, (12) landscape operator, and (13) pediatrician. We could have easily talked about how the cosmetologist doesn't use a ruler when a client says something like, "Take two inches off the back"! Rather, the cosmetologist uses sight and hands to estimate the length of two inches of hair to cut off. We could have included the many ways the cook used estimation to prepare meals in the restaurant with little time to consider using measurement utensils to prepare the dishes. In fact, sometimes, the amount of an ingredient is so small that a measurement tool would be cumbersome and so is replaced by a "pinch" of the ingredient in the cook's fingers. Furthermore, we could have included the estimation skills used by the basketball referees, with one of the most common estimation task being the counting of seconds to monitor players whose team has possession of the basketball but who might stand "in the paint" for more than the allowed three seconds. The referees do not use a stop watch to measure three seconds, but instead they estimate the passing of three seconds by saying "one, one thousand, two, one thousand, three, one thousand" at a practiced pace and waving their arms in sync with the counting. Finally, we could have mentioned the estimation practices of the butcher who often grabs a handful of ground beef to begin a customer's order, attempting to get as close to the amount of pounds a customer wants as possible because good estimates in this job saves the butcher time and quickly meets the customer's request.

We were very engaged in the work of interviewing and observing these professionals, and we found great pleasure in making them aware of how important the skill of estimation was to their work. Some of the professionals were working in such an automatic way that they hadn't given much thought to the role of estimation in their job duties. Hence the experience, while it was enlightening to us to learn about the ways estimation is used by various professionals, it was also very enlightening to many of the professionals.

We believe that this mutual enlightening about the role of estimation in the work place would still exist if we were to replicate our study with other professionals. We believe that we would become more informed about the ways professionals use estimation and that likewise, some, if not all of the professionals, would be informed when given the guidance to stop and think about the ways they use estimation. If we were to replicate the study today, some of the professionals we would select could be as follows: (1) flight attendant, (2) cell phone repairperson, (3) painter, (4) tattoo artist, (5) pool service manager, (6) optometrist, (7) crane operator, (8) newspaper deliverer, (9) car salesman, and (10) golf course designer.

Our approach today would be the same, to observe the professionals at work and to follow up with discussions on how and why they used estimation to engage in their work duties. Such an exercise would continue to refine our perspective about the intersection of mathematics and culture, particularly as it relates to professionals and their work duties.

The Changes

Since writing the article, we've continued to be enlightened about the rich perspectives that surround mathematics in the context of the ways people live, work, and play. We became very interested in a teaching of mathematics that was inclusive of all students, which in fact gave consideration to the culture of the learner as a means empowering them to learn. I (Adams) and several doctoral students followed up on this interest by spending almost a year observing, visiting, and interviewing a teacher of mathematics that we determined was masterful at facilitating a teaching of mathematics that acknowledged the presence of students' culture in the classroom. One of the doctoral students, Emily P. Bonner, conducted her dissertation research in Mrs. Merriex's classroom and penned a book about the experience: "Unearthing Culturally Responsive Mathematics Teaching: The Legacy of Gloria Jean Merriex" (2011).

Mrs. Merriex was remarkable in her approach to students and to teaching mathematics. She used her personal context and historical reference to ground her instruction. She grew up in the community and knew almost all of the people that lived near the school. She had taught many of her students' grandparents, parents, and siblings. She lived in the community, and she worshipped at the church that was located just across the street from the school. She walked the community, and for the students, who mostly walked to school, she shared the same neighborhood experiences, good and bad. She would often interject bits about happenings in the neighborhood into the mathematics instruction as a way of connecting with students' real-life experiences. To further bolster her connection with students and to make them feel valued and to motivate them, she would often share stories from her childhood memories and build in similarities to her own students.

When we would ask students in mathematics class what they thought about Mrs. Merriex, students would respond with statements like "She doesn't play!", "She means business", or "She's hard, but its fun!" In fact, Mrs. Merriex's approach to teaching mathematics included demanding all students to be attentive and participatory, requiring all students to try, and securing a classroom environment that welcomed all learners of all

NOTES

levels of ability. Mrs. Merriex told the story once of how a young boy was not allowed to be in her mathematics class because he was "learning disabled", but she used her influence with the boys' parents and school leadership to get the boy placed in her mathematics class. Eventually, the boy grew academically, and under Mrs. Merriex's tutelage, the boy advanced from elementary school. This young man has since graduated from a major university, and he continues to excel in the work place.

This look back at Mrs. Merriex and her artful and skillful manner of teaching mathematics with such attention to the students, their varying abilities, talents and interests, their life's circumstances, and their aspirations for the future fueled us to continue considering the value of intersecting mathematics and culture. Now when we look at the literature, we see more dissemination of scholarship related to the intersection of culture and the teaching and learning of mathematics. This shows up in various ways, for example, the National Council of Teachers of Mathematics (NCTM) provides voice via its Equity Principle (NCTM, 2000) which, in part, states that it is vital for teachers of mathematics to appreciate the cultural diversity that is present in the classroom and to be aware of their (teachers') own conceptions, perspective, and biases about and toward others, particularly, students.

A few years ago, I (Adams) was invited to attend a small meeting of mathematics educators held at the University of Michigan to discuss formative assessment in mathematics classrooms. When I arrived at the meeting, I realized that formative assessment was the larger topic and that there were several sub topics chosen as a means of making formative assessment more robust and relevant to today's diverse classroom. One sub-topic was culturally relevant pedagogy (Ladson-Billings, 1995), a framework of instruction that promotes academic excellence without dismissing elements of culture in the classroom. The concept of the meeting was to discuss how the sub-topics impacted or interacted with formative assessment in the mathematics classroom. I was invited to participate in the meeting discussions using the culturally relevant pedagogy framing. We've had several meetings and conference presentations since that initial meeting, and I've concluded the following regarding how culturally relevant pedagogy intersects with formative assessment.

First, formative assessment is grounded in having clarity about the mathematics content and using good tasks in instruction. Our initial idea for the article, that it would be valuable to contribute to the literature on the ways people use estimation at work relates to this idea of clarity about mathematics content. Authentic estimation experiences rest on good measurement tasks. From each of the interviews we conducted, teachers can use those contexts to either select or create good measurement tasks for students: tasks that are meaningful, mathematically-sound and would engage students in cognitive thinking and open-ended problem solving.

Secondly, formative assessment involves using a variety of assessment methods to gather information about what students are or are not learning. One such assessment method is that of classroom discourse, including asking and answering questions. Culturally relevant pedagogy involves positive and robust communication between teachers and students, where students' answers, responses, comments, questions, etc. are first of all, valued, and secondly, acknowledged in the classroom. As we were interviewing and observing the professionals while they engaged in their work duties, it was very evident that communication, for example, between the professional and a customer, was very important to the entire work duty process. Now when we think of mathematics and culture, we have to give consideration to how mathematics is communicated and how discourse in the mathematics classroom is set up to support student learning, particularly in a diverse classroom environment.

Volume 1

A third, but not necessarily final, thought in regards to culturally relevant pedagogy and formative assessment is that the context that situates the mathematics in a task should be meaningful to students. While the abstract nature of mathematics will lend itself to abstract contexts, all of mathematics does not have to be taught from this perspective and it definitely should not be so in the early years of students' mathematics learning. The 14 professionals we interviewed and the four professionals that were presented in our previously published article is just a very small example of ways teachers can make mathematics instruction, and subsequently, formative assessment more effective and more interesting to students.

The Challenges

Since we published our original article, there have been great challenges on the mathematics education landscape. We will discuss two of these challenges: the introduction of the Common Core State Standards for mathematics and the preparation of preservice teachers of mathematics. In 2010, the Common Core State Standards for mathematics (CCSSM) (National Governors Association Center for Best Practices & Council of Chief State School Officers) were introduced across the United States. The standards actually included content standards for grades K-12 as well as a set of eight Mathematical Practices that are applicable across grades K-12. Great debates about national adoption of and standardized assessment of the CCSSM occurred across the country. There were strong voices in support of the CCSSM and strong voices in opposition of the CCSSM. Voices for the CCSSM saw these new standards as one way of leveling the playing field for students so that, for example, students in New Jersey would experience the same high quality mathematics content as students in Utah. Voices in opposition of the CCSSM saw these new standards as one way the federal government was intervening in the education practices and policies at the state and local level. Collectively, both positions included voices of parents as well as educators at every level. The one element that most people could agree on were the Mathematical Practices, given that they were not driven by content and they reflected best practices for learning mathematics for novice as well as advanced learners of mathematics.

In regards to our original work, we still valued estimation as an important activity of measurement. In light of this, when we considered how estimation was treated in the CCSSM, we found that "estimate" makes its first explicit appearance in Grade 2, where a standard supports students estimating measurement of length in standard units, such as inches, feet and meters. Through a process of unpacking this standard, the underlying expectation is that students have already had experience estimating lengths using informal units of length such as paper clips, straws, string, etc. Secondly, the standard also has embedded in its intent that students have experiences with contexts in which estimation of length is necessary or meaningful to a situation that involves the attribute of length.

However, what we find in our continued work of teaching mathematics and working with teachers of mathematics is that real life contexts often elude the teaching and learning experience of measurement. We believe, therefore, that continued efforts such as those

NOTES

begun as described in our original article are important to classroom teachers as well as curriculum developers.

Secondly, I (Harrell) have found teacher education to be a major challenge area in terms of communicating effective practice to future mathematics teachers. Many future teachers still view mathematics as obtaining "the" correct answer to a problem. Estimation, with its virtual guarantee of many possible solutions, is seen as inaccurate and, therefore, problematic. Based on the findings in the estimation study, we have incorporated many measurement and estimation activities into instruction for preservice teachers. Students estimate length, area, and volume by using referents, visually estimating a variety of objects, and measuring with a formal tool to check the accuracy of their estimations and thereby refine their estimation skills. Similarly, students hold objects to estimate their weight, and then measure using a formal scale in order to refine their skills. During this process, I (Harrell) share that this methodology is "how the professionals do it", and I am able to use examples from the article.

Furthermore, these activities have opened new venues for potential research as I (Harrell) have developed hands-on activities for future teachers and grade K-8 students that use proportional reasoning skills within the context of weighing common objects on a balance scale. For example, by holding a penny in her hand, a student can estimate its weight. Then placing the penny on a balance scale, she can check her estimate and discover that three one-gram weights appear to balance the penny. Using proportional reasoning, if we quadruple the number of pennies and quadruple the number of weights, the scale should still balance. However, since the penny does not weigh exactly three grams, the scale no longer balances. Using estimation to help students learn concepts beyond measurement provides a rich learning environment as well as an interesting area of research.

In addition to using estimation as a robust framework for understanding proportional reasoning, the study has also shifted my thinking in preparing future teachers to teach algebra. Again, the goal is to estimate the weight of common objects. However, instead of placing gram-weights on the scale until the scale balances to estimate the weight of the object, the scale is already balanced with a given number of like objects on each side of the scale as well as a given number of gram weights on each side of the scale. Students are asked to remove objects and weights from the scale while maintaining balance in order to estimate the unknown weight of the given object. These classroom ideas have shaped my writing as can be seen in the algebra chapter of the book that I (Harrell) co-authored (Moch, Wares & Harrell, 2014).

The Commitment

Our scholarship following the publication of our original article includes two major projects. First, I (Adams) expanded our work by engaging in funded research to study the positionality of African American girls (grades 5-8) in relation to mathematics and science (Pringle, Brkich, Adams, West-Olatunji, & Archer-Banks, 2012). For the effort, the team of researchers observed instruction in mathematics classes, interviewed teachers, school guidance counselors, and students. One of the findings from this work attended to the discontent between African American girls and the contexts presented in mathematics and science classrooms. From interviews, we learned that the African American girls in the study often perceived mathematics and science curriculum and instruction to be far removed from their interests and ways of learning. We observed a lack of mathematics experiences, such as estimation activities, that allowed for flexible thinking among the students. Another key development of this work was the revealing of the critical role that school counselors can play in encouraging girls of color to study mathematics and science.

Volume 1

Secondly, a more recent engagement in this work involved a presentation on building students' capacity to learn (Adams, 2014). A tenant from this presentation is that to build students' capacity to learn requires providing students with experiences "to be in the mathematics". Estimation, particularly through the lens of how people apply this skill in the marketplace, provides many examples for how situate learning of estimation so that students are in the mathematics rather than observers of others doing the mathematics.

When we engaged in the observation and interviewing of the diverse group of professionals engaged in their work to show how mathematics, specifically estimation, is versatile and reflective of the cultural context in which it is used, our underlying aim was to provide examples that teachers of mathematics could use in the classroom to involve students in the various ways people actually use the skill of estimation to do their work. One of the ways we can contribute to improving this matter is to translate our article into a tool that teachers can use in the classroom. To that aim, we offer the following guide for creating a project that students can engage in:

1. Arrange students in teams of 3-4 students.

2. Share one example from the article to kick-start the project for the students.

3. Provide an opportunity for students to brainstorm a list of professionals they might observe and/or interview to determine how these professionals use the skill of estimation to perform their work duties.

4. Collaborate with the students to determine how to gain access to one or more professionals to carry out the project.

5. Assist students in developing a list of questions to ask during the interview of the professional(s) about the application of estimation.

6. Assist students in developing a list of points to look for during observation of the professional(s) using estimation in the work place.

7. Instruct students to prepare an oral and/or visual presentation of their experience.

8. Collect the students' experiences and build a library of estimation possibilities to reinforce students' learning and to inform future students.

NOTES

Conclusion

As mathematics educators, we are very fortunate to be part of a discipline that is quite reflective of its past. One reason for this reflection is that new thinkers are constantly emerging in the field. There are also new happenings, such as the strong presence of technology in classrooms, which cause us to pause and think about what has been, what is and what should be for the benefit of teachers and learners of mathematics. In reflecting on our article, we find that the topic is still relevant to and meaningful for the teaching and learning of mathematics. We hope that teachers will find the examples we use beneficial, and also motivating for seeking other examples of professions that will fuel students' interests and needs to see mathematics in action.

References

Adams, T. L., & Harrell, G. (2010). *A study of estimation by professionals at work. Journal of Mathematics and Culture, 5*(2) (ISSN 1558-5336).

Adams, T. L. (2014, October). *Building students' capacity to learn.* Presentation invited by and delivered at the annual conference of the North Carolina Council of Teachers of Mathematics, Greensboro, NC.

Bonner, E. P. (2011). *Unearthing culturally responsive mathematics teaching: The legacy of Gloria Jean Merriex*. Lanham, MD: Hamilton Books.

Ladson-Billings, G. (1995). Toward a theory of culturally relevant pedagogy. *American Educational Research Journal,* 32, 3, 465-491.

Moch, P. L., Wares, A., & Harrell, G. (2014). *Creating a positive change in elementary teacher mathematics*. Dubuque, IA: Kendall/Hunt Publishing.

National Council of Teachers of Mathematics. *Principles and standards for school mathematics*. Reston, VA: National Council of Teachers of Mathematics.

Pringle, R. M., Brkich, K. M., Adams, T. L., West-Olatunji, C., & Archer-Banks, D. A. (2012). Factors influencing elementary teachers' positioning of African American girls as science and mathematics learners. *School Science and Mathematics, 112*(4), 217-229.

Van de Walle, J. A., Karp, K. S., & Bay-Williams, J. M. (2013). *Elementary and middle school mathematics: Teaching developmentally.* Boston, MA: Pearson.

National Governors Association Center for Best Practices & Council of Chief State School Officers. (2010). *Common core state standards (mathematics).* Washington DC: National Governors Association Center for Best Practices & Council of Chief State School Officers.

Volume 1

Ethnomathematics: Cultural Assertions and Challenges Towards Pedagogical Action

Volume 1

Daniel Clark Orey
orey@iceb.ufop.br
Departamento de Matemática
Instituto de Ciências Exatas e Biológicas
Campus Morro de Cruziero
Universidade Federal de Ouro Preto
Ouro Preto, MG
35000-000
Brasil

Milton Rosa
milrosa@hotmail.com
Encina High School
1400 Bell Street
San Juan Unified School District
Sacramento, California 95825 USA

Abstract

In this article, the authors discuss ethnomathematical concepts that were found useful in the interpretation of assertions and challenges faced by the Ethnomathematics Program for teacher professional development, as described by D'Ambrosio. The authors share the student's teaching-learning process, and for mathematics education in our ongoing work in Brazil and the United States. The discussion begins with a brief examination of where ethnomathematics stands today in respect to the process of teaching and learning of mathematics.

Resumo

Neste artigo, os autores discutem alguns conceitos etnomatemáticos que podem ser considerados úteis para a interpretação das asserções e desafios enfrentados pelo Programa Etnomatemática, para o desenvolvimento profissional dos professores, para o processo de ensino-aprendizagem dos alunos, e para a educação matemática; conforme descrito por D'Ambrosio; em nosso contínuo trabalho realizado no Brasil e nos Estados Unidos. Esta discussão começa com um breve exame sobre o posicionamento atual do Programa Etnomatemática em relação ao processo de ensino e a prendizagem em matemática.

Introduction

We begin with the assumption that those who study and use mathematics academically operate from a unique paradigm. That is, as mathematicians, we see the world from a certain point of view, and most importantly, we operate from within a certain cultural perspective. Modern Western academic mathematics[1] has come to represent a grammar and language that originated in the Mediterranean Region in the 16th Century. Hence, like many of the other cultural vehicles, that have come to dominate the planet, the Western mathematical way of thinking spread and grew along with civilizations in that region. At the same time, this particular form of mathematics also came into contact with other ways of thinking and interpreting the world. The conquest and domination of the planet, along with and through the use of modern mathematics by European-North American colonizers was not without its consequences. Arrogance coupled with a not so very subtle sense of entitlement and a scientific cultural hegemony has spread faster than it has sought to understand or come to terms with the thousands of traditions and time honored forms of thinking, calculating, and problem solving. D'Ambrosio (1997) has stated that as "individuals become integrated into humankind as solidarity, we will be able to contribute to the preservation of a common inheritance" (p.12). Sharing this point of view "we hope that, in this phase of the evolution of our species, that the respect for diverse peoples will not be replaced by our arrogance, envy, prepotency, and great power and so that when people began to work with quantity, measure, shape, and modeling" (D'Ambrosio, 1997, p. 12).

Many people, the authors included, have been attracted to Western-academic mathematics for its elegant, yet powerful sense of logic, form, as well as its scientific, cultural, and artistic qualities. It goes without saying that the mathematics of the 21st century has allowed impressive scientific achievements to immerge in all parts of the planet, yet this came with incalculable costs to the millions of people, their cultures and civilizations as European/North American ideas, morality, and science have come to dominate and control, and capture our imagination. It has, as well, enabled some of the most horrific, social, scientific, ecological and cultural disasters in the history of the planet. The Ethnomathematics Program considers a concept of mathematics which includes a critical, moral, holistic, and global perspective. That is, a mathematics curriculum that "walks the mystical way with practical feet[2]".

D'Ambrosio's Approach to Ethnomathematics

The authors have chosen D'Ambrosio's approach of ethnomathematics as our philosophical basis to our work in ethnomathematics and school mathematics. However, other researchers and educators have used a diversity of approaches to demonstrate the dualistic concepts between both approaches in mathematics

1 The "scientific way of thinking."

2 Attributed to Former President of Stanford University D. S. Jordan, quoted in *Bahá'í World,* VI, p.480. and found at: http://bahai-library.com/books/quest/index.html

NOTES

education. For example, Gerdes (1997b) stated that these concepts were indigenous mathematics (Gay & Cole, 1967; Lancy, 1976); sociomathematics (Zaslavsky, 1973); informal mathematics (Posner, 1978); spontaneous mathematics (D'Ambrosio, 1982); oral mathematics (Carraher et al., 1982); oppressed mathematics (Gerdes, 1982); hidden or frozen mathematics (Gerdes, 1982); folk mathematics (Mellin-Olsen, 1986); non-standard mathematics (Carraher et al. 1982, Gerdes, 1982; Harris, 1987); and mathematics codified in know-how (Ferreira, 1991). Gerdes (1997) also stated that these "concepts have been gradually united under the more general "common denominator" of D'Ambrosio's ethnomathematics" (p. 339).

In this context, we feel that D'Ambrosio's (1993) approach is broader in concepts, ideas, and objectives. It considers mathematical modeling as a methodology that may be used as a tool in the ethnomathematics program. The authors also consider this perspective as a new trend for the pedagogical actions of the ethnomathematics program for its agenda for the 21st century.

Ethnomathematics

Ethnomathematics is the application of mathematical ideas and practices to problems that confronted people in the past or are encountered in present contemporary culture (D'Ambrosio, 2001). Much of what we call *modern* mathematics came about as diverse cultural groups sought to resolve unique problems such as exploration, colonization, communications, and construction of railroads, census data, space travel, and other problems-solving techniques that arose from specific communities. More than ten years ago, the National Academy of Sciences (1994) reported that cultural variables have strongly influenced how students consider themselves as human beings, how they came to understand the world, and interpret their own and others experiences. Simply put, culture affects the ways we acquire and use our own mathematical knowledge.

In attempting to create and integrate mathematical materials related to different cultures that draw on students' own experiences in an instructional mathematics curriculum, we believe that it is possible to apply ethnomathematical strategies in teaching and learning mathematics. These strategies include, but are not limited to, the:

> *...historical development of mathematics in different cultures (e.g. the Mayan numeration system). Another aspect could be prominent people in different cultures that use mathematics (e.g. an African-American biologist, an Asian-American athlete). Mathematical applications can be made in cultural contexts (e.g. using fractions in food recipes from different cultures). Social issues can be addressed via mathematics applications (e.g. use statistics to analyze demographic data (ICME-7, 1992, WG-10, Multicultural Mathematics, p.3 - 4).*

Seen in the above context, Gilmer (1990) stated "...there is an urgent need to multiculturalize the mathematics curriculum" (p.4). A growing body of literature on ethnomathematics has been stimulated by concerns for equity, equality, and excellence in a context of diversity. Many teachers worldwide have come to realize that students become motivated when they are involved in their own learning. This is especially true when dealing directly with issues of greatest concern to themselves (Freire, 1970).

The challenge that many societies face today is to determine how to shape a modernized, national culture integrating selected aspects of traditional culture that coexist in an often delicate balance with those of westernized globalization. An increased cultural, ethnic, and racial diversity provides both an opportunity and challenge to societies and institutions, with many questions related to schooling forming an integral part of this question.

Ethnomathematics as a Program

The inclusion of moral consequences to mathematical-scientific thinking, mathematical ideas and experiences from different cultures around the world, the acknowledgment of contributions that individuals from diverse cultural groups make to mathematical understanding, the recognition and identification of diverse practices of a mathematical nature in varied cultural procedural contexts, and the link between academic mathematics and student experiences should become a central aspect to a complete study of mathematics. This is one of the most important objectives of an ethnomathematics perspective in mathematics curriculum development. Within this context, D'Ambrosio (1990) describes ethnomathematics as:

> *The prefix ethno is today accepted as a very broad term that refers to the social-cultural context, and therefore includes language, jargon, and codes of behavior, myths, and symbols. The derivation of mathema is difficult, but tends to mean to explain, to know, to understand, and to do activities such as ciphering, measuring, classifying, ordering, inferring, and modeling. The suffix tics is derived from techne, and has the same root as art and technique (p.81).*

In this case, ethno refers to groups that are identified by cultural traditions, codes, symbols, myths and specific ways used to reason and to infer. Mathematics is therefore more than counting, measuring, classifying, inferring, and modeling. Ethnomathematics forms the intersection set between cultural anthropology and institutional mathematics and utilizes mathematical modeling to solve real-world problems, and translate them into modern mathematical language systems. Eglash (2002) stated that mathematical modeling is a tool, which "provides a translation from indigenous knowledge systems to Western mathematics. This perspective is crucial to give minority students a sense of cultural ownership of math, rather than an empty gesture toward inclusiveness" (p. 4).

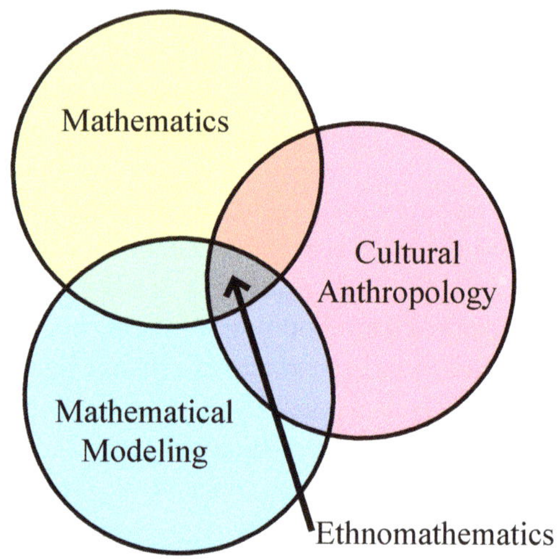

Figure 1: Ethnomathematics as an intersection of three disciplines

An essential aspect of the program includes an ongoing critical analysis of the generation and production of mathematical knowledge and the intellectual processes of this production (D'Ambrosio, 1993; Rosa, 2000). This process also includes the social mechanisms of the institutionalization of knowledge (academics), and its transmission (education).

NOTES

A General Assertion

Ethnomathematics is a contemporary global pedagogical trend in education (Scientific American-Brasil, Edição Especial, 2005). This approach has allowed a number of mathematicians and educators to criticize (Greene, 2000; Ravitch, 2005) ethnomathematics and express fears that this trend may represent a pulling away from certain cultural norms, even though some social realities underlie the need for many multicultural efforts to reform curricula.

The world's economy is globalized; yet, traditional mathematics curricula neglects, indeed often rejects, the contributions made by the world's non-dominant cultures. Given these conditions, an ethnomathematical perspective in the curriculum may be seen as giving new, and expanded, often extremely complicated definitions of a given society's unique scientific contributions. This context allows school mathematics to be seen "as the process of inducting young people into mathematical aspects of their culture" (Gilmer, 1990, p. 4).

Ethnomathematics can reshape our greater cultural identity in a positive way (D'Ambrosio, 1995; Zaslavsky, 1996) by requiring the inclusion of a greater representation of the true mathematical practices and problems of a student's own community (D'Ambrosio, 1998; Zaslavsky, 1993, 1996). It helps both educators and students alike to understand the universality of mathematics, in the context of mathematical practices of day-to-day life, preliterate cultures, professional practitioners, workers, and academic or school mathematics. It can do this by taking into account the historical evolution, and the recognition of natural, social and cultural factors that shape human development (D'Ambrosio, 1995).

A Brief Summary of Assertions

There are numerous favorable arguments as related to the implementation of ethnomathematics as a program towards pedagogical action. We summarize these arguments in the following categories:

Education

The Ethnomathematics Program promotes the rights of all people, no matter their sexual orientation, gender, ethnicity, race, and socio-economic status. It does this in order to allow learners to understand the issues and problems of all the members of our increasingly globalized society (D'Ambrosio, 1990, 1995; Croom, 1997; Fasheh 1982; Zaslavsky, 1991, 1996, 1998). The students will learn to make contributions and learn to appreciate the achievements of their own and other cultures (D'Ambrosio, 1990, Joseph, 1991, Zaslavsky, 1996).

Mathematics

The Ethnomathematics Program deals with both the content and the process of curriculum, classroom management, teacher expectations, professional development, and relationships among teachers, administrators, students, and the community (Borba, 1990; D'Ambrosio, 1985, 1990, 1995, 1998; Zaslavsky, 1998). This approach allows students to make connections with historical developments of mathematics ideas, practices, and the contributions made by diverse cultural groups and individuals.

Ethnomathematics posses important characteristics of being able to help develop the concept of mathematics. In other words, ethnomathematics may be seen as a transformational endeavor because it transforms what mathematics really is. Thus a possible purpose of the analysis of an ethnomathematical perspective as an addition the school curriculum could be tracing the developments or transformations of mathematical ideas and practices throughout history.

Ethnomathematics also develops the study of mathematical ideas and practices of socio-cultural groups, but how this is realized in the classroom is still problematic because much of ethnomathematical research and investigations identify ethnomathematical forms of mathematics but do not continue to develop the pedagogical actions for this program. Classrooms research about ethnomathematics and its role is a crucial aspect of this perspective because this program should be implemented in classrooms (D'Ambrosio, 1993).

Ethnomathematics emphasizes the communal and tends to connect mathematics with its contexts. If these two components are to be brought together then ethnomathematics may be conceived as an overarching aspect of the curriculum because mathematics may be humanized, that is, ethnomathematics may be viewed as a philosophical, contextual, affective, and attitudinal approach to the curriculum.

From a perspective of using mathematical modeling towards pedagogical action for the ethnomathematics program, teaching is much more than the transference of knowledge because teaching becomes an activity that introduces the creation of knowledge (Freire, 1998). This approach in mathematics education is the antithesis of turning containers to be filled with information (Freire, 1970). Bassanezi (2002) characterizes this process as "ethno-modeling" (p. 208), and defines ethnomathematics as "the mathematics practiced and elaborated by different cultural groups, and involves the mathematical practices that are present in diverse situations in the daily lives of members of these diverse groups" (p. 208).

Students

Students need to develop skills in critical thinking and analysis that can best be applied to all areas of life. These skills include vital issues involving health, environment, race, gender, and socioeconomic class (D'Ambrosio, 1990, 1998; Freire, 1970; Zaslavsky, 1998). Bassanezi (1990, 1994), Borba (1990), D'Ambrosio (1990, 1998), and Zaslavsky (1998) agree that ongoing contact of students with diverse and different ways of thinking and practicing mathematics, will raise interest in learning the necessary and required content, by having students apply mathematical concepts to future professional contexts and by facilitating student performance (Bassanezi, 1990, 1994; NCTM, 1989; Zaslavsky, 1990).

In an Ethnomathematics Program, students develop abilities, increased creativity, and a sound set of research habits, as they learn the required mathematics. They develop the capacity to create a hypothesis (Bassanezi, 1990, 1994; D'Ambrosio, 1990, 1993; Biembengut, 1999; Hogson, 1995). Ethnomathematics contributes to the development of student capacity by selecting data and the subsequent adaptation to their needs (Biembengut, 1999, Croom, 1997; Hodgson, 1995), by encouraging contact with biology, chemistry, physics, geography, history, and language (Bassanezi, 1990, 1994; D'Ambrosio, 1995; Zaslavsky, 1991, 1993, 1996, 1998), and by developing work in groups, sharing tasks, learning how to appreciate both criticism and alternate opinions, respecting the decisions of others

NOTES

and of the group, and by facilitating student interactions in a globalized society, students share global and interactive visions necessary to develop successful mathematical content (D'Ambrosio, 1993; Bassanezi. 1990, 1994; Freire, 1970, Gerdes, 1988a, 1988; Zaslavsky, 1998). In our opinion, a major objective of ethnomathematics as a program can be to raise student self-confidence, to enhance creativity, and to promote cultural dignity (D'Ambrosio 1990; Rosa, 2000). These factors are essential for students to learn all school subjects.

Teachers and Educators

As societies become increasingly urbanized, diverse and globalized, the application of an ethnomathematical perspective in the curriculum encourages further and ongoing intellectual and professional development of teachers. It also encourages long-term learning through a diversity of experiences. Teachers are characterized as facilitators, advisers, and mediators of the mathematics learning process (Bassanezi, 1990, 1994; Biembengut, 1999; Hogson, 1995). Biembengut (1999), Hogson (1995), Orey (2001), and Rosa (2000) all have shown that teachers and students can discover a process of understanding mathematics together if given proper support and encouragement. This alternative allows students to learn mathematics content through varied experiences that are related to the cultural, historical, and scientific evolution of mathematics.

School Mathematics

Ethnomathematics turns perceived conceptual poverty into conceptual richness. This process can go wrong, for example when non-Western knowledge systems are idealized, or when educational processes trap students in old ways of thinking. However the process is likely to be positive provided that ethnomathematics:

- looks forward, not backward; that is ethnomathematics as an expression of contemporary thought, is not just a recording of historical ideas and practices;
- assumes a sophisticated knowledge system because it is not just mathematical skills and drills.

When a mathematical system is actively used as a system based on theory that can solve real problems, this can be described as modeling. In such modeling activity, both conventional mathematics and the mathematical system of the cultural group are used. The point of this activity as ethnomathematics is not just to solve problems, nor just so that people understand alternative mathematical systems, but also it is so that people understand more about the role that mathematics play in a societal context (Orey, 2000). It can enable a better understanding of mathematical systems through the use of mathematical modeling.

According to D'Ambrosio (2000), mathematical modeling is a process of translation and elaboration of problems and questions growing from any real situation (system), that forms an image or sense of an idealized version of the system. By analyzing their role in the contextual reality as a whole, this holistic context allows those engaged in the modeling process to study the systems of reality in which there is an equal effort made by them to create an understand of all components of the system as well as the interrelationships among them (D'Ambrosio, 1985, 1989, 1990, 1993).

In 2002, D'Ambrosio commented about an ethnomathematical example that naturally comes across as a mathematical modeling methodology. For example, a group of Brazilian teachers (1989-1990) studied the cultivation of vines brought to Southern Brazil by Italian immigrants in the early twentieth century. They chose to investigate this theme because, in their opinion, the problem to be analyzed should be linked with the culture of the people in that region. Both

Bassanezi (2002) and D'Ambrosio (2002) have sited this is an excellent example of the connection betwen ethnomathematics and modeling.

Teachers started their project by interviewing wine producers and doing an ethnographical and historical research about their theme. One of the producers of wine was also involved in the building of their own barrels. The barrel used to store the wine formed a truncated cone. The teachers realized that the process used by the producer to find the volume of the barrel was different from the process used in academic mathematics, but worked perfectly well, and had been used for centuries in Italy. The producer's process was based on techniques that were learned from his father and transmitted to the members of his family through generations by his ancestors long before arriving in Brazil. The teachers in this study used mathematical modeling techniques (academic mathematics) to model the process used by the producer (ethnomathematical practice).

Through modeling (Rosa, 2000; Rosa & Orey, 2003), it is possible to mix school mathematics and ethnomathematics. Hence, students practice academic mathematics while learning to model situations which are generated by ethnomathematics (D'Ambrosio, 1993). This is especially true when they are encouraged to solve problems that are related to their own communities. Mathematical modeling acts as a bridge between ethnomathematics and Western-academic mathematics which will be required for the student's success in today's globalized technological society.

A General Challenge

The traditional mathematical curriculum may not allow implementation by ethnomathematical perspectives. This challenge depends on how ethnomathematics aspects are selected from student experiences or from their own environment and community, because as a paradigm, it is culturally rooted, and often there is no traditional *syllabus* or assessment standards. The selection of mathematical content is dependent on situations that are interesting to students themselves because motivation and creativity are key components in the teaching and learning process in mathematics. In this curriculum model, teachers must learn to be both flexible and knowledgeable in the content because they be the ones who select cases which are related to student cultural background and their environment.

Seen in this context, part of this challenge is to adopt a methodology that deals with ethnomathematical perspectives in the mathematics curriculum. It is feasible to believe that one methodology that is adequate to this perspective is the use of mathematical modeling strategies. The selection of mathematical content can be directed to cover specific topics, standards or benchmarks in required mathematics content. However, many teachers are not trusted or allowed to work away from required texts and curriculum, nor are they sufficently prepared or supported to be able to connect mathematical modeling and ethnomathematics. This most certainly limits the possibility for an ethnomathematics perspective to enter in classrooms in the United States and other countries.

NOTES

To work with this perspective it is necessary that a school and community have an open curriculum where by the mathematics content that must be covered depends on the development of the class and how students become motivated. In this context, mathematical modeling becomes a methodology which is intrinsic to the Ethnomathematics Program. A careful look into the history of mathematics reveals that mathematical modeling is the root out of which much of modern mathematics has developed, often from a process of abstraction, which represents the real world. This can also be true in traditional school and academic mathematics.

A Summary of the Challenges to Ethnomathematics

There are numerous unfavorable arguments as related to the implementation of ethnomathematics as a program. We outline some of these challenges in the following categories:

Education

For several years it has been argued that traditional uses of the school curriculum do not foster a genuine disposition for realistic mathematics in students (Davis, 1989). Yet the same argument can be used against many attempts to make ethnomathematics useful to educators and students. Other concerns are related to the lack of enough time to develop content that enables teachers to execute pre-established pedagogical plans. Concerns related to the application of ethnomathematics as pedagogical action generally include:

- Few, if any, textbooks and other pedagogical materials about ethnomathematics are in use in classrooms, this is especially true in regards to materials in English.
- A scarcity of university ethnomathematics courses leaves teachers and researchers unprepared to argue this issue, and leaves the door open to further criticism of the field.
- Most research being done in ethnomathematics is being published in languages other than English (the most recent example being the 2005-Scientific American-Brasil Special Edition on Ethnomathematics).
- Few, if any, assessment instruments are appropriate to this new curriculum model; this is especially true in the United States.
- There is a real danger of ethnomathematics being taught as folkloristic and "primitivist" (Eglash, 2002) introductions to "real" or academic mathematics.
- Much of the published ethnomathematics activities represent a shallow, superficial understanding of the field with a sense of "multicultural experiences" based only upon "exposure to diversity", and an inability to look at the deeper mathematics present.

Students

Many students have difficulty with working in groups or in cooperative learning styles. Many teachers and parents are unsure about how their children should work without traditional classroom structures. Because of the traditional, often passive aspect of schooling, many students and teachers have not developed the habit of formulating questions and mathematizing[3] ideas and hypotheses, and have difficulty assimilating several subjects simultaneously (Burak, 1994; Pedroso, 1998), or seeing mathematics outside of the traditional classroom context.

Teachers and Educators

Often because of the lack of their own personal experiences, and their inability to comprehend languages other than English, many educators in the United States and other countries are reticent to try cross-cultural methods in their academic training or professional growth in mathematics. Many educators are reticent to de-emphasize their traditional authority in favor of group work (Burak, 1994; Cross & Moscardini, 1985; Pedroso, 1998). Other reasons may relate to issues of time for planning lessons (Burak, 1994; Pedroso, 1998; Zaslavsky, 1998). Many educators are not familiar with the interface between mathematics and other subjects, that is, the interdisciplinary fashion of the mathematics curriculum, which is required when teachers work in this perspective. In a generally mathematics phobic culture, certainly the reverse is equally true. As well, many mathematics educators are not always prepared to employ practices that will enable underserved and underrepresented groups to learn, understand, and comprehend mathematics (Burak, 1994, Zaslavsky, 1994).

[3] Mathematization is the development of a given problem, that is, the transformation of the problem into mathematical language. This can be accomplished by the formulation of the hypotheses; classification of the data and information as important and non-important to the hypotheses, selection of important variables, selection of symbols adapted for these variables and description of those relationships in mathematical terms.

School Mathematics

It has been extremely difficult to mix the goals and philosophy of ethnomathematics, academic benchmarks, standards, and goals related to passing of standardized exams that are based in traditional school mathematics (Rosa, 2000). Many teachers and educators feel that their students will not learn mathematics if they use an ethnomathematical approach in school curricula. They believe that this perspective makes students and teachers accountability in standardized examinations extremely problematic.

We believe that one way for ethnomathematics to be taken seriously in North America and other countries, is for ethnomathematicians to develop rigorous research in regards to linking ethnomathematics and modeling activities to the mandated mathematics curriculum as being undertaken in places such as Brazil. Like it or not, educators are less likely to use ethnomathematics if they cannot see at this time how this would allow them to succeed in improving overall student achievement.

Challenges towards Pedagogical Action

The pedagogical action of the ethnomathematical program underscores the importance of doing the ethnomathematical work first. In this context, coming to a good understanding of the mathematical aspects of culture and having a clear purpose to the educational activity are vital aspects of this perspective. The consequence of these ideas is that ethnomathematical work in the schools is not a simplistic presentation of cultural examples or simply situating mathematics in cultural contexts. Rather it

NOTES

requires considerable background work, complete understanding, and pedagogical sophistication. This is a complex task, takes time, and is difficult to get right. For example, it is convenient to state that teachers may interpret an ethnomathematical approach by starting with the students' outside socio-cultural-economic realities, but the students may refuse to study their own realities because they consider them oppressive. In this case, they may not identify this contextualization as mathematics because they already have a grounded mathematical conception as a previous knowledge. In this case, one possibility is to start with the students' existing mathematical conceptions, even if they are traditional, because this is another way to develop the mathematical contents followed by a critical examination of these conceptions. The consequence of this educational aspect for teacher education is significant. It means that teachers must know more about mathematics and additional pedagogical skills in order to help students undertake a critical examination of the mathematical contents.

Final Considerations

An eventual inclusion of ethnomathematics in the school curriculum could happen simply because current trends in many urban schools (on-growing migration, immigration and the increasing complex, urban diversity of our populations) will demand it. However, ethnomathematics can be considered negative when it restricts ethnic groups to stereotypes and leaves them unprepared to participate in greater academic endeavors. It can be considered equally negative if it is seen to "water down" the necessary mathematics content in general. Ongoing work needs be undertaken by ethnomathematicians that shows that we can and do integrate rigorous mathematics such as modeling, into the curriculum.

As stated earlier, ethnomathematics seeks to recognize the contributions, values, rights, and the equality of opportunities of all cultural groups that compose any given society. Educators can begin to develop an ideal equality among students and build a foundation for promoting academic excellence for all their students (Croom, 1997) by implementing ethnomathematics and modeling.

Ethnomathematics may come to exert some influence on mathematics education in the United States, as it has in other countries (most notably Brazil) if researchers develop rigorous activities that are linked to standards and curriculum that makes mathematics a living subject for both teachers and students. Traditional mathematics education aims at transmitting a certain amount of content and uses it in artificial situations presented as problems. The problems are artificially formulated, in such a way that they can only help memorization skills, at best. These techniques and problems are usually boring, uninteresting, obsolete, and unrelated to the modern reality of students. These characteristics of traditional mathematics education are responsible for downgrading of school satisfaction and achievement in many countries. Ethnomathematics through modeling may restore a sense of pleasure in the doing of mathematics.

The use of calculators and computers must be integrated into the concept of ethnomathematics and mathematical modeling curriculum (D'Ambrosio, 2001). The emphasis in ethnomathematics can be considered conceptual while the emphasis in modeling can be thought of as performance. And both, concepts and performance are aided in significant ways by advanced technology (D'Ambrosio, 2000; Rosa, 2000).

In relation to the ongoing development and implementation of ethnomathematics and modeling in the Untied States and other countries, debate should continue as part of ongoing development and research that creates inclusive paths for the learning and teaching of mathematics in schools, communities, and societies. Sociological questions about the relationships between institutionally dominant majority and minority cultures needs be reflected by ethnomathematics research in this, our increasingly interconnected and globalized, world.

In closing, since the vast amount of research in the field is being published outside of the United States, it is necessary for more research to be translated and published into English. A recent special of Scientific American - Brasil was published in Portuguese with out any plans what so ever for Scientific American to do the same in English. NASGEm might consider assisting researchers in other countries to translate and publish their work in journals in the United States in order to establish a better reputation for our own work the field.

References

Bassanezi, R.C. (1990). A modelagem como estratégia de ensino-aprendizagem [Mathematical modeling as teaching-learning strategy]. Campinas: UNICAMP.

Bassanezi, R.C. (1994). Modeling as a teaching-learning strategy. *For the Learning of Mathematics, 14*(2), 31-35.

Bassanezi, R. C. (2002). *Ensino-aprendizagem com modelagem matemática* [Teaching and learning with mathematical modeling]. São Paulo: Editora Contexto.

Biembengut, M.S. (1999). *Modelagem matemática e implicações no Ensino - aprendizagem de matemática* [Mathematical modeling and its implications in teaching and learning mathematics]. Blumenau: Editora da FURB.

Borba, M.C. (1990). Ethnomathematics and education. *For the Learning of Mathematics, 10*(1), 39-43.

Burak, D. (1994). Critérios norteadores para a adoção da modelagem matemática no ensino fundamental e secundário [Basic criteria for the adoption of mathematical modeling in elementary, middle and high schools]. *Zetetiké, 2*(2), 47-60.

Croom, L. (1997). Mathematics for all students: Access, excellence, and equity. In J. Trentacosta & M. J. Kenney (Eds.), *Multicultural and gender equity in the mathematics classrooms: The gift of diversity* (p.1-9). Reston, VA: NCTM.

Cross, M. & Moscardini, A.O. (1985). *Learning the art of mathematical modeling.* West Sussex: Ellis Horwood Limited.

D'Ambrosio, U. (1985). Ethnomathematics and its place in the history and pedagogy of mathematics. *For the Learning of Mathematics, 5*(1), 44-48.

D'Ambrosio, B. (1989). Como ensinar matemática hoje? [How to teach mathematics today?]. *Temas & Debates, 2*(2), 11-13.

D'Ambrosio, U. (1990). *Etnomatemática* [Ethnomathematics]. São Paulo: Editora Ática.

NOTES

D'Ambrosio, U. (1993). *Etnomatemática: Um programa* [Ethnomathematics: A program]. A Educação Matemática em Revista, 1(1), 5-11.

D'Ambrosio, U. (1997). *Transdisciplinaridade* (Trandisciplinarity). São Paulo, SP, Brazil: Editora Palas Athena.

D'Ambrosio, U. (1995). Multiculturalism and mathematics education. International Journal on Mathematics, Science. *Technology Education, 26*(3), 337-346.

D'Ambrosio, U. (1998). Ethnomathematics: The art or technique of explaining and knowing (P. B. Scott, Trans.). Las Cruces: ISGEm. (Original work published in 1990).

D'Ambrosio, U. (2000). Etnomatemática e modelagem [Ethnomathematics and modeling]. In. Domite, M. C. (Ed.). *Anais do Primeiro Congresso Brasileiro de Etnomatemática – CBEm-1.* São Paulo: FE-USP, 142.

D'Ambrosio, U. (2001). *Etnomatemática: Elo entre as tradições e a modernidade* [Ethnomathematics: A link between traditions and modernity]. São Paulo, SP: Editora Autêntica.

D'Ambrosio, U. (2002). Prefácio [Forward]. In Bassanezi, R. C. Ensino-aprendizagem com modelagem matemática [Teaching and learning with mathematical modeling]. São Paulo: Editora Contexto.

Davis, R.B. (1989). The culture of mathematics and the culture of schools. *Journal of Mathematical Behavior, 8*, 143-160.

Eglash, R. (2002). Learning ethnomathematics: A software environment for teacher profession development and students' classroom use. Non-published project in progress. Rensselaer Polytechnic Institute.

Fasheh, M. (1982). Mathematics, culture, and authority. *For the Learning of Mathematics, 3*(2), 2-8.

Ferreira, E. S. (1991). Por uma teoria da etnomatemática [For an ethnomathematical theory]. *BOLEMA, 7*, 30-35.

Freire, P. (1970). *Pedagogia do Oprimido* [Pedagogy of the oppressed].Rio de Janeiro: Paz e Terra.

Freire, P. (1998). *Pedagogy of freedom: Ethics, democracy, and civic courage.* New York: Rowman and Litttlefield.

Gay, J. & Cole. M. (1967). *The new mathematics and an old culture: A study of learning among the Kpelle of Liberia.* New York, NY: Holt, Rinehart and Winston.

Gerdes, P. (1982). *Mathematics for the benefit of the people.* Paramaribo: CARIMATH.

Gerdes, P. (1988). On culture, geometrical thinking and mathematics education. *Educational Studies in Mathematics, 19*(3), 137-162.

Gerdes, P. (1988a). On possible uses of traditional Angolan sand drawings in the mathematics classroom. *Educational Studies in Mathematics, 19*(1), 3-22.

Gerdes, P. (1997). On culture, geometrical thinking and mathematics education. In A. B. Powell & M. Frankenstein (Eds.), *Ethnomathematics: Challenging Eurocentrism in Mathematics* (pp. 223-248). Albany: State University of New York.

Gerdes, P. (1997a). Survey of current work on ethnomathematics. In A. B. Powell & M. Frankenstein (Eds), *Ethnomathematics: Challenging Eurocentrism in mathematics education.* (pp.331-371). Albany, NY: State University of New York Press.

Gilmer, G. (1990). An ethnomath approach to curriculum development. *ISGEm Newsletter, 5*(2), 4-5.

Greene, E. (2000). Good-bye Pythagoras? *Chronicle of Higher Education, 47* (6), A16-A18.

Harris, M. (1987). An example of traditional women's work as a mathematics resource. *For the Learning of Mathematics, 7*(3), 26-28.

Hodgson, T. (1995). Secondary mathematics modeling: Issues and challenges. *School Science & Mathematics, 95*(7), 351-357.

ICME 7 – 7th International Congress in Mathematics Education (1992). Working Group 10 on Multicultural/Multilingual Classrooms. Topic subgroup 1: Curriculum, Resources, and Materials for Multicultural/Multilingual Classrooms, organized by Rick Patrick Scott. Quebec, Canada. In R.P. Scott (Ed.), ICM-7 Working Group 10 on Multicultural/Multilingual Classrooms (p.3-5). ISGEm Newsletter, Las Cruces, NM: ISGEm.

Joseph, G.C. (1991). *The crest of the peacock: Non-European roots of mathematics*. London: I.B. Taurus.

Lancy, D. (Ed.). (1978). The indigenous mathematics project. *Papua New Guinea Journal of Education, 14*, 1-217.

Mellin-Olsen, S. (1986). Culture as a key theme for mathematics education. Post seminar reflections. *Mathematics and Culture: A seminar report*. Radal: Caspar Forlag, 99 -121

National Council of the Teachers of Mathematics (1989). *Curriculum and evaluation standards for school mathematics.* Reston, VA: National Council of the Teachers of Mathematics.

Orey, D.C. (1989). Ethnomathematical perspectives on the NCTM Standards. In R.P. Scott (Ed.). *ISGEm Newsletter – International Group on Ethnomathematics* (p. 5-7). Las Cruces, NM: ISGEm.

Orey, D. C. (2000). The ethnomathematics of the Sioux tipi and cone. In Selin, H. (Ed.). *Mathematics Across Culture: the History of Non-Western Mathematics*. Dordrecht, Netherlands: Kulwer Academic Publishers, 239 – 252.

Orey, D.C. (2001). The algorithm collection project. Available at http://www.csus.edu/indiv/o/oreyd/ACP.htm_files/Alg.html

Pedroso, S.R. (1998). Modelagem matemática como método de ensino e aprendizagem: Relato de uma experiência em sala de aula [Mathematical modeling as a teaching and learning method: a report of a classroom experience]. *ETECAP, 1*(1), 1-10.

Posner, J. (1982). The development of mathematical knowledge in two West African societies. *Child Development, 53*, 200-208.

Ravitch, D. (June, 20th, 2005). Ethnomathematics: Even math education is being politicized. Wall Street Journal. Editorial Page.

Rosa, M. (2000). From reality to mathematical modeling: A proposal for using ethnomathematical knowledge. Unpublished master thesis. California State University, Sacramento, California.

NOTES

Rosa, M. & Orey, D.C. (2003). Vinho e queijo: Etnomatemática e Modelagem [Wine and cheese: Ethnomathematics and Modeling!]. *BOLEMA, 16*(20), 16.

Scientific American Brasil – Edição Speical (2005). Etnomatemática [Ethnomathematics]. In Laura Knapp (Ed.). São Paulo, SP, Brazil: Ediouro, Segmento-Duetto Editorial Ltda. The National Academy of Sciences. (1994). *Cultural Diversity and Early Education: Report of a Workshop.* Washington DC: National Academy Press.

Zaslavsky, C. (1973). *Africa counts: Number and patterns in African culture.* Boston, MA: Prindle, Weber and Schmidt.

Zaslavsky, C. (1990). Symmetry in American folk art. *Arithmetic Teacher, 38*, 6-12.

Zaslavsky, C. (1991). Multicultural mathematics education for the middle grades. *Arithmetic Teacher, 38* (6), 8-13.

Zaslavsky, C. (1993). *Multicultural mathematics: Interdisciplinary cooperative-learning activities.* Portland, Maine: J. Weston Walch.

Zaslavsky, C. (1994). *Multicultural math: Hands-on activities from around the world.* Jefferson City, MO: Scholastic Professional.

Zaslavsky, C. (1996). *The multicultural math classroom: Bringing in the world.* Portsmouth, NH: Heinemann.

Zaslavsky, C. (1998). Ethnomathematics and multicultural mathematics education. *Teaching Children Mathematics. 4*(9), 502-503.

The Role of Language in Ethnomathematics

Volume 1

Tamsin Meaney
Senior Lecturer in Education
Charles Sturt University
Wagga Wagga NSW Australia 2678
tmeaney@csu.edu.au

Uenuku Fairhill
Te Kuro O Te Koutu
uen_fai@koutu.school.nz

Tony Trinick
University of Auckland
t.trinick@auckland.ac.nz

Abstract

Drawing on experiences from Mäori immersion schools in New Zealand, this paper examines the impact of language when ethnomathematical practices are discussed. The inclusion of ethnomathematical perspectives into the mathematics education of indigenous students is often described as being beneficial. In New Zealand, it is also perceived as supporting the revitalisation of the Mäori language, te reo Mäori. This paper will argue that cultural practices including ethnomathematical ones cannot be separated from the language in which they were developed. Changing the language or the linguistic register in which the practices are discussed will have an impact on how the practices are perceived by students. This could result in a loss in the fundamental values that would normally accompany the practices. Without proper consideration of this issue many of the benefits aligned with using these practices may be nullified.

Introduction

Ethnomathematics began as 'the study of mathematical ideas of non-literate people' (Ascher & Ascher, 1986) but soon broadened to be the mathematical practices of a particular group of people whether they be carpenters (Masingila, 1994) or cardiovascular surgeons (Shockey, 2002). The mathematics used and developed by Western mathematicians is, therefore, one form of ethnomathematical practices (Borba, 1990). Borba (1990) suggested that ethnomathematics is '[m] athematical knowledge expressed in the language code of a given sociocultural group'. Other researchers in this area have also made connections to language. D'Ambrosio who is considered by many as the 'father of ethnomathematics' (Stillman & Balatti, 2000) described a research programme in ethnomathematics as 'the study of the generation, organisation, transmission, dissemination and the use of jargons, codes, styles of reasoning, practices, results and methods' (D'Ambrosio, 1992, p. 1183).

Given that ethnomathematics is concerned with the mathematical practices of particular cultural groups, it is not surprising that descriptions of it are linked to language. Geertz (1975) suggested that when an experience is described in language, meaning is added. However, this meaning is constrained by an individual's linguistic repertoire that developed from previous discussions with others. By describing an experience in words, aspects of it are highlighted whilst others are downplayed, often unintentionally (Meaney, 2005). For example, in te reo Mäori, *runga* can mean 'up' but in different contexts can also mean 'above', 'top' and 'south' and are linked to 'life', 'standing' and 'Sky Father' (Salmond, 1976). These are marked by different linguistic structures but these other meanings will stay with them. These connotations are considerably different to those of the English word 'up' but reflect how the concepts are related in Mäori culture.

Specific cultural practices such as ethnomathematical ones are embedded within the language of the culture. Wardhaugh (1998, p. 216) stated that individuals from a particular culture 'come to use their language in ways that reflect what they value and what they do'. In a discussion of mathematics concepts found within the languages of Ojibway and Inuit hunters, Denny (1986) showed that in Inuktitut the context in which counting occurs will be signalled by the addition of a suffix. For example, 'if events rather than things are being counted, the verb suffix –iqtaq-, "do so many times", can be added to the number root' (Denny, 1986, p. 145). Denny believed that this was consistent with people who live a hunting existent prefer 'to include together in a single representation a lot of information from the situation being dealt with' (p. 147). In te reo Mäori, numerical prefixes are context specific and/or determined by a person's relationship with the objects that are being quantified. For instance, a mountain can be quantified using the same descriptors as those for people only when there is a relationship between the speaker and the mountain. This relationship is more important than the quantities of the item being discussed. Changing the language which is used to describe the experience, whether it is a change in natural language, such as English to te reo Mäori, or a change in register within a language will have an impact on how that experience is described and therefore what is valued in this experience.

NOTES

The mathematics register consists of the words and grammatical expressions that are used consistently to describe mathematical ideas (Roberts, 1998). Any group of people who regularly talk about mathematical ideas will have specific ways to succinctly convey their meanings. Halliday (1978) stated '[w]e can refer to a "mathematics register", in the sense of the meanings that belong to the language of mathematics (the mathematical use of natural language that is: not mathematics itself), and that a language must express if it is being used for mathematical purposes' (p. 195). It is only when the mathematical ideas within an ethnomathematical activity are valued by discussants that the mathematics register will be deployed. However, cultures that have not labelled a body of knowledge and skills as 'mathematics' may not have a developed mathematics register to describe their ethnomathematical practices in this way. When a new mathematics register is developed, the choice of terms and grammatical expressions will have an impact on the connotations that accompany the mathematical ideas. The development of the mathematics register in te reo Mäori, as an example of this, is discussed later in this paper.

Although the relationship between cultural practices and language has been recognised for some time, the use of ethnomathematics within the classroom has rarely been discussed in relationship to the language of instruction. Instead, discussions are more likely to be centred around the potential benefits to students of using ethnomathematical practices (see Gerdes, 1996 and Stillman & Balatti, 2000). Stillman and Balatti (2000) suggested that in recognising the mathematics in real-life situations students would gain skills that would motivate them to learn about abstract concepts that could otherwise be considered as being devoid of meaning. These experiences could also support students to critique the assumptions on which these mathematical practices were based. The process of using ethnomathematics practices would also result in links being made between the students' community and the classroom. As was described by Meaney (2002), these benefits are seen as being particularly useful to indigenous students. The inclusion of ethnomathematical perspectives is believed to support indigenous students to gain better results because these students feel that their backgrounds and experiences were valued in the classroom and that mathematics can be developed by others outside of Western culture (Howard, 1995). As a consequence of the belief that using ethnomathematical practices in classrooms will benefit students, there are numerous accounts of how to integrate different activities into mathematics classrooms (see for example Powell & Temple, 2001). Yet there is very little research to show whether the potential benefits have manifested themselves as real benefits (Vithal & Skovsmose, 1997). The work by Lipka, Hogan, Webster, Yanez, Adams, Clark, and Lacy (2005) would be the exception to this. Their work has shown that the use of culturally-based mathematics teaching with Yup'ik students in Alaska resulted in significant improvement in test results. Given that the incorporation of ethnomathematical practices into classrooms can only be done through language, any investigation of the benefits needs to include consideration of language issues.

Ethnomathematical practices in kura kaupapa Mäori

This paper explores the language issues around introducing ethnomathematical practices, including those of Western mathematics, into the mathematics classrooms of a Mäori immersion school (known as a kura kaupapa Mäori). In kura kaupapa Mäori, 'the pedagogy of these schools is based on, but not exclusively, Mäori preferred teaching and learning methods' (Smith, 1990 p147-148). The document which is the guiding principle for kura kaupapa Mäori, *Te Aho Matua* (Kura Kaupapa Working Group, 1989), states that the child's own home is their first school, intimating that school should be homelike with a shared culture and values. Ethnomathematical practices of the local community would, therefore, seem to be appropriate activities to incorporate into the kura's mathematics classrooms. However, there are several issues that need to be considered in relationship to the introduction of ethnomathematical practices because of the students' bilingual/bicultural position within their community.

The particular kura kaupapa Mäori that is the focus of our discussion is in a provincial city of New Zealand. Students who attend this kura use only te reo Mäori whilst at kura and are encouraged to use it when communicating with their friends outside of school hours. Although the parents made active decisions to place their children in a Mäori immersion situation, they themselves may not be fluent in te reo Mäori. As well, te reo Mäori is a minority language in New Zealand and so many of the children's outside experiences would be conducted in English. These would include activities that would use their mathematical understandings such as going to fast food restaurants and playing computer games. Although the children at this kura would also participate in some traditional Mäori cultural experiences both inside and outside the kura, much of their lives would not be very different from their non-Mäori (Pakeha) peers. The situation that the students at this kura are in is linguistically quite complex and so issues of language cannot be separated from considerations about how to teach mathematics.

The data for the paper is drawn from the experiences and opinions of the six staff involved in teaching mathematics in the kura. Information came from conversations with staff and also from responses to a questionnaire. The questionnaire was used as part of a larger project that was investigating language issues in the mathematics classrooms of this kura (see Fairhall, Meaney and Trinick, 2007 for further details). The researchers all had a long term relationship with the kura and its teachers. Uenuku Fairhall is the principal of the kura. Tony Trinick had been involved in the preservice mathematics education of some of the teachers, whilst Tamsin Meaney had done her PhD research at this kura in 1998 and 1999.

The following discussion first looks at teachers' opinions about using ethnomathematical experiences in the classroom and then considers the language issues in regard to this.

Many of the gains from using ethnomathematical practices in classrooms are presumed to be made in relationship to the acquisition of the standard mathematics curriculum. Learning about ethnomathematical practices becomes the entry point for gaining a better understanding of school mathematics concepts (Gerdes, 1996). As a result, ethnomathematical practices tend to be described in the mathematics register of the language of instruction. As Stillman and Balatti (2000) warned, this process potentially 'divorces the cultural practices from their context and trivializes and fragments them from their real meaning in context' (p. 325).

In an earlier paper, Meaney and Fairhall (2003) described how some of the parents had seen some of their own experiences, such as building fish traps and smoking fish, as being useful contexts for their children to learn mathematics. However, they believed that the teachers' professional experience was a more valuable guide to choosing appropriate contexts for teaching mathematics. In

NOTES

responses to the questionnaire, the current teachers mostly suggested counting and money experiences as outside-school mathematical experiences that learners and their parents would consider to be useful in the classroom. When asked specifically about the traditional Mäori knowledge and practices which should be included in mathematics classrooms, suggestions ranged from looking at time and other measurements to looking at number patterns in traditional arts such as tukutuku (woven decorative panels), whakairo (carving), raranga harakeke (weaving flax) as well as whakatere waka (navigation skills). Uenuku Fairhall, who is the principal of the kura, also suggested that traditional practices should be used that showed how Mäori ancestors used mathematics.

One activity that he has used successfully with various groups of children and adults is that of land division. Land belongs to extended family groups and a person's share in the land will depend upon the number of generations since the original title was recognised. Generally family members are unclear of how they came to have their proportion of land. At a family night in 2004, parents came as students to attend various classes with their children acting as assistant teachers. Describing land division using the language of fractions meant that for the first time many parents understood how their land share came about and understood how fractions operated. Previously, many parents had described their own poor experiences of learning mathematics at school (Meaney & Fairhall, 2003). To find an activity that had meaning for them and gave them an understanding of fractions was something that was very much appreciated by these parents.

This activity exhibited many of the points suggested as being valuable in using ethnomathematics practices with indigenous students. It relates to the students' backgrounds as it shows that mathematical concepts were incorporated into traditional activities and makes the concept of fractions more than just a set of abstract ideas, 'devoid of meaning'. It also seemed that in this situation, discussion of the activity using the mathematics register in te reo Mäori added another dimension that had not been apparent to the parents previously.

Concerns have been raised about the use of ethnomathematical practices as tokenistic add-ons to a mathematics curriculum (Meaney, 2002). By discussing a cultural practice using the mathematics register, it is the mathematics which was focussed on. The importance of knowing about how to perform a traditional art or appreciate it aesthetically can become lost when the focus becomes mathematics (Roberts, 1996). In the land division activity, added meaning was gained by the participants when they talked about it using the mathematics register. If ethnomathematics practices are included into classrooms, then discussion of them through the mathematics register needs to add something to the meaning rather than take something away.

The mathematics register in te reo Mäori

Discussion of ethnomathematical practices in te reo Mäori is complicated by the newness of much of the vocabulary. Although some textbooks had been translated into Mäori in the 19th century, the development of the Mäori mathematics register has been a fairly recent endeavour (Barton, Fairhall & Trinick, 1995). Transliteration of expressions and terms into Mäori are of limited value as they neither express the mathematical concept or have appropriate baggage that comes with using a term which has derived from cultural experiences. In the 1980s a series of meetings (hui) were held with primary and secondary teachers who were teaching mathematics in Mäori. Two of the authors, Tony Trinick and Uenuku Fairhall were at these hui. Teachers and others shared the terms that they were using to discuss mathematical ideas and a list of terms for particular mathematical concepts were compiled (see Barton & Cleave, 1989).

Volume 1

Eventually, the Mäori Language Commission became involved and a set of 600 terms and some grammatical structures was agreed upon (Barton, Fairhall & Trinick, 1995). These were then first as *Nga Kupu Tikanga Pangarau: Mathematics Vocabulary* (Learning Media, 1991) and then an expanded version was included as a glossary at the back of the new Pängarau (mathematics) curriculum document (Ministry of Education, 1994). A dictionary of mathematical terms was published in 2004 and made available to all schools (Christensen (Ed), 2004).

However, the lack of people who continually discuss mathematical ideas in te reo Mäori has impacted on the standardisation of the mathematics register terms. The small number of teachers who teach mathematics in te reo Mäori is spread over a large area of New Zealand. As a result, few terms have become solidly integrated into the lexicon with many kura and even individual teachers using local expressions. Words that are used less frequently, such as whenu (cosine), are more likely to be standardised, whereas more frequently used words such as multiplication are less likely to be. In the 1980s, whakarau was initially coined for multiplication. In recent years, there has been a shift to using whakarea but some teachers have resisted this change. Teachers need a strong understanding of mathematics and te reo Mäori, if they are to explore with students how best to make use of the language for learning mathematics. There could be problems in exploring other mathematical ideas with non-standardised language. Very few teachers learnt mathematics in te reo Mäori and so many are learning the mathematical terms in Mäori at the same time as they are teaching the concepts to the students. Teacher in the earlier grades are also unlikely to know the mathematical terms that their students will need to use in the following years of study. Although this situation will improve as more teachers have themselves been educated in kura, it is difficult at the present time for many teachers to utilise the resources of te reo Mäori in teaching mathematical concepts.

If Western mathematics is considered to be one example of ethnomathematical practices, then its discussion in te reo Mäori also needs examination. When responding to the questionnaire, almost all the teachers mentioned that new vocabulary had an effect on students' ability to learn mathematics. Some of the teachers felt that because students only came in contact with the mathematics register at school their ability to fluently discuss mathematical ideas was reduced. There was a recognition that a student's ability to acquire the mathematical register would depend on their teacher's own fluency in te reo Mäori in general as well as in the mathematics register (te reo tätatai).

Uenuku Fairhall suggested that if the class and its teacher were confident in discussing mathematics and had a sophisticated understanding of the mathematics then more culturally appropriate terms would develop. He gave an example of his senior class' change of terms for the expression 'like terms' in algebra, when identifying them in expressions such as the following $-4x - 3x + 3x^2 -5x^2$. The term used in the new dictionary of mathematical terms, *Te Reo Pängarau* (Christensen, 2004) is *röpü* meaning 'grouping'. Students in Uenuku's senior classes coined the term, *whakawhänau*, meaning 'making families'. It was through discussion of what was happening when 'like terms' were gathered together that the students felt that *whakawhänau* was more

NOTES

appropriate. This is because the connotations which this term invoked more readily fitted their understanding of what was happening. Discussion of Western mathematics in te reo tätaitai (mathematics register) needs to be done in a way which makes use of the cultural connotations of the Mäori words. Traditionally, whänau (family) was rarely used to describe inanimate objects. Therefore, the term whakawhänau may need to bc considered as a step to helping students understand the mathematical idea that will then be replaced with the standardised term röpü.

Development of te reo tätaitai needs to be done in a way which minimises damage to te reo Mäori and Mäori culture (Barton, Fairhall & Trinick, 1998). One teacher in responding to a question on how te reo Mäori is changing as a result of using it to teach mathematics stated 'ma tou rourou, ma toku rourou, ka ora te iwi' (two heads are better than one). This suggests that the resources of both languages, English and te reo Mäori, can be made use of and it may be that destructive damage can be minimalised if thought is put into how te reo tätaitai is developed (Roberts, 1998). Another teacher suggested that archaic terms should be given new mathematical meanings rather than making up new terms as one way of holding onto the traditional cultural values connected to these older terms. For example, *hangai* traditionally meant to take a shortcut or to go directly to a location, rather than going around the outside. In recent years it has gained the mathematical meaning of diagonal, thus sharing the idea of cutting across in the shortest manner.

Conclusion

The use of ethnomathematical practices in mathematics classrooms warrants thoughtful consideration. This is because every activity is embedded within the language of the culture in which it arose. This language will highlight those features of the activity which the culture values. Changing the way an activity is discussed either by using the mathematics register or by changing the language will have implications for how the activity is perceived by the students. It may also change the actual practice itself by focussing on other ideas.

This paper used examples from a kura kaupapa Mäori to illustrate the need for the mathematics register to add rather than to subtract meaning from an ethnomathematical practice. If this is not done then the value of the activity in its own right could be lost in the eyes of the learners. This paper also looked at how changing languages from English to te reo Mäori to discuss mathematical ideas is not a simple procedure. The development of indigenous languages for describing Western mathematics needs to respect the cultural values of the users of the language and that means using terms which have appropriate connotations both for the mathematics but also for the culture.

Acknowledgement

This research was funded by the New Zealand Ministry of Education through a Teaching and Learning Research Initiative, administrated by the New Zealand Council of Educational Research.

References

Ascher, M. & Ascher, R. (1986). Ethnomathematics. *History of Science 14*, 125-144.

Barton, B. & Cleave, P. (Eds) (1989) *He kupu tikanga tau ähuatanga: Mathematics vocabulary*. Special Edition of *Ngä Mauranga* Auckland: MECA, Auckland College of Education.

Barton, B., Fairhall, U. & Trinick, T. (1995). History of the development of Mäori mathematics vocabulary. In A. Begg et al. (eds), *SAMEpapers 95*, Hamilton, NZ:CSMER, 144-160.

Barton, B., Fairhall, U. & Trinick, T. (1998). Tikanga re tatai: Issues in the development of a Maori mathematics register. *For the learning of mathematics 18*(1), 3-9.

Borba, M. C. (1990). Ethnomathematics and Education. *For the learning of mathematics 10*(1), 39-43.

Christensen, I. (Ed) (2004). *Te reo Pängarau* Wellington: Ministry of Education.

D'Ambrosio, U. (1992). Ethnomathematics: A research program on the history and philosophy of mathematics with pedagogical implications. *Notices of the American Mathematical Society, 39*(10), 1183-1185.

Denny, J. P. (1986). Cultural ecology of mathematics: Ojibway and Inuit hunters. In M. Closs (Ed.) *Native American Mathematics* Dallas: University of Texas.

Eglash, R. (1997). Bamana sand divination: Recursion in ethnomathematics. *American Anthropologist, 99*(1), 112-123.

Fairhall, U., Trinick, T., Meaney, T. (2007). Grab that kite! Acquiring the mathematics register in te reo Mäori. *Curriculum Matters, 3*, 48-66.

Geertz, C. (1975). *The Interpretation of Cultures.* London: Hutchinson & Co.

Gerdes, P. (1996). Ethnomathematics and Mathematics Education. In A. J. Bishop, M. A. Clements, C. Keitel, J. Kilpatrick and C. Laborde (Eds.) *International Handbook of Mathematics Education*, (pp. 987-1023). Dordrecht: Kluwer Academic Publishers.

Halliday, M. A. K.: 1978, *Language as Social Semiotic*, London: Edward Arnold.

Howard, P. (1995). Social and cultural factors in the mathematics classroom: empowering Indigenous students *The Proceedings and Papers of Bridging Mathematics Network Conference*, (pp. 31-35). Batchelor, NT: Batchelor College.

Kura Kaupapa Maori Working Group, Katarina Mataira chair (1989). *Te Aho Matua* Wellington: Ministry of Education.

Learning Media (1991). *Nga kupu tikanga pangarau: Mathematics vocabulary*. Wellington: Ministry of Education.

Lipka, J., Hogan, M. P., Webster, J. P., Yanez, E., Adams, B., Clark, S. & Lacy, D. (2005). Math in a cultural context: Two case studies of a successful culturally based math project. *Anthropology and Education Quarterly, 36*(4), 367-385.

NOTES

Masingila, J. O. (1994). Mathematics practice in carpet laying. *Anthropology and Education Quarterly, 25*(4), 430-462.

Meaney, T. (2002). Symbiosis or cultural clash? Indigenous students learning mathematics. *Journal of Intercultural Studies, 23*(2), 167-187.

Meaney, T. (2005). Mathematics as text. In A. Chronaki and I. M. Christiansen (Eds.) *Challenging Perspectives in Mathematics Classroom Communication.* (pp. 109- 141). Westport, CT: Information Age.

Meaney, T. & Fairhall, U. (2003). Tensions and possibilities: Indigenous parents doing mathematics education curriculum development". In L. Bragg, C. Campbell; G. Herbert & J. Mousley (Eds.), *Mathematics Education Research: Innovation, networking, opportunities* (pp. 507-514). Melbourne: Deakin University.

Ministry of Education (1994). *Pängarau, mathematics in the New Zealand Curriculum.* Wellington: Learning Media.

Powell, A. B. & Temple, O. L. (2001). Seeding ethnomathematics with oware: Sankofa. *Teaching children mathematics, 7*(6) 369-373.

Roberts, T. (1996). Whose mathematics do we teach? *The Australian mathematics teacher, 52*(4), 42-46.

Roberts, T. (1998). Mathematical registers in Aboriginal languages. *For the learning of mathematics, 18*(1), 10-16.

Salmond, A. (1976). Te ao tawhito: A semantic approach to the traditional Mäori cosmos. Reprinted in 1986 in *Nga kete waananga: A selection of papers relating to Äkonga Maori* Auckland: Auckland College of Education.

Shockey, T. L, (2002). Etnomatematica de uma Classe Profissional: Cirurgioes Cardiovasculares. *BOLEMA, 15*(17), 1 – 19.

Smith, G. H. (1990). Ki te whai ao, ki te ao marama: Crisis and change in Maori education. In A. Jones, G. McCulloch, J. Marshall, G. H. Smith, L. Tuhiwai- Smith (eds) *Myths and Realities: Schooling in New Zealand* (pp. 123-155) Palmerston North: Dunmore Press.

Stillman, G. & Balatti, J. (2000). Contribution of ethnomathematics to mainstream mathematics classroom practices. In B. Atweh, H. Forgasz & B. Nebres (Eds.) *Sociocultural research on mathematics education: An international perspective*, (pp. 201-216). Mahwah, NJ: Lawrence Erlbaum Associates.

Vithal, R. & Skovsmose, O. (1997). The end of innocence: A critique of "ethnomathematics", *Educational Studies in Mathematics, 34,* 131-157.

Wardhaugh, R. (1998). *An introduction to sociolinguistics.* Malden, Mass: Blackwell Publishers.

Ethnomathematics in a European Context: Towards an Enriched Meaning of Ethnomathematics

Volume 1

Karen François
Centre for Logic and Philosophy of Science (CLWF) Free University Brussels (VUB)

Abstract

This paper considers the field of enquiry called ethnomathematics and its role within mathematics education. I elaborate on the shifted meaning of 'ethnomathematics'. This "enriched meaning" impacts on the philosophy of mathematics education. Currently, the concept is no longer reserved for the so called 'nonliterate' people, but also includes diverse mathematical practices even within Western European classrooms. Consequently, mathematics teachers are challenged to handle people's cultural diversity occurring within every classroom setting. Ethnomathematics has clearly gained a prominent role, within Western European curricula, becoming meaningful in the exploration of various aspects of mathematical literacy. I discuss this enriched meaning of ethnomathematics as an alternative, implicit philosophy of school mathematical practices.

Keywords:

- Ethnomathematics
- Diversity
- Politics
- Philosophy
- Values.

Introduction

Until the early 1980s, the notion 'ethnomathematics' was reserved for the mathematical practices of 'nonliterate' – formerly labeled as 'primitive' – peoples (Ascher & Ascher, 1997). What was needed was a detailed analysis of the sophisticated mathematical ideas within ethnomathematics, which it was claimed were related to and as complex as those of modern, 'Western' mathematics. D'Ambrosio (1997), who became the "intellectual father" of the ethnomathematics program proposed "a broader concept of 'ethno', to include all culturally identifiable groups with their jargons, codes, symbols, myths, and even specific ways of reasoning and inferring". Currently, as a result of this change within the discipline of ethnomathematics, scientists collect empirical data about the mathematical practices of culturally differentiated groups, literate or not (Gordon, 2004; Pica, Lemer, Izard & Dehaene 2004; Stathopoulou, 2006; Moreira & Pires). The label 'ethno' should thus no longer be understood as referring to the exotic or as being connected with race. This changed and enriched meaning of the concept 'ethnomathematics' has had its impact on the philosophy of mathematics education. From this point on, ethnomathematics became meaningful to every classroom since multicultural classroom settings are generalized all over the world. Every classroom today is characterized by diversity (ethnic, linguistic, gender, social, cultural, etc.). Teachers in general but also mathematics teachers have to deal with the existing cultural diversity since mathematics is defined as human and cultural knowledge, like any other field of knowledge (Bishop 2002). The shifted meaning of ethnomathematics into a broader concept of cultural diversity became meaningful within the community of researchers working on the topic of ethnomathematics, multicultural education and cultural diversity. In the European context, the topic was absent at the first two conferences of the Conference of European Research in Mathematics Education (CERME 1, 1998; CERME 2, 2001), the topic appeared at CERME 3 (2003) as "Teaching and learning mathematics in multicultural classrooms". At CERME 4 (2005) and CERME 5 (2007) the working group was called "Mathematics education in multicultural settings". At CERME 6 (2009) the working group now was called "Cultural diversity and mathematics education" and at the CERME 7 (2011) the working group was called "Diversity and mathematics education: Social, cultural and political challenges/ issues." Since then, and in line with the theoretical development of the concept of ethnomathematics, there has been an explicit consideration of the notion of cultural diversity. Looking at the spaces beyond Europe one can observe a similar evolution in the research field of ethnomathematics. It became a research field within mathematics education that is concerned with the social and political aspects of the learning of mathematics. In this sense the difference between Critical Mathematics Education (CME) and ethnomathematics became vague as mentioned by Ole Skovsmose in a recent interview:

> *I see CME as represented by many different approaches in mathematics education, and certainly by many approaches that do not use the label of CME. For instance, the whole movement of mathematics education for social justice I see as an example of CME. The*

NOTES

approach suggested by Renuka Vithal talking about "a pedagogy of conflict and dialogue" can also be an example. Different examples are found in many discussions of mathematical modelling. **Much work in ethnomathematics shares the interest and concern of CME.** *And today several people in Brazil contribute explicitly to the development of CME.*

Such different approaches represent concerns with respect to mathematics education that to me indicates a critical position. (bold my own emphasis, Alrø, Ravn & Valero, 2010: 4).

In this quote one can immediately observe the common interest of CME and ethnomathematics. One approach cannot be reduced to the other although both approaches share a critical position that can enrich the theory and practice of mathematics education and the specific concern on social, cultural and political challenges.

Dealing With Cultural Diversity in the Classroom

Ethnomathematics applied in education had a Brazilian origin, but the research field of ethnomathematics eventually became a common interest all over the world. The research field has been extended from a so called exotic interpretation (connected to race and origin) to a way of intercultural learning that is applicable within any learning context where the teacher is dealing with the social and cultural diversity of all children and their background. Dealing with cultural diversity in the classroom is the universal context within which each specific context has its place. The common and universal concern is the fact that children's background needs to be taken into account. The specificity of each learning process is the diversity that appears at each context related to the diverse backgrounds of the children. The meaning of the ethno concept has been viewed as an ethnical group, a national group, a racial group, a professional group, a group with a philosophical or ideological basis, a socio-cultural group and a group that is based on gender or sexual identity (Powell 2002: 19). This list could still be added to but since lists will always be deficient, all the more because some distinctions are relevant only in a specific context, we use the all-embracing concept of *cultural diversity*. The meaning of this concept of cultural diversity in the field of mathematics is also used by Bishop (2002) to identify the nature of mathematics as a human and cultural knowledge. This view enables us to see the comparative culture studies regarding mathematics that describe the different mathematical practices, not only as revealing the diversity of mathematical practices but also to emphasize the complexity of each system. In addition there is interest in the way that these mathematical practices arise and how they are used in the everyday life of people who live and survive within a well-defined socio-cultural and historical context. Consequently there has to be a translation of these studies (see e.g. Stathopoulou, 2006; Pinxten & François, 2007) to mathematics education where the teacher is challenged to introduce the cultural diversity of pupil's mathematical practices in the curriculum since pupils also use mathematical practices in their everyday life. This application exceeds the mere introduction in class of the study of new cultures or – to put it dynamically – new culture fields (Pinxten 1994:14). In the following paragraph I will illustrate some of these first 'ethno mathematical' moves that are made, even before dealing with cultural diversity arose.

Diversity within mathematical practices was considered as a practice of the 'other', the 'exotic'. It was not considered relevant to mathematics pupils from a Westernised culture. That is why the examples regarding mathematics (and adjacent sciences) are an enquiry of all kinds of so called 'exotic traditions' such as sand drawings from Africa, music from Brazil, games such as

Patience the way it is played in Madagascar, the arithmetic system of the Incas or the Egyptians, the weaving of baskets or carpets, the Mayan calendar, the production of dyes out of natural substances, drinking tea and keeping tea warm in China, water collection in the Kalahari desert, the construction of Indian arrows, terrace cultivation in China, the baking of clay bricks in Africa, the construction of African houses. The examples are almost endless (Bazin & Tamez 2002). Notwithstanding the good intentions of these and similar projects, referring to Powell & Frankenstein (1997) I would like to emphasize that these initiatives may well turn into some kind of folklore while originally intending to offer intercultural education.

> *However, I also stress that I am not advocating the curricular use of other people's ethnomathematical knowledge in a simplistic way, as a kind of "folkloristic" fiveminute introduction to the "real" mathematics lessons. (Powell & Frankenstein 1997: 254)*

In line with the empirical research by Pinxten & François (2007) on mathematical practices in classroom settings, one can verify many appropriate examples of pupils' mathematical practices that may be used in class, not as some kind of exoticism but as the utilization of a mathematical concept. Starting from pupils' mathematical knowledge, and their everyday mathematical practices, is a basic principle of the new orientation towards realistic mathematics education and the development of innovative classroom practices (Prediger 2007). A central question that one needs to address here is how the teacher acquires knowledge of the students out of school and how they can bring the students' daily mathematical practices into the learning process. In other words, how one can move from a teacher-centered learning process towards a pupil-centered learning process where pupils' mathematical practices can enter the classroom? The answer is initially given by Cohen & Lotan (1997). They describe how cooperative interactive working can be structured and they also explain the benefits of interactive learning in groups to deal with diversity. For that purpose the Complex Instruction theory (as a specific variant of cooperative learning) was developed which they implemented in education. Meanwhile this didactical approach has had an international take-up in Europe, Israel and the United States and it has been elaborated to the didactics of Cooperative Learning in Multicultural Groups (CLIM) (Cohen 1997: vii). This teaching method has been tested in a number of settings, in distinct age groups and with regard to different curricula (Cohen 1997: 137, Neves 1997: 181, Ben-Ari 1997: 193). The acquisition of mathematical content was also part of this research. Complex Instruction is a teaching method with the equality of all pupils as its main objective. It tries to reach all children and tries to involve them in the learning process, irrespective of their diverse backgrounds (François & Bracke 2006). In order not to peg cultural diversity down to a specific kind of diversity, in this context Cohen (1997: 3) speaks of working in heterogeneous groups. Heterogeneity can be found in every group structure. Even a classroom is characterized by a diverse group of pupils where every pupil has in some way his or her everyday mathematical practices.

The Complex Instruction teaching method implies the integration of the daily experiences of the children since it uses

NOTES

open tasks that can be worked on based on the practices children bring from home. It is not rare that even parents are involved in the learning process. During the interactive learning process children have to negotiate how to solve the problem, e.g. how to measure, what to use for measuring, what kind of reference is used and which example is worked on (see e.g. Pinxten & Francois, 2007 on Navajo children in a North American context and Turkey immigrants in Flemish schools). If pupil-centered learning is taken seriously, teachers are challenged to deal with the present mathematical practices while teaching mathematics. This way, ethnomathematics becomes a way of teaching mathematics where cultural diversity of pupils' everyday mathematical practices are taken into account (François 2007).

Ethnomathematics in Every Classroom

The extended notion of ethnomathematics as dealing with pupils' everyday mathematical practices has equality of all pupils as its main objective. Ethnomathematics becomes a philosophy of mathematics education where mathematical literacy is a basic right of all pupils. The teaching process tries to reach all pupils and tries to involve them in the learning process of mathematics, irrespective of their cultural diversity. All pupils are valued and treated as equal. This notion of mathematics for everyone fits in with the ethical concept of pedagogic optimism that is connected with the theory of egalitarianism. This ethical-theoretical foundation on which the project of equality within education is based, assumes that the equality is measured at the end of the line. As reported by the justice theories of John Rawls (1999) and Amartya Sen (1992), pupils' starting positions can be dissimilar in such a way that a strictly equal deal will prove insufficient to achieve equality. A meritocratic position – which measures the equality at the start of the process – thus cannot fully guarantee equal chances (Hirtt, Nicaise & De Zutter 2007: 61-84). An egalitarian position starts from a pedagogic optimism and it needs to take into account the diversity of those learning in order to give equality maximum chances at the end of the line.

By extending the notion of ethnomathematics to cultural diversity and mathematics education, the distinction between mathematics and ethnomathematics seems to disappear. If every mathematical practice is a cultural embedded and thus contingent practice (which the author agrees with), the concept of ethnomathematics seems to become superfluous. Hence the critical question can be raised whether the achievements of ethnomathematics will not then become lost. On the contrary, the distinction between ethnomathematics and mathematics can only disappear by acknowledging and implementing the achievements of ethnomathematics in mathematics education. The issue of the distinction between ethnomathematics and mathematics has been raised before within the theory development of ethnomathematics (Setati 2002). Being critical of the dominant Western mathematics was the basis out of which the discipline of ethnomathematics has developed and now the time is right to also raise the critical questions internally, within the field of ethnomathematics itself. What exactly distinguishes ethnomathematics from mathematics? Setati raises this question in a critical review on the developments within ethnomathematics as a theoretical discipline that is dissociated and distinguished from mathematics (Setati 2002). Setati sees mathematics as a mathematical practice, performed by a cultural group that identifies itself based on a philosophical and ideological perspective (Setati 2002: 31). Every mathematics teacher is supposed to use a series of standards that are connected with the profession and with obtaining qualifications. The standards are philosophical (about the way of being), ideological (about the way of perceiving) and argumentative (about the way of expressing). Both mathematics and ethnomathematics are embedded in a normative framework. So the question can be raised as to whether the values of mathematics and ethnomathematics are indeed that distinctive.

It cannot be denied that ethnomathematics is based on an emancipatory and critical attitude that promotes the emancipation and equality of discriminated-against groups (Powell & Frankenstein 1997). This general idea of emancipation can also be found in the UNESCO's view of education. Moreover one can see in its mission a tight connection with socio-economic development, with working on an enduring and peaceful world, while respecting diversity and maintaining human rights. In this mission statement, education is obviously connected with the political factor. One can question if this mission statement is also characteristic of mathematics education in general. Research on social justice in mathematics education (e.g. Apple 1992; Burton 2003; Gutstein 2003, 2006; Gutstein & Peterson 2005) shows that this is not even close to the case. This is why ethnomathematics and CME are valuable research programs to this objective of equality in mathematics education.

> *UNESCO believes that education is a key to social and economic development. We work for a sustainable world with just societies that value knowledge, promote a culture of peace, celebrate diversity and defend human rights, achieved by providing education for all. The mission of the UNESCO Education Sector is to provide international leadership for creating learning societies with educational opportunities for all populations; provide expertise and foster partnerships to strengthen national educational leadership and the capacity of countries to offer quality education for all. (UNESCO 1948)*

Taking into account these general stipulations one can conclude that the explicit values of the general education objective connect to the values of equal chances for all pupils which are central within ethnomathematics. Consequently the expansion of ethnomathematics as a way of teaching mathematics which takes the diversity of pupils' mathematical practices into account can be justified. There is a kind of inequality in every group and the real art is to learn to detect the boundaries of inequality and the boundaries of cultural diversity. Instead of a depreciation of the concept 'ethnomathematics' this extended notion could mean a surplus value in situations where heterogeneity and cultural diversity are less conspicuous.

Within ethnomathematics education two aspects are highlighted. First there is the curriculum's content. Often this is the first step when implementing ethnomathematics. Besides the mathematics that can be found in the traditional curriculum, there will now be additional space to be introduced to more exotic or traditional mathematics practices. Powell & Frankenstein (1997) also emphasize this aspect in their definition of the enrichment of a curriculum through ethnomathematics. Stressing other mathematical practices offers the opportunity to gain a better perception in their own mathematical practice and its role and place in society (D'Ambrosio 2007a: 33). It also offers the opportunity to philosophize and critically reflect on their own mathematical practice. In language teaching it goes without saying that it is better to learn more than one language. It broadens

NOTES

the outlook on the world and offers a better adaptation to dealing with other people in this globalized world. Knowledge of several languages is undoubtedly an advantage and besides it broadens the knowledge of the mother tongue. This comparison could even be extended to the mathematics education where knowledge of mathematical practices of several cultural contexts and throughout time proves to be advantageous. A second aspect within ethnomathematics is the didactics, the way that the learning process is set up. Here an interactive and cooperative approach is crucial (Cohen 1997, César 2009). The two aspects obviously have mutual grounds. An interactive approach results in contents being defined also by the learning with an active participation in the learning process. This aspect is strongly emphasized by researchers who investigate the integration of so-called traditional groups within the academic context. This is expressed as one of Graham's key questions in his enquiry into mathematics education for aboriginal children: what do the children bring to school? (Graham 1988: 121). With the extended notion of ethnomathematics as cultural diversity and mathematics education and with the emphasis on dealing with pupils' everyday mathematical practices, ethnomathematical practice is now closer to the social environment of the pupil and disconnected from its original (exotic) cradle. Both the theory and practice of ethnomathematics have opened eyes and broadened minds. It immediately answers the question as to what exactly could be of benefit to the 'highly-educated countries' – with their outstanding results in international comparative investigations – regarding ethnomathematics as it originally developed, as a critical and emancipatory theory and as a movement that aimed to give all pupils equal chances. In a final section about ethnomathematics I would like to link mathematics education, politics and human rights.

Ethnomathematics as Human Right

D'Ambrosio, who is the mathematician and educationalist of the mathematics on which ethnomathematics is based, situates mathematics education within a social, cultural and historical context. He can also be considered the first to explicitly link mathematics education and politics. Mathematics education is a lever for the development of the individual, national and global wellbeing (D'Ambrosio 2007a, 2007b). In other words the teaching and learning of mathematics is a mathematical practice with obviously a political grounding. D'Ambrosio advances the political proposition that mathematics education should be accessible to all pupils and not only to the privileged few. This proposition has been registered in the OECD/PISA report, which is the basis for the PISA-2003 continuation enquiry.

> *Mathematical literacy is an individual's capacity to identify and understand the role that mathematics plays in the world, to make well-founded judgements and to use and engage with mathematics in ways that meet the needs of that individual's life as a constructive, concerned and reflective citizen. (OECD, 2004: 37)*

This specification of mathematical literacy clearly implies that this form of literacy is a human right for every child, such that the child gets a chance to participate to the world in a full, constructive, relevant and thoughtful way. This proposition recurs later in the essays of Alan J. Bishop (2006) where he demonstrates the link between mathematics, ethnomathematics, values and politics (Bishop 2006: 7).

Mentioning mathematics education and values education in one and the same breath does not sound unambiguous because mathematics is undeniably perceived as non-normative.

It is a widespread misunderstanding that mathematics is the most value-free of all school subjects, not just among teachers but also among parents, university mathematicians and employers. In reality, mathematics is just as much human and cultural knowledge as any other field of knowledge, teachers inevitably teach values [...]. (Bishop 2002: 228)

We will give some examples from literature how this values come to play in school curricula and in the classroom. Bishop (1997) presents different sets of ideals and values associated with mathematics. He differentiates six categories: rationalism and objectism (dimension of ideology), control and progress (dimension of sentiment), openness and mystery (dimension of sociology). In addition Ernest (2003) inquires on the inherent values in mathematics. Abstract is valued over concrete, formal over informal, objective over subjective, justification over discovery, rationality over intuition, reason over emotion, general over particular, theory over practice, the work of the brain over the work of the hand, and so on. These examples are recognized and reflected in empirical research on classroom practice as François (2007) shows for the case of rationalism and objectivism in Flanders secondary mathematics curriculum.

It is predominantly within D'Ambrosio's' ethnomathematics research program that the link of mathematics and mathematics education with values is extended to the political domain. According to D'Ambrosio still too many people are convinced that mathematics education and politics have nothing in common (D'Ambrosio 2007a: 27). He challenges this cliché. In his recent work D'Ambrosio (2007a, 2007b) takes as his starting point the Universal Declaration of Human Rights where articles 26 and 27 highlight the right to education and to share in scientific advancements and their benefits.[1] This declaration concerning education is further developed and confirmed within the UNESCO's activities by means of the World Declaration on Education for All in 1990 and ratified by 155 countries. Finally the declaration has been applied in mathematical literacy in the OECD/PISA declaration of 2003. D'Ambrosio regrets that these declarations are not well-known by mathematics teachers since they play a key role in the emancipation process. In line with the World Declaration, 'mathematics education for all' implies a critical reflective way of teaching mathematics. According to D'Ambrosio, this way of teaching does not receive sufficient opportunities. Following Bishop (1997) he criticizes the technically-oriented curriculum with its emphasis on technique and drill and where history, philosophy and critical reflection are not given a chance. D'Ambrosio develops three concepts to focus on in a new curriculum regarding the usage of the international (UNESCO) emancipatory objectives - *literacy, matheracy* and *technoracy*. We first will explain what D'Ambrosio means by this concepts. Afterwards we make the comparison with the work of Skovsmose (1994).

Literacy has to do with communicative values and it is an opportunity to contain and use information. Here both spoken and written language is concerned but so are symbols and meanings, codes and numbers. Mathematical literacy is undoubtedly a part of it. *Matheracy* is a tool that offers the chance to deduce, to

NOTES

develop hypotheses and to draw conclusions from data. These are the base points for an analytical and scientific attitude. Finally there is *Technoracy* which offers the opportunity to become familiar with technology. This does not imply that every pupil should or even could get an understanding of the technological developments. This elementary form of education needs to guarantee that every user of a technology should get to know at least the basic principles, the possibilities and the risks in order to deal with this technology in a sensible way or deal not at all with it.

With these three forms of elementary education, which can be developed throughout the ethnomathematics research program, D'Ambrosio wants to meet the Universal Declaration of the Human Rights that relate to the right to education and the right to the benefits of the scientific developments.

One can observe a tight connection with the analysis Skovsmose (1994) makes in his research on mathematical knowledge. He provides us a well defined meaning of mathemacy when he explains how mathematics education in general can be organized so as to develop different types of knowledge. He makes the distinction between mathematical knowing, technological knowing and finally reflective knowing. The first type of knowing is associated with the skills developed in traditional teaching; the second type is associated with the competence in mathematical model building and finally the reflective knowing is associated with the competence in evaluating applications of mathematics. To Skovsmose, a CME must integrate the three types of knowledge; the mathematical knowing, the technological knowing as well as the reflective knowing. One example Skovsmose (1994) gives us to better understand his notion of reflective knowing is the relation between technology and mathematics education. In a standard curriculum of mathematics education, this connection is understood as the use of computers in the classroom. However the connection between technology and mathematics education can be understood in a broader sense, connected to societal and political concerns about the impact of technology on society. Here we enter a critical stance in mathematics education. Besides the development of the mathematical and the technological knowing, a reflective knowing is developed with the mathematics curriculum.

If we compare Skovsmose's (1994) analysis of the concept *mathemacy* with D'Ambrosio's (1990) analysis of the concept ethnomathematics we can identify an interesting comparison. D'Ambrosio's (1990) analyses the concepts of ethnomathematics by looking at the three constitutive parts of the concept, namely *mathema, technés* (or *tics*) and *ethno*:

> *I call* ***mathema*** *the actions of explaining and understanding in order to survive. Throughout all our own life histories and throughout the history of mankind,* ***technés*** *(or* ***tics****) of* ***mathema*** *have been developed in very different and diversified cultural environments, i.e. in the divers ethnos. So, in order to satisfy the drives towards survival and transcendence, human beings have developed and continue to develop, in every new experience and in diverse cultural environments, their* ***ethno-mathematics.*** *(D'Ambrosio 1990: 369)*

In this analysis we can identify the traditional knowledge, a way of explaining and understanding in order to survive. Secondly we can observe the techniques, the applications and the models by which the mathematical knowledge is handled and practiced. Thirdly we enter the critical part that, in the analysis of D'Ambrosio, refers to the existence of the diverse ways in which all diverse mathematics are constituted, used and applied. This is D'Ambrosio's reflective part, the notion of *ethnos* that refers to the diverse nature of mathematics.

Conclusion

This paper considered the shifted meaning of ethnomathematics and its role within mathematics education. Ethnomathematics is no longer reserved for so-called nonliterate people; it now refers to the cultural diversity in mathematics education. Mathematics teachers are therefore challenged to handle pupils' diverse everyday mathematical practices. In line with the UNESCO declaration (1948) on education and the OEDC declaration (2004) on mathematical literacy, ethnomathematics clearly gained a more prominent role. Within Western curricula, ethnomathematics became meaningful to explore as an alternative, implicit philosophy of school mathematical practices. The extended notion of ethnomathematics as dealing with pupils' cultural diversity and with their everyday mathematical practices brings mathematics closer to the social environment of the pupil. Ethnomathematics is an implicitly value-driven program and practice on mathematics and mathematics education. It is based on an emancipatory and critical attitude that promotes emancipation and equality (Powell & Frankenstein 1997). Where the so-called academic Western mathematics still is locked in the debate on whether it is impartial or value-driven, the ethnomathematics' purposes stand out clearly right from the start. The historian of mathematics Dirk Struik postulated the importance of ethnomathematics. He validates ethnomathematics as both an academic and political program. There mathematics is connected to its cultural origin as education is with social justice (Powell & Frankenstein 1999: 418). D'Ambrosio even puts it more sharply: "Yes, ethnomathematics is political correctness (D'Ambrosio 2007a: 32)".

The implication for research is threefold. First, research has to reveal the (explicit and implicit) values within mathematics, mathematical practices and mathematics education. Second, research has to investigate thoroughly the use and integration of pupils' mathematical practices in the curriculum. Third, pupils' daily mathematical practices have to be studied.

Notes

[1.] Article 26. (1) Everyone has the right to education. Education shall be free, at least in the elementary and fundamental stages. Elementary education shall be compulsory. Technical and professional education shall be made generally available and higher education shall be equally accessible to all on the basis of merit. (2) Education shall be directed to the full development of the human personality and to the strengthening of respect for human rights and fundamental freedoms. It shall promote understanding, tolerance and friendship among all nations, racial or religious groups, and shall further the activities of the United Nations for the maintenance of peace. (3) Parents have a prior right to choose the kind of education that shall be given to their children. Article 27. (1) Everyone has the right freely to participate in the cultural life of the community, to enjoy the arts and to share in scientific advancement and its benefits. (2) Everyone has the right to the protection of the moral and material interests resulting from any scientific, literary or artistic production of which he is the author. (United Nations Educational, Scientific and Cultural Organization. 1948)

NOTES

References

Alrø, H., Ravn, O. & Valero, P. (eds.) (2010). *Critical Mathematics Education: Past, Present, and Future. Festschrift for Ole Skovsmose.* Rotterdam: Sense Publishers.

Apple, M. W. (1992). Do the Standards go far Enough? Power, Policy, and Practice in Mathematics Education. *Journal for Research in Mathematics Education, 23*, 412-431.

Ascher, M. & Ascher, R. (1997). Ethnomathematics. In A. Powell & M. Frankenstein (eds.), *Ethnomathematics, Challenging Eurocentrism in Mathematics Education* (pp. 25-50). Albany: State University of New York Press.

Bazin, M. & Tamez, M. (2002). *Math and Science A cross Cultures. A ctivities and Investigations from the Exploratorium.* New York: The New Press.

Ben-Ari, R. (1997). Complex Instruction and Cognitive Development. In E. Cohen & R. Lotan (eds.), *Working for equity in heterogeneous classrooms: sociological theory in action* (pp. 193-206). New York: Teachers College Press.

Bishop, A. J. (1997). *Mathematical Enculturation, A Cultural Perspective on Mathematics Education.* Mathematics Education Library, Vol. 6., Dordrecht: Kluwer Academic Publishers.

Bishop, A. J. (2002). Research Policy and Practice: The Case of Values. In P. Valero & O. Skovsmose (eds.), Mathematics Education and Society. *Proceedings of the Third International Mathematics Education and Society Conference MES3*, (sec. ed.), 2 Vols (pp. 227-233). Denmark: Centre for Research in Learning Mathematics,

Bishop, A. J., Clarke, B., Corrigan, D. & Gunstone, D. (2006). Values in Mathematics and Science Education: Researchers' and Teachers' Views on the Similarities and Differences. *For the Learning of Mathematics, A n International Journal of Mathematics Educatio, 26*(1), 7-11.

Burton, L. (ed.). (2003). *Which way social justice in mathematics education?* Westport, Conn.: Praeger.

César, M. (2009). Listening to different voices: Collaborative work in multicultural maths classes. In M. César, & K. Kumpulainen (eds.), *Social Interactions in Multicultural Settings* (pp. 203-233). Rotterdam: Sense Publishers.

Cohen, E. & Lotan, R. (eds.). (1997). *Working for Equity in Heterogeneous Classrooms: Sociological Theory in Action.* New York: Teachers College Press.

D'Ambrosio, U. (1990). The History of Mathematics and Ethnomathematics. How a Native Culture Intervenes in the Process of Learning Science. *Impact of Science on Society, 40*(4), 369-377.

D'Ambrosio U. (1997). Ethnomathematics and its Place in the History and Pedagogy of Mathematics. In A. Powell & M. Frankenstein (eds.), *Ethnomathematics, Challenging Eurocentrism in Mathematics Education* (pp. 13-24). Albany: State University of New York Press.

D'Ambrosio, U. (2007a). Peace, Social Justice and Ethnomathematics. *The Montana Mathematics Enthusiast, Monograph 1*, 25-34.

D'Ambrosio, U. (2007b). Political Issues in Mathematics Education. *The Montana Mathematics Enthusiast, Monograph 3*, 51-56.

Ernest, P. (2003) *The Philosophy of Mathematics Education, Studies in Mathematics Education*, London: Routledge Falmer.

François, K. & Bracke, N. (2006). Teaching Statistics in a Critical Way: Historical, Philosophical and Political Aspects of Statistics. *Proceedings of the 7th International Conference on Teaching Statistics ICOTS-7*. 2-7 July 2006, Brazil.

François, K. & Van Bendegem, J.P. (eds.). (2007). *Philosophical Dimensions in Mathematics Education*. New York: Springer.

François, K. (2007). The Untouchable and Frightening Status of Mathematics. In K. François & J.P. Van Bendegem (eds.), *Philosophical Dimensions in Mathematics Education* (pp 13-39). New York: Springer.

Gordon, P. (2004). Numerical Cognition Without Words: Evidence from Amazonia. Science, 306, 496-499. First appeared in *Science Express*, online publication August 16th 2004.

Graham, B. (1988). Mathematical Education and Aboriginal Children. In A.J. Bishop (ed.), *Mathematics Education and Culture*. Educational Studies in Mathematics. 19 (pp. 119-135). Dordrecht: Kluwer Academic Publishers.

Gutstein, E. (2003). Teaching and Learning Mathematics for Social Justice in an Urban, Latino School. *Journal for Research in Mathematics Education, 34*(1), 37-73.

Gutstein, E. (2006). *Reading and Writing the World with Mathematics: Toward a Pedagogy for Social Justice*. New York: Routledge.

Gutstein, E. & Peterson, B. (eds.). (2005). *Rethinking Mathematics: Teaching Social Justice by the Numbers*. Milwaukee: Rethinking Schools.

Hirtt, N., Nicaise, I. & De Zutter, D. (2007). *De School van de Ongelijkheid.* Berchem: EPO.

Moreira, D. & Pires, G. (2006). Aspectos Etnomatemáticos na Educação das Crianças Ciganas. *Actas das Jornadas Caminhos da A lteridade: Os Ciganos na Península Ibérica* (pp.123-143). Lisboa: Câmara Municipal de Lisboa.

Neves, A. H. (1997). The Relationship of Talk and Status to Second Language Acquisition of Young Children. In E. Cohen & R. Lotan (eds.), *Working for Equity in Heterogeneous Classrooms: Sociological Theory in Action* (pp. 181-192). New York: Teachers College Press.

Organisation for Economic Co-operation and Development. (2004). *Learning for Tomorrow's World - First Results from PISA 2003*. Paris: OECD.

Pica, P., Lemer, C., Izard, V. & Dehaene, S. (2004). Exact and approximate arithmetic in an Amazonian Indigene Group. *Science, 306*, 499-503.

Pinxten, R. (1994). *Culturen Sterven Langzaam. Over Interculturele Communicatie*. Antwerpen-Baarn: Hadewijck.

Pinxten, R. & François, K. (2007). Ethnomathematics in Practice. In K. François & J.P. Van Bendegem (eds.), *Philosophical Dimensions in Mathematics Education* (pp 213-227). New York: Springer.

NOTES

Powell, A. (2002). Ethnomathematics and the Challenges of Racism in Mathematics Education. In P. Valero & O. Skovsmose (eds.), *Mathematics Education and Society. Proceedings of the Third International Mathematics Education and Society Conference MES3*, (sec. ed.), 2 Vols. (pp. 17-30). Denmark: Centre for Research in Learning Mathematics.

Powell, A. & Frankenstein, M. (eds.). (1997). *Ethnomathematics, Challenging Eurocentrism in Mathematics Education.* Albany: State University of New York Press.

Powell, A. & Frankenstein, M. (1997). Ethnomathematical Praxis in the Curriculum. In A. Powell & M. Frankenstein (eds.), *Ethnomathematics, Challenging Eurocentrism in Mathematics Education* (pp. 249-259). Albany: State University of New York Press.

Powell, A. & Frankenstein, M. (1999). In His Prime: Dirk Jan Struik Reflects on 103 Years of Mathematical and Political Activities. *Harvard Educational Review, 69*(4), 416-446.

Prediger, S. (2007) Philosophical Reflections in Mathematics Classrooms. In K. François & J.P. Van Bendegem (eds.), *Philosophical Dimensions in Mathematics Education* (pp. 43-58). New York: Springer.

Rawls, J. (1999). *A Theory of Justice.* (Rev. ed.), Cambridge: Harvard University Press.

Setati, M. (2002). Is Ethnomathematics = Mathematics = Antiracism? In P. Valero & O. Skovsmose (eds.), *Mathematics Education and Society. Proceedings of the Third International Mathematics Education and Society Conference MES3*, (sec. ed.), 2 Vols. (pp. 31-33). Denmark: Centre for Research in Learning Mathematics.

Sen, A.K. (1992). *Inequality Reexamined.* Cambridge: Harvard University Press.

Skovsmose, O. (1994). Towards a Critical Mathematics Education. *Educational Studies in Mathematics, 27*(1), 35-57.

Stathopoulou, C. (2006). Cultural and Cognitive Conflicts in School Context: Mathematics Education as an Acculturation Process for Romany Students. *Themes In Education, 7*(2), 131-143.

United Nations Educational, Scientific and Cultural Organization. (1948). *Universal Declaration of Human Rights*. Retrieved August 14, 2011, from http://www.unesco.org/education/information/50y/nfsunesco/doc/hum-rights.htm.

Valero, P. & Skovsmose, O. (eds.). (2002). Mathematics Education and Society. *Proceedings of the Third International Mathematics Education and Society Conference MES3*, (sec. ed.), 2 Vols. Denmark: Centre for Research in Learning Mathematics.

Bridging Policy and Practice with Ethnomathematics

Volume 1

Linda H.L. Furuto, Ph.D.
University of Hawai'i – West O'ahu
Mathematics/Science Subdivision
96-129 Ala Ike C104E, Pearl City, HI 96782
Email: lindafuruto@post.harvard.edu
Tel.: 808-454-4830
Fax: 808-453-6176

Abstract

In an effort to bridge policy and practice in diverse populations, research was conducted at U.S. higher educational institutions (Harvard, UCLA, University of Hawai'i) in the field of ethnomathematics. Ethnomathematics addresses first, how cultural values affect teaching, learning, and curriculum; and second, how mathematics education affects schooling process dynamics. A macroperspective provides a foundation for exploring the policy framework that will be bridged by ethnomathematics to explore practices at the school level (equity and quality education, building successful partnerships, mentoring in diverse populations). This has been accomplished primarily through the Hōkūle'a, a voyaging canoe internationally renowned for rekindling the Pacific tradition of celestial navigation to locations such as Tahiti, Rarotonga, Micronesia, Japan, and the U.S., and is preparing for a worldwide voyage in 2013 of which the author will be participating in. It is a vehicle to explore real-world mathematics applications in global communities, and represents resourcefulness, inventiveness, wisdom grounded in the past, and hope for the future.

He lawaiʻa no ke kai papaʻu, he pōkole ke aho;

he lawaiʻa no ke kai hohonu he loa ke aho.

A fisherman of the shallow sea uses only a short line;

a fisherman of the deep sea has a long line.

This ʻŌlelo Noʻeau (proverb) was the guiding saying of the research study. One whose knowledge is shallow does not have much, but one whose knowledge is deep has the ability to deepen perspectives through partnerships with culture and traditions that permeate pedagogy, practices, and values.

In an effort to bridge policy and practice in diverse populations, research focuses on a study conducted at U.S. higher educational institutions (Harvard University, UCLA, and the University of Hawaiʻi) in the field of ethnomathematics. Ethnomathematics refers to a broad cluster of ideas ranging from distinct numerical and mathematical systems to multicultural education, including race/ethnicity, socioeconomic class, and special needs (D'Ambrosio, 2001). It allows us to address how cultural values can affect teaching, learning, and curriculum; and how mathematics education can then affect political and social dynamics in the schooling process.

A macroperspective of mathematics at the national, state, and local levels provides a foundation for exploring the policy framework. This will then be bridged by ethnomathematics in order to view practices at the grassroots, school level. As examples of ethnomathematics, the University of Hawaiʻi – West Oʻahu has implemented the following: (1) equity and quality education, (2) building successful partnerships and practices, and (3) mentoring in diverse populations. This has been made possible by a National Science Foundation grant and has been accomplished primarily through the Hōkūleʻa, "star of gladness," a double-hulled Hawaiian voyaging canoe that is an integral part of research.

Policy Framework

A number of studies show Hawaiʻi's reference to national and international populations. The Programme for International Student Assessment (PISA) is a triennial survey of the knowledge and skills of 15 year olds. It is the product of collaboration between participating countries and economies through the Organisation for Economic Cooperation and Development (OECD). More than 400,000 students in 57 countries making up close to 90% of the world economy took part in PISA 2006. The focus was on science but the assessment also included reading and mathematics and collected data on student, community, and institutional factors that could help to explain differences in performance such as culture and family background (PISA, 2006).

The U.S. currently ranks 25th out of 30 OECD countries in mathematics. In the 2003 PISA study, although the U.S.'s ranking remained fairly static, 24th out of 29 OECD countries, the mean score dropped from 483 in 2003

NOTES

to 474 in 2006. By contrast, top ranking countries Finland and Korea increased (Finland 544 to 548; Korea 542 to 547). Mexico, ranked last both times, improved its score from 385 in 2003 to 411 in 2006. Although the U.S. relative ranking has stayed constant, the overall mean scores of the countries at the top and bottom increased (OECD, 2009).

We know the U.S. has challenges to overcome, but some states are performing extremely well such as Massachusetts. Near the other end of the spectrum is Hawai'i. The National Assessment of Educational Progress (NAEP) mathematics assessment is administered periodically to 4th and 8th graders. NAEP results serve as a common metric for all states and selected urban districts. In a state comparison of NAEP 8th grade mathematics average scaled scores from the 2007 assessment, Hawai'i's average scale score of 269 is significantly lower than the national average of 280. In fact, Hawai'i is ranked 47 out of 50 states (NCES, 2007). According to the University of Hawai'i Vice President of Community Colleges John Morton, "This [mathematics] is not a new problem, what is new is that we're working together" (Morton, J., Personal Conversation, October 10, 2011).

Hawai'i is the only statewide school district in the country, and it has functioned this way since before it became a state. Its first public school law in 1840 included a centralized structure. According to the University of Hawai'i Institutional Research Office, 77.0% of the 7,754 public, parochial, and independent school students go on to attend one of the University of Hawai'i 10 campuses (UH IRO, 2012). In spite of these illuminating statistics, the first statewide Mathematics Summit "Defining the Mathematics Crisis", designed to facilitate discussion and action on raising mathematics achievement levels of Hawai'i students to ensure college and career success was held in 2008. This is a collaborative effort between the major stakeholders in the state, the University of Hawai'i Systemwide Office and State of Hawai'i Department of Education. Each of these perspectives will be examined in further detail.

The goals of the continuing State of Hawai'i Mathematics Summit series are to: (1) improve the mathematics pipeline leading to career and college ready mathematics, (2) improve alignment of courses so that students transition smoothly between institutions and courses, and (3) prepare more qualified and effective mathematics teachers/instructors.

The State of Hawai'i Department of Education is doing its part to raise mathematics preparation of all graduates by revising mathematics benchmarks, quality education, and graduation requirements. Specifically, this includes an increase of 3 to 4 credits of mathematics for the graduating class of 2013. The goal is to have 50% of students achieve the DOE recognition diploma in 2013, and 80% by 2018. Students will need to achieve, at the minimum, the level of algebra II. Incentives for completing the DOE recognition diploma include scholarships, college admission with honors, placement into college level mathematics courses, and carpentry or dry wall apprenticeship program application test waivers. State Superintendent Patricia Hamamoto has clearly expressed, "We need to figure out how to reach our students that are different learners. Drill and kill doesn't work anymore. Our future teachers have to be solid in content and pedagogy. We spend too much money in the DOE on re-teaching our mathematics teachers" (Hamamoto, P., Personal Conversation, January 19, 2009).

In the past, many students taking algebra 2 went on to major in STEM fields. However, now, the students in algebra 2 classrooms comprise of a more diverse group of students who have differing learning styles. We need teachers to teach to the strengths of the students, and not rely habitually on methods that may have worked for the advanced students. Thus, given various career paths, how do we help students transition into University of Hawai'i colleges and universities? For students entering the

Volume 1

remedial and development courses, how does policy help them transition into college-level courses?

We need to improve the mathematics pipeline leading to career and college ready mathematics so that we not only increase the number of students earning degrees, but more specifically increase the number of STEM degrees. The University of Hawai'i systemwide strategic goal for degrees and certificates of achievement earned is an increase of 3-6% per year and an increase in the number of University of Hawai'i STEM degrees by 3% per year. Not enough students performing well in University of Hawai'i pipeline to higher level mathematics courses. Only 1/5 of entering University of Hawai'i community college students successfully completed a remedial/developmental mathematics course in Fall 2006. 5,538 students began in the entering cohort, and only 1,149 successfully completed at least one remedial/development course in their first academic year (UH IRO, 2012). There is a huge leak in the pipeline.

University of Hawai'i production falls significantly short in terms of annual teacher vacancies, and this has been a major area of focus in recent years. In AY2007-2008, there were 418 qualified mathematics teachers. However, in 2007-2015 the state is projected to need 1,432 qualified mathematics teachers, and last year the total University of Hawai'i systemwide production of mathematics education majors was 28. Only 47.8% of secondary mathematics courses are being taught by "highly qualified" teachers, as deemed by the Hawai'i State Teachers Standard Board (UH IRO, 2012). According to University of Hawai'i Vice President of Academic Planning and Policy Linda Johnsrud, the mental ability to do the process that comes with doing mathematics is paramount. "We don't need more human calculators; we need people who can think" (Johnsrud, L., Personal Conversation, October 10, 2011).

The major stakeholders in the State of Hawai'i are committed to taking action to improve the pipeline leading to "career and college ready" mathematics, improve the alignment of courses so that students transition smoothly between institutions and courses, and prepare more qualified and effective mathematics teachers. Given the policy framework outlined, ethnomathematics has been implemented as a tool to bridge policy with practice in order to accomplish these goals.

Overview of Ethnomathematics

The term *ethno* describes "ingredients that make up the identity of a group: language, codes, values, beliefs, community, class, food and dress, habits, and physical traits" (D'Ambrosio, 2001, p. 308). *Mathematics* expresses a "broad view of mathematics which includes arithmetic, classifying, ordering, inferring, and modeling" (D'Ambrosio, 2001, p. 308). An understanding of these terms allows teachers to expand their mathematics perceptions and more effectively instruct their students in a growing school climate of diversity.

Teachers and the public in general do not commonly say that mathematics and identity are connected. When teachers do acknowledge a link, they often engage their students in multicultural activities merely out of curiosity. Such activities usually

NOTES

refer to a culture's past and to cultures that are removed from that of the students in the class. An important component of mathematics education today should be to reaffirm the individuality and identity of students. As students experience mathematics in this manner, they not only learn to value mathematics but also develop a greater respect for those who are different from themselves.

Just as literacy has come to mean much more than reading and writing, mathematics must also be thought of as more than counting, calculating, sorting, and comparing. Today's children are living in a civilization that functions by mathematically based technology and unprecedented means of communication. A very clear picture of future employment opportunities is given by Robert Reich (1992) in *The Work of Nations*. This picture includes the need for a technologically capable work force whose members participate in the global economy and are able to create solutions to problems that currently do not exist, with technologies that have not yet been invented. A goal of mathematics education should be to build upon students' abilities to successfully use a diverse array of knowledge to solve problems and communicate their thinking as they gain an awareness of the capabilities and limitations of an interconnected global community (Ascher, 1987, 1991; Thomas, 1985, 1987).

Ethnomathematics encourages us to witness and attempt to understand how mathematics continues to be adapted and used by people around the world. Students should be encouraged to construct personal mathematical understandings and be able to explain their work. When all students' invention, experience, and application of mathematics are realized and respected, they are given equal opportunity for access and achievement (Harvard University Achievement Gap Initiative, 2008; Kyselka, 1987; UCLA Center X, 2008).

The mathematical ideas of the peoples of the Pacific region have often been overlooked, particularly in science, technology, engineering, and mathematics. Every cultural group develops its own ways and styles of explaining, understanding, and coping with their environment. In recent years, there has been recognition that among indigenous peoples there were other non-Western forms of knowledge, responses to survival, and wisdom. There is tremendous value in knowledge systems that back different societal structures (Harvard University Achievement Gap Initiative, 2008; Stroup, 1985; UCLA Center X, 2008).

Particularly in Pacific communities, it is central to understand how space and time are integrated. This is critical when we study peoples whose behavior is intimately related to the sea and to navigation as highlighted in the opening ʻōlelo noʻeau (proverb). Pacific communities relied heavily on canoeing and navigation to survive in the past, produced a technology that still resonates today, and exemplify the importance and processes of spatial and linear conceptualizations (Babayan, Finney, Kilonsky, & Thompson, 1987; Chauvin, 2000; Finney, 1979, 1986).

Ethnomathematics was forged in the experiences, reflections, and hopes for a better quality of life. In the next section, I will show how ethnomathematics may be used as a tool to connect policy and practice in U.S. higher educational institutions in the following three main areas: (1) equity and quality education, (2) building successful partnerships and practices, and (3) mentoring in diverse populations. Before delving into these, an overview of the University of Hawaiʻi – West Oʻahu (UHWO) is provided.

University of Hawaiʻi – West Oʻahu

Within the University of Hawaiʻi system there are 10 campuses, including three universities and seven community colleges. Headcount enrollment at the University of Hawaiʻi

10 campuses measured 53,526 in Fall 2011, and the University of Hawai'i - West O'ahu (UHWO) campus posted the largest enrollment increase in percentage terms of any campus in Fall 2011 at 21.3%. This is a product of UHWO moving into a new phase of growth in Fall 2007, with the transition from a two-year upper division campus to a four-year, comprehensive university with an emphasis on the liberal arts, serving professional, career-related, and applied fields (UHWO Mission Statement, 2011). This is an especially critical period to support and retain traditionally underrepresented populations, particularly in the science, technology, engineering, and mathematics disciplines.

To provide some general information, UHWO is the only four-year baccalaureate degree granting degree institution on the west side of the island of O'ahu. The new campus in the city of Kapolei is scheduled to open its doors in Fall 2012. UHWO is accredited by the Accrediting Commission for Senior Colleges and Universities of the Western Association of Schools and Colleges, and offers more than 20 bachelor and certificate programs. UHWO provides distance learning programs for residents on six of the eight inhabited islands, Hawai'i, Kaua'i, Lāna'i, Maui, O'ahu, and Moloka'i.

The average age of students is 31 years old, and the ethnic breakdown is as follows: Asian/Pacific Islander 11%, Caucasian 15%, Filipino 15%, Chinese 3%, Japanese 13%, Native Hawaiian 25%, Mixed 13%, and Other 5%. UHWO has the highest percentages of Native Hawaiians and Filipinos at any four year university, public or private, in the State of Hawai'i. Of the academic achievement statistics, the most striking differences occur in Native Hawaiian and Filipino groups. The ethnically diverse student body at UHWO will continue to grow as highlighted by projections that the population will increase by 216.1% in seven years from 866 students in 2006 to 1,872 students in 2013 (UH IRO, 2012). This provides an ideal setting to nurture, develop, and celebrate cultural heritage and identity in mathematics.

Equity and Quality Education

First, in terms of equity and quality education, the most effective programs address issues of support and retention with a systematic, multi-faceted approach involving a variety of teaching/learning strategies, integrated mentoring programs, and course-based learning assistance (Astin & Oseguera, 2005; Bok, 2006; Hurtado, 2001). In Spring 2008, UHWO opened its first Mathematics Center with the support of a National Science Foundation grant. The Mathematics Center has a three-fold purpose of providing academic mentoring, personalized tutoring, and research experiences, all of which are critical to achieve the goal of increased student success in college, particularly for traditionally underrepresented students.

The National Science Foundation funding has allowed UHWO to effectively address and work toward accomplishing the strategic plan in terms of immediate, intermediate, and long term mathematics goals. Tutoring has serviced the immediate needs of assisting students enrolled in mathematics courses. The UHWO Mathematics Center plays an instrumental role on campus since all UHWO students are required to take one college level mathematics course to fulfill graduation

NOTES

requirements. The intermediate and long term goals are to provide a bachelor's degree in middle and secondary mathematics education, along with relevant experiences, to address one of the State of Hawai'i's paramount challenges. The UHWO Mathematics Center is the central hub for mathematics.

The mission of the UHWO Mathematics Center is to uniquely: celebrate cultural heritage and diversity, honor common values, foster a global community of students and educators, encourage global sustainability, and catalyze positive change. There are four main principles that reflect the culture and traditions of the communities we serve. These principles inform and direct educational efforts, and focus on promoting a positive learning environment:

1. Navigation: Vision, exploration, education.
2. Honoring Elders: Tradition, heritage, kupuna, family, our first people.
3. Environment: Our sacred earth.
4. Island Living: Our Pacific ʻohana and our special multiethnic community.

According to ethnomathematics scholar Marcia Ascher, "Mathematical ideas involve number, logic, or spatial configuration, and in particular, the combination or organization of these into cultural systems or structures" (1998, p. 185). We have learned through ethnomathematics how deeply cultural values can impact teaching, learning, and curriculum; and how mathematics education can then affect political and social dynamics and policies in the schooling process.

Within the areas of navigation, honoring elders, environment, and island living, the central values are:

- Mālama: To care for
- Aloha: To love
- ʻImi ʻIke: To seek knowledge
- Lokomaikaʻi: To share with each other
- Naʻau Pono: To nurture a deep sense of justice
- Olakino Maikaʻi: To live healthy

Specifically, through mālama, each student is cared for individually and collectively as part of a larger organization. In aloha, love is the foundation to capitalize on unique contributions in the process of ʻimi ʻike or seeking knowledge. Lokomaikaʻi is necessary to share with and learn from each other in a mutual process of continual growth. In doing so, a bond is created between past, present, and future generations by nurturing a deep sense of justice, equity, and fairness in naʻau pono. Finally, the planet earth, like Hawaiʻi, is an island which is cared for and sustained through olakino maikaʻi. These are Hawaiian values, but they are also universal values. Their impact crosses state, national, and international boundaries and most importantly, instills in students a sense of global responsibility.

The UHWO Mathematics Center fosters an environment where diverse students learn to think critically, receive tutoring and mentoring support, hone skills, and prepare for jobs to ensure the future mathematics development of the State of Hawaiʻi and beyond.

Building Successful Partnerships and Best Practices

Second, ethnomathematics has provided a framework for building successful partnerships and practices. Collaboration has taken place between campus administration, the Pacific Islands Mathematical Association of Two-Year Colleges, Hawaiʻi Council of Teachers of Mathematics, Polynesian Voyaging Society, University of Hawaiʻi Systemwide Office and State of Hawaiʻi Department of Education. Each understands the importance of quality learning, teaching, and personal development.

For example, over the past five years, UHWO has been involved with the Polynesian Voyaging Society's Hōkūle'a. The Hōkūle'a, "star of gladness," is a double-hulled Hawaiian voyaging canoe that is an integral part of the UHWO Mathematics Center and connects students to cultural and historical traditions. The Hōkūle'a is internationally renowned for the role it has played in rekindling the Pacific Island tradition of non-instrument way-finding techniques that include celestial navigation to locations such as Tahiti, Rapa Nui, Rarotonga, Marquesas Islands, Samoa, Micronesia, Japan, and the U.S. mainland, and is preparing for a voyage around the world in 2013 of which the author will be participating in. Thousands of children around the world have connected to the Hōkūle'a in the last 35 years. She is a powerful vehicle to explore real-world applications of mathematics in the Hawaiian and global communities, and represents resourcefulness, inventiveness, wisdom grounded in the past, and hope for the future (Finney, Kilsonky, Somsen, & Stroup, 1986; Honolulu Star-Bulletin, December 26, 2007; Jonsson, 1988; Pacific Business News, December 20, 2007).

According to Hōkūle'a navigator Nainoa Thompson, "Our ancestors sailed across a vast ocean, one third of the earth's surface, and to accomplish this great feat they needed mathematical vision and tools to see islands over the horizon, the ability to plan intentional voyages of discovery, the discipline to train physically and mentally, the courage to take risks, and a deep sense of aloha to bind the crew together during the voyage. The values worked in the past and they will work today" (Thompson, N., Personal Conversation, October 29, 2011).

Ethnomathematics scholar Ascher continues, "Mathematics has no generally agreed-upon definition; it means to some degree what was included in their formal Western models of schooling" (1998, p. 2-3). Some attempts to define mathematics emphasize its objects of study and others its methods; some definitions are extremely narrow and others exceptionally vague and broad. Concerns about what mathematics is are generally in the domain of philosophers and the historians who write its history. Many societies, including Native Hawaiian, did not have a formal system of writing until Western contact. As a result, the category "mathematics" is Western and not necessarily found in traditional cultures. Such is the case in Hawai'i, where people for centuries performed complex calculations of trigonometric angles, wind speed, and geometrical star houses, yet there was no word for "mathematician" or "scientist" (Babayan, Finney, Kilonsky, & Thompson, 1987; Best & Hongi, 1906; Lewis, 1977).

In order to inculcate the value of their culture, UHWO students and future mathematics teachers are actively involved with voyaging and lesson planning. While sanding surfaces and varnishing rails of the Hōkūle'a, students learned about the connectivity of mathematics and ancient pathways based on celestial navigation. The Hawaiian star compass is the foundational element in navigation. The canoe is oriented to the rising and setting points of stars, and the distant horizon is divided into a compass comprising 32 equidistant directional points of 11.25°. Each point is the midpoint of a directional house, or hale, hence the 32 houses of 11.25° each divide and organize the 360° circular horizon into equally placed hale that the stars reside in (see figure 1).

NOTES

The star compass is like a mirror. Excluding the names of the four main cardinal points, the seven remaining names reflect each other in all four of the divisional quadrants. A star rising in one house travels a path that parallels the celestial equator - never crossing it and set in the same named house on the opposite side of the compass. For example, a star rising in the house of ʻĀina Koʻolau sets in the house of ʻĀina Hoʻolua. Ocean swells, however, always travel in a direction that moves from one horizon passing through the canoe and exiting in a direction that is directly opposite. For example, a swell that travels from the compass house ʻĀina Koʻolau moves in a direction that passes through the canoe and exits the horizon in the compass house of ʻĀina Kona.

Knowing the rising and setting points of the many different celestial bodies (sun, moon, stars, planets) combined with knowledge of swell patterns and its placement within the star compass helps the navigator orient the canoe on long journeys. Understanding the relationship between the 32 different houses of the star compass and the canoe is the fundamental first step in non-instrument, celestial navigation.

> *Upon learning about the mathematical relationships, one student commented: I didn't realize how much math had to be factored into going on a voyage…the amount of supplies, the dimensions of each canoe, location, and distance and travel time. It was interesting to learn about the historical aspects of the voyage. I want to learn math so I can learn more about the great feats of my ancestors (Bello, M., Personal Conversation, October 27, 2010).*

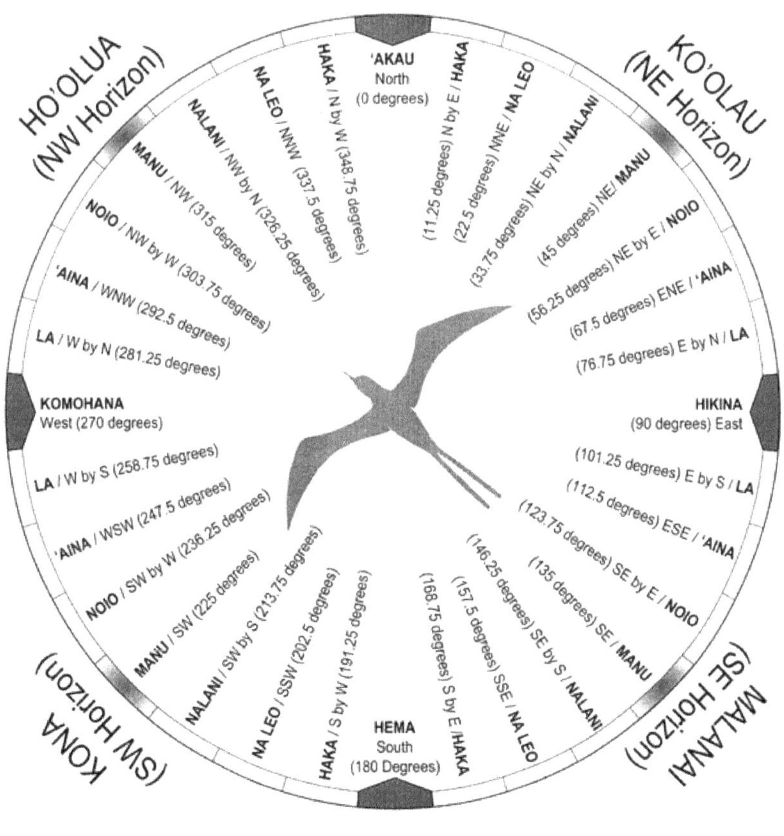

Figure 1

Through working together on real-world mathematics applications in addition to traditional class assignments, the student experience is enhanced through a culturally-sensitive mathematics curriculum in geometry, algebra, statistics, precalculus, and calculus. For example, vectors are discovered through sailing and navigating voyaging canoes first-hand; the Hawaiian Star Compass is used as a guide to study the trigonometric unit circle and apply half-angle and double angle formulas; geometry is investigated by maintaining and caring for Hawaiian fishponds; algebraic functions are studied while playing traditional Hawaiian Makahiki games (season to honor the Hawaiian god Lono) such as throwing spears at banana stalks and rolling stones between various configurations of pegs on the ground.

While it is very important to learn about textbook mathematics, it is equally critical to understand that there were great mathematicians that emerged from Hawai‘i and the Pacific Islands, including the students' ancestors.

Mentoring in Diverse Populations

Third, in a widely cited report, the National Center for Public Policy and Higher Education (2004) indicates that only 18 out of 100 students complete any type of postsecondary education within six years of graduating from high school. The outcomes of the Hawai‘i pipeline are below the national average with only 13 completing a postsecondary degree (UH IRO, 2012). The same study shows that mentoring and building relationships is one of the most effective ways to prevent attrition. Based on ethnomathematics, mentoring activities are a builtin component of developing mathematics at UHWO through cultural and historical-based field studies, and projects between future mathematics teachers in local high schools.

For example, the Polynesian Voyaging Society has invited UHWO prospective as well as current high school teachers to participate in designing educational curriculum for the Hōkūle‘a. With a legacy of ocean exploration as its foundation, the Polynesian Voyaging Society and UHWO reaffirm our commitment to undertake voyages of discovery, respect and learn from our heritage and culture, and strengthen learning which integrates voyaging experiences through mentoring experiences. There is a firm commitment to nurturing communities that value learning and share knowledge in order to foster living well on islands. Through mentoring and voyaging, we have come to celebrate cultural heritage and diversity, while honoring common values.

Among the products of the mentoring program between UHWO prospective high school teachers, current high school teachers, and the Polynesian Voyaging Society is an array of educational curriculum for future voyages included in the Starter Pack for Teachers. For example, these include development of a Hōkūle‘a education website, satellite phone calls, crewmember visits, canoe tours, curriculum resources, and teacher sails. The Starter Pack for Teachers was created to increase involvement with the voyage both on the canoe and in the classroom. There are multiple educational opportunities for teachers, students, and schools throughout the world. Through the Hōkūle‘a Education Program weblog online forums, a student space and teacher space, students and teachers from around

NOTES

the world can meet and get to know each other, and communicate and exchange ideas on important themes connected to the voyage such as environmental stewardship, cultural heritage, and sustainability related to mathematics.

Whenever possible, students and teachers have the opportunity, via satellite phone, to speak with and ask questions directly to crew members on the canoes as they travel across the Pacific. Upon arriving in each port, Hōkūle'a crew members visit with teachers, students, and schools for canoe tours and educational lectures. UHWO prospective teachers and current high school teachers also sail together on the Hōkūle'a to experience navigation first-hand. These travels have gone to extraordinary places in the Hawaiian Islands and have enabled a committed network of teachers to help Polynesian Voyaging Society design culturally-sensitive mathematics curriculum.

We seek to form a global community of students and educators to create a more sustainable world and cultivate positive change. In the process, this provides prospective UHWO mathematics teachers with the opportunity to make this dream a reality and accomplishes State of Hawai'i mathematics goals. The Hōkūle'a Education Program website serves as a communication platform to bring students and teachers worldwide together to share their ideas, knowledge, and mathematics perspectives that help foster a more integrated and sustainable world (http://pvs.kcc.hawaii.edu/hoonaauao/intro_hoonaauao.html). An additional resource is the Hōkūle'a Worldwide Voyage website (http://www.hokuleawwv.org/).

Hawai'i, our special island home, is a place where the land and sea are cared for, and people and communities are healthy and safe. As another example of mentoring in diverse populations, each semester UHWO mathematics students visit the Hawai'i Institute of Marine Biology, a world-renowned research institute situated on Coconut Island in Kāne'ohe Bay. This ideal location is just 15 miles from the main campus of the University of Hawai'i and downtown Honolulu, and allows students to learn more about intersections of culture, environmental awareness, conservation, tropical marine science, and mathematics in their Hawaiian backyard.

Coconut Island is surrounded by 64 acres of coral reef, designated by the State of Hawai'i as the Hawai'i Marine Laboratory Refuge. The island itself covers approximately 29 acres, with six acres enclosed in lagoons that are used for keeping organisms in captivity for study. Coconut Island provides research facilities for its faculty and students, as well as visitors, who come from all over the world. Through field studies to Coconut Island, students form mentoring relationships with researchers that allow them to consider STEM related fields for careers, future graduate work and studies, and lifelong learning.

This is evidenced by the comments of a student in a reflection following the field study to Coconut Island:

> *What made this trip interesting is the fact that we were taught how to use math formulas that we had learned in class in a real world situation, such as determining the population growth of a certain species through functions or mapping the reefs surrounding Coconut Island through linear equations. Overall this was probably one of the most practical ways for us as a class to see how math can be used in a real life situation, and personally it definitely beats a classroom setting any day of the week. However, what topped this experience off is the fact that we were able to interact with the ocean's sea creatures and learn about its previous inhabitants from our mentors. All in all, this was definitely a once in a lifetime opportunity that*

I'm glad I didn't miss (Pasion, M., Personal Communication, April 13, 2011).

Examples of student work may be found in research and practicum-based textbooks that have been distributed throughout the state, "Ethnomathematics Curriculum Textbook: Precalculus, Trigonometry, and Analytic Geometry" (2011) and "Ethnomathematics Curriculum Textbook: Algebra, Geometry, and Number Foundations" (2010). The textbooks serve to engage current and prospective mathematics teachers and mathematicians by continuing dialogue on critical themes in mathematics, supplementing mathematics curriculum, and enriching teacher training materials.

According to the National Council of Teachers of Mathematics' Position Statement on Equity in Mathematics Education, a culture of equity depends on the joint efforts of all participants in the community of students, educators, families, and policymakers, among which includes, "The school community acknowledges and embraces all experiences, beliefs, and ways of knowing mathematics… High expectations, culturally relevant practices, attitudes that are free of bias, and unprejudiced beliefs expand and maximize the potential for learning…All students have access to and engage in challenging, rigorous, and meaningful mathematical experiences" (January 2008). Such practices empower all students to build a relationship with mathematics that is positive and grounded in their own cultural roots and history. All members of the classroom group must accept the responsibility to engage with and support one another throughout the learning experience, which is a critical component of mentoring and fulfilling State of Hawai'i mathematics goals.

Further Discussion

The successes have been tremendous, and we continue to meet and overcome challenges. In the past couple of years, we have worked to develop a bachelor of arts degree in mathematics education to work towards fulfilling the teacher shortage in the State of Hawai'i at the middle and secondary levels (Harvard University Achievement Gap Initiative, 2008; Kyselka, 1987; UCLA Center X, 2008). Through implementing the best practices strategies, we have seen the proportion of underrepresented students double in mathematics courses, particularly for Native Hawaiian and Filipino populations. The overall passing rate has increased and almost doubled from 42.8% to 80.6%, far above the average passing rate for the University of Hawai'i system (UH IRO, 2012). Furthermore, more students are taking mathematics courses, necessitating additional courses in precalculus elementary functions and analytic geometry.

Challenges have included securing additional funding for development of the UHWO Mathematics Center with the increasing student body population. However, based on the overwhelmingly positive statistics, we have built partnerships to facilitate this growth, including local and statewide organizations. The National Council of Teachers of Mathematics states, "Mathematics performance and learning of groups that have traditionally been underrepresented in mathematics fields can be improved by interventions that address social, affective,

NOTES

and motivational factors. Recent research documents that social and intellectual support from peers and teachers is associated with higher mathematics performance for all students, and that such support is especially important for many African American and Hispanic students" (2008, p. 6).

There has never been a more important time to cultivate sustainable conditions that advance student success in college. Building relationships based on students' backgrounds is one of the most effective ways to increase student retention and success. This study began with a macroperspective analysis based on collaborative research with major stakeholders and policymakers in the State of Hawai'i. There are tremendous benefits of solidifying values, overcoming challenges, and working together with the University of Hawai'i Systemwide Office and State of Hawai'i Department of Education. An overview of policy provided a framework for understanding mathematics at the national, state, and local levels, including the importance of quality learning and teaching in personal development.

Ethnomathematics has proven to be an effective bridge between policies introduced at the macroperspective level and hands-on practices and learning experiences in the communities we are serving. As examples of best practices, this study covered three: (1) equitable and quality education, (2) successful partnerships and practices, and (3) mentoring in diverse populations. Based on principles of ethnomathematics, researchers at Harvard University, UCLA, and the University of Hawai'i explored examples of work at the UHWO Mathematics Center funded by the National Science Foundation; the Hōkūle'a, a double-hulled Hawaiian voyaging canoe that is an integral part of the center and connects students to cultural and historical traditions; and the Hawai'i Institute of Marine Biology's Coconut Island's fostering environment of student success, personal achievement, retention, and support (D'Ambrosio, 2001; Harvard University Achievement Gap Initiative, 2008; Honolulu Star-Bulletin, December 26, 2007; Pacific Business News, December 20, 2007; UCLA Center X, 2008).

It is hoped that students, educators, and communities will engage in communication that will allow us to collectively examine how universal values bind us together. The planet earth, like Hawai'i, is also an island which must be cared for and sustained instilling in each of us a sense of global responsibility. In the words of Polynesian Voyaging Society President and navigator Nainoa Thompson, "It is not about the canoe. The canoe is a piece of it. It is those beliefs and principles that you are willing to sail for. It is about the minds and hearts of children. I hope your voyages are a catalyst to change how to help children embrace the future, and that is why we are sailing around the world" (Thompson, N., Personal Conversation, October 29, 2011). Whether the journey is at the local, state, or national level, may we work together to be catalysts of positive change.

References

Ascher, M. (1987). "Mu Torere: An Analysis of a Maori Game." *Mathematics Magazine,* Vol. 60(2), pp. 90-100.

Ascher, M. (1991). Ethnomathematics: A Multicultural View of Mathematical Ideas. CA: Brooks/ Cole Publishing Company.

Ascher, M. (1995). "Models and Maps from the Marshall Islands: A Case in Ethnomathematics." *Historia Mathematica,* Vol. 22, pp. 347-370.

Ascher, M. (1998). Ethnomathematics: A Multicultural View of Mathematical Ideas. NY: Chapman & Hall/CRC.

Astin, A. & Oseguera, L. (2005). Degree Attainment Rates at American Colleges and Universities. CA: Higher Education Research Institute, Graduate School of Education, University of California, Los Angeles.

Babayan, C., Finney, B., Kilonsky, B., & Thompson, N. (1987). "Voyage to Aotearoa." *The Journal of the Polynesian Society*, Vol. 96(2), pp. 161-200.

Beckwith, M. (1932). Kepelino's Traditions of Hawaii. HI: Bernice P. Bishop Museum Bulletin 95.

Best, E. & Hongi, H. (1906). *Maori Numeration.* Paper presented at the Auckland Institute Conference, Auckland, New Zealand. October 3, 1906.

Bok, D. (2006). Our Underachieving Colleges: A Candid Look at How Much Students Learn and Why They Should be Learning More. NJ: Princeton University Press.

Chauvin, M. (2000). "Useful and Conceptual Astronomy in Ancient Hawaii." *Astronomy Across Cultures: The History of Non-Western Astronomy*. Vol. 1, pp. 91-125.

D'Ambrosio, U. (1984). The Intercultural Transmission of Mathematical Knowledge: Effects on Mathematical Education. Campinas: UNICAMP.

D'Ambrosio, U. (2001). Ethnomathematics Link between Traditions and Modernity. Rotterdam: Sense Publishers.

Economic Modeling Specialist Inc. (EMSI). (2007). Website: http://www.economicmodeling.com/.

Emory, K. (1924). The Island of Lanai: A Survey of Native Culture. HI: Bernice P. Bishop Museum Bulletin 12.

Finney, B. (1979). Hōkūle'a The Way to Tahiti. NY: Dodd, Mead, & Company.

Finney, B., Kilonsky, B., Somseon, S., & Stroup, E. (1986). "Re-Learning a Vanishing Art." *The Journal of the Polynesian Society*, Vol. 95(1), pp. 41-90.

Goetzfridt, N. (2008). Pacific Ethnomathematics A Bibliographic Study. HI: University of Hawai'i Press.

NOTES

Gutstein, E. & Peterson, B. (2006). Rethinking Mathematics: Teaching Social Justice by the Numbers. WI: Rethinking Schools, Ltd.

Hamamoto, P., Personal Conversation, January 19, 2009.

Harvard University Achievement Gap Initiative. (2008). Raising Achievement and Closing Gaps in Whole School Systems: Recent Advances in Research and Practice. MA: Harvard University Graduate School of Education Press.

Honolulu Star-Bulletin. *Hōkūle'a Helps in Teaching Math,* Website: http://starbulletin.com/2007/12/26/news/briefs.html, December 26, 2007.

Honolulu Star-Bulletin. *Hōkūle'a Powered by Math*, Website: http://archives.starbulletin.com/2008/03/16/news/story02.html, March 16, 2008.

Hurtado, S. & Schoem, D. (2001). Intergroup Dialogue: Deliberative Democracy in School, College, Community, and Workplace. MI: University of Michigan Press.

Johnsrud, L., Personal Conversation, October 10, 2011.

Jonsson, E. (1988). "The Carolinian Cane as a Navigation Instrument." *Pacific Island Focus*, Vol. 1(1), pp. 14-33.

Kyselka, W. (1987). An Ocean in Mind. HI: University of Hawai'i Press.

Lewis, D. (1977). "Mau Piailug's Navigation of Hōkūle'a from Hawai'i to Tahiti." *Topics in Culture Learning*, Vol. 5, pp. 1-23.

Morton, J., Personal Conversation, October 10, 2011.

National Center for Public Policy and Higher Education. (2006). Measuring Up 2006. Washington, D.C.

National Center for Education Statistics. (2007). National Assessment of Educational Progress Report. Washington, D.C.

Organisation for Economic Co-operation and Development. (2009). Annual Report. Paris, France.

Pacific Business News. *Math Students Will Use Voyaging Techniques*, Website: http://www.bizjournals.com/pacific/stories/2007/12/17/daily36.html, December 20, 2007.

Powell, A. & Frankenstein, M. (1997). Ethnomathematics: Challenging Eurocentrism in Mathematics Education. NY: State University of New York Press.

Programme for International Student Assessment. (2006). Science Competencies for Tomorrow's World, Volume I: Analysis. Washington, D.C.

Selin, H. (2001). Mathematics Across Cultures: The History of Non-Western Mathematics. MA: Kluwer Academic Publishers.

Stroup, E. (1985). "Navigating without Instruments: The Voyages of Hōkūle'a. *Oceanus,* Vol. 28(3), pp. 69-75.

Taylor, C. (1957). Hawaiian Almanac. HI: Tongg Publishing Company.

Thomas, S. (1987). The Last Navigator. NY: Henry Holt and Co.

Thomas, S. (1985). "The Sons of Palulap: Navigating without Instruments in Oceania." *Oceanus*, Vol. 28(1), pp. 52-58.

Thompson, N., Personal Conversation, October 29, 2011.

University of California, Los Angeles Center X. (2008). The Role that Teachers' Beliefs About Mathematics Play in Bringing About Change in the Elementary Mathematics Classroom: A Professional Development Model. CA: UCLA Graduate School of Education and Information Studies Press.

University of Hawai'i Institutional Research Office. (2012). MAPS Enrollment Projections, Website: http://www.iro.hawaii.edu/maps/mltitles.asp, January 15, 2012.

University of Hawai'i. (2007). Remedial and Developmental Education in the University of Hawai'i Community College System White Paper Group, Website: http://www.hawaii.edu/offices/cc/docs/remedialdev% 20task%20force/Remedial_Dev_Ed_Paper.pdf, October 29, 2007.

University of Hawai'i Council of Chief Academic Officers and the Office of Academic Planning and Policy. (2006). Measuring Our Progress, Website: http://www.hawaii.edu/ovppp/mop/mop06.pdf, October 29, 2007.

Zaslavsky, C. (1973). Africa Counts; Number and Pattern in African Culture. MA: Prindle, Weber, & Schmidt, Incorporated.

Public Good or Private Commodity? Mathematics Education in Japan and Implications for the U.S.

Linda H.L. Furuto
University of Hawai'i
Pearl City, HI
lfuruto@hawaii.edu

Abstract

Given the flow of global policy and educational reform, a number of countries are approaching junctures on issues of equal access and opportunity. This study begins with the major role supplemental mathematics education plays in Japan. Research was conducted as a visiting scholar at the University of Tōkyō, and supported by classroom experience. The contextual background then leads to a discussion on sector relationships in the U.S., where a culture has emerged of relying on privatized education such as academic tutoring and test preparation. Although educators in both countries question the impact of growing sector disparities, the reality is that they have just begun to address issues of equity and accessibility to ensure education is provided as a public good.

Keywords:

- Mathematics Achievement
- Cultural Diversity
- Equity

Introduction

Educational Equity

Given the flow of global policy and educational reform, a number of countries are approaching significant educational junctures on issues of equal access and opportunity. According to the National Council of Teachers of Mathematics (2008), further research is needed to prepare the next generation of scholars because "ensuring America's stature as the world's leader hinges on our ability to educate *all* students as we move into an era with increased technical demands" (p. 2). In order to examine the U.S. case, a cross-national comparison of mathematics education as a public good versus private commodity with Japan will be presented.

Theoretically, Japan provides a system of education open to all citizens. However, there is a growing tendency for the wealthy to opt for private schools and supplemental education that provide children greater opportunity to pass national examinations and eventually entrance to one of the prestigious universities. Rather than measuring the abilities of students in more than one endeavor, the Japanese system bases most of the selection process on a test score. *Juku* and supplemental education in Japan supply not only educational choice but a form of competition to the public school system as well (Russell, 2002). *Juku* are defined as special types of private cram schools that offer academic lessons conducted after regular school hours and on the weekends. These cram schools prepare students for higher levels of education, as well as rigorous national examinations (Russell, 2002).

According to recent Japanese government surveys, nearly 70% of students enroll in the private sector by the time they leave the ninth grade. The largest *juku* companies are traded on stock exchanges, and the industry generates 1.4 trillion yen (approximately $12 billion) in revenue annually (Yano Research Institute, 1994; LeTendre, 2001; DeCoker, 2002). Some 6.5 million children are estimated to attend *juku* in Japan, which has become an important consumer commodity for wealthier families. The growing acceptance of market-driven approaches in education continues to strengthen this trend.

This research comes at an opportune moment for Japanese and U.S. educators, policymakers, and researchers. Equity-based reform has consumed discussions of academic achievement in both societies, especially given that U.S. schoolchildren continue to lag behind international standards. The 2003 Third International Mathematics Science Study (TIMSS) resulted from the need for reliable and timely data on the mathematics and science achievement of U.S. students compared to that of students in other countries. TIMSS is the most comprehensive and rigorous assessment of its kind ever undertaken with a total of 41 participating countries. TIMSS director James Stigler concluded that instructional practices and expectations of all children regardless of cognitive factors is one of the leading variables in increasing academic achievement. (National Center for Education Statistics, 2003).

Even though almost every state in the U.S. is working to develop higher standards for what students should be learning, along with the means for assessing their progress, the quick-fix solutions implemented so far

NOTES

have not had a noticeable impact. In light of this, the U.S. has developed an emergent culture of relying on privatized education. A booming industry offers online diagnostic tests, private academic tutoring, SAT prep courses, and personalized guidance on how to navigate the college-application process. National Council of Teachers of Mathematics literature asserts that although the number of U.S. studies has grown in recent years due to changes in policies and priorities at federal agencies, these studies are only beginning to yield findings on privatized education and their number remains comparatively small (NCTM, 2008).

In this study, I begin with an introduction to the Japanese test preparation culture, followed by an in-depth presentation of supplemental education with qualitative and quantitative data. Research was conducted over several years as a visiting scholar at the University of Tōkyō, and complemented with hands-on experience at four public and private schools. I will focus on West Cram School, which is one of the largest providers of supplemental education in the country thus yielding valuable information. Data analysis raises issues of educational equity and the impact of sectors on the Japanese system. The contextual background leads to a discussion on implications for public and private relationships in the U.S., and the future of access to education. This research is a cooperative project, supported by the University of Tōkyō, Waseda University, Ōbirin University, University of California, Los Angeles, and the Japanese Ministry of Education, Culture, Sports, Science, and Technology (MEXT).

Supplemental Education

Japan supports a wide-range of academic institutions outside of the public school system. This dense network consists of home tutors, correspondence courses, *juku*, and examination preparation schools (*yobiko*), among others. These forms of extra schooling have been described as "shadow education" because their private sector curriculum tends to shadow and enhance the curriculum offered in the public schools (Rohlen, 1983; Stevenson & Baker, 1992).

There are two major types of *juku*, individual enrichment courses and academic *juku*. The academic *juku* are further divided into review *juku* which cater to students who need remedial assistance, and advancement *juku* which cater to students preparing for the entrance examinations. Review *juku* are popular with upper-elementary and junior high school students. Students in the latter *juku* receive basic help in subjects that they find difficult. Many of these *juku* are run out of private homes by teachers who use the same texts used in the classroom. The advancement *juku* are the ones that most closely match Western stereotypes. These schools are largely attended by junior high and high school students seeking to prepare for the upcoming entrance examinations. Moreover, there are schools that specifically prepare students for the college entrance examination (Stevenson & Baker, 1992; Tsukada, 1991). Students who fail to get into the college of their choice may spend a year or two studying at *yobiko* after graduating from high school. This is not uncommon since attendance at an esteemed university may be a life-altering course.

Individual enrichment courses, primarily non-academic in nature, are most popular among elementary and nursery school students. These *juku* teach activities such as swimming, piano, arts and crafts, dance, and calligraphy. Some Japanese consider them to be distinct from the academic *juku* because of the lack of rigor and discipline.

In a survey of over 60,000 students conducted by the Ministry of Education, Culture, Sports, Science, and Technology (MEXT, 2006), the percentage of students attending *juku* was documented over a twenty-year period. Enrollment in individual enrichment courses tended to decrease over time while enrollment in advancement courses increased. Enrollment also varied markedly according to the size and location of cities, with a much higher participation rate of students in large, metropolitan areas such as Tōkyō. Figure 1 shows the trends of percentage of students enrolled in *juku* by grade level.

Figure 1. Percentage of students enrolled in *juku* by grade level.
SOURCE: MEXT, 2006, p. 9.

Grade Level	Advancement *juku*		Enrichment *juku*	
	1985	2005	1985	2005
Elementary				
1	6.2%	12.1%	60.1%	70.1%
2	10.1	14.1	69.1	79.4
3	12.9	17.5	76.8	81.2
4	15.4	23.6	78.0	82.5
5	21.1	31.1	73.9	77.5
6	29.6	41.7	65.7	70.6
Junior High				
1	41.8	52.5	35.7	36.7
2	44.5	59.1	26.4	29.3
3	47.3	67.1	19.6	18.8

Impact on the Japanese Educational System

In a 1994 MEXT survey, students said the second most important benefit of attending *juku* was that they could "go beyond public school lessons" (p. 76). If students can surpass the standards, classroom study becomes a place of academic review. Academic *juku*, such as West School, provide lessons not offered by public schools. For example, some parents enroll their elementary school age children to study English, which doesn't become compulsory in public schools until later years. When *juku* accelerates students to the extent that school becomes a place for review, *juku* begin to impeded on public education, and the state system is no longer a place of educational equity.

As long as entrance examinations are key to academic and future career success, and *juku* can help students improve their scores, private educational services will continue to be in demand. They emphasize the tenuous authority of national standards when unofficial standards, such as a high-stakes entrance examination, set the learning bar higher.

The growing presence of private educational services in Japan may be seen in the trend toward private schooling in Japan since the mid-1990s. The most sought after schools combine the junior high and high school years into a combined six-year program, enabling students to attend the same school uninterrupted by an entrance exam that normally is taken at the end of Japanese junior high school. Such private schools claim that, in the absence of the test, they are freer to offer augmented curricular and cultural activities (i.e., food festivals, sports tournaments, and New Year's traditional events) and respond more sensitively to individual needs. Such schools also have a track record of placing their graduates in competitive universities. Amano (1996) demonstrates that the composition of students who gain admission to the most selective university in Japan has changed dramatically over the course of 25 years from predominantly public school graduates to predominantly private school students. Moreover, "of the top 20 high schools in the country that sent students to the nation's most prestigious University of Tōkyō in 1964, five were private schools. By 1979, the private school figure had grown to 11, and by 1989, 25 years later, 75% of the

NOTES

schools sending graduates to the University of Tōkyō were private and only five were public" (Amano, 1996, pp. 279-280). Based on Amano's data, if one accepts the premise that attending a test-focused *juku* is a requirement for successful admission to a private high school or selective university, it seems clear that studying at *juku* is an integral component of educational success in Japan (Amano, 1996).

Methodology

Research was conducted at North Elementary School, South Elementary School, East Elementary School, and West Cram School, with permission to use data granted by the respective schools and districts. The upper-elementary students and schools were selected because of the turning point in educational paradigms that occurs during the schooling and early- youth development of preadolescents (10-12 year olds). In upper-elementary school, students experience dramatic shifts in the type of instruction received, become aware of the expectations adults have for them, and undergo psychological and physiological changes. Japanese students are also subjected to powerful social pressures from peers and the media that mirror wider concerns in Japanese society. The complexity of these educational concerns, such as the tension between the ideal of learning as an exploration and learning as examination preparation, at this juncture makes change difficult. Thus, the experiences of Japanese students provide critical information on education as a public good versus a private commodity.

Survey analysis was the first methodological strategy employed to answer the underlying research question, what factors of public and private education affect mathematics achievement as observed in upper-level Japanese elementary schools? The quantitative survey took approximately 30 minutes to complete, and was based on the work of Helen Astin (2000), Alexander Astin (1987, 2000), Astin, Green and Korn (1987), and Derek Bok (1996) at the UCLA Cooperative Institutional Research Program. The UCLA Cooperative Institutional Research Program is an interdisciplinary center for research, evaluation, information, policy studies, and research training in education. I incorporated this body of knowledge into my survey because of the tools and resources they provided to conduct research at the institutional level, and the partnerships they developed with educational organizations based on longitudinal data.

My survey included demographic, attitudinal, family/community-related, and future-oriented objective questions. Data analysis of ordered logistic regression focused on the logit and probit versions of the ordered regression model, in terms of an underlying latent variable, selected to fit the types of data and variables under consideration. Figure 2 provides a description of major variables and measurements.

Figure 2. Description of major variables and measurements.

Variable Name	Description	Measurement
q00	School	1 = North Elementary 2 = South Elementary 3 = East Elementary 4 = West Cram School
q01	Grade Level	1 = Fifth Grade 2 = Sixth Grade 3 = Other
q02	Gender	1 = Male 2 = Female
q03	Age	1 = < 9 Years 2 = 9 Years 3 = 10 Years 4 = 11 Years 5 = 12 Years 6 = > 12 Years
q04	Anticipated Highest Academic Degree	1 = Junior College 2 = University 3 = Graduate/Professional School 4 = Other
q05	School-related Experiences	1 = Frequently 2 = Occasionally 3 = Never
q06	Future Occupation	1-18 = Various Occupations 19 = Other 20 = Haven't Decided
q07	Hours Spent Doing Weekly Activities	1 = None 2 = 1-2 Hours 3 = 3-5 Hours 4 = 6-10 Hours 5 = 11-15 Hours 6 = 16-20 Hours 7 = 20+ Hours
q08	Personal Importance of Values/Goals	1 = Essential 2 = Somewhat Important 3 = Not Important
q09	Items Possessed by Respondent	1-12 = Various Items

NOTES

Variable Name	Description	Measurement
q10	Rating of Respondent's Characteristics/Traits	1 = Upper 10% 2 = Above Average 3 = Average 4 = Below Average 5 = Lower 10%
q11	Elementary School Mathematics Grades	1= High 2 = Upper Middle 3 = Middle 4 = Lower Middle 5 = Lower 6 = Don't Know, Can't Remember
q12	Teacher Expectations and Views	1 = Frequently 2 = Occasionally 3 = Never
q13	Number of Years Attending *Juku*	1 = < 1 Year 2 = 1 Year 3 = 2-3 Years 4 = 4-5 Years 5 = 6-7 Years 6 = > 7 Years
q14	Grade Level Began Attending *Juku*	1 = First Grade 2 = Second Grade 3 = Third Grade 4 = Fourth Grade 5 = Fifth Grade 6 = Sixth Grade

Features of Japanese schooling were investigated secondarily with qualitative techniques. The focus groups were semi-structured, open-ended interviews, involving 4-5 participants. They were complementary to quantitative methods, providing further understanding of the organizational history and structure of schools, curriculum and instruction, assessment and evaluation, and their influences on mathematics achievement (Fetterman, 1998; Bickman, 1993, 1998; Rog, 1998). Some of the focus group questions included "What role do public and private school teachers play in the lives of students?" and "Is there a noticeable difference mathematics achievement between students who attend *juku* and those who don't?" Focus group interviews were conducted with 25 fifth and sixth grade students and 25 teachers/administrators at the four research sites, for a total of five student focus groups and five teacher/administrator focus groups.

Major Findings

Descriptive Analysis

Descriptive analysis illustrates general characteristics of the sample population of 278 fifth and sixth grade students at four schools. There was no missing data for any of the variables in the analyses. Figure 3 is a contingency table of grade level (q01) by school (q00). The chi- square test is not statistically significant, indicating there is no association between grade level and school attended. There is a fairly even representation of fifth and sixth graders across all four schools. Overall, 52.52% of the population is in the fifth grade (146 students) and 47.48% is in the sixth grade (132 students). North Elementary is coded as school 1, South Elementary is school 2, East Elementary is school 3, and West Cram School is school 4. 24.82% of the sample is from North Elementary School (69 students), 30.58% is from South Elementary School (85 students), 28.42% is from East Elementary School (79 students), and 16.19% is from West Cram School (45 students).

Figure 3. Contingency table of grade level (q01) by school (q00).

grade level	school				Total
	1	2	3	4	
grade 5	38 55.07	42 49.41	42 53.16	24 53.33	146 52.52
grade 6	31 44.93	43 50.59	37 46.84	21 46.67	132 47.48
Total	69 100.00	85 100.00	79 100.00	45 100.00	278 100.00
Pearson chi2(3) = 0.5347 Pr = 0.911					

Contingency tables of gender by school and age by school indicate that gender and age composition are relatively homogeneous across schools. Approximately half of the population sample is male (48.56%, 135 students), and the other half is female (51.44%, 143 students). Interestingly, there is a noticeable difference between male and female numbers at West Cram School, where 57.78% of the students are male and 42.22% are female. The majority of students are 11-12 years of age (97.84%), with a few 10 and 13 year olds. The ages are generally similarly distributed across schools. There tend to be slightly more 11 year olds (53.60%, 149 students) than 12 year olds (44.24%, 123 students).

The intraclass correlation of mathematics achievement (acadachvt) by school (q00) also provides useful information. Based on review of the literature and the UCLA Cooperative Institutional Research Program, the intraclass correlation (0.56) is substantial and shows that schools are significantly different across populations and fairly homogeneous within, as expected (Aldrich & Nelson, 1984; Astin & Astin, 2000; Bickman, Hedrick, & Rog, 1993; Bok, 1996). Figure 4 provides a contingency table of cross-tabulation and chi-square of mathematics achievement (acadachvt) by school (q00) that support intraclass correlation findings. The chi-square test is statistically significant at the 0.001 α-level ($p = 0.000$), indicating there is a strong association between mathematics achievement and school attended.

NOTES

Figure 4. Contingency table of mathematics achievement (acadachvt) by school (q00).

acadachvt	school				Total
	1	2	3	4	
1	3 4.62	3 3.90	35 52.24	0 0.00	41 16.27
2	19 29.23	55 71.43	29 43.28	3 6.98	106 42.06
3	43 66.15	19 24.68	3 4.48	40 93.02	105 41.67
Total	65 100.00	77 100.00	67 100.00	43 100.00	252 100.00
Pearson chi2(6) = 168.0325 Pr = 0.000					

Juku Data Analysis

The trends of students attending *juku* by school are revealing about the types of students that have access to supplemental education. Figure 5 is a cross-tabulation and chi-square of the number of years attended *juku* (q13) by school (q00). In each cell within the table, the top number is the frequency, the second number is the row percentage, and the third number is the column percentage. The Pearson chi-square value is given at the bottom, along with the significance probability. The chi-square test is significant at the $p < 0.001$ level, indicating that there is a relationship between schools (1 = North Elementary School, 2 = South Elementary School, 3 = East Elementary School, 4 = West Cram School) and the number of years students attend *juku*.

Figure 5. Cross-tabulation and chi-square of *juku* attendance (q13) by school (q00).

years *juku*	school				Total
	1	2	3	4	
0-1 yrs	51 24.88 73.91	66 32.20 77.65	68 33.17 86.08	20 9.76 44.44	205 100.00 73.74
2-5 yrs	17 26.15 24.64	18 27.69 21.18	11 16.92 13.92	19 29.23 42.22	65 100.00 23.38
6-7 yrs	1 12.50 1.45	1 12.50 1.18	0 0.00 0.00	6 75.00 13.33	8 100.00 2.88
Total	69 24.82 100.00	85 30.58 100.00	79 28.42 100.00	45 16.19 100.00	278 100.00 100.00
Pearson chi2(6) = 37.8335 Pr = 0.000					

The number of years attended *juku* is divided into three categories (0-1 years, 2-5 years, and 6-7 years). Attending *juku* for 2-5 years at the elementary school level is a good indicator of time, resources, and financial investment of parents in their children.

Most of the students in this category began attending *juku* when they were in the first or second grade. 26.15% of students at North Elementary School, 21.18% of students at South Elementary School, 13.92% of students at East Elementary School, and 42.22% of students at West Cram School attended *juku* 2-5 years.

In Figures 6 and 7, mathematics achievement (acadachvt) is the outcome variable and number of years attended *juku* (q13) is the predictor variable. Mathematics achievement is based on three point categories from 1 point (low) – 3 points (high) that reflect the elementary school mathematics grades of the student measured in the quantitative survey. The number of years attended *juku* (q13) proves a statistically significant predictor of mathematics achievement at the 0.05 α-level ($p = 0.049$). A one point change in the number of years attended *juku* is associated with a 0.94 increase in the logistical odds of being in a higher category of mathematics achievement. Thus, a one point change in number of years attended *juku* results in a significant difference on the outcome of mathematics achievement.

Figure 6. Ordered logistic regression of *juku* attendance (q13) and mathematics achievement (acadachvt).

Ordered logistic regression

Number of obs	=	252
Wald chi2(1)	=	3.87
Prob > chi2	=	0.0490
Pseudo R2	=	0.0285

Log pseudolikelihood = -250.81947

(Std. Err. adjusted for 4 clusters in q00)

acadachvt	Coef.	Robust Std. Err.	z	P>|z|	[95% Conf. Interval]	
years *juku*	.9392696	.4771544	1.97	0.049	.0040643	1.874475
/cut1	-.5067169	.8429937			-2.158954	1.14552
/cut2	1.54194	.4376586			.6841453	2.399736

The odds ratio provides information to determine the effect of mathematics instruction on mathematics achievement. A student's ability to move to the next higher category of mathematics achievement is 2.558, with a one point increase in number of years attended *juku*. Recall, mathematics achievement is a three point system reflecting elementary school mathematics grades, so a more than two point increase is extremely significant (Aldrich & Nelson, 1984; Astin & Astin, 2000; Bickman, Hedrick, & Rog, 1993; Bok, 1996).

Ordered logistic regression may also be expressed as a probability. Figure 7 provides predicted probabilities of mathematics achievement given the number of years attended *juku*. The predicted probability of a student with 0-1 years of *juku* having a low level of mathematics achievement is 0.1906,

NOTES

while a student with 6-7 years of *juku* having a low level of mathematics achievement is much less at 0.0347. On the opposite end of the spectrum, the predicted probability of a student with 0-1 years of *juku* having a high level of mathematics achievement is 0.3537, while a student with 6-7 years of *juku* having a high level of mathematics achievement is more than double that amount at 0.7817. Overall tests of the parallel regression/proportional odds and other assumptions were met.

Figure 7. Predicted probabilities of mathematics achievement (acadachvt) based on number of years attended *juku* (q13).

Number of years attended *juku*	Predicted Probability of Outcome 1	Predicted Probability of Outcome 2	Predicted Probability of Outcome 3
1	0.1906	0.4556	0.3537
2	0.0843	0.3323	0.5834
3	0.0347	0.1835	0.7817

The analysis confirms that educational equity in Japan is affected by the ability of the middle- and upper-classes to pay for the extra schooling in private *juku* that less wealthy parents are unable to provide. *Juku* have become such an indispensable part of a student's college preparation that those who can afford it attend. The most successful *juku* (measured by the number who pass the elite universities' exams) have become so profitable that they are able to hire the best teachers and use the newest technology to evaluate and monitor performance.

West Cram School

West Cram School is among the most prestigious and reputable cram schools in all of Japan. Over a period of several years, I worked at the West Cram School in Shibuya primarily with the Director of Operations and the General Manager. As one of the largest providers of supplemental education in the country, research at West Cram School has much to offer in terms of understanding the intersections of private and public sectors in Japan. I will begin with a general overview of the West Cram School and its educational strategies.

There are 37,500 fourth through sixth grade students enrolled in West Cram School nationwide. Approximately 12,000 are in the fourth grade, 13,000 are in the fifth grade, and 12,500 are in the sixth grade. Of that figure, 29,000 of the students reside in the Tōkyō area. West Cram School is divided into three main regions: Tōkyō (Chiba, Kanagawa, Saitama), Kansai (Ōsaka, Kobe), and Kyūshū (Fukuoka, Kagoshima). In Tōkyō, more than 20% of students in the fourth through sixth grades attend *juku*. Each year, West Cram School provides 33.5% of the students to the top ten private junior high schools in the Tōkyō area (Gonzales, Calsyn, Jocelyn, Mak, Kastberg, Arafeh, Williams, & Tsen, 2001; Nichinōken, 2007). This is the highest and most distinguished figure in the country. Most students attend public elementary schools, but their parents send them to private junior high schools, high schools, and universities.

Students generally attend West Cram School three times per week (Monday, Wednesday, Friday) from 16:50-20:45. There are three 70-minute periods with a five minute break between Periods 1 and 2, and a 15-minute dinner break between Periods 2 and 3. The subject content mainly focuses on mathematics, social studies, science, and Japanese national language (kokugo). Tuition is based on the number of subjects taken (two or four), grade (fourth, fifth, or sixth), ability tracking group (red—highest, white—middle, or gray—lowest), and the number of months students enroll in West Cram School. For example, a sixth grade student in the red group, enrolled in four classes, and attending classes year-round (there are 40 days of summer vacation and students attend West Cram School 20 of those days), pays a monthly tuition of approximately U.S. $800-1000. The tuition covers items such as registration, test fees, textbooks, science lab fees, handouts, and other resources.

The diagram in Figure 8 represents themes and periods of change within the West Cram School education strategy as described by the General Manager. In 1980, the focus was on passing the examination, followed by a shift in the 1990s to preparing students for a successful higher education transition. The drop in 1995 is an effect of shifting to a business model that was based on accruing as many students as possible. In 2000, West Cram School aspired to increase learning and the quality of education. That is the trend West Cram School has been continuing through 2009. The slight decrease toward the end is because West Cram School sees room for improvement in education for all students rather than the overreliance on entrance examination-driven preparation.

Figure 8. Education strategies and foci of West Cram School

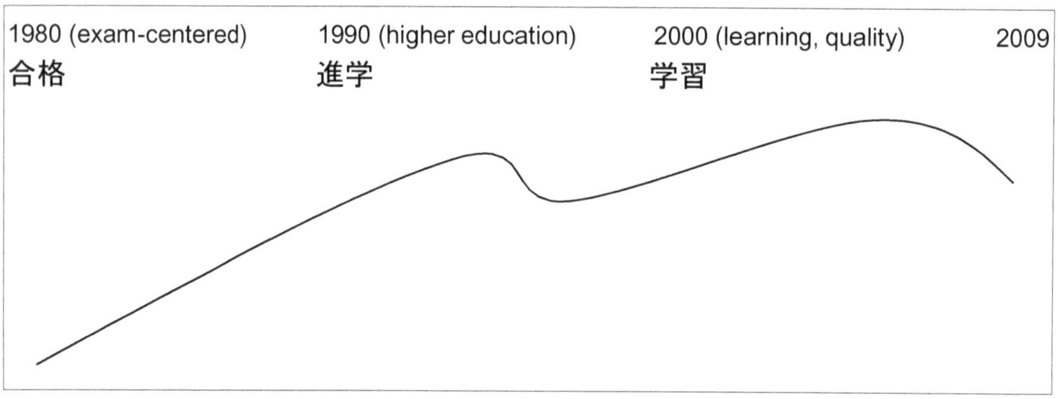

In each period, the West Cram School aimed to fulfill its mission of "meeting the needs of students through complementary education to their six years of elementary schooling" (Asai, M., personal communication, March 1, 2006). Today, West Cram School realizes there is still a strong focus on examination preparation, however, in light of global pressures to make education accessible to all there is an emphasis on quality. Teachers and staff strive to promote the character of students in relation to the greater society through structural and pedagogical strategies, such as the "Inventing the Future" Project where students discuss values, personal philosophies, and how these will affect their future careers and life decisions. As a leader in academic achievement, especially mathematics and sciences, developments in Japan have an impact on U.S. education. Several applications will be explored in the next section on sector relationships in the U.S.

NOTES

Sector Relationships to U.S. Education

The common response of *juku* and other forms of supplemental education is to cater to parents' concerns that official standards are growing more distant from the entrance examination realities of Japanese and U.S. universities. The rising attendance at the Japan's most prestigious universities by students from wealthier families undermines the egalitarian thrust of the educational system, and gives us reason to examine the U.S. case more closely (Beauchamp, 1991, 2003; Rosenbaum & Kariya, 1989). The result is that parents bridge the gaps of public schooling through the private purchase of supplemental education lessons. Privatization has created a system for the purpose of passing entrance examinations in Japan, and is forging a growing market in the U.S. Three areas that may be applicable to education in the U.S. are curriculum and assessment, teacher/parent expectations, and mathematics resources.

Curriculum and Assessment

The public sector, following a uniform curriculum through grade nine, allows for the outward appearance of equity, while the private sector, in its parallel curriculum of supplemental study, allows for the reality of a competitive world where students strive to get ahead. Since there is no direct connection between the private sector and the MEXT, the curriculum, textbooks, and lessons are designed and created by West Cram School teachers and staff generally based on entrance examinations into prestigious junior high and high schools. Curriculum is designed to push students to excel, rather than catering to the lower-achieving students. Classes at West Cram School are about one semester (minimum) - one year (maximum) faster than their public elementary school peers. For example, the fifth graders in the West Cram School class I observed were studying the same materials on division and multiplication of fractions as sixth graders at their counterparts at a nearby public elementary school.

Mastering the curriculum taught in supplemental education very often determines a student's future employment and lifestyle. There is intense pressure to succeed on achievement tests, culminating in the university entrance examination. To test well is critical for Japanese students, as universities in Japan almost exclusively base their admissions decisions on test scores. West Cram School and most privatized systems in Japan utilize weekly and monthly tests that result in the shifting of ability groups. The achievement tests assess proficiency and knowledge content of national entrance examinations, in addition to West Cram School curriculum.

The U.S. phenomenon follows this example in some respects. As in Japan, private tutoring has taken off in the United States due to the competitive pressures of high-stakes examinations. The two leading test-preparation firms in the United States, Princeton Review and Kaplan Test Prep, profit from the weight given to SAT scores in college admissions offices. U.S. parents see higher education as key to the future economic success for their children, and tutoring services benefit from the inability of public schools to address a controversial issue such as test preparation with the single-mindedness of a for-profit business.

However, there are some unique differences in the private-tutoring phenomenon in the U.S. Firms sell enrichment, remedial, and test-preparatory services, but also strengthen their product lineup with services that meet the demands of a U.S. market. Princeton Review, Kaplan, and other test preparation companies offer college guidance packages. These take advantage of parental anxiety and the inability of over-burdened high school counselors to provide close assistance in a process that is growing more competitive and strategic.

Teacher/Parent Expectations

Teachers and supervisors at West Cram School frequently take on responsibilities in addition to their regular course load. Through various methods of observations, teachers know the capabilities, learning styles, and personalities of each student. Meetings are regularly scheduled to discuss the character, problem-solving style, attitude toward learning, and behavior of individual students. After combining this assessment with the performance on the four core subjects (mathematics, social studies, science, and Japanese national language), teachers determine how to best help students. The strength of this relationship transfers into personalized support and tailored expectations. On a personal level, teachers know where students live, their family backgrounds, academic challenges, and future goals. For example, a map in the main room at West Cram School pinpoints to where each student's home is located in the vicinity. Parents trust the teachers and staff at West Cram School because their children are given such close, supervised attention.

The impact of family expectations and influence on student performance should not be underestimated. Almost immediately after birth, West Cram School parents generally take a strong, vested interest in the academic development of their children. According to General Manager Ōta Katsumi, "Mothers often spend hours on academic games and other activities supportive of learning" (Ōta, K., personal communication, January 13, 2006). It is common for West Cram School parents to buy books for preschoolers, including those written in English to cultivate language skills at an early age. Most West Cram School children know how to read and write before they enter elementary school, and can do simple computations.

Once a child enters elementary school, the maternal involvement increases significantly as evidenced by quantitative and qualitative data collection. There are classes for mothers called *mama juku* that prepare the mothers in subjects their children are studying at West Cram School (Stevenson, Azuma, & Hakuta, 1986; White, 1987). Collaborative homework is considered a means for developing a sense of responsibility in the child and for molding character. This constant striving by mothers for improved performance is present throughout a child's schooling years up until university, and it is intensified by students in the supplemental education system.

In contrast to Western theories of achievement which tend to emphasize individual effort and ability, the Japanese consider academic achievement to be an outgrowth of a cooperative effort and planning. In other words, if students work hard and stay on task, they can learn. Teachers and parents at West Cram School consistently uphold the belief that "individual effort is of greater significance than individual ability" (Asai, M., personal communication, March 1, 2006). Students generally strive for, and meet, these high expectations that are not as deep-rooted in the U.S. The National Mathematics Advisory Panel (2008) advocates increased research in collaborative instruction and learning, especially through the involvement of parents, teachers, and other educators (p. 1).

NOTES

Mathematics Resources

Access to resources within the West Cram School and supplemental education system is another area where differences between public and private sectors are brought to the forefront. West Cram School provides textbooks, specialized libraries, facilities, and other materials that aren't present at public elementary schools many of the children attend. These resources give students space to link the lessons to previous knowledge, control what parts of learning to use, and develop the skills to explain concepts intuitively.

In the field of mathematics, students at West Cram School have the opportunity to pursue the study of logic and critical thinking in this type of environment. Fifth grade mathematics teacher Mr. Asano regularly allows his students time to brainstorm, probe arguments, and discuss solutions. For example, one day the topic was calculating rate based on distance and time. Mr. Asano opened the class with, "Tell me what you know about rate and connect this to real-world applications" (Asano, T., personal communication, January 27, 2006). As previously explained, there are three 70 minute periods at West Cram School per evening. The first 70-minute period was solely focused on the formula, rate = distance ÷ time (速さ ＝ 距離 ÷ 時間). This solidified the foundation for proceeding to the average speed of objects during the second 70-minute period of mathematics. Students began by choosing seven moving objects and ranking them from slowest to fastest. Then, the students were allowed to use resources within West Cram School— computers, books, their notes, the internet, compasses, rulers, protractors, teachers/staff, and other mathematical tools—to calculate the speed. The groups creatively chose animals (in order, snail, sloth, dog, grasshopper, horse, rabbit, cheetah), and transportation modes (in order, cow, horse, car, boat, bullet train, airplane, space shuttle). To measure speed, the first group combined innovative resources with mathematical reasoning. They took a polygon shape to represent a snail, and moved it slowly along a ruler to measure speed and distance. One member of the group timed the movement on a West Cram School stopwatch and another confirmed distance based on Cartesian coordinates. This is an example of the types of engaging and thought- provoking experiments conducted by students under the supervision of their teachers. Students were encouraged to develop a deeper comprehension of the elementary concept.

In the National Council of Teachers of Mathematics Address "Foundations for Success" there are a number of references to mathematics resources, with an emphasis on textbooks. U.S. mathematics textbooks are extremely long—often 700-1,000 pages (p. 7). Excessive textbook length often detracts from applications such as those present at West Cram School, and can contribute to a lack of coherence. Mathematics textbooks are much smaller in many nations such as Japan with higher mathematics achievement than the U.S., thus demonstrating that the great length of our textbooks is not necessary for high achievement. Representatives of several publishing companies who testified before the panel indicated that one substantial contributor to the length of the books was the demand of meeting varying state standards for what should be taught in each grade. It is recommended that publishers make every effort to produce shorter and more focused mathematics textbooks.

Similar to the Japanese system, access to resources is another arena where the public and private sector disparities are highlighted. Because of their private nature, U.S. supplemental education programs can delve further into subject matter. Unlike public schools, they can be non-ideological about controversial topics such as preparing for high-stakes exams, for example, and teaching to the test.

Raising the level of achievement through curriculum and assessment, teacher/parent expectations, and mathematics resources are three applications that form a basis for Japanese and U.S. sector comparisons. There

Volume 1

is much the U.S. can learn from trends in the Japanese educational system. In Japan, *juku* and the form of specific, packaged education they provide are so essential to preparatory education that some educators have gone so far as to propose ending public support of education at the elementary and secondary levels, allowing the *juku*-type education to replace public schools (Cummings, 1980; Horio, 1994; Okano & Tsuchiya, 1999). While this is not expected to occur in the U.S., the benefits of supplemental education based on the three areas of comparisons are consistently documented in the records of entrance examinations into junior high school, high school, and university (Aso & Amano, 1983; Cummings, Beauchamp, Ichikawa, Kobayashi, & Ushiogi, 1986; Tsukada, 1991).

Further Discussion

In order to explore the research question of mathematics education as a public good or private commodity, I began with a portrait of the major role supplemental education plays in ensuring the success of Japanese students, both on tests administered within the country and international comparisons made on the basis of mathematics achievement test scores. The growing pressure to succeed on examinations is taking a toll in Japan, with educators questioning its impact, both on growing disparities between the wealthy and poor and the fact that tests are starting to determine curriculum.

Data collected at West Cram School, one of the largest and most reputable providers of supplemental education in the country, shows the degree to which the private sector is influential in shaping Japanese education. Upper level elementary school students (fourth – sixth grades) were selected based on the experiences in the rite of passage into adolescence and adulthood. Understanding how children balance culture, peers, academics, and other processes offers a valuable opportunity to examine their impact on educational achievement. This was accomplished by observing academic classes, extracurricular activities (i.e., sports practices, music rehearsals, tea ceremonies, language clubs, etc.), supplementary education institutions, and school cultural festivals.

The trends of students attending *juku* by school were very revealing about the types of students with access to supplemental education. Number of years attended *juku* is a good indicator of time, resources, and financial investment of parents in their children. 26.15% of students at North Elementary School, 21.18% of students at South Elementary School, 13.92% of students at East Elementary School, and 42.22% of students at West Cram School had attended *juku* 2-5 years. Ordered logistic regression showed that there was a statistically significant relationship between mathematics achievement and the number of years students attended *juku* at the 0.05 α-level ($p = 0.049$). The analysis highlighted the effect of socioeconomic class and the ability to pay for the extra schooling in the private sector on mathematics achievement.

Qualitative data also sharpened recurring questions about social inequality. Participant observations and focus group interviews provided a lens to view three examples of the symbiotic relationship between the public and private sectors in the Japanese and U.S. educational systems: curriculum and assessment, teacher/parent expectations, and mathematics resources. Although there

NOTES

appears to be growing concerns about the *juku* in Japan and the incredible pressure placed on Japanese students, there is still widespread support among the population for these schools. Even with varying enrollment according to the size and location of cities, with a much higher participation rate of students in metropolitan areas, over 40% of sixth graders nationwide attend *juku* (MEXT, 2006). Situations are developing in which private tutoring businesses may help to support official standards in the U.S., following Japan's lead. Private U.S. companies see a potential market in the nation's public school systems whose students perform poorly on standardized tests. Such tests are increasingly administered by state education departments to monitor school instruction and student achievement. Business firms hope to sell test preparation programs that will strengthen U.S. student performance, particularly in mathematics and science fields.

Throughout research, concurrent enrollment as a Visiting Scholar at the University of Tōkyō and maintaining connections with the four research sites provided complementary macroperspective analysis and access to databases at the national level, as well as hands-on learning within the classroom environment. This study was a cooperative project, supported by the University of Tōkyō, Waseda University, Ōbirin University, University of California, Los Angeles, and the MEXT.

Parents in both Japan and the U.S. seem committed to spending the money and time having their children enrolled in supplemental education classes to give them a competitive advantage in the testing to determine which high schools, colleges, and universities they may attend. Although the MEXT and U.S. Department of Education seem to be concerned about the greater pressure being placed on students to succeed on tests, the reality is that they have just begun to address issues of equity and accessibility as seen through curriculum, assessment, teacher/parent expectations, and increased mathematics resources, in a system that will ensure education is provided as a public good (NAEP, 2008; MEXT, 2006).

Volume 1

References

Aldrich, J. & Nelson, F. (1984). *Linear Probability, Logit, and Probit Models: Quantitative Applications in the Social Sciences*. Newbury, California: SAGE Publications, Inc.

Amano, I. (1990). *Education and Examination in Modern Japan*. Tōkyō, Japan: Tōkyō University Press.

Aso, M., & Amano, I. (1983). *Education and Japan's Modernization*. Tōkyō, Japan: Japan Times.

Astin, A., & Astin, H. (2000). *Leadership reconsidered: engaging higher education in social change*. Battle Creek, MI: W.K. Kellogg Foundation.

Baker, D. (2001). "TIMSS-R: Innovation in international information for American educators." *Education Statistics Quarterly, 3*(1), 17-19. Washington, D.C.: National Center for Education Statistics.

Beauchamp, E. (ed.). (1991). *Windows on Japanese Education*. Westport, CT: Green Press.

Beauchamp, E. (2003). *Comparative Education Reader*. New York, NY: Routledge.

Bennett, W. (1987). "Looking at Japanese Education: Implications for American Education." *NASSP Bulletin, Vol. 71*(499), pp. 102-108.

Bickman, L., Hedrick, T., & Rog, D. (1993). *Applied research design: a practical guide*. Thousand Oaks, CA: Sage Publications.

Bok, D. (1996). *The state of the nation: government and the quest for a better society*. Cambridge, MA: Harvard University Press.

Boyatzis, R. (1998). *Transforming qualitative information: thematic analysis and code development*. Thousand Oaks, CA: Sage Publications.

Buraku Kaihō Kenkyūsho. (ed.). (1983). *The Road to a Discrimination-free Future*. Ōsaka, Japan: Buraku Liberation Research Institute.

Clune, W., & Witte, J. (eds.). (1990). Choice and Control in American Education, *Volume 1: The Theory of Choice and Control in American Education*. New York, NY: Routledge.

Cozzens, M., & Fuhrman, S. (2001). "Invited Commentary: Lessons from the Third International Mathematics and Science Study-Repeat." *Education Statistics Quarterly, 3*(1), 17-19. Washington, D.C.: National Center for Education Statistics.

Cummings, W. (1980). *Education and Equality in Japan*. Princeton, NJ: Princeton University Press.

Cummings, W., Beauchamp, E., Ichikawa, S., Kobayashi, V., & Ushiogi, M. (1986). *Educational Policies in Crisis, Japanese and American Perspectives*. New York, NY: Praeger in association with the East-West Center.

DeCoker, G. (2002). *National Standards and School Reform in Japan and the United States*. New York: Teachers College Press.

NOTES

Duke, B. (1986). *The Japanese School: Lessons for Industrialized America*. Tōkyō, Japan: Tōkyō University Press.

Duke, B. (1989). *The Great Educators of Modern Japan*. Tōkyō, Japan: Tōkyō University Press.

Fetterman, D. (1998). *Qualitative approaches to evaluation in education: the silent scientific revolution*. New York: Praeger.

Hanushek, E., & Luque, J. (2003). Efficiency and Equity in Schools around the World. *Economics of Education Review, 22*(5), 481-502.

Hess, R., Holloway, S., & McDevitt, T. (1986). Family influences on achievement in Japan and the United States. In Harold Stevenson, Hiroshi Azuma, and Kenji Hakuta, eds., *Child Development and Education in Japan*, pp. 181-194. Tōkyō, Japan: Shibundō.

Horio, T. (1994). *Education in Modern Japanese*. Tōkyō, Japan: Tōkyō University Press.

Ishida, H. (1993). *Social Mobility in Contemporary Japan: Educational credentials, class and the labour market in a cross-cultural perspective*. London, England: Macmillan Press.

Kariya, T., & Rosenbaum, J. (1987). Self-Selection in Japanese Junior High Schools: A Longitudinal Study of Students' Educational Plans. *Sociology of Education, 60*(3), 168-180.

LeTendre, G., Baker, D., Akiba, M., Goesling, B., & Wiseman, A. (2001). "Teacher's Work: Institutional Isomorphism and Cultural Variation in the U.S., Germany, and Japan." *Educational Researcher, 30*(6), 3-15.

Levin, H. (1990). "The theory of choice applied to education." In William Clune and John Witte, eds., *Choice and Control in American Education, Volume 1: The Theory of Choice and Control in American Education*, pp. 247-284. London: Routledge.

Lewis, C., & Tsuchida, I. (1988). Teaching is Cultural. *American Educator, 22*(4), 12-17.

Monbushō. (1997). Gakkō Kyōin Tōkei Chōsa Hōkokusho Heisei 7 Nendo. Tōkyō, Japan: Monbushō.

Nakajima, K. (1989). "Narusa Jinzo." In Benjamin Duke, ed., *The Great Educators of Modern Japan*, pp. 67-85. Tōkyō, Japan: Tōkyō University Press.

Nakauchi, T. (1986). *The Modernization of Japanese Education*. Tōkyō, Japan: International Society for Educational Information.

National Assessment for Educational Progress (NAEP). (2001). *The Nation's Report Card Mathematics 2000*. Washington, D.C.: U.S. Department of Education.

National Assessment for Educational Progress (NAEP). (2008). *The Nation's Report Card Mathematics 2007*. Washington, D.C.: U.S. Department of Education.

National Center for Education Statistics (NCES). (2003). *Teaching Mathematics in Seven Countries: Results from the TIMSS 1999 Video Study*. Washington, D.C.: U.S. Department of Education.

National Mathematics Advisory Panel. (2008). *Foundations for Success: The Final Report of the National Mathematics Advisory Panel*. Washington, D.C.: U.S. Department of Education.

Nichinōken. (2007). General Catalog. Tōkyō: Nichinōken Press.

Office of Educational Research and Improvement (OERI). (1988). *The educational system in Japan: Case Study Findings*, pp. 41, 218-236. Washington, D.C.: U.S. Department of Education.

Okano, K., & Tsuchiya, M. (1999). *Education in Contemporary Japan: Inequality and Diversity*. London: Cambridge University Press.

Reich, R. (1992). *The Work of Nations. Preparing Ourselves for Twenty-First Century Capitalism.* New York: Vintage Books.

Rog, D., & Bickman, L. (1998). *Handbook of applied social research methods*. Thousand Oaks, CA: Sage Publications.

Rohlen, T. (1983). *Japan's High Schools*. Berkeley, CA: University of California Press.

Rosenbaum, J., & Kariya, T. (1989). From high school to work: Market and institutional mechanisms in Japan. *American Journal of Sociology, 16*(1), 134-165.

Sasagawa, K. (1993). Gaikokujin no gakushūken mondai to 'posutokokuminkyōiku-jidai' no kyōiku-gaku. *Kyōiku, 43*(2), 16-27.

Sōmuchō Tōkeikyoku. (2006). *Japan Statistical Yearbook 2005/2006*. Tōkyō, Japan: Sōmuchō.

Stevenson, H., Azuma, H., & Hakuta, K. (1986). *Child Development and Education in Japan.* Tōkyō: Shibundō.

Stigler, J., & Hiebert, J. (1988). Teaching is Cultural. *American Educator, 22*(4), 4-11.

Stigler, J., & Hiebert, J. (1999). *The Teaching Gap: Best Ideas from the World's Teachers for Improving Education in the Classroom*. New York: Free Press.

Tsukada, M. (1991). *Yobiko Life: A Study of the Legitimation Process of Social Stratification in Japan.* Berkeley, CA: Institute of East Asian Studies.

Tyack, D. (1974). *The One Best System*. Cambridge, MA: Harvard University Press.

U.S. Department of Education. (1997). *International Comparisons of Entrance and Exit Examinations*. Washington, D.C.: Office of Educational Research and Improvement.

Vogel, E. (1979). *Japan as number one: lessons for America*. Cambridge, MA: Harvard University Press.

White, M. (1987). *The Japanese Educational Challenge*. New York: Free Press.

Whitty, G., & Edwards, T. (1998). School choice policies in England and the United States: An exploration of their origins and significance. *Comparative Education, 34*(2), 211-227.

Culturally Relevant Pedagogy as an Ethnomathematical Approach

Milton Rosa
CEAD/UFOP
milton@cead.ufo.br

Daniel Clark Orey
CEAD/UFOP
oreydc@cead.ufo.br

Volume 1

Abstract

The implementation of a culturally relevant pedagogy in the mathematics curriculum helps to develop students' intellectual, social, emotional, and political learning by using their own cultural referents to impart their previous knowledge, skills, and attitudes. A culturally relevant pedagogy provides ways for students to maintain their identity while succeeding academically. In this context, there is a need to examine the embeddedness of mathematics in culture by drawing from an ethnomathematical perspective that takes on the cultural nature of knowledge production into the mathematics curriculum. Ethnomathematics and culturally relevant pedagogy-based approaches to mathematics curriculum are intended to make school mathematics relevant and meaningful as well to promote the overall quality of students' educational experience.

Keywords:

- Culturally Relevant Pedagogy
- Ethnomathematics
- Ethnomathematics Curriculum
- Culturally Relevant Curriculum
- Cultural Aspects of Mathematics

Introduction

In the last two decades, the theory of Culturally Relevant Pedagogy was developed. It examines the teaching-learning process within a critical paradigm through explicit connections between the culture of the students and school subject matter. Culturally relevant pedagogy "can be defined as using cultural knowledge, prior experiences, frames of reference, and performance styles of ethnically diverse students to make learning encounters more relevant and effective for them" (Gay, 2000, p. 29). In this regard, Ladson-Billings (1995) stated that cultural relevant pedagogy as an oppositional pedagogy in which collective empowerment is its center of focus. According to this perspective, the overall goal of culturally relevant pedagogy is to empower students through learning activities that help them to develop their literacy, numeracy, technological, social and political skills in order to be active participants in a democratic society (Ladson-Billings, 1995). However, it is necessary to emphasize that these activities, including mathematical activities, must be relevant to the students' cultural backgrounds.

In relation to the pedagogical work developed in schools, Hart (2003) affirmed that "the views of pedagogy within the literature on ethnomathematics are compatible with work on culturally relevant pedagogy" (p. 42), which examines the cultural congruence between students' community and school. However, it is paramount that educators and teachers understand that cultural congruence indicates respect for the cultural background of the students. According to Zeichner (1996), if educators and teachers want to implement the principle of cultural congruence in their classrooms, then they should be knowledgeable of and respect the various cultural traditions of their students. In order to reach this goal, Rosa (2010) affirmed that these professionals should develop a clear sense of their own ethnic and cultural identities to be able to understand and appreciate those of their students.

On the other hand, since mathematics usually tends to be presented as a set of objective and universal facts and rules, this subject is often viewed as *culture free* and not considered as a socially and culturally constructed discipline (Lee, 2003). However, to change this perception, it is necessary that educators and teachers understand what counts as knowledge in mathematics as well as how knowledge may be related to norms and values of diverse cultures. In other words, if educators and teachers deal with integrating diverse cultures in the classroom, then they need a conceptual framework to make coherent decisions regarding to the curricular activities concerning the mathematics curriculum. Rosa (2010) argued that this approach may help educators and teachers understand how their own cultural biases influence their judgments about students' performance and obstruct students' ability to learn mathematics.

The aim of this paper is to show that there is a need to examine the embeddedness of mathematics in culture by drawing from an ethnomathematical perspective that takes on the cultural nature of knowledge production into the mathematics curriculum. The argumentation is that culturally relevant pedagogy may be considered as an ethnomathematical approach to mathematics curriculum because it intended to make school mathematics relevant and meaningful regarding the promotion of the overall quality

NOTES

of students' educational experience. In order to do so, it is necessary to provide a summary of a related literature on culturally relevant pedagogy and ethnomathematics to show that these two pedagogical approaches help to develop students' intellectual, social, emotional, and political learning by using their own cultural referents to impart their previous knowledge, skills, and attitudes.

Theoretical Framework

The theoretical framework used in this article is Ethnomathematics and Culturally Relevant Pedagogy. In the context of Culturally Relevant Pedagogy, Gay (2000) stated that it was developed out of concern for the serious academic achievement gap experienced by low- income students, students of color, and students from linguistically and culturally diverse environments. According to Klotz (2006), culturally relevant pedagogy uses the cultural knowledge, prior experiences, frames of reference, and learning styles of ethnically diverse students to make learning more relevant and effective with the objective to strengthen their connectedness with schools and as consequence reduce behavior problems and enhance learning.

In this regard, Ladson-Billings (1995) affirmed that educators and teachers benefit from being culturally relevant by contextualizing instruction and schooling practices while maintaining academic rigor and helping students to achieve their academic potential. According to Ogbu and Simons (1998), a culturally relevant pedagogy provides ways for students to maintain their cultural identity while succeeding academically. This means that a culturally relevant pedagogy is designed to fit school culture with students' culture to help them to understand themselves and their peers, develop and structure social interactions, and conceptualize mathematical knowledge.

On the other hand, ethnomathematics studies the cultural aspects of mathematics. It presents mathematical concepts of the school curriculum in a way in which concepts are related to the students' cultural backgrounds (D'Ambrosio, 1990), thereby enhancing their abilities to make meaningful connections and deepening their understandings of mathematics. Thus, there is a need to examine the embeddedness of mathematics in culture, drawing from a body of literature that takes on the students' cultural root of knowledge production into the mathematics curriculum. Rosa (2010) argued that the application of ethnomathematical approaches and culturally relevant pedagogy into the mathematics curriculum are intended to make school mathematics more relevant and meaningful to students and to promote the overall quality of their education.

Mathematics and Culture

Mathematics was, for a long time, regarded as neutral and culturally free, and also considered free of social values (D'Ambrosio, 1990). It was always taught in schools as a culturally neutral or free subject that involved learning supposedly universally accepted facts, concepts, and contents. In other words, Western or academic mathematics consists of a body of knowledge of facts, algorithms, axioms, and theorems. Ethnomathematicians such as D'Ambrosio (1990), Powell and Frankenstein (1997), and Rosa and Orey (2007) argued that mathematics is a cultural product and that every culture has developed its own forms of mathematics. For example, Gerdes (1996) stated that:

> *Ethnomathematicians emphasize and analyze the influences of sociocultural factors of the teaching, learning, and development of mathematics (...). They look for cultural elements [that] reveal mathematical and other scientific thinking. [In addition,] ethnomathematicians also look for (...) activities that may serve as a* ***starting point*** *for doing and elaborating mathematics in the classroom (p. 917-918).*

These ethnomathematicians also argued that the pervasive view of mathematics as Eurocentric and value-free misrepresents the evolution of modern mathematics. This perception is also reinforced by the students' experience of the way mathematics is taught in schools. Brown, Cooney, and Jones (1990) suggested that educators and teachers' view of mathematics is transmitted to the students in their instruction and this fact helps to shape students' views about the nature of mathematics. Even though the universality of mathematical truths is not questioned, it is only in the last three decades that mathematics as *culture free* has been challenged (Rosa & Orey, 2006).

According to this context, Bishop, Hart, Lerman, and Nunes (1993) argued that "there is no sense in regarding mathematics learning as abstract and culture free" (p. 1) because the learning process cannot be abstract and context free. This means that learning cannot be free of environmental and societal influences. The studies conducted by Bandeira and Lucena (2004) and Chieus (2004) that looked at mathematical practices in a variety of contexts confirm this assertion. Orey (2000) and Rios (2000) argued that it is worth noting that the contextualization of mathematics has been described as the identification of mathematical practices developed in different cultural groups.

Because mathematics is a cultural construct; it is a product of cultural development. In this regard, this claim of mathematics contradicts the claims that are made by modern mathematics as universal, objective, and culturally neutral. According to Nasir, Hand, and Taylor (2008), this is an important issue that leads to the development and inquiries of culturally relevant pedagogy and ethnomathematics According to this perspective, Rosa and Orey (2006) affirmed that ethnomathematics as a program was developed "to confront the taboos that the mathematics as a field of study is universal and acculturated" (p. 20).

Ethnomathematics and the Cultural Aspects of Mathematics

The study of ethnomathematics describes the mathematical ideas and practices of identifiable cultural groups found in any culture (D'Ambrosio, 1985). The search for solutions for specific problems that help the development of mathematics are always imbedded in a cultural context. In so doing, in order to understand how mathematics (*tics*) is created, it is necessary to understand the problems *(mathema)* that precipitate it. In D' Ambrosio' (1993) point of view, in order to understand those problems (*mathema*), it is necessary to consider the cultural context (*ethnos*) that drives them.

The essence of the ethnomathematics program is to acknowledge that there are different ways of doing mathematics by considering the appropriation of the academic mathematical knowledge developed by different sectors of the society as well as by considering different modes in which different cultures negotiate their mathematical ideas and practices (D'Ambrosio, 2001). In this conception, ethnomathematics is a program that investigates the ways in which different cultural groups comprehend, articulate, and apply ideas and concepts that can be identified as mathematical practices.

NOTES

Moreover, ethnomathematics as a program may be described as a way in which people from a particular culture use mathematical ideas and concepts for dealing with quantitative, relational, and spatial aspects of their lives. This way of viewing mathematics validates and affirms people's experience of mathematics because it demonstrates that mathematical thinking is inherent to their lives. Further evidence of this assertion is given by Orey (2000) who stated that "the paradigm that diverse cultures use or work within involves out of unique interactions between their language, culture, and environment" (p. 248). Within this context, D'Ambrosio (1993) argued that in an ethnomathematical perspective, mathematical thinking is developed in different cultures in accordance to common problems that are encountered within a cultural context. In this perspective, in order to solve specific problems, *ad hoc*[1] solutions are created, generalized methods are developed from those solutions to solve similar problems, and theories are developed from these generalized methods.

In the context of ethnomathematics, many cultural differentiated groups *know* mathematics in ways that are quite different from academic mathematics as taught in schools (D'Ambrosio, 1990). The tendency has been to consider these *ad hoc* mathematical practices as non-systematic and non-theoretical knowledge systems. In contrast, the study of ethnomathematics underlies a structure of inquiry in *ad hoc* mathematical practices by considering how these practices and problem solving can be developed into methods and theories. Since different types of problems are common in different cultures, the kinds of solutions, methods, and theories that are developed may differ from culture to culture. In this regard, D'Ambrosio (1993) stated that what is recognized as a problem and a solution in one culture may have no meaning in another culture.

Mathematical ideas and practices are identified in cultural activities in traditional and non-traditional societies (Orey, 2000). This means that ethnomathematics refers to the study of mathematical concepts embedded in cultural practices by recognizing that all cultures and all people develop unique methods and sophisticated explications to understand, comprehend, and transform their own reality (Rosa & Orey, 2007). It also recognizes that the accumulated methods of these cultures are engaged in a constant, dynamic, and natural process of evolution and growth in every society. D'Ambrosio (1990) stated that ethnomathematics has come to mean the study of how people within various cultural groups develop techniques to explain and understand their world in response to problems, struggles, and endeavors of human survival, including material needs as well as art and spirituality through the use of the development of artifacts, which are objects created by members of a specific cultural group that inherently give cultural clues about the culture of its creator and users. Rosa and Orey (2008) affirmed that this perspective "provides an important opportunity for educators to link current events and the importance of these artifacts in the context of ethnomathematics, history, and culture" (p. 33).

Another presupposition of ethnomathematics is that it validates all forms of mathematical explaining and understanding formulated and accumulated by different cultural groups (D'Ambrosio, 1993). This knowledge is regarded as part of an evolutionary process of change that is part of the same cultural dynamism present as each group comes into contact with each other (Rosa, 2005). A study of the different ways in which people resolve problems and the practical algorithms on which they base these mathematical perspectives becomes relevant for any real comprehension of the concepts and the practices in the mathematics that they have

[1] Ad hoc is a Latin expression that means *for this purpose*. It generally means a solution designed for a specific problem or task, non-generalizable, and which cannot be adapted to other purposes.

developed over time (Rosa & Orey, 2008). In other words, ethnomathematics refers to forms of mathematics that vary as a consequence of being embedded in cultural activities whose purpose is other than doing mathematics. In this perspective, Orey (2000) affirmed that "ethnomathematics might be characterized as a tool to act in the world" (p.250) and as such, it provides insights into the social role of academic mathematics.

Culturally Relevant Pedagogy

A culturally relevant pedagogy focuses on the need for teachers to not only value the diverse home cultures of students they teach, but explicitly address the home culture and students' understanding of it in the classroom, to teach mathematics in a culturally appropriate manner situated within students' funds of knowledge (Moll, Amanti, Neff, & Gonzalez, 1992) that they bring to school, and to be educated with an understanding of the historical legacy of racism in society and its current impact on education (Ladson-Billings, 1995). The understandings and processes of culturally relevant pedagogy underscore students' connections to their home culture, and provide ways for school leaders and teachers to support cultural connections in school and use them to scaffold learning.

The application of culturally relevant pedagogy in classroom validates and incorporates students' cultural background, ethnic history, and current societal interests into teachers' daily instruction. It addresses students' socio-emotional needs and uses ethnically and culturally diverse materials for its pedagogical action in classrooms (Gay, 2000). In this perspective, Ladson-Billings (1994) stated that culturally relevant pedagogy is an educational approach that empowers students intellectually, socially, emotionally, and politically through the use of cultural and historical references to convey knowledge, impart academic skills, and change students' attitudes towards academic instruction.

Gay (2000) and Howard (2003) have taken their studies a step further with the development of culturally relevant pedagogy that attempts to integrate the culture of different racial and ethnic groups into the overall academic program. In this point of view, Scheurich and Skrla (2003) argued that the basic premise of culturally relevant pedagogy "is that teachers should teach using philosophies and methods that respect, value, and use positively the strengths of students' home cultures, contexts, and languages" (p.48). Richards, Brown, and Forde (2000) affirmed that the use of culturally relevant pedagogy delineates and promotes the achievement for all students because effective teaching and learning take place in an environment that is culturally supported, learner-centered context, whereby the strengths students bring to school are identified, nurtured, and utilized to promote their achievement.

The results of the studies conducted by Gay (2000) and Ladson-Billings (1994) showed that the academic achievement of students who come from culturally and linguistically diverse backgrounds improve if schools and teachers ensure that classroom instruction is conducted in a manner that is relevant to their home and community cultures. For example, the study of Lipka and Adams (2001) clearly addressed the application of culturally relevant pedagogy. The purpose of

NOTES

their study was to determine the effectiveness of a culturally-based unit of instruction on mathematics, which dealt with the learning of perimeter and area by sixth grade students. Their conclusion was in favor of the effectiveness of the culturally relevant based curriculum treatment as implemented in their study.

Irvine and Armento (2001) suggested that culturally relevant pedagogy allows teachers to provide and use meaningful learning materials; create environments, which include cultures, customs, and traditions that are different from their own; and include lessons that assist students in making meaningful connections between their lives and school-related experiences. In other words, participating in culturally relevant teaching essentially means that teachers create a bridge between students' home and school lives, while they still meet the expectations of the district, state, and federal curricular requirements because this kind of instructional pedagogy utilizes the backgrounds, knowledge, and experiences of the students to inform the teacher's planning mathematical lessons and methodology. Ogbu (1992) stated that a culturally relevant pedagogy must provide a way for students to maintain their cultural identity while succeeding academically. In other words, it is important that teachers contextualize instruction and schooling by applying culturally relevant pedagogy in their pedagogical practices as well as embodying relevance and rigor to the educational process. In this regard, Rosa (2010) affirmed that this is one of the goals of the ethnomathematics as a program.

A Culturally Relevant Pedagogy Curriculum

It is paramount that educators and teachers make use of culturally diverse curriculum content in the mathematics classroom because effective teaching and learning for culturally diverse students may be expedited through the use of instructional materials that recognize the contributions that people of diverse ethnic groups have made to the current bank of knowledge. When students see visions of people who look and live like those of their culture, both currently and historically, it not only breeds interest but also provides motivation for students to believe they can achieve, and that their achievement is not only expected but valued. In a culturally relevant pedagogy, the curriculum bridges the academic learning with students' prior understanding, knowledge, native language, and values (Waxman & Tellez, 2002). This perspective into the school curriculum allows teachers to learn from and about their students' culture, language, and learning styles to make instruction more meaningful and relevant to their students' lives. For example, Demmert and Towner (2003) have examined the research based on Native education and found six critical elements of culturally relevant education, which suggested an impact on academic achievement of Native-American students. These critical elements are the recognition and use of Native languages; the implementation of a pedagogy that uses traditional cultural characteristics; the application of teaching strategies and curriculum that are congruent with traditional culture and traditional ways of knowing; strong Native community participation in education; and knowledge and use of political mores of the community.

In the context of culturally relevant pedagogy, there is a need to examine the embeddedness of mathematics in culture, drawing from a body of literature that takes on the cultural nature of knowledge production into the mathematics curriculum (D'Ambrosio, 1990). Mathematics as part of the school curriculum must reinforce and value cultural knowledge of students rather than ignore or negate it because a culturally relevant curriculum should fully integrate students' cultural mathematics knowledge. Rosa and Orey (2007) argued that this mathematics curriculum must be grounded in a constructivist approach to learning and seeks to change the way mathematics teachers construct their learning environments by producing teachers who are able to facilitate a mathematics learning environment grounded in real life experiences and to support students in the social construction of mathematics.

An Ethnomathematics Curriculum

The results of the studies conducted by Eglash (1997) and Rosa and Orey (2007) show evidences that including cultural aspects in the curriculum will have long-term benefits for mathematics learners; that is; cultural aspects contribute to recognizing mathematics as part of daily life, enhancing the ability to make meaningful connections, and deepening the understanding of mathematics. In this regard, Chieus (2004) stated that the pedagogical work towards an ethnomathematics perspective allows for a broader analysis of the school context in which the pedagogical practices transcend the classroom environment because these practices embrace the sociocultural context of the students. Damazio (2004) agreed with this perspective by suggesting that the pedagogical elements necessary to develop the mathematics curriculum are found in the school community. This means that the field of ethnomathematics presents some possibilities for educational initiatives that help to reach this goal. In D'Ambrosio's (1990) point of view, it is important to recognize that ethnomathematics is a research program that guides educational pedagogical practices. However, according to Monteiro, Orey, and Domite (2004), it is necessary to point out that the incorporation of the objectives of the ethnomathematics program as pedagogical practice in the school curricula and its operationalization and transmission in the field of education is a recent field of study that is still developing its own identity in the pedagogical arena.

The trend towards ethnomathematical approaches to mathematics curriculum and pedagogy reflects a comprehensive development in mathematics education. Ethnomathematical approaches are intended to make school mathematics more relevant and meaningful to students, and to promote the overall quality of education. It is necessary to plead for a more culturally sensitive view of mathematics to be incorporated into the school curriculum. For example, Powell and Frankenstein (1997) proposed the elaboration of a mathematics curriculum that is based on students' knowledge, which allows teachers to have more freedom and creativity to choose academic mathematical topics to be covered in the lessons. They suggested that through dialogue with students, teachers can apply mathematical themes that help them to elaborate the mathematics curriculum. In their point of view, teachers can engage students in the critical analysis of the dominant culture as well as the analysis of their own culture through an ethnomathematical perspective. In this context, it is necessary to investigate the conceptions, traditions, and mathematical practices of distinct cultural groups with the intention to incorporate them into the mathematics curriculum.

Further, Rosa and Orey (2003) stated that a culturally relevant mathematics curriculum based on an ethnomathematical perspective infuses the students' cultural backgrounds in the learning environment in a holistic manner. One possibility for an ethnomathematical curriculum may be labeled as mathematics in a meaningful context in which students are given opportunities to relate their new

NOTES

learning experiences to knowledge and skills they have previously learned (Bandeira & Lucena, 2004). In this regard, it is particularly important that the mathematical learning experiences of students acknowledge their cultural backgrounds and experiences in the learning of mathematics. This mathematical approach is presented as a cultural response to students' needs by making connections between their cultural background and mathematics (Rosa, 2010). This approach supports the view that "mathematics is conceived as a cultural product which has developed as a result of various activities" (Bishop, 1988, p. 182) with the objective of making mathematics more relevant to students because every culture is assumed to have such mathematical responses and these responses are valid content for a mathematics classroom. A classroom using this type of ethnomathematical curriculum would be full of examples that draw on the students' own experiences and on experiences that are common in their cultural environment. These examples would be vehicles for communicating mathematical ideas, which themselves would be relatively unchanged. Rosa and Orey (2008) affirmed that ethnomathematics aims to draw from the students cultural experiences and practices of the individual learners, the communities, and the society at large, in using them as vehicles to not only make mathematics learning more meaningful, but more importantly, to provide students with the insights of mathematical knowledge as embedded in their social and cultural environments.

It is assumed that this kind of curriculum motivates students to recognize mathematics as part of their everyday life and enhances students' ability to make meaningful mathematical connections by deepening their understanding of all forms of mathematics. For example, Duarte (2004) investigated the uniqueness of mathematical knowledge produced by workers in the home construction industry through the study of mathematical ideas and practices that they develop in the construction sites. In this study, there was a reflection on the mathematical knowledge possessed by the members of this working class in order to academically legitimate their knowledge and determine the pedagogical and curricular implications that are inferred in the process of productions of this knowledge.

The objective of developing an ethnomathematical curriculum model for classrooms is to assist students to become aware of how people mathematize and think mathematically in their culture, to use this awareness to learn about formal mathematics, and to increase their ability to mathematize in any context in the future (Duarte, 2004; Rosa & Orey, 2006). This ethnomathematical curriculum leads to the development of a sequence of instructional cultural activities that enables students to become aware of potential practices in mathematics in their culture so that they are able to understand the nature, development, and origins of academic mathematics (Rosa & Orey, 2007). Students also value and appreciate their own previous mathematical knowledge, which allows them to understand and experience these cultural activities from a mathematical point of view, thereby allowing them to make the link between school mathematics and the real world and daily life in this globalized society (Rosa & Orey, 2003).

An ethnomathematical curriculum brings a broader understanding about mathematics into the classroom mathematics program. Most mathematics curricula focus on mastery of skills, accumulation of facts, rules, and algorithms that are necessary for examinations. The curriculum is experienced as mathematical content, so most students leave school thinking that mathematics is something to be done only at school and that it has no relevance to their real lives. According to Rosa and Orey (2003), an ethnomathematical curriculum helps students understand the nature of mathematics, they acquire the tools to better comprehend the relevance of mathematics in the various aspects of their everyday life.

Volume 1

Rosa and Orey (2003) argued that an ethnomathematics curriculum offers to the students, especially minority students, motivation to perceive mathematics as an important cultural tool that facilitates their mathematical learning. They also affirmed that the establishment of cultural connections is a fundamental aspect in the development of new strategies to the process of teaching and learning mathematics because it allows students to perceive mathematics as a significant part of their own cultural identity. Warschauer (1999) argued that the use ethnomathematics in the school curricula is an effective tool that contributes to improve the learning of mathematics of minority students.

This curriculum focuses on mathematics as a process, rather than a collection of facts, and it is based on the idea that mathematics is a human creation that emerges as people attempt to understand their world. Therefore, mathematics is seen as a process, and as a human activity, rather than just as a set of academic content (Rosa, 2010). This implies that an ethnomathematical curriculum is not just about the application of relevant contexts in learning and teaching mathematics, but is also about generating formal mathematics from cultural ideas. Thus formal mathematics is better understood, appreciated, and made more meaningful to its learners.

Ethnomathematics and Culturally Relevant Pedagogy

An important change in mathematical instruction needs to take place in order to accommodate continuous and ongoing change in the demographics of students in mathematics classrooms. According to Torres-Velasquez and Lobo (2004), it is necessary to integrate a culturally relevant pedagogy into the existing mathematics curriculum because it proposes that teachers contextualize mathematics learning by relating mathematical content to students' real life-experiences.

In this context, the guidelines of both the National Council of Teacher of Mathematics (NCTM, 1991) and the Brazilian Ministry of Education and Culture (BRASIL, 1996) highlighted the importance of building connections between mathematics and students' personal lives and cultures. Along with this line, Rosa and Orey (2006) affirmed that "When practical or culturally-based problems are examined in a proper social context, the practical mathematics of social groups is not trivial because they reflect themes that are profoundly linked to the daily lives of students" (p. 34). According to Rosa and Orey (2008), the culturally relevant pedagogy should focus on the role of mathematics in a socio-cultural context that involves the ideas and concepts associated with ethnomathematics, which uses an ethnomathematical perspective for solving problems. In this regard, Ladson-Billings (1995) stated that teachers working from a culturally relevant perspective demonstrate a belief that students can be competent regardless of race or social class, provide students with scaffolding between what they know and what they do not know, focus on instruction during class rather than busy-work or behavior management, extend students' thinking beyond what they already know, and exhibit in-depth knowledge of students as well as subject matter. In this perspective, students may be successful in mathematics when their understanding of it "is linked

NOTES

to meaningful cultural referents, and when the instruction assumes that all students are capable of mastering the subject matter" (Ladson-Billings, 1995, p. 141).

According to this perspective, Rosa (2010) affirmed that mathematical knowledge in the culturally relevant pedagogy may be viewed as an approach of an ethnomathematics program. In this perspective, D'Ambrosio (1990) stated that ethno is defined as all culturally identifiable groups with their jargons, codes, symbols, myths, and even specific ways of reasoning, inferring, and modeling; mathema is defined as categories of analysis; and tics is defined as methods or techniques developed by the members of diverse cultural groups to solve problems they face daily. Gustein, Lipman, Hernandez, and de los Reyes (1997) argued that, in a culturally relevant mathematics classroom, educators and teachers build from the students' cultural background (*ethno*) by orienting mathematical lessons toward their culture and experiences (*mathema*) while developing their critical thinking skills (*tics*).

Final Considerations

Culturally relevant pedagogy is a teaching methodology that is designed to fit the school culture of the students' culture as the basis for helping them to understand themselves and their peers, develop and structure social interactions, and conceptualize knowledge (Ladson-Billings, 1994). It builds on and values the cultural experiences and knowledge of all students regardless of whether they are represented by dominant or non-dominant cultural systems and empowers students intellectually, socially, emotionally, and politically by using cultural referents to impart their knowledge, skills, and attitudes in the pedagogical work in schools. Culturally relevant pedagogy combines an examination of the cultural and socioeconomic influences on teaching and learning. As well, it includes knowledge, with a commitment to challenging social injustices and reflecting on educational injustices by identifying obvious and subtle individual, institutional, and cultural actions that perpetuate social structures.

According to this perspective, ethnomathematics studies the cultural aspects of mathematics and presents the mathematical concepts of the school curriculum in a way that are related to the students' cultural backgrounds by enhancing their ability to make meaningful connections and deepening their understanding of mathematics. Ethnomathematics links student's diverse ways of knowing and learning and culturally embedded knowledge with academic mathematics because it explores academic and culturally rich ways to provide more inclusive developmental programs for the diverse populations served at educational institutions. It is a program that includes curricular relevance and which builds a curriculum around the local interests and culture of the learners (Rosa, 2010). Teaching mathematics through cultural relevance and ethnomathematical perspective helps students to know more about reality, culture, society, environmental issues, and themselves by providing them with mathematics content and approaches that enable them to successfully master academic mathematics.

References

Bandeira, F. A., & Lucena, I. C. R.(2004). *Etnomatemática e práticas sociais* [Ethnomathematics and social practices]. Coleção Introdução à Etnomatemática [Introduction to Ethnomathematics Collection]. Natal, RN, Brazil: UFRN.

Bishop, A. J. (1988) *Mathematics enculturation: A cultural perspective on mathematics education.* Dordrecht, Netherlands: Kluwer.

Bishop, A. J., Hart, K., Lerman, S.; & Nunes, T. (1993). *Significant influences on children's learning of mathematics*. Paris, France: UNESCO.

BRASIL (1996). *Lei das Diretrizes e Bases da Educação Nacional* [Law of Guidelines and Bases of National Education]. Brasília, DF: MEC.

Brown, S. I., Cooney, T. J., & Jones, D. (1990). Mathematics teacher education. In Houston, W. R. (Ed.). *Handbook of research on teacher education* (pp. 639-656). New York, NY: Macmillan.

Chieus, G. J. (2004). Etnomatemática: Reflexões sobre a prática docente [Ethnomathematics: Reflections on teaching practice]. In Ribeiro, J. P. M., Domite, M. C. S., & Ferreira, R. (Eds.). *Etnomatemática: Papel, valor e significado* [Ethnomathematics: Role, value, and meaning] (pp. 185-202). São Paulo, SP, Brazil: ZOUK, 2004.

Damazio, A. (2004). *Especifidades conceituais da matemática da atividade extrativa do carvão* [Conceptual specifications of mathematical activities of coal extraction]. Coleção Introdução à Etnomatemática [Introduction to Ethnomathematics Collection]. Natal. RN, Brazil: UFRN.

D'Ambrosio, U. (1985). Ethnomathematics and its place in the history and pedagogy of mathematics. *For the Learning of Mathematics, 5*(1), 44-48.

D'Ambrosio, U. (1990). *Etnomatemática* [Ethnomathematics]. São Paulo, SP, Brazil: Editora Ática.

D'Ambrosio, U. (1993). Etnomatemática: Um programa [Ethnomathematics: A program]. *A Educação Matemática em Revista, 1*(1), 5-11.

D'Ambrosio, U. (2001). What is Ethnomathematics, and how can it help children in schools? *Teaching Children Mathematics, 7*(6), 308-310.

Demmert, W. J. Jr.; & Towner, J. C. (2003). *Final paper: A review of the research literature on the influences of culturally based education on the academic performance of native American students*. Portland, OR: Northwest Regional Educational Library.

Duarte, C. G. (2004). *Implicações Curriculares a partir de um olhar sobre o mundo da construção civil* [Curricular implications concerning the world of civil construction]. In Knijnik, G., Wanderer, F, & Oliveira, C. J. (Eds.). Etnomatemática: Currículo e Formação de Professores [Ethnomathematics: Curriculum and Teacher Education] (pp. 195-215). Santa Cruz do SL, RS, Brazil: EDUNISC.

Eglash, R. (1997). When math worlds collide: Intention and invention in ethnomathematics. Science. *Technology and Human Values, 22*(1), 79-97.

NOTES

Gay, G. (2000). *Culturally responsive teaching: Theory, research, and practice*. New York, NY: Teachers College Press.

Gerdes, P. (1996). Ethnomathematics and Mathematics Education. A. Bishop et al. (Eds.). *International Handbook of Mathematics Education* (pp. 909-943). Utrecht, Netherlands: Kluwer Academic Publishers.

Gutstein, E.; Lipman, P.; Hernández, P.; & de los Reyes, R. (1997). Culturally relevant mathematics teaching in a Mexican American context. *Journal for Research in Mathematics Education, 28*, 709-737.

Hart L. E. (2003). Some Directions for Research on Equity and Justice in Mathematics Education. In L. Burton (Eds.). *Which Way Social Justice in Mathematics Education? International Perspectives on Mathematics Education* (pp.25-50). Westport, CT: Praeger Publishers.

Howard, T. C. (2003). Culturally relevant pedagogy: Ingredients for critical teacher reflection. *Theory Into Practice, 42*(3), 195-202.

Irvine, J. J., & Armento, B. J. (2001). *Culturally responsive teaching: Lesson planning for elementary and middle grades*. New York, NY: McGraw-Hill.

Klotz, M. B. (1994). Culturally competent schools: Guidelines for secondary school principals. *Principal Leadership*. Retrieved from http://www.nasponline.org/resources/principals/.

Ladson-Billings, G. (1994). *The dreamkeepers: Successful teachers of African American children.* San Francisco: Jossey Bass Publishers.

Ladson-Billings, G. (1995). Toward a theory of culturally relevant pedagogy. *American Educational Research Journal, 32*(3), 465-491.

Lee, O. (2003). Equity for linguistically and culturally diverse students in science education: A research agenda. *Teachers College Record, 105*(3)465–489.

Lipka, J., & Adams, B. (2001). Improving rural and urban students' mathematical understanding of perimeter and area. Fairbanks, AK: University of Alaska.

Moll, L. C., Amanti, C., Neff, D., & Gonzalez, N. (1992). Funds of knowledge for teaching: Using a qualitative approach to connect homes and classrooms. *Theory into Practice, 31*(2), 132-141.

Monteiro, A.; Orey, D. C.; & Domite, M. C. (2004). Etnomatemática: Papel, valor e significado [Ethnomathematics: Role, value, and meaning]. In Ribeiro, J. P. M., Domite, M. C. S.; Ferreira, R. (Eds.). *Etnomatemática: Papel, valor e significado* [Ethnomathematics: Role, value, and meaning] (pp.13-37). São Paulo, SP, Brazil: ZOUK.

Nasir, N. S., Hand, V., & Taylor, E. V. (2008). Culture and mathematics in school: boundaries between cultural and domain knowledge in the mathematics classroom and beyond. *Review of Research in Education, 32*, 187-240.

NCTM. (1991). *Professional Standards for Teaching Mathematics*. Reston, VA: National Council of Teachers of Mathematics.

Ogbu, J. U. (1992). Understanding cultural diversity and learning. *Educational Researcher, 21*(8), p. 5-14.

Ogbu, J. U., & Simons, H. D. (1998). Voluntary and involuntary minorities: A cultural- ecological theory of school performance with some implications for education. *Anthropology and Education Quarterly, 29*(2), 155-188.

Orey, D. C. (2000). The ethnomathematics of the Sioux tipi and cone. In Selin, H. (Ed.). *Mathematics across culture: the History of non-Western mathematics* (pp. 239-152). Dordrecht, Netherlands: Kulwer Academic Publishers.

Powell, A. B. & Frankenstein, M. (1997). Ethnomathematics praxis in the curriculum. In Powell, A. B. & Frankenstein, M. (Eds.). *Challenging Eurocentrism in mathematics education* (pp. 249-259). New York, NY: SUNY.

Richards, H., Brown, A., & Forde, T. (2000). *Addressing diversity in schools: Culturally responsive pedagogy. Practitioner Brief. NCCREST.* Retrieved from www.nccrest.org/Briefs/Diversity_Brief.pdf.

Rios, D. P.(2000). Primero etnogeometría para seguir con etnomatemática [First ethnogeometry to follow with ethnomathematics]. In. Domite, M. C. (Ed.). *Anais do Primeiro Congresso Brasileiro de Etnomatemática – CBEm-1* (pp. 367-375). São Paulo, SP, Brazil: FE-USP, 2000.

Rosa, M. (2005). Curriculo e matemática: Algumas considerações na perspectiva etnomatemática [Curriculum and mathematics: Some considerations in a ethnomathematical perspective]. *Plures Humanidades, 6*, 81-96.

Rosa, M. (2010). *A mixed-methods study to understand the perceptions of school leaders about English language learners (ELL): The case of mathematics*. Unpublished doctorate dissertation. California State University, Sacramento, California, USA.

Rosa, M., & Orey, D. C. (2003). Vinho e queijo: Etnomatemática e Modelagem! [Wine and cheese: Ethnomathematics and modelling!]. *BOLEMA, 16*(20), 1-16.

Rosa, M., & Orey, D. C. (2006). Abordagens atuais do programa etnomatemática: delinenando-se um caminho para a ação pedagógica [Current approaches in the ethnomathematics as a program: Delineating a path toward pedagogical action]. *BOLEMA, 19*(26), 19-48.

Rosa, M.; & Orey, D. C. (2007) Cultural assertions and challenges towards pedagogical action of an ethnomathematics program. *For the Learning of Mathematics, 27*(1), 10-16.

Rosa, M.; & Orey, D. C. (2008). Ethnomathematics and cultural representations: Teaching in highly diverse contexts. *Revista Acta Scientiae, 10*, 27-46.

Scheurich J. J. & Skrla, L. (2003). *Leadership for equity and excellence: Creating high- achievement classrooms, schools, and districts*. Thousand Oaks, CA: Corwin Press, Inc.

Torres-Velasquez, D., & Lobo, G. (2004). Culturally responsive mathematics teaching and English language learners. *Teaching Children Mathematics, 11*, 249-255.

Warschauer, M. (1999). *Electronic literacies: Language, culture, and power in online education.* Mahwah, NJ: Lawrence Erlbaum Associates.

Waxman, H. C., & Tellez, K. (2002). *Research synthesis on effective teaching practices for English language learners*. Philadelphia, PA: Mid-Atlantic Laboratory for Student Success.

Zeichner, K. (1996). Educating teachers to close the achievement gap: Issues of pedagogy, knowledge, and teacher preparation. In Williams, B. (Ed.). *Closing the achievement gap: A vision to guide change in beliefs and practice* (pp. 55-77). Alexandria, VA: Association for Supervision and Curriculum Development.

A Study of Estimation by Professionals at Work

Thomasenia Lott Adams
University of Florida
Gainesville, FL
tla@coe.ufl.edu

Gregory Harrell
Valdosta State University
Valdosta, GA
gharrell@voldosta.edu

Abstract

Estimation is a learned skill that supports learning and applications in measurement as well as other quantitative contexts. Estimation is an important skill to learn in that it provides a foundation for measurement – it helps one determine the most appropriate tools for measuring specific items. Estimation is equally important as it helps one determine whether or not an actual measurement obtained is reasonable for a given context. It can be beneficial for learners to explore the application of estimation so that they have context and motivation for learning to estimate. The authors engaged with selected professionals during the course of their work to obtain information about when they use estimation, how they use estimation, how they determine if estimation is helpful and when they decide not to use estimation. The authors share estimation activities of the professionals, which result in "estimation declarations" and suggestions for the teaching of estimation.

Introduction

Estimation provides the foundation for measurement because estimation skills help us to determine the specific units or tools people will need to obtain a measurement. For example, an estimate of a length will help one to determine which tool (e.g., ruler, meter stick, measuring tape, trundle wheel, etc.) is most appropriate for determining the actual measurement. Estimation also helps us make decisions about whether or not a measurement obtained is reasonable for a given context (National Council of Teachers of Mathematics, 2000). Hence, estimation is both a precursor to and a foundation for measurement. In addition, "when people do not have specific information or need a precise answer, they often make an educated guess. An estimate is an educated or reasonable guess based on information, prior knowledge, and judgment. Even when information is known, the situation may call for an estimation" (Kennedy, Tipps, & Johnson, 2008, p. 175).

Estimation

Reys, Lindquist, Lambdin, and Smith (2009) define estimation as "...a process of producing answers that are close enough to allow for good decisions without performing elaborate or exact computations" (p. 225). In addition, Kennedy, Tipps and Johnson (2008) indicate that while "estimation is a reasonable guess, hypothesis, or conjecture based on numerical information", the estimate itself also "... communicates numerical information" (p. 173). While it is important to provide strategies to empower students to estimate, it can be very challenging to incorporate meaningful and real-life experiences from which these strategies emerge. To many learners, estimation is abstract and unrelated to "real" activity. "To produce an estimate is...a very difficult task for young children. They do not easily grasp the concept of "estimate" or "about" (Van de Walle, Karp, & Bay-Williams, 2010, p. 141). To this end, it is beneficial for learners to engage in discussions around issues that use such terms and or phrases as "about", "more or less than ______?, "closer to ______ or to ______?" (Van de Walle, Karp, & Bay-Williams, 2010, p. 141). Another perspective of estimation is the determination of a "...range within which the solution [or estimate] might fall" (Hatfield, Edwards, Bitter & Morrow, 2008, p. 107).

Many students struggle with building the skill of estimation because it is more complex than responding with a suggestion isolated from conceptual understanding of the mathematics. In fact, "students who are proficient at written computations are not necessarily good estimators" (Reys, Lindquist, Lambdin, & Smith, 2009, p. 225). According to Bassarear (2008)

> *The availability of inexpensive calculators has profoundly changed the way we compute. Most people now rely on calculators when exact answers are needed. However, estimating skills are still very important in cases where an exact answer is not needed and to the check the reasonableness of results obtained on the calculator. Estimation, in turn, requires good mental arithmetic skills, which come from an understanding of the nature of the operations, a firm understanding of place value, and*

NOTES

> *the ability to use various properties (for example, the associative and distributive properties) (p. 143).*

Recognizing this deficiency students often have with estimation, the authors determined to uncover how the concept and skill of estimation are addressed by professionals who estimate in the context of their work. The intent of the work is to disseminate these contexts so that teachers and teacher educators can have thoughtful examples of the application of estimation to provide for students and to use as contexts for instruction regarding estimation. The framework for the exploratory study rests with Ubiratan D'Ambrosio's notion of Ethnomathematics. According to D'Ambrosio (2007), "the great motivator for the research program known as ethnomathematics is to seek to understand mathematical knowing/doing throughout the history of humanity, in the contexts of different interest groups, communities, peoples and nations" (p. 8). This was the impetus for the authors exploring professionals engaged in their work that proposed to involve the act of estimation.

Methodology

Selecting Professions and Participants

In considering the work of estimating, the authors discussed professions thought to use estimation as a function of their work and collaborated on determining the professions to include. The authors concluded deliberations with 16 possible professions. Once the authors identified the professions that would receive focus, the authors scanned the community for representative persons for those professions and proceeded to call each professional to request each professional to engage in an on-the-job interview to observe the professionals at work. Of primary concern was to be able to observe professionals at work. The authors held the position that "the natural setting is the place where the researcher is most likely to discover, or uncover, what is to be known about the phenomenon of interest" (Maykut & Morehouse, 1994, p. 45).

All of the professionals were adults over the age of 21. There were 12 males and 5 females. The authors interviewed and observed 17 professionals engaged in the work of their profession (two shared the task of one profession.). Several of the professionals were known to one of the authors but not to the extent that the professionals needed to be removed from exploration. This current paper includes only four participants (White males, average age of 40) from the larger study.

Data Collection

The authors developed an interview guide that consisted of the following questions:

1. In what ways do you use mathematics on a daily basis to perform your work duties?
2. Specifically, how do you employ the skill of measurement?
3. In measuring, when do you use the skill estimation?
    a. What do you estimate?
    b. How do you estimate?
    c. When do you estimate?
4. What makes an estimate good or acceptable for particular professional tasks?
    a. Why don't you estimate? (That is, what situations lead to a decision not to estimate?)

The authors audiotaped the interview as well as collected the responses and field notes in writing as the participants spoke. Each audiotaped interview was scheduled for one hour at the place or site of employment for the profession so that the author could also observe the professionals at work with the expectation to observe the professional engage in estimation. The sessions lasted an average of 90 minutes. Several interactions lasted longer because of the nature of the profession and also several professionals preferred to talk as they worked rather than stop working to just engage in the interview. The author engaged the professionals in the interview as they worked, with the professionals often demonstrating something they were saying; this also provided prompts for questions of clarification about the professional activity. Visual observation data was recorded by still photography.

Data Analysis

One author transcribed each videotaped interview, using written notes for clarification when needed and possible. Both authors independently reviewed the participants' responses and chunked the data from each participant by responses to the interview question. This naturally led to the following categories of data.

1. Estimation tasks the participants engaged in for a task
2. Estimation strategies the participants used for a task
3. Participants' reasons for estimating and determining a good estimate for a task
4. Participants' reasons for not estimating for a task

The authors then collaborated to identify the salient points from the participants' voices regarding their engagement in estimating for their work tasks. Furthermore, the authors engaged in reading the participants' responses to search for patterns related to the participants' actions of measurement and estimation during the work experience. The authors coded data that related to identified patterns and then searched for sub-themes. Using these sub-themes, the authors revisited the data and condensed the sub-themes to result in a set of "estimation declaration" that would have implications for teaching estimation.

Results

The Traffic Patrolman

The traffic patrolman attended a two-day radar school to learn how to estimate the speeds of traveling vehicles. His instructor drove vehicles past the patrolman and fellow trainees at a set speed. They would then estimate the speed of each vehicle. After the simulation, the instructor would reveal the actual speed of each vehicle. The patrolman admitted: "You mess up, but over the two days, you practice all day long. You eventually get to the point to where you can make a guess." During work hours, the patrolman will often park his patrol car, watch a couple of vehicles pass by and make estimates about the speed of the vehicles. If he gets several vehicles that "look like" they are going 35 mph and the radar supports that, then he knows what 35 mph looks like in that particular setting.

According to the patrolman, white vehicles, larger vehicles, and vehicles traveling through a curve will look like they are going faster than they actually are. During the interview the patrolman estimated a school bus' speed at 40-43 mph, but the radar speed registered at 38-39 mph. For him, his estimate was acceptable because it was within 5 mph. He is confident that he has developed his estimation of speed of vehicles through

NOTES

practice and experience to the extent that on a "good day", he can usually get within 3-4 mph of a vehicle's actual speed.

The Timber Sales Consultant

The timber sales consultant estimates several factors such as (1) size of tracts of trees in acres, (2) quarter-acre circular plots within tracts, (3) diameters of individual trees, (4) heights of individual trees, and (5) the number of board feet within a tract. Here are two of his estimation experiences:

To estimate the height of individual trees, he has to consider that "most sawmills want you to cut off the height estimate at a six-inch diameter top." Since the tree tapers in size as it grows, he stops to get his height estimate at the point where he thinks the diameter of the tree is down to six inches. For a tree that fell in the 14-inch diameter class, he calls "this tree a fourteen three. Three meaning there are three 16-foot cuts that I can get up to a six-inch top. So I'm assuming that up there at 48 feet, there is a six-inch top, so that gives me three 16-foot cuts." In terms of height, he classifies the trees in terms of half logs and logs, where every 16 feet is one log. So, fourteen three also provides the information that the tree is three logs in height.

"I had a little tool, a hypsometer, that you could back off a chain (66 feet), shoot the bottom of the tree, then turn it and shoot to where you think the tree has a six-inch top. It's pretty accurate in giving you a height measurement. That was back in 1972 that I used it." Through repetition, he no longer uses a hypsometer. He is accurate enough to visually estimate the height of the trees. When he estimates the heights of trees, he states, "I'm not even thinking of it in terms of feet. I'm thinking about it in logs. I'm looking at it thinking three, three and a half, two and a half." In other words, he looks at the tree and immediately classifies the tree as a three-log (48 foot) tree, 3.5-log (56 foot) tree, etc., and he uses how he classified other trees as a way to check his current estimate. As he "cruises" a tract of land, he will categorize trees on a tally sheet

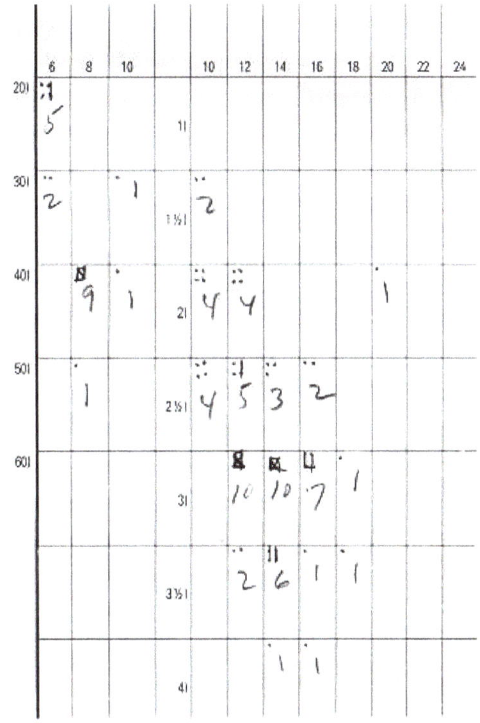

Photograph 1. Tally sheet used for placing trees in categories during the "cruise" of acreage

The final cruise report (Photograph 2) has the same tree categories as the tally sheet used during each cruise. At the end of the year, all of the cruises performed by him are tallied. His aim is that his total estimate of volume (board feet) for the timber cruised comes within 3% of the actual total board feet cut out from the trees. He emphasizes getting a good acreage estimate and a good diameter classification. The good acreage estimate is most important because "missing the acreage will blow the entire cruise." He focuses on getting a good diameter classification because an error in diameter classification throws off the board feet volume much more than an error in height classification. For example, a three-log, 16-inch diameter tree will have 211 board feet, but if it is incorrectly placed in the 14-inch diameter class, then the estimate will

Volume 1

be 146 board feet, or 65 board feet too little. On the other hand, if the same three-log, 16-inch diameter tree is incorrectly placed in the 2.5 log category, then the estimate will be 185 board feet, or only 26 board feet too little. As he drives up to the tract, he gets a general sense for how many board feet per acre the tract of trees will have. Then, "when I'm going through there, I get a feel for how much there is per acre." As he is cruising the tract of land, he starts thinking, for example, "I'm on 7000 board feet per acre." According to him, "if it's normal South Georgia timber, there's going to be 6000-8000 board feet to the acre." He uses this up-front "feel for it" to verify his estimate in the end, to sense that he has a good estimate.

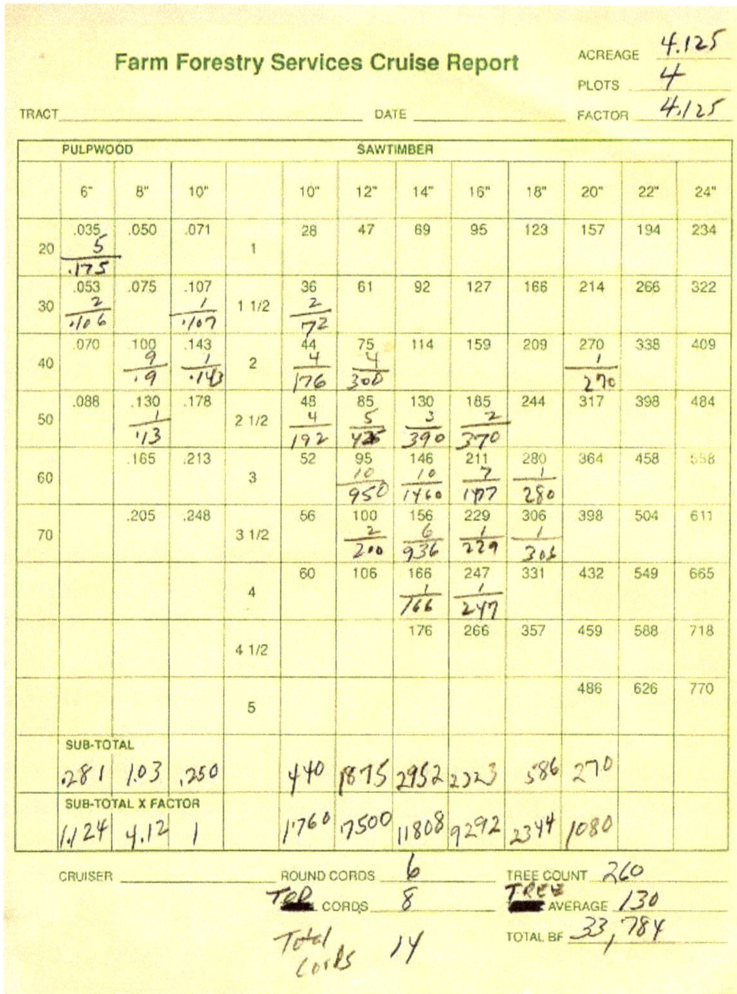
Farm Forestry Services Cruise Report

ACREAGE 4.125
PLOTS 4
FACTOR 4.125
TRACT ______ DATE ______

	PULPWOOD				SAWTIMBER							
	6"	8"	10"		10"	12"	14"	16"	18"	20"	22"	24"
20	.035 5 .175	.050	.071	1	28	47	69	95	123	157	194	234
30	.053 2 .106	.075	.107 1 .107	1 1/2	36 2 72	61	92	127	166	214	266	322
40	.070	.100 9 .9	.143 1 .143	2	44 4 176	75 4 300	114	159	209	270 1 270	338	409
50	.088	.130 1 .13	.178	2 1/2	48 4 192	85 5 425	130 3 390	185 2 370	244	317	398	484
60		.165	.213	3	52	95 10 950	146 10 1460	211 7 1477	280 1 280	364	458	558
70		.205	.248	3 1/2	56	100 2 200	156 6 936	229 1 229	306 1 306	398	504	611
				4	60	106	166 1 166	247 1 247	331	432	549	665
				4 1/2			176	266	357	459	588	718
				5						486	626	770
SUB-TOTAL	.281	1.03	.250		440	1875	2952	2323	586	270		
SUB-TOTAL X FACTOR	1.124	4.12	1		1760	7500	11808	9292	2344	1080		

CRUISER ______
ROUND CORDS 6
TOP CORDS 8
Total cords 14
TREE COUNT 260
TREE AVERAGE 130
TOTAL BF 33,784

Photograph 2. Example of final cruise report

NOTES

The Farmer

According to the farmer, "when you have to work with Mother Nature, the conditions really do make a difference in how you estimate." Yet, he still has to do the best he can for his job. He estimates factors like (1) the amount of seed to use, (2) sizes of planting fields, (3) when to plant, and (4) when to harvest. In order to determine the amount of seed to use, he uses lessons he learned when he farmed with his father. "My daddy knew how many acres were in the field and we would plant about half of it. We had to plant at least a quarter to half of the field to know. That would still give a guess, because you had to guess where half of the field was." From this planting, they could then determine how much seed would be needed to finish planting the field. He now has an acre counter (wheel) mounted on his planter. It is a wheel that rolls out digits like an odometer. He multiplies the distance on the acre counter by the row width. This gives him the number of acres planted, and he uses this tool as part of his seed estimate. He fills the planter "level full" with seeds and starts planting the field by making one round through the field. "If that round is one acre or a half-acre, then you estimate how much seed you have used." To determine the amount of seed used, he fills the planter level full again and estimates how much is left in the fifty-pound bag. This gives how much seed is being used per acre after making just one round in the field. Since he knows the size of the field, he can then estimate how much seed is needed to finish planting the field.

"Most all fields have been measured by the [state farming] office. You can get a number there, but it's not always accurate. For example, fence lines grow into the field. The trees and bushes growing out into the field will make the field a lot smaller even in a few years. You put fertilizer out there and the trees really grow fast. So if you have the acreage, you still have to estimate. You take two or three acres off a 50-acre field. After planting one time, I go with what the acre counter says. I do all of my own work, so I'm out in the open all of the time looking at land. I know what size my fields are and I'm looking at them all of the time."

He also estimates when to plant based on the ground temperature. He plants at about 65°F, and he obtains the ground temperature from reports. He must also look at the weather that is developing. An upcoming cold front will delay planting. "You just know when to plant. My daddy used to say when you kicked the covers back it was time for planting."

He knows how long it takes to harvest a set number of acres, but conditions vary, so he still has to estimate. If peanuts have been rained on, for example, then it takes a lot longer to harvest. His equipment now gathers six rows at a time, but other factors such as the condition of the equipment impact the rate of harvesting individual fields. "...You have to estimate down time, because you can count on that. For new equipment, you can estimate less down time." All of these factors impact how long the harvesting process will take.

"I used to think that you needed to get within 10 or 15 percent of my goal when estimating. The way things have been going lately on the farm, if I get within 25% of my goal then I'm doing really good."

The Automotive Technician

The automotive technician started off using measuring devices, but then through experience and instruction, learned to estimate. He credits his estimation abilities to the repetition of measuring with a device early in his career. He also credits instruction from others.

Rather than go grab a measuring device, he was instructed to "look at this, or feel this." Early on in his job, he "relied on things like a gauge, the actual measuring device, until you get to the point where you can look at it and say 'this is where you are at.' Now I can go actually measure it, but I'm telling you this is where you are at." He does not need to measure any more. It took him "at least two years" to get to this point, but "that doesn't encompass the entire spectrum of what you are doing. I would say at least a third of what you are doing in that two- year period you would be comfortable enough to make an estimate without going to get a measuring device." He uses other common objects at times. He demonstrated that if you put a penny upside down in the tire tread and "can see all of Lincoln's head (Photograph 3), then you have less than 3/32 of an inch and your tire is worn out."

Photograph 3. From the top of Lincoln's head to the top of the coin is 3/32nd of an inch.

Estimation Declarations: Implications for Teaching Estimation

When the authors analyzed the interview data in search of specific estimation enactments, the authors found many and determined to classify them as "estimation declarations". These declarations emerged from the shared work of estimation by the traffic patrolman, timber sales consultant, farmer, and automotive technician and can be used to support the teaching of estimation.

1. There are distinct and real-life contexts for exercising estimation skills and these contexts are fruitful for exploration beyond abstract examples of estimation. Consider the four professionals.

2. People who are skilled at estimation have spent a great deal of time in practice and repetition of their skills, and they use these prior experiences to develop estimations.

3. Watching and receiving instruction from others with more estimation experience are important for building estimation skills.

4. Estimation within particular contexts relies on factors of that context (e.g., variations, average or usual measurements, characteristics of objects being measured).

5. Different estimation skills exercise different senses and body activities (e.g., intuition, sight, walking paces).

6. Practice and comparison with actual measurements first helps build estimation skills until one "gets a feel" for the task at hand.

7. Acceptable or good estimates vary with the context in which the estimates are developed.

NOTES

8. Some estimates are more critical for a given situation than others. That is, some estimates, if too high or too low, may have a greater impact on a situation than other estimates that are too high or too low (e.g., planting an appropriate amount of seed).

9. In many instances, factors that cannot be controlled (e.g., nature or natural phenomena) impact the outcome of estimates.

10. Within given contexts, estimates may change as other factors change (e.g., size of farming fields).

11. Personal and cultural traditions carry valuable information that can be useful for estimation (e.g., the farmer's father's advice about planting).

12. Knowing informal tools that are useful for estimation (e.g., coins) can benefit estimation activities.

Knowing and doing mathematics is manifested in many ways, including in the work of people as they perform required tasks. The estimation declarations presented represent just some of the insight gained from the interviews and from accompanying the professionals on their jobs. Some implications for pedagogical practice in teaching estimation include the following:

- Share information about how professional use measurement estimation to perform their job duties as a way of connecting classroom instruction to real-life experiences. Just being about to share these interviews with students have helped to capture their interests in estimating in the context of measurement.

- Invite professionals to speak about and demonstrate their estimation experiences and skills. This will help students see the usefulness of estimation in real-life contexts.

- Seek opportunities for students to "shadow" professionals as they perform job duties that involve estimation.

- Provide ample opportunities for students to repeatedly practice estimation.

- Students can conduct interviews of professionals in their local communities to gather information about how these professionals use estimation in the course of their work. Students can follow up with written reports and demonstrations.

- Encourage students to develop informal estimation benchmarks. For example, the automotive technician uses a penny to support his work.

- Design simulations that model the ways professionals estimate on their jobs. For example, the estimation practices of the traffic patrolman can be modeled inside or outside using moving objects (e.g., students pushing toy cars or navigating remote cars, students riding bicycles, etc.) and stopwatches.

Conclusion

The skill and enactment of estimation comes to life across many professions as when examining the act of doing mathematics when professionals engage in their work. Missing from students' mathematical experiences is often a sense of what doing mathematics means outside of the classroom context. Approaching mathematics via the lenses of ethnomathematics reveals ways in which teachers can present more meaningful mathematics contexts to students and empower them to engage in the ways of doing mathematics.

References

Bassarear, T. (2008). *Mathematics for elementary school teachers*. Boston, MA: Houghton Miflin.

D'Ambrosio, U. (2007). *Ethnomathematics: Link between traditions and modernity*. Rotterdam, The Netherlands: Sense Publishers.

Hatfield, M. M., Edwards, N. T., Bitter, G. G., & Morrow, J. (2008). *Mathematics methods for elementary and middle school teachers*. Hoboken, NJ: John Wiley & Sons.

Kennedy, L. M., Tipps, S., & Johnson, A. (2008). *Guiding children's learning of mathematics*. Belmont, CA: Thomson.

Maykut, P. & Morehouse, R. (1994). *Beginning qualitative research: A philosophical and practical guide*. Washington, DC: The Falmer Press.

National Council of Teachers of Mathematics. (2000). *Principles and standards for school mathematics*. Reston, VA: NCTM.

Reys, R., Lindquist, M. M., Lambdin, D. V., & Smith, N. L. (2009). *Helping children learn mathematics*. Hoboken, NJ: John Wiley & Sons.

Van de Walle, J. A., Karp, K. S., & Bay-Williams, J. M. (2010). *Elementary and middle school mathematics: Teaching developmentally.* Boston, MA: Allyn & Bacon.

Volume 1

Culture That Counts

Volume 2 ◆ Community

Volume 2 Community

Volume 2
Community

Sense and Simplicity: Ethnomathematics of Negev Bedouins

Ada Katsap
Mathematics Education Department
Kaye Academic College of Education
Beer-Sheva, 84536
Israel
03-571-4784
adak@013.net

Fredrick "Rick" Silverman
School of Teacher Education
College of Education
University of Northern Colorado
Greeley, Colorado 80639
USA
970-351-2760
flsilver@gmail.com

Abstract

This article presents examples of the mathematics inherent in lives of the Negev Bedouins and examines interlacing of ethnomathematics in school mathematics.

The particular descriptions of the Bedouin women's craftwork in embroidery, weaving, and similar pursuits is based on theories of frieze and wallpaper groups and symmetries of color to elucidate the mathematics that occurs naturally in those pursuits. Another domain refers to the Bedouins' heritage, as expressed in various games played by the Negev Bedouins, one of which this article describes fully, and with an eye toward using it as a math class activity. Brief discussion suggests a process for introducing ethnomathematics into a mathematics curriculum to promote likelihood of students' understanding of mathematics that is present in current practices originating in the cultural tradition of the learners.

Keywords:

- Negev Bedouins
- Culture and Mathematics
- Women's Craftwork
- Teaching

Introduction

A Case Study of the Role of Ethnomathematics among Teacher Education Students from Highly Diverse Cultural Backgrounds (Katsap and Silverman, 2008) was published in JMC (3, 1). It describes the research methodology and findings with mathematics prospective teachers (MPT), Jewish and Bedouin alike, who came together in the Ethnomathematics Program (EP) in the Kaye College of Education in Beer Sheva, the Negev, Israel. The population demographics from which Kaye College derives its students leads to a student body of Negev Arab-Bedouin Muslim and Jewish students in approximately equal numbers. The Bedouin MPT attending the College from the Negev settlements come from families that follow the traditional ways of life and preserve the ancient customs and traditions, including the religious imperatives of Islam. Some Jewish MPT have religious backgrounds, others have secular. These variations, Jewish and Bedouin alike, among all the students in the course, explain the differences in the traditional practical mathematical knowledge from their own heritages.

Moreover, that article (Katsap and Silverman, 2008) describes the program's impact and shows examples of MPT's learning of naturally occurring mathematics in the cultures of their own people and that of their peers. It also reveals various ethnomathematics activities derived directly from cultural experiences, outlines mathematics content of the course, and indicates mathematics pedagogy. An exposure to ethnomathematics assisted MPT to better understanding of mathematics rooted in socio-cultural settings and pointed to a new way for them to view mathematical concepts, one that can and should be used for linking school mathematics and naturally occurring mathematics in people's lives. The article gives an account of academic mathematics topics brought to light in the course, on the one hand, and on the other, practical mathematics identified by each culture, Jewish and Bedouin. The intercultural dialogue that took place in the course was unique because it is not common in Israel that Arabic and Jewish mathematics students will listen, question, and conduct discussions across the cultural separation with such a high level of respect for one another as was evident in this course. The summary indicated that the research demonstrated that Ethnomathematics in mathematics teacher education can influence the MPT's professional development in such a way that they became more likely to incorporate humanistic and social aspects of mathematics in their transition from teacher education students to contractually employed teachers in schools.

How did that research we presented in 2008 impact us in the ensuing years? The ethnomathematics revealed in the Bedouin culture during the research period had a strong appeal to us, leading to greater desire to learn it in depth, study, understand, and bring the ethnomathematics of the Negev Bedouins to the eyes of many. We toyed with the idea of writing a book. And so began Ada Katsap's visits with many Bedouins of the Negev, Israel, to interview them and to collect original materials from them about their lives. She took numerous photographs and wrote extensive notes that became the cornerstone for our book, published in 2016: *Ethnomathematics of Negev Bedouins' Existence in Forms, Symbols, and Geometric Patterns*. Below is an image of the book's cover and as it appears on the website of Sense Publishers, providing a glimpse into

NOTES

the book's content. One of the topics in the book deals with a question, "Why and how might it be possible to use the naturally occurring mathematics of the Bedouins for enhancing the teaching and learning of school mathematics?"

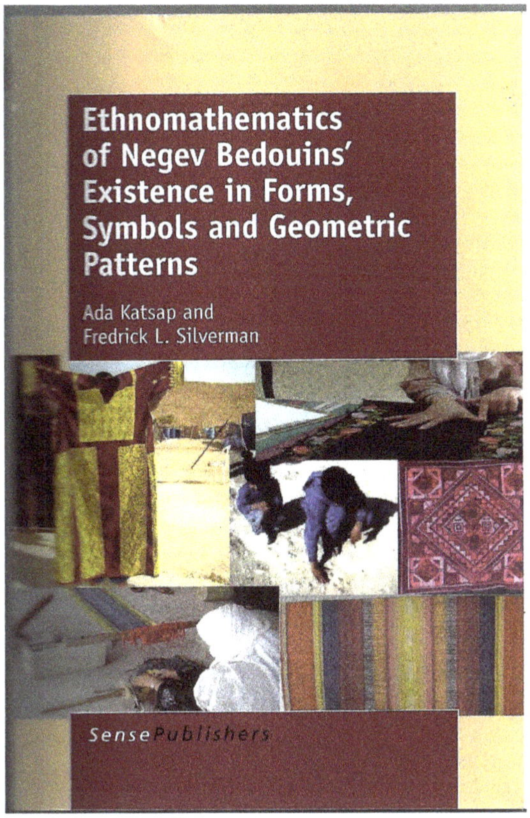

https://tinyurl.com/NegevBedouins

The book *Ethnomathematics of Negev Bedouins' Existence in Forms, Symbols, and Geometric Patterns* overview

The book presents an abundance of mathematics inherent in the lives of Bedouins of the Negev. The math can be explained within the fields of ethnomathematics and academic mathematics, thus being useful to explore interlacing of ethnomathematics motifs in school mathematics. The reader will be able to recognize the ethnomathematical narrative gleaming throughout the Negev Bedouin culture; it can provoke a journey into their world, as well as attest to the beauty and sophistication of mathematics that occurs naturally especially in the women's craftwork, and also in built structures, games, and in many other aspects of Bedouin life.

The book opens with a narrative of Bedouin history followed by a brief overview of ethnomathematics. The essence of the book is an in-depth view of Bedouin women's traditional handicraft through which they reflect social and cultural activities in weaving, embroidering, and similar pursuits. Their creations reveal mathematical ideas incorporated in embroidery compositions of repeated patterns of flowers and geometric figures in varying scales. Bedouin men construct dwellings, tents, desert wells, and such. They and their children play games attuned to sand and other features of the desert environment.

The collected empirical data about the mathematical practices of the Bedouins of the Negev was examined. The structures and systems that we discerned in the data are the bases for within which our work presents the various activities. Those organizing structures and systems include symmetry groups, color symmetry groups, and others. Photographs document the examples of Bedouin ethnomathematics, are the subject of considerable analysis, and appear throughout the book. The book's two concluding chapters include discussion about bridging the gap between mathematics experiences in school and those outside school. Also considered toward the end of the book are mathematically problematic situations embedded in Bedouin sociocultural heritage likely to appeal to teachers for use with school students from basic school level. The book encourages readers to attend to the effort to utilize its findings to facilitate learning and teaching mathematics in the precollege grades. The following paragraphs recount some aspects of the subjects from our book.

Ethnomathematics of Bedouin women craftwork

The search for Bedouin embroideries in the Negev led to focusing on three scenes, the first one being the Desert Embroidery Factory, which commenced operation in the Negev Bedouin town of Laqye, and gives expression to a unique combination of art, culture, and employment for Bedouin women. Embroidery motifs created there are based on uniquely traditional Bedouin embroidery from the Negev area. The factory has a small yard and includes a visitor center and store, and serves as the gathering place for social meetings and lectures for both the embroidery and those who take an interest in it.

The second scene was the Negev markets where women bring their merchandise: dresses, cushions, and decorations whose most prominent characteristic is an embroidery of many colors and patterns.

And the third scene is the dwellings in the villages unrecognized by Israel and in the Bedouin towns in the Negev. Of course, in reality those settlements exist. "Unrecognized villages" is a political statement about some Bedouin settlements in the Negev that the government of Israel does not accept as legal. Although the embroideries were the product of work of women from various settlements, we have found a common denominator characterizing the Bedouin embroidery in the Negev as a whole, one that sets it apart from those of their Bedouin counterparts in other regions, such as in Northern Israel and in Hebron with the Palestinian Bedouins. Our book provides a detailed description of the uniqueness of the Negev region embroidery and compares it with embroidery patterns originating among Bedouins from other regions.

Following are several examples from "Transformations, shapes and patterns analysis in the Negev Bedouins' Embroideries," Chapter 6 of our book. The symmetries of frieze and wallpaper groups theories come into use for particular descriptions of Bedouin women's handicraft.

We began by displaying frieze group features of the patterns of the commonly occurring strips in the Bedouins' embroideries we collected. A frieze group is a mathematical concept used to classify designs that are

NOTES

characterized by repetition in one direction, based on the symmetries in the pattern. There are seven symmetry groups of Frieze patterns (Washburn and Crowe, 1988). We use the names of frieze groups and the notation for the symmetry group that were standardized and adopted by the International Union of Crystallography (IUC) in 1952. The symmetry groups of frieze patterns are named in the four-symbol format of the IUC notation: *p, m* or *a*, and numbers *1* or *2*.

All labels begin with "p," which denotes pattern, followed by "m" or "a," and conclude with "1" or "2." For example, in **pmm2**, "p" means pattern, the first "m" means vertical reflection, the second "m" means horizontal reflection, and the last number "2" means a 180° rotation. The different quartet combinations indicate the kinds of motions that create the patterns.

Examination of the frieze strips revealed that the most prevalent pattern in the Bedouin strips belongs to the frieze group denoted *p1m1*. These strips have a horizontal reflection symmetry. Whether imaginary or real, a horizontal line in the middle of each strip acts as an axis of the symmetry that enables creation of equal twin parts. See the figures 1-4 below.

Figure 1. Strip with symbolic notation p1m1- horizontal reflection symmetry

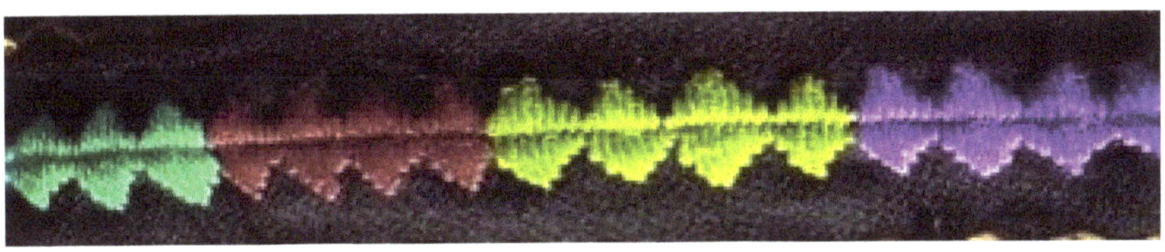

Figure 2. Strip with symbolic notation p1m1 - horizontal reflection symmetry

Figure 3. Strip with symbolic notation p1m1- horizontal reflection symmetry

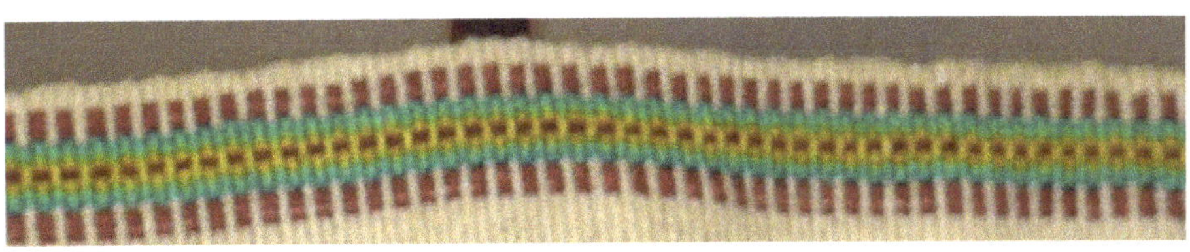

Figure 4. Strip with symbolic notation p1m1 – horizontal reflection symmetry

Two examples of kinds of frieze symmetry groups appear in the next two figures, 5 and 6. Figure 5 contains a combination of vertical and horizontal reflections (denoted *pmm2*), that is to say, the design simultaneously has horizontal and vertical reflection symmetry (without attention to the colors of the cells). This means that it is possible to imagine two perpendicular lines in the middle of the strip, and two axes of symmetry intersecting at 90°.

Figure 5. Strip with symbolic notation pmm2 – horizontal reflection, vertical reflection, & 180° rotation

In Figure 6, the frieze group of the strip has vertical reflection symmetry only (denoted *pm11*). This means that it is feasible to imagine a vertical line, an axis of symmetry, in the middle of the strip.

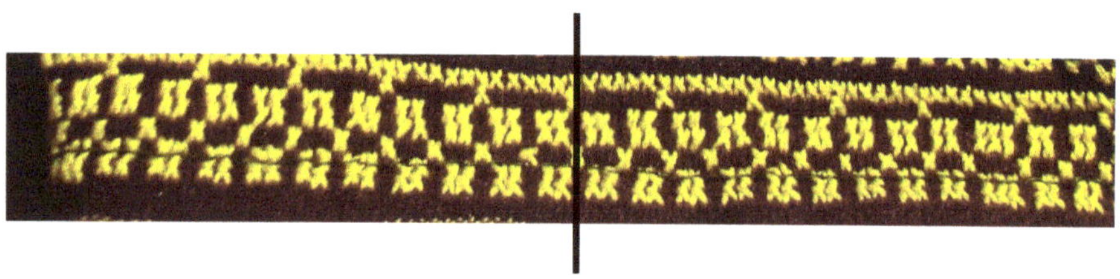

Figure 6. Strip with symbolic notation pm11 - translation, vertical reflection only, no rotation

A very rich and complex world comes into view from Bedouin women's embroideries on the dresses and cushion pillows. We characterized the components of them through the features of the 17 wallpaper symmetric groups, according to IUC notation (Fedorov, 1891, completed by Polya, 1924/ published in English in 1971). The emphasis was on the chest part (called *toubs*) of the dresses, and cushion pillows, as shown in Figures 7-9.

NOTES

In mathematics, a *wallpaper pattern* is any subset of the plane whose translational symmetry group is repetitive in two independent directions. A plane symmetry group, termed *wallpaper group*, is a mathematical classification of two-dimensional repetitive patterns, based on the symmetries in the patterns and having two independent translations. The symmetry groups of wallpaper patterns are denoted, according to the Hermann–Mauguin or International crystallographic notation (Hermann, 1928, and Mauguin, Ch. 1931) in the form of a two-symbol to four-symbol system that combines the following letters and numerals: c, g, m, p, 1, 2, 3, 4, 6. In the standard crystallographic notation, the numerals 2, 3, 4, and 6 refer to rotations through angles of 180°(2), 120°(3), 90°(4), 60°(6). The crystallographic theorem in its basic form was based on the observation that the rotational symmetries of a crystal are limited to 2-fold, 3-fold, 4-fold, and 6-fold, hence, the numeral 5 does not preserve the translational symmetry and therefore does not appear in the system. All 17 groups have been used in the art and craft of the Muslim world for generations. Our book presents details and references for this system.

Figure 7. A chest part of dress for a woman

Figure 8. A full dress

NOTES

Figure 9. A cushion pillow

Figures 7 and 10 show the chest part of dresses, and corresponding IUC notation *p4m* (Washburn and Crowe, 1988). Their lattice type is square and has a rotation of order 4; the reflection axes intersect at **45°**, and all rotation centers lie on reflection axes.

Figure 10. A chest part of dress

Figure 11 shows the chest part of a dress for which the corresponding IUC notation is *p1*. The lattice type is rectangular, without rotation, and with translational symmetry only.

Figure 11. A chest part of dress

The study of color symmetry groups has occupied, and still occupies, many researchers. An example of two-color symmetry groups is provided by Ascher (1991), where analysis revealed symmetry groups in strip decorations that have been created by the Inca. Figure 12 is an example of Bedouin handmade knitted decoration. That knitting uses two types of patterns, right isosceles triangles and diamonds. Negev Bedouin women invite the observer's direct attention to the motif in which the particular coloring of the patterns in the embroidery influences the symmetry groups that one can recognize in the motifs. Figure 12 shows a decoration that enables viewers to see in two different images, according to the selection of the background of the motif.

Figure 12. A handmade knitted decoration

Chapter 7, "Patterns, colors, and designs in Bedouins rugs and decorations" of our book presents other Bedouin crafts, rugs, and decorations made by weaving. Resources for these products were markets and also Bedouin lodgings, the same as for products described earlier. The *Laqye Negev Weaving Project* in Laqye was an additional resource for information.

Figure 13. Carpets and cushions in Laqye Negev Weaving showroom

Our book presents a collection of original stripe patterns found in the Negev Bedouin rugs, carpets, and other belongings, a sampling of which appears in Figure 13 that were organized systematically by product type and available for purchase. Below we present some examples of the weaving products with explanations. Katsap and Silverman (2016, pp. 178-180) present a table that delineates the Bedouin strip patterns and shows their variations. Figure 14 shows a carpet piece comprised of 5 patterns that appear in the table: (a) 3 color segments with translation symmetry, (b) ladder pattern with bars, (c) ladder pattern without bars, (d) one-side comb, and (e) vertical line.

NOTES

Figure 14. A portion of a carpet that the purchaser uses as desired

The Bedouins who settled throughout the Negev desert wove many practical items which served their needs in the nomadic environment. Figure 15 shows a woven cushion that reveals different combinations of the same colors. The weaver knew how to divide the colors between the combinations of groups and choose the appropriate color for the middle of the design to create a symmetrical and harmonious design.

Figure 15. A woven cushion

Figures 16 and 17 show a donkey bag and junk bag. Such accessories are very common in the unrecognized villages across the Negev.

Figure 16. A donkey bag

Figure 17. A junk bag

The foregoing section portrayed the embroidering and weaving products of the Negev Bedouin weavers and embroidery women. Sole patterns and combinations of some patterns and colors in the artifacts expose the suggestive language of symbols of Bedouin society, descriptions of which appear in our book.

Bedouins desert games as an educational tool

Games are susceptible to mathematical and logical analyses, some relationships being evident, others not obvious. "Culture arises and unfolds in as play" (Huizinga, 1984, p. ix). Hence, play is an activity integrally connected to cultural heritage and to the world that surrounds the players. Many games induce competition, along with a drive to improve one's play, and as a result, players may apply analytical thinking to further their understandings of games for which strategy might be possible. Games may relate to players' cultural heritage, which is an integral part of the society to which they belong. Various games played by the Negev Bedouins, elders and children's alike are a major part of leisure culture in the desert. Figure 18 shows a Bedouin child playing with stones.

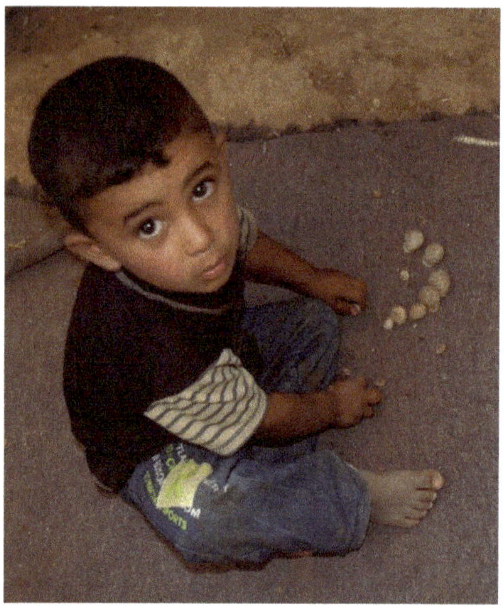

Figure 18. This picture was published on 6/4/2008 in the section "Bedouin Children"

Another domain draws upon Bedouin heritage, as expressed in various games played by the Negev Bedouins, one of which this article describes fully and with an eye toward using it as a math class activity. Six games played by the Negev Bedouins are presented in our book in a form suitable for implementing as a math class activity with children aged 6 to 14 years, according to the level of the class. Play can be outdoors, at a sand station indoors, or simply indoors with the pieces on a classroom table or the floor. Following is one of the six games described in our book, chosen because the game fascinated us, is common, as are the other 5, is challenging, and has this photo that engaged us (Katsap and Silverman, 2016, pp. 264-267).

Noked or Mozkat - 5 Stone Game

The Bedouin game Mozkat uses 4 small stones of equal size and 1 large stone, as in Figure 19. The large stone, known as *Mozkat*, which is in the top center of the figure, gave the game its name.

Figure 19. Bedouin 5 Stones in Mozkat Game

NOTES

Game description

The players use their judgment to discriminate between "little" and "large," comparatively sized as in the photo. The game is intended for two or more players, each of whom, in his turn, throws the Mozkat and distributes the little stones according to the rules of each stage. The description below applies to two players, but by turns it can include more than two players. The objective is to do two things simultaneously: catch the Mozkat and arrange the stones, each stage offering different combinations. Each player tries to go through all the six stages, and if the player accomplishes them successfully, he wins 4 points. The winning player is the one who scores the most points. Before they begin the game, the players decide on the winning number of points. We did not collect information about whether the number of players has any effect on determining the number of points required to win the game. It was said that the number of points required to win the game does have implications for the number of rounds to play. One complete round includes attempts at all six stages, unless players decide otherwise. Our interviews led us to conclude that games usually consist of 12 to 16 points.

Six point-scoring stages in stone arrangement and Mozkat throwing

1. Stage one: 1+3 stone division means pick up 1 and then pick up 3, and throwing and catching the Mozkat.
2. Stage two: 2+2 stone grouping and throwing and catching the Mozkat.
3. Stage three: 3+1 stone grouping and throwing and catching the Mozkat.
4. Stage four: 4 stones grouping and throwing and catching the Mozkat.
5. *Touza* Stage
6. *Gababa* Stage

Game Rules

Starting the game: Four little stones are scattered on the sand.

1. The first player puts the stones into a 1+3 grouping, throws the Mozkat into the air, and immediately throws 1 stone into the air. The players adjust height for their skill level and according to their perceptions of fairness. He catches the stone, and while doing that must also catch the falling Mozkat with the same hand. If he misses, the turn goes to his opponent; if he succeeds, he continues to stage 2. We do not know one way or the other whether the stones are replaced to the original positions after each player's turn, but we speculate that the players likely decide that question for themselves. Furthermore, we speculate that the players decide whether one who did not succeed at any given stage has the chance next time at that same stage or must start over from the beginning.
2. The player – kneeling, sitting, or whatever comfortable position - puts the stones into a 2 + 2 grouping, throws the Mozkat into the air, and immediately throws two stones into the air. He catches the two stones, and while the Mozkat is in the air catches two stones in one hand and while doing that must also catch the falling Mozkat with the same hand. Counting the Mozkat, 3 stones will be in the air. If he misses, the turn goes to his opponent; if he succeeds, he continues to stage 3.
3. The player puts the stones into a 3+1 grouping, throws the Mozkat into the air, and immediately throws 3 stones into the air. He catches the three the stones, and while doing that must also catch the falling Mozkat with the same hand. Counting the Mozkat, 4 stones will be in the air. If he misses, the turn goes to his opponent; if he succeeds, he continues to stage 4.

4. The player throws the Mozkat into the air, and immediately throws 4 stones into the air. He must catch the four stones, and while doing that must also catch the falling Mozkat with the same hand. Counting the Mozkat, 5 stones will be in the air. If he misses, the turn goes to his opponent; if he succeeds, he continues to stage 5.

5. This stage is called *Touza.* Two fingers, the thumb and any other finger, of one hand are arranged to create a semi-circle on the sand. The distance between any two stones is approximately one *shib'r*, about the span of one hand. The four stones are cast upwards behind the hand laid on the ground with the half-circle of the finger shape. Once the stones touch the ground, the player throws the Mozkat into the air. He must sweep the four stones on the ground with one stroke through the half-circled fingers of the hand laid on the ground and must immediately catch the falling Mozkat before it hits the ground. If he misses, the turn goes to his opponent; if he succeeds, he continues to stage 6.

6. This final stage is called *Gababa.* At this stage, the player takes all four stones into both hands with open palms, throws the four stones into the air, and quickly turns his hands over, and tries to catch the stones on the back of the hands. Apparently, *Gababa* does not involve the Mozkat. Neither of our two sources mentioned it for this stage. Completion of all the 6 stages scores points.

The game continues until one of the players completes the pre-determined number of rounds, each of the 6 stages, unless decided otherwise at the start, and thus scores the number of points determined at the beginning as sufficient for winning, 4 points per round, and with 12 or 16 points usually the agreed total to win.

Game objectives in the area of mathematics:

1. Counting and adding of numbers.

2. Grouping the four pebbles into possible, allowable permutations, which are a subset of the complete set of partitions that sum to 4.

Obviously, this game requires considerable concentration ability.

Ethnomathematics in math curriculum

The brief discourse below is to suggest possibilities for incorporating mathematics, curriculum, teachers, and students with an ethnomathematical approach. Izmirli (2011) explained broadly the connections between ethnomathematics and its potential for learning and teaching mathematics:

> *The ethnomathematical perspective is cognizant and receptive of the impact of various cultural conventions and inclinations, including linguistic practices, social and ideological surroundings, for doing and learning mathematics. It regards mathematics education as a potent structure that helps learners and teachers to critique and transform personal, social, economic, and political constructs and other cultural patterns (p. 39).*

NOTES

Introducing ethnomathematical approaches into mathematics curriculum and pedagogy is intended to make school mathematics more relevant and meaningful, and to develop *ethnoknowledge*, a term coined by Borba (1990), which is "acquired by students in the pedagogical action process of learning mathematics in a culturally relevant educational system" (Rosa and Orey, 2011, p. 48). Integration of ethnomathematics in mathematics curriculum can lead to higher regard for mathematics among students, given its "inherent mathematical value in special cultures and societies" (Zhang and Zhang, 2010, p. 157). Taking that standpoint into account, education must enhance cultural dignity (D'Ambrosio, 2002), and it should base learning and understanding of mathematics on recognition that students learn better than otherwise when they can use sources created by their communal peers to complement other materials, including textbooks (Darling-Hammond & Bransford, 2005).

One of the classroom activities that Katsap and Silverman (2008) explained pertains to preservice teachers reproducing the process that Bedouin women use to weave a specific carpet pattern. These teacher education students described and showed the Bedouin women's process as follows: "measure the width of each part of carpet, starting at the right-hand edge of the carpet and proceeding successively… to the midline of the image;" [the actions] "require careful measuring and counting of stitches." In this activity, the MPTs teachers identified with Bedouin women who do such carpet weaving, as expressed by this comment from one of the MPTs:

> *The conversation conducted inside the college class about the geometric figures and the kinds of geometric transformations [that] appeared in the carpets built a bridge between academic mathematics and outside classroom life, and it permitted mathematical-socio-cultural dialogue, thus facilitating better understanding of several transformations. (p. 86).*

Our research indicates that creation of a linkage between the prospective teacher's culture and academic mathematical concepts in school curricula can lead to a special sense of satisfaction with their learning of mathematics, as well as with their teaching it. The teachers' personal stories of naturally occurring mathematics in their own life experiences can be a model for the same to arise in the math learning of their students. Ethnomathematics can serve teachers well for eliciting students' authentic examples and problems couched, first, in common language, and later in the language and symbolism of formal mathematics associated with mathematics content courses.

A Summary

We advocate that the purpose of education is to create a sound basis for individual growth and development of the learners, so that learning will remain an inseparable part of their lives. Ethnomathematics opens doors, new to some, into the understanding of mathematics with respect to culture and personal experience. Thus, heritage - in the form of enduring practices, traditions, and general cultural elements of the learners - gains a substantive place in the teaching and learning of mathematics.

Our book brings forth examples of the mathematics embedded unobtrusively, that is, naturally occurring, in the daily lives of the Bedouins of the Negev. It furthers the realization that mathematics, the academic discipline, has its roots in the everyday experiences of everyone, across time and cultures, and is evident, in particular, in the Negev of Israel among the Bedouins who inhabit those glorious desert expanses. Our book's examples may support, or even inspire, teachers of school mathematics, university faculty in mathematics education and content, and in some instances educators in other subject areas. In particular, though, a desirable outcome is to motivate educators to do more than is common with respect to linking mathematics and culture in teaching and learning mathematics in ways that benefit both students and teachers, thereby utilising the complementary relationship between ethnomathematics and academic mathematics.

NOTES

References

Ascher, M. (1991). *Ethnomathematics: A Multicultural View of Mathematical Ideas*. Pacific Grove: Brooks/Cole, pp. 203. Chapter six: Symmetric Strip Decoration, pp. 156-179.

Borba, M. C. (1990). Ethnomathematics and education. *For the Learning of Mathematics, 10*(1), 39-43, cited in Rosa, M. & Orey, D. C. (2011). Ethnomathematics: the cultural aspects of mathematics. *Revista Latinoamericana de Etnomatemática, 4*(2). 32-54.

Bradley, C. and Cracknell, A. (1972). The Mathematical Theory of Symmetry in Solids. Oxford, Clarendon press.

D'Ambrosio, U. (2002). Ethnomathematics: An overview. This text is the translation of the closing lecture delivered to the II Congresso Internacional de Etnomatemática, 05 a 07 de agosto de 2002, in Ouro Preto, MG, Brazil.

Darling-Hammond, L. & Bransford, J. (Eds.) (2005). *Preparing Teachers for a Changing World: What Teachers Should Learn and Be Able to Do*. San Francisco. Jossey-Bass.

Fedorov, E. S. (edition from 1971). *Symmetry of Crystals.* Translated from the 1949 Russian Edition. American Crystallographic Association, New York.

Hermann, C. (1928). Zur systematichen Structurtheorie 1. Eine neue Raumgruppensymbolik, Z. Kristallogr. 68, 257-287. Cited in Bradley, C. and Cracknell, A. (1972).

Huizinga, J. (1984). *Homo Ludens: a study of the play-element in culture*. Hebrew edition: The Bialik Institute (Mosad Bialik): from original Dutch edition: Huizinga, J. (1938). Homo Ludens: Proeve Ener Bepaling Van Het Spelelement Der Cultuur. Groningen, Wolters-Noordhoff. Reprinted in 1980 in Great Britain by Redwood Burn Ltd Trowbridge & Esher. Pp.220.

Izmirli, I. M. (2011). Pedagogy on the Ethnomathematics - Epistemology Nexus: A Manifesto. *Journal of Humanistic Mathematics*, Vol. 1 Issue 2 (July 2011), pp. 27-50.

Katsap, A., Silverman R. (2016). *Ethnomathematics of Negev Bedouins' Existence in Forms, Symbols, and Geometric Patterns*. Rotterdam: Sense Publishers.

Katsap, A., Silverman R. (2008). A Case Study of the Role of Ethnomathematics among Teacher Education Students from Highly Diverse Cultural Backgrounds. *The Journal of Mathematics and Culture*. June 2008, V.3 (1) ISSN - 1558-5336.

Mauguin, Ch. (1931). Sur le symbolisme des groupes de répétition ou de symmétrie des assemblages cristallins. Z Kristallog. 76, 542-558. Cited in Bradley C. and Cracknell A. (1972).

Pólya, G. (1924). Über die Analogie der Kristallsymmetrie in der Ebene. *Zeitschrift für Kristallographie*, 60, 278-282, cited in Schattschneider D. (1990). *Visions of Symmetry: Notebooks, Periodic Drawings, and Related Work of M. C. Escher*, W. H. Freeman and Co., New York.

Rosa, M. & Orey, D. C. (2011). Ethnomathematics: the cultural aspects of mathematics. *Revista Latinoamericana de Etnomatemática*, 4(2). 32-54.

Washburn D. K. and Crowe D. W. (1988). *Symmetries of Culture: Theory and Practice of Plane Pattern Analysis*, Seattle: University of Washington Press. (Page 63). p. 299. *books.google.co.il/books?isbn=0295970847*

Zhang, W., Zhang, Q. (2010). Ethnomathematics and Its Integration within the Mathematics Curriculum. *Journal of Mathematics Education*. June 2010, Vol. 3, No. 1, pp.151-157.

Hñähñu (Otomi) Mathematics Revisited

Thomas E. Gilsdorf
Department of Mathematics
Central Michigan University
Mount Pleasant, MI 48859 USA

Abstract

English: This article represents reflections on the motivations, consequences, pedagogical considerations, and possible future developments related to the article *Mathematics of the Hñähñu: The Otomies*, previously published in the *JMC*.

Español: Este artículo representa reflexiones sobre las motivaciones, consecuencias, consideraciones pedagógicas, e indicaciones para posibilidades de investigaciones, conectadas con el artículo *Mathematics of the Hñähñu: The Otomies*, previamente publicado en la revista *JMC*.

Hñähñu (Otomi):

Hñähñu u Otomí: Nuna t'ukant'ofo uni ma mfeni dige'a ya ot'e ya jä'i, ra hanja ra nt'utuate ne ra mfädi, ñ'udi te dra t'ot'e pa dra thoni ya ra r'ayo mfädi, ri njuni nu'a thuts'i ko ra t'ukant'ofo ra thuhu *ra munts'amfädi dig'ea ya b'ede, te ri boni ya ñ'udi, ne ya mfädi hingi nthuts'i dige'a ya Otomi*, ge xa nthutsí pa dra unga mfädi ha ya nt'ofo ngu gi nuti *JMC*.

Keywords:

- Ethnomathematics
- Otomi Mathematics
- Cultural Mathematics

Motivations for the original study

My motivations for writing the article Gilsdorf (2009) originated in 2001. At that time I was on sabbatical leave in Mexico and there was to be an international conference on the history of science and mathematics. Ubiratan D'Ambrosio was one of the scheduled speakers. I wanted to submit something to present at that conference. At first I thought about discussing the mathematics of the Maya, but some people I talked to told me that Mayan mathematics is a fairly common topic for such events and that it would be more interesting if I decided to pursue something that would be less familiar. I began investigating some of the lesser-known indigenous groups of Mexico and decided to study the mathematics of the Otomies – which in their own language, are referred to as the Hñähñu. See Wright (1997) for variations and explanations of the terms Hñähñu and Otomi. I will generally use the term *Otomi*. The research I did in order to prepare for the presentation eventually grew into what became the article that appeared in the JMC. It is worth pointing out that before the JMC article, I wrote an article for the journal *Estudios de Culturas Otopames*, Gilsdorf (2008). That article differs from what I wrote for the JMC in that the journal *Estudios de Culturas Otopames* is an anthropological journal.

In what follows I would like to describe some details of how I investigated the subject of Otomi mathematics.

Obstacles and progress of the original study

For starters, I found that virtually no one had studied the mathematics of the Otomi. I found a couple of references to lists of number words that had been published, such as in Barriga Puente (1998), Merrifield (1968), and in Urbano and Acuña (1990), but no other studies beyond that. Hence, the first obstacle I faced was to find the information through other means. An important resource for me was, and continues to be, the Otopame Study Group at the National University of Mexico (UNAM), which consists mainly of anthropologists, but also historians and others, who are interested in the cultures of the Otomangue linguistic family in Mexico that contains the Otomi as one of its cultures. Specifically, the focus is on the Otopame linguistic subfamily, which includes the Otomi, Mazahua, Pame, Chichimeca-Jonaz, Matlatzinca, and Ocuilteco cultures. Numerous discussions I had with various people from the Otopame Study Group led me to important resources such as Bartholomew (2000), Carrasco (1950), Galinier (1990), Lastra and Bartholomew (2001), Soustelle (1993[1937]), and later Albores (2006), Castillo Escalona (2000), Hermann Lejarazu (2001), Lastra (2006), and Wright (1997), among others. These references represent the foundation of the research I did for writing the paper. I was able to create an analysis of Otomi mathematics from the point of view of mathematics of Otomi numeration, some indications of the ritual use of number in Otomi culture, the Otomi calendar, the Aztec mathematical symbols that were also used by the Otomi, as well as general historical an anthropological considerations relevant to Otomi mathematics.

NOTES

Another obstacle related to finding references is that there have been significantly fewer investigations into Otomi culture compared to the better-known cultures of the Aztecs (also referred to as the Mexicas) and Maya. This seems unusual, considering that the Otomies are known to be one of the largest and oldest cultures of Mesoamerica, comparable in size to the Aztecs and Maya (Lastra, 2006, p. 25), (Wright, 1997, p. 439 – 441). One reason for the lower level of interest has to do with negative images that have been associated with the Otomies over time – see Wright (1997) and Wright (2005, p. 19). Another consideration is that, in general, the Otomies are not considered to have been builders of an empire, which distinguishes them from the two better-known cultures of the Aztecs and the Maya. There is a certain Western influence to this. One finds in Wikipedia (2016), for example, under 'Major Civilizations' the statement "Ancient Mexico can be said to have produced five major civilizations: the Olmec, Maya, Teotihuacan, Toltec, and Aztec. Unlike other indigenous Mexican societies, these civilizations (with the exception of the politically fragmented Maya) extended their political and cultural reach across Mexico and beyond." Meanwhile, on the History Channel's website *Mexico Timeline* (n.d.), under the topic heading "From Ancient Mesoamerica to the Toltecs", the Maya are described in ways like (referring to the time period centered about 250 CE), "The Mayan civilization, centered in the Yucatán peninsula, becomes one of the most dominant of the area's regional groups, reaching its peak around the sixth century A.D., during the Classic period of Mesoamerican history." The Aztecs are described with statements like, "The mighty Aztecs conquer their chief rivals in the city of Azcapotzalco and emerge as the dominant force in central Mexico." and "In the great cities of the Aztec empire, magnificent temples and palaces and imposing stone statues decorating most street corners, plazas and landmarks …" The Otomies are not mentioned in either of these contexts, and the statements from Wikipedia and the History Channel insinuate a popular Western belief that the most successful cultures are those that create empires in order to conquer, dominate, and build impressive structures.

A third type of obstacle that I encountered is the blurring of Otomi and Aztec cultures. As I indicated in the original paper, these two cultures have a long a complicated history of inhabiting the same general area of what is now central Mexico, and thus having a long history of cultural interaction. In contexts such as numeration and calendars, there are connections and similarities between the Otomi and Aztec cultures, and it is not always easy to determine what information is originally Otomi as opposed to what information is originally Aztec.

Some of the outcomes of publishing the JMC article

The biggest outcome is that the content of Chapter Nine of Gilsdorf (2012) is based almost entirely on the JMC article. Between writing the JMC article and that chapter, I had a chance to polish up the content.

One of the aspects of cultural mathematics is that of analyzing art and decoration, including symmetry patterns, such as is done in Ascher (1991), and I was interested in including this kind of analysis of Otomi art and decoration in my research. In Soustelle (1993[1937], p. 84 - 95), there is a detailed discussion of how Otomi weavers create their products by counting threads and reversing orientations to create symmetry patterns. Nevertheless, at one point one of the persons of the Otopame group informed me that there are traditional weavers of Otomi culture in the state of Querétaro in Central Mexico and that I might be able to interview some of them. This information allowed me to visit San Miguel Tolimán in the state of Querétaro, and interview two traditional Otomi weavers. The interviews turned out to be extremely useful because I was able to ask the weavers how they constructed their products, how they formed symmetric patterns, and about managing large thread counts. The information they gave me directly confirmed what I had read in resources such as Soustelle (1993[1937]). Looking back on that original work, I consider the interviews to have been a crucial part of putting together a quality research product. This is another outcome of writing the article, and it has motivated me to try to include more of an ethnographic aspect to my research in cultural mathematics, such as I have done in Gilsdorf (2011).

There were other outcomes that came out of writing the article, mainly from a pedagogical point of view, which I discuss next.

Pedagogical considerations

The JMC article had a substantial influence on my perceptions of how to teach mathematics, in particular, cultural mathematics.

One aspect of this influence relates to how students perceive learning about cultures that they have never heard of. When I have taught courses in cultural mathematics, in the part in which I teach about Otomi mathematics, it usually turns out that few students have heard of the Otomi culture. I have found, on one hand, that students' minds are open to learn about a culture that is new to them. On the other hand, there is a tendency for some students to think that if a culture did not follow a path of developing mathematics similar to Western mathematics, then the mathematics of that culture does not offer interesting information to understanding mathematics. As we go through the different mathematical considerations of Otomi culture, the students can see for themselves how people from distinct cultures understand, interpret, and express mathematics in distinct ways. For example, students learn that the Otomies use a numeration system that is a combination of bases 5, 10, and 20 (a so-called 5-10-20 Mesoamerican system), which of course, is very different than the decimal system. Experiences like this, of going through Otomi mathematics in the classroom, motivated me to reflect on how to orient students toward an adequate perspective of culture and mathematics for learning ethnomathematics and related topics. Such reflections led to the main ideas of Chapter One of Gilsdorf (2012), which includes a careful description of cultural tendencies, interpretations of mathematics, and so on.

A second pedagogical consideration related to the work I did for the JMC article has to do with having students discover concepts and ideas on their own, rather than me dictating such information to them.[1] I employ this teaching mentality in every course I teach. Thus for instance, when the time comes to study the topic of calendars, I ask the students questions like, "Is it necessary for a culture to calculate Leap Year in order for the calendar of that culture to be useful?" Later, while studying Otomi mathematics, we encounter a statement from the *Huichapan Codex* (Lastra and Bartholomew, 2001) in which a local Spaniard had described how he thought the Otomies had calculated Leap Year by adding six hours to each yearly calendar, so that after four years, a total of twenty-four hours would have been added to the calendar. This description leads me to ask the students a number of questions that prompt them to think about mathematics and culture, such as, "How would the Otomies divide a day

[1] This kind of activity goes by names such as *Inquiry Based Learning and Self Discovery*, but I am not sure that what I promote in class fits exactly with how these terms are defined.

NOTES

into smaller units such as hours?" (e.g., by observing shadows of fixed objects), and "How likely is it that Otomies would divide their days into twenty-four units just like in Western culture, especially considering that the Otomies count using a 5-10-20 system?" In the end, students learn that the Otomies did not include a specific calculation of Leap Year in their calendar, yet adjusted their calendar so as to make it useful for agricultural cycles. I find these kinds of questions to be particularly useful in enlightening students as to how mathematics and culture interact.

Related developments since then

My interest in Otomi mathematics and related topics has progressed since the writing of the article. I have continued to consider topics connected to Otomi mathematics. One such topic is that of the connection between gender and cultural mathematics. In particular, in the area of textile art there are strong connections between the anthropological concept of gender[2] and the skills of women in activities such as weaving and embroidery. One finds that in Otomi and Mazahua[3] cultures, there is an importance placed on the skills of women as textile artists that indirectly shows a cultural evidence and support for the abilities of women to apply mathematical concepts to their processes of creating textile products. This cultural importance of the weaving and embroidering skills of women in the Otomi, Mazahua, and other cultures, differs from cultural perceptions of mathematical skills of women in Western mathematics. See Berlo (1996), Brumfiel (1996), Gilsdorf (2014, 2015), Hendon (2006), Romaní (2005), Vizcarra and Guadarrama (2006), for more details.

I would like to describe an example that illustrates my use of the phrase "the indirect cultural evidence of mathematical abilities". Consider the photo below of an embroidered cloth done by a Mazahua woman.

Figure 1. An embroidered cloth, Mazahua culture. Photograph © 2009, Thomas E. Gilsdorf.

It is clear from the embroidery pattern that the artist who made this cloth understands the concepts of vertical and horizontal (and perhaps also 180° rotational) symmetry, as verified for example, in Soustelle (1993[1937]). However, when I asked the artist how she created the patterns on the cloth, she responded by stating, "No sé, nada más arreglo los pétalos para que se ven correctos." which translates to, "I don't know, I just arrange the petals so they look correct." The artist's response indicates an important example of how people from distinct cultures can have a solid understanding of a mathematical concept, even though they do not express that understanding in terms equivalent to Western mathematical definitions and terminology. Indeed, according to Ascher (2002, p. 3), "Most cultures do not set mathematics apart as a distinct, explicit category." This tells us that a person such as the woman of Mazahua culture

[2] The concept of gender has to do with cultural interpretations and expectations of activities that are typical of women and men. It is not simply observations of the biological differences between female and male humans.

[3] The Mazahua culture is part of the Otopame linguistic family. The Mazahua and Otomi cultures are distinct.

who made the above embroidered cloth is not likely to view her work as mathematics, rather, as a cultural product that, from the point of view of ethnomathematics, includes the utilization of mathematical skills.

A second development that resulted from the writing of the article occurred when I was re-reading the reference *Instituto Lingüístico del Verano* (1979), regarding Otomi numeration. On page 74 of that reference, referring to Otomi constructions of large numbers, appears the statement, "Aunque es posible lograr cifras bastante elevadas por este sistema, la yuxtaposición se alarga tanto que es más fácil usar los números de español.", which translates to, "Although it is possible to obtain higher level expressions using this system, the juxtaposition gets so long that it is easier to [just] use Spanish numbers." I am not sure how I had missed it earlier, but when I saw it again, it was clear that this is an ethnocentric statement. It motivated me to determine how higher values are expressed in the Otomi number system. For example, one million is expressed as: *?na mahuąhi de ga mahuąhi* ("a thousand thousands"). I later included this construction in my discussion of Otomi mathematics in Chapter Nine of Gilsdorf (2012, p. 184). After this observation, I looked through some of the references of lists of number expressions in resources such as Barriga Puente (1998), which includes lists of number expressions for most major cultures in the Western hemisphere[4]. I noticed that many such lists include examples of number expressions up to a point where patterns would likely repeat, with few examples of larger values. The assumption is that constructions of number expressions will indeed repeat, depending on the counting base or bases in use. However, keeping those lists relatively short gives a false impression that the corresponding cultures were unable to use or understand large number values. The effect that had on my viewpoint of number systems is that when I now discuss that topic, I remind listeners that the lists of number expressions are not meant to be exhaustive and that no assumption should be made that people in the corresponding culture were unable to comprehend large number values.

[4] Barriga Puente's lists of number expressions are mostly taken from references by anthropologists, missionaries, historians, and so on, who describe numeration systems with a wide variety of detail and information.

Possibilities for further research

A) Studies *of other cultures; the use of ethnographic methods*. As I mentioned at the beginning, when I first began investigating Otomi mathematics I found that their mathematics had not previously been studied. This is the case for many cultures around the world. Hence, one possibility for further research is to conduct investigations into the mathematics of cultures whose mathematics has not been carefully studied. There are plenty of cultures for which this possible. A particular path that could be incorporated into such research is the inclusion of ethnographic techniques typically associated with anthropologists. One finds for example in Byram and Feng (2004, p. 155) a description of how ethnographic methods have proven to be quite useful in the area of second language acquisition. I believe such methods could also be useful for learning about the mathematics of other cultures.

NOTES

B) Pedagogy *and mathematics education*. One consequence of considering pedagogical aspects of Otomi mathematics is that I have found a number of ways that the inclusion of cultural aspects of mathematics enriches the experiences of students with regard to understanding and appreciation of mathematics, as I indicated in section 4. Thus, a second possibility for further research is to investigate more ways that cultural mathematics/ethnomathematics can be included in mathematics classrooms. People who study ethnomathematics are aware of this, and studies are currently being done. Nevertheless, I believe it bears repeating that cultural mathematics should be an important part of general mathematics education.

C) Otopame *mathematics*. With respect to specific areas of investigation connected with Otomi mathematics, there are several possibilities. Specifically, of the cultures of the Otopame linguistic group, the only culture whose mathematics has been studied is that of the Otomies. Hence, one could investigate the mathematics of the Mazahua, Pame, Chichimeca-Jonaz, Matlatzinca, and Ocuilteco cultures.

Acknowledgement:

Part of the research for this article was done while the author was at the Instituto Tecnológico Autónomo de México (ITAM). The author would like to acknowledge research support for this paper by the Asociación Mexicana de Cultura, A. C. I also wish to thank Mr. Raymundo Isidro Alavez for his important translation of the abstract into Hñähñu (Otomi).

References

Albores, B. (2006). Los graniceros y el tiempo cósmico en la región que ocupó el Matlanzinco. *Estudios de Cultura Otopame,* **5**. Mexico City: Instituto de Investigaciones Antropológicas, UNAM,
71 - 117.

Ascher, M. (1991). *Ethnomathematics: A Multicultural View of Mathematical Ideas*. Pacific Grove: Brooks/Cole.

Ascher, M. (2002). *Mathematics Elsewhere: An Exploration of Ideas Across Cultures*. Princeton: Princeton University Press.

Barriga Puente, F. (1998). *Los Sistemas de Numeración Indoamericanos; Un Enfoque Areotipológico*. Mexico City: UNAM.

Bartholomew, D. (2000). Intercambio lingüístico entre Otomi y Náhuatl. *Estudios de Cultura Otopame,* **2**, Y. Lastra and N. Quezada, Eds. Mexico City: UNAM, 189 - 201.

Berlo, J. C. (1996). Beyond Bricolage: Women and Aesthetic strategies in Latin American textiles. In *Textile Tradition of Mesoamerica and the Andes*. Austin: University of Texas Press, 437 – 479.

Brumfiel, E. (1996). The quality of tribute cloth: The place of evidence in archaeological argument. *American Antiquity,* **61** (3): 453-462.

Byram, M. and Feng, A. (2004). Culture and language learning: teaching, research and scholarship. *Language teaching,* **37**, no. 3, 149-168.

Carrasco Pizana, P. (1950). *Los otomíes: Cultura e Historia Prehispánicas de los Pueblos Mesoamericanos de Habla Otomiana* (No. 15). Mexico City: Biblioteca enciclopédica del Estado de México.

Castillo Escalona, A. (2000). *Persistencia Histórico – Cultural San Miguel Tolimán*. Querétero: Universidad Autónoma de Querétero.

Galinier, J. (1990). *La Mitad del Mundo: Cuerpo y Cosmos en los Rituales Otomíes*. Mexico City: Universidad Nacional Autónoma de México (UNAM), Centro de Estudios Mexicanos y Centroamericanos, Instituto Nacional Indigenista.

Gilsdorf, T. E. (2008). Etnomatemáticas de los Otomíes. *Estudios de Cultura Otopame,* **6**, Mexico City: UNAM, 165 - 181.

Gilsdorf, T. E. (2009). Mathematics of the Hñähñu: The Otomies. *Journal of Mathematics and Culture,* **4** (3), 84 - 105.

Gilsdorf, T. E. (2011). Ethnomathematics of the Purhépecha – Interviews regarding mathematics and culture, with indigenous artists. Everett: *Jacobs Research Fund Archives*.

Gilsdorf, T. E. (2012). *Introduction to Cultural Mathematics with Case Studies in the Otomies and Incas*. Hoboken: John Wiley & Sons.

Gilsdorf, T. E. (2014). Arte textil Otomí y Mazahua como una expresión femenina de las matemáticas. *Estudios de Cultura Otopame*, Mexico City: UNAM, In press.

Gilsdorf, T. E. (2015). Gender, Culture and Ethnomathematics. In *Proceedings of the Eighth International Conference of Mathematics Education and Society*. S. Mukhopadhyay, & B. Greer (Eds.), Portland, OR: Portland State University, 531-542.

Hendon, J. (2006). Textile production as craft in Mesoamerica: Time, labor and knowledge. *Journal of Social Archaeology,* **6**, 354-378.

Hermann Lejarazu, M. (2001). Códices tributaries de Mezquiahuala. In *Códices del Estado de Hidalgo, State of Hidalgo Codices*. Sotelo Santos, L. E., Ballesteros García, V. M., Luvián Torres, E. (Eds.) Pachuca: UAEH, 88 – 99.

History of Mexico. (2016, May 24). In *Wikipedia, The Free Encyclopedia*. Retrieved 17:37, May 25, 2016, from https://en.wikipedia.org/w/index.php?title=History_of_Mexico&oldid=723145978.

Instituto Lingüístico del Verano. (1979). *Luces Contemporáneas del Otomí*. Mexico City: Secretaría de Educación Pública.

Lastra, Y. (2006). *Los Otomies, su Lengua y su Historia*. Mexico City: Instituto de Investigaciones Antropológicas, UNAM.

Lastra, Y. and Bartholomew, D., Eds. (2001). *Códice de Huichapan. Paleografía y traducción* (of notes by Lawrence Ecker). Mexico City: Instituto de Investigaciones Antropológicas, UNAM.

Mexico Timeline. (n.d.). *History*. Retrieved from http://www.history.com/topics/mexico/mexico-timeline. April 13, 2016.

Merrifield, W. R. (1968). Number names in four languages of Mexico. In *Grammars for Number Names*, H. Brandt Corstius, Ed. Dordrecht: D. Reídle Publishing Company.

NOTES

Romaní C. C. (2005). *Bordado Tradicional Mazahua de Michoacán*. Zitácuaro: PACMYC.

Soustelle, J. (1993). *La famille Otomi – pame du Mexique Central*, Travaux en Mémoires de l'Institut d'Ethnologie, Paris, 1937. Translation into Spanish: *Centro de Estudios Mexicanos y Centroamericanos*, Fondo de Cultura Económica, 1993.

Urbano, A. and Acuña, R. (1990). *Arte Breve de la Lengua Otomí y Vocabulario Trilingüe: Español-Náhuatl-Otomí*. UNAM.

Vizcarra Bordi, I. and Guadarrama, N. M. (2006). Las niñas a la casa y los niños a la milpa: La construcción social de la infancia mazahua. *Convergencia,* **13**, no. 40, Toluca: Universidad Autónoma del Estado de México.

Wright, D. (1997). Manuscritos Otomies del Virreinato. In *Códices y Documentos sobre México, Segundo Simposio, Volumen II*, Rueda Mithers, S., Vega Sosa, C., Martínez Barca, R. (Eds.), Mexico City: Instituto Nacional de Antropología e Historia (INAH), 437 – 462.

Wright, D. (2005). Hñahñu, Ñuhu, Ñhato, Ñuhmu. *Arqueología Mexicana,* **13**, no. 73, May - June.

Volume 2
Community

Finding the 'Space Between' in Developing Mathematical Curriculum and Pedagogy across Cultures: The Mathematical Ways of the Northern Ute

Dr. Jim Barta, Dean
Tift College of Education
Mercer University
1501 Mercer University Drive
Macon, Georgia 31207
barta_jj@mercer.edu

Tod Shockey
University of Toledo
tod.shockey@utoledo.edu

Abstract

In this article, the authors describe changes in their insights a decade after they wrote about how mathematics was used traditionally in the culture of an indigenous people, the Northern Ute. The transition involved a significant reframing of perception shifting their notion of math as a noun to one where math is a verb. The personal, professional, and instructional implications of this paradigm shift invited the authors to consider where meaning was taking place as the instigator of the math acted with an object. They discovered the Ute perspective illustrated that the important event happened 'in the space between'. The authors discuss the importance of teachers understanding this position and its effect on teaching and learning.

Keywords:

- Mathematics
- American Indian
- Culture
- Ute
- Critical Pedagogy

Math is a part of our lifestyle too but we don't have it in the way that the White man has it, say... math like times tables, and addition or subtraction. We have those things also in our traditional ways of beading, our raising of our horses and cattle, or building fences, putting up teepees, etc. If you don't have your culture, if you don't have your language, your arts and crafts you are just another person out there in the world. If we have our beadwork, our songs and dances, if we believe in our old ways, our traditional ways then that is going to uplift us someday while the world fails around us. We will have something to fall back on. Right now we can still live in a White world; we are in a White world. My dad says we're considered White people because we speak their language; we go to work and pay bills. We are not like Indian people the way we used to be but we still have a part of our Indian ways and that belongs to us. We should hang on to that. It's good to have your culture and your language to fall back on. It gives you a good feeling to know it is a part of you and that is where you came from.

(Personal conversation with Fabian Jenks, Ute Tribal Elder, February 2001)

Introduction

We (Barta and Shockey, 2006) opened with this quote a decade ago and use it again as a stepping off point to now reconsider how our thinking has changed concerning ethnomathematics.

In the years since those words were first spoken much has indeed transformed. Today, fortunate scholars and interested students and teachers can easily locate a plethora of books on ethnomathematics with a click of a finger. Yet, we remember a time not so long ago when as new students to the study few if any single book existed. Rather, to consider the field, we had to read eclectically. Knowledge was constructed as we considered where overlap existed in the research and literature involving anthropology, mathematics, sociology, linguistics, and psychology, cognition, etc. Overlap was important but just as vital were the perceived 'gaps' or the 'spaces between' where different subjects being compared helped inform the learner pressed to seek emerging vantage points striving to understand the implications of such coexistence. This reciprocity of influencing and being influenced by two or more disparate viewpoints was to us a critical aspect of better understanding the "space between" and is a main topic we will discuss further in this paper. We now expand on this topic to discuss our personal transformations related to our scholarship and in particular that article written. Additionally, we will explain how our new perspectives shaped and continue to shape the work we do, the insights gained, the new stories told and the pedagogy and curriculum that we create.

NOTES

Positioning

Ten years ago, we admit we hardly considered the positionality we expressed through our research on the mathematics of the Ute. Looking back and judging now we assumed an 'us vs. the other' perspective sharing what we had learned through our scholarship in an act of respect and appreciation. We see things differently today and will try to describe the transition that has occurred within us. Today we are much more aware of the critical need to consider one's position before striving to understand what is occurring through interactions within and between others. For instance, both the authors were educated within the traditional Eurocentric paradigm. Neither of them are American Indian. The authors speak no native languages nor practice indigenous traditions. They do not live on a reservation or with the communities comprised of human beings who are American Indian. As scholars, they traveled to the reservations and developed professional working relationships with the community elders who understood their interest in learning about traditional uses of mathematics in native cultures and trusted their sincerity in using these understandings to try to improve the mathematical educations their sons and daughters, nieces and nephews received. Yet, we seem to have been granted temporary entry gaining a brief glimpse of the lived applications of mathematics in the lives and meaning of our Native informants. We were transformed by our interactions, professional and personal with our American Indian or First Nations colleagues. It is our belief that one can never return to being the same person they were following any diverse interaction. We can never return to that same place we once occupied nor think about things in entirely the same way as we once did.

As the comparison between cultural influences of knowing and using math was made between our western ways of mathematical understanding and that of other cultures, we discovered a new space. It was in this 'space between' that we came to realize much more profoundly we saw others and ourselves using mathematics to think and do.

Border Crossings and Singularity of Space

In traditional education, the mathematics and science taught in schools seldom includes overt connections with culture. This education positioning, whether intentional or unintentional is often non-contested and rather is provided as the superior standard against which to judge all others. Many math students and teachers today still do not even know the term ethnomathematics; it is no wonder their instruction and curriculum in relation to culture is more a reflection of what has been than what could be.

It really should come as no surprise. The majority of university teacher educators were educated in similarly consistent settings. While many do value culture and diversity few have any real practice or experience in multiculturalizing their mathematical instruction or their curriculum. It is nearly impossible to recognize what has never been seen. Consequently, many students view mathematics as a spectator sport rather than one in which they can participate. Perso (2002) contends that the mathematics taught in schools exclusively reflects a "western-techno mathematics". Its roots lie in the creation of a thought pattern resulting from a Cartesian worldview developed in the 15th century that has since been accepted as the standard for not only mathematical and scientific endeavors, but has been infused into virtually every aspect of behavior of modern western-based societies around the world. This perspective includes a perception of linear deductive logic and separate objectification of the world and things found in it.

For the Native American student, this cultural disconnection poses additional obstacles for achievement. It is as if the child were being asked to see through two different pairs of glasses; one which includes a perception of "reality" based on a dependence on the western worldview, a subject, whose concepts, content, and information are encased in a cloak of western impression and definition, and the other, a mathematical expression of life and animation. One term is based on a definition of math as a noun and the other instead understood as a verb.

Glen Akinhead, University of Saskatchewan uses the phrase "crossing the border" to describe the boundaries students must navigate to obtain new scientific knowledge. I believe the same can be said for mathematics. (p. 87). Soloman (1998) adds that coming to feel comfortable within this new territory is, for most of young students, as long and slow a process as joining a new tribe might be. To make it even more difficult, the students find that while they learn, their own culture of everyday life and meaning continues to intrude on every side because "their own culture of everyday life and meaning continues to intrude on every side" (Soloman, 1998 p. 171).

Native students, who may already feel separated from the culture of school, may feel doubly removed from the subculture of math and science classes, and, may find it even more difficult to negotiate "meaning from one domain to another" (Cajete, 1999, p. 97). Saxe (1982) in his review of mathematics studies concerning indigenous students found that many Native American students experience particular difficulties in solving mathematical problems when the problems are not perceived as being culturally relevant. The National Council of Teachers of Mathematics (NCTM) has articulated the necessity for all students to more fully understand the many connections that exist between mathematics and its real world applications (NCTM, 2000).

What we describe is a singularity of space constructed in classrooms that limit true student understanding of what mathematics is and the role it plays in their lives. From such a position majority students mistakenly conceive the mathematical superiority of this one type of ethnomathematics. For the minority student, the cultural ways of knowing and doing of their community are devalued as they must strive to learn from a different cultural paradigm often without any overt or intentional instruction to help them gain those necessary skills. Developing a third space, or the space that exists between we suggested is of paramount importance in helping our students truly succeed.

The Third Space or the Space Between

We accept the position of opportunity as opposed to an "us and them" approach when considering Jenks' quote above. Seminal research conducted by Lipka and Mohatt (1998) in collaboration with Ciulistet (a group of Yup'ik educators) began to reveal strategies to reconcile two cultures, the culture of school, and the Yup'ik culture. The reconciliation process offered insights into "third space" described by Likpa, Sharp, Adams and Sharp (2007) citing Gee (1996):

> *Gee's (1996) concept of borderlands describes the space where two cultures or linguistic styles meet but co-evolve into a practice that is not strictly either and becomes a new creation. This is theoretically*

NOTES

> *important. Classrooms have the potential for being these "third spaces:" belonging to not one or the other, but both. These third spaces have the potential to become productive uncharted zones between school and local cultural knowledge and norms. From a critical pedagogical perspective, this third space has the potential for changing historically situated authority structures. (Lipka, et al., 2007, p. 97)*

The third space, a term often used in informal science literature as well, may also be considered as an intellectual, abstract space where research attempts to situate findings in beneficial ways for individuals. This space may be viewed through "Two-Eyed Seeing." Albert Marshall, a Mi'kmaw Elder, describes Etuaptmumk (Mi'kmaw word for Two-Eyed Seeing): "It refers to learning to see from one eye with the strengths of Indigenous knowledges and ways of knowing, and from the other eye with the strengths of Western knowledges and ways of knowing…and learning to use both eyes together, for the benefit of all." (http://www.integrativescience.ca/Principles/TwoEyedSeeing/). We contend that through this 'two-eyed seeing, an educational system may embrace Native students through respect and affirmation of the importance of the individual, culture, and language.

Looking Back

In 2006, our paper was not informed by the work of Pike (1967) and his introduction of the concepts of emic and etic. We did not discuss the possible implications for ethnolinguistics and the possible implications for the Ute language of numbers. Vandendriessche (2016) highlights interdisciplinary opportunities that extend into the Ethnomathematics community with ethnolinguistics, another view that we had not considered. We had omitted the seminal work of Bassanezi (1994) and Rosa and Orey (2013a, 2013b) on ethnomodeling. In our defense, as shown by the dates of some of the cited research, we were not informed as the work was not yet available. In the past ten years more scholarship is crossing disciplinary boundaries, we take advantage of these boundary crossings as we look back and more importantly look forward. In the time since, we have more thorough studied the research of Pike, Vandendriessche, Bassanezi, and Rosa and Orey, and others. Leontiev was yet another highly influential scholar helping to guide our thinking.

Alexei Leontiev, a highly influential Soviet scholar of the 20th Century and a student colleague of Vygotsky, was instrumental in developing a socio-historical" approach to psychology. He professed human activity as a central focus of understanding human psychology. He described the importance of the activity involving real individuals working in groups or in situations expressly interacting with physical objects and activities in the world. He discussed the relationship that is formed as a human being interacts with others or with objects. It is within this interaction, this third space, or this place between where a new meaning is realized. Acknowledging this space between is profoundly important in developing thought and understanding.

How then does one begin to discover such space? Alan Bishop (1991) describes ways to access mathematical interaction in multiple human activities that involve counting, measuring, designing, locating, explaining, and playing. Bishop's six as many refer to them has served a framework for many of us working in the Ethnomathematics program. In 2006, we employed Bishop's activities as the lens to consider the mathematical ways of the Ute people. This framework exposed the space between mathematics and Ute culture has it illuminated the effect of mathematics on the people and reciprocally the influence of their cultural community on mathematics. Today, as suggested by Shockey and Silverman (2016) we include calendar as a seventh element to be included in Bishop's (1991) framework.

Mathematical Ways of an Aboriginal People: The Northern Ute

In the Ute language, there is no one specific word for mathematics. Ute Elder Fabian Jenks professed,

> *"You white men make it a noun, for us it is a verb; we name it as we use it. When we hunt, cook, chop wood, etc. it is math. In Ute there are many names for math!" (Personal conversation – May 2004)*

We consider this single statement to be one of the most profound of all that indigenous colleagues have shared with us. It demonstrates numbers are alive and that shows itself to us through its use. It confirmed what we had learned in our educational instruction as teachers, that for students to be successful they must participate in relevant and active engagement in such instruction. It is insufficient for the teaching to be framed in some vague or ambiguous word problem or in the lived experience of the instructor. Rather for Ute and other Native children on the Rez: to learn and understand us, we had to connect with their authentic interactions and those of the cultural community in which they were enculturated. For children to succeed they had to dance with the numbers.

The Third Space in Aboriginal Measuring and Designing

In returning to our original article we again witness the third space in what we had described.

Rather than creating a set of universal measurement standards Ute measurement was very contextualized and individual. A stick or a length of rawhide for instance possessed its own length that could then be used as a standard or comparison with another object. One Elder stated, "Everything was built with a measurement within itself." Parts of the body such as hands, length of an outstretched arm, paces, etc. were used and were specific to the person to which they belonged. In essence, each person carried with them their own standard units and thereby what one constructed was proportionately appropriate for the user. Hide shirts, buckskin dresses, moccasin footwear, and even tools and weapons were made with the body of the person acting as a measurement template.

The following examples illustrate that in many indigenous communities what was measured or designed was implicitly framed within the relationship between the maker/being and his or her intent, the object, and the act involved situated in the context in which each existed as an active participant. This relationship is illustrated in the discussion of Ute measurement, "…Ute measurement was very contextualized and individual" (Barta & Shockey, 2006, p. 84) and in descriptions of the use of body dimensions Lipka, Shockey, and Adams (2003) for constructing a Yu'pik kayak. Measurement for the Ute people is important in cooking. Beyond the notion of a 'pinch of this' the Ute's understood that "a woven bowl" (Barta & Shockey, 2006, p. 85) satisfied a volume requirement for cooking. For example, if a mom or dad prepares a certain dish for the family they would use a bowl that they knew would properly hold the ingredients.

NOTES

In effect anytime after the same dish could be properly prepared by using the bowl itself as a kind of measuring device. The preparer would know from experience (and interaction with the bowl) how full or how much fuller it would become as ingredients were added. The volume of the bowl (the amount in which it was filled) allowed for a certain number of people to be fed. The cook could increase or decrease for more or less guests by considering that effect on the volume of the bowl. One Elder stated, "When we had guests Mom would always use the brown bowl to make the meal because she knew that it would feed the number of people coming." (Personal conversation, April 2005)

On a different scale, Lipka, and Mohatt (1998) found that fish racks built by Yu'pik people were built with respect to the body measures of the builder that led to an outcome related the amount of fish the rack would hold. The Yu'pik builder understood that the rack needed to hold enough fish to get through the long arctic winters. Over time and through both observing and learning from others, Yu'pik people knew how much fish was typically needed by those they fed to carry them through the winter. They also knew the optimal capacity of their racks. A calculation occurred considering variables of fish needed for winter and the amount of fish a rack could hold for drying. The builder/fisherman was very conscious of the calculations they were making and the importance of being right.

With respect to spatial reasoning, "The beadworker typically used their sense of what looks good rather than making strict mathematical calculations to produce work" (Barta & Shockey, 2006, p. 85). The design that was imagined and then translated into a physical pattern was inextricably reliant on who it was for and the message or dream the beadworker strove to communicate. Many examples of this mind to fingertip reasoning have been revealed in Ethnomathematics scholarship (Shockey & Pollington, 2005; Shockey, 2006).

Various calendars were used again reflecting the important relationship between the system of measurement (calendric) and the importance of such timing or measurement that in part determining when important cultural and social events throughout the year were to take place.

"The Ute calendar was organized in relation to the moon and the various positions and shapes of it" (Barta & Shockey, 2006, p. 97). Calendars have been important for groups for centuries, see for example Bjarnadottir (2010). Penobscot basketmaker Paula Thorne (personal communication) shared the importance of the calendar with respect to harvesting sweet grass. Sweet grass is harvested after it blooms thus assuring that the plant thrives and will be there year after year, harvesting before the bloom would lead to the plants demise. When to hunt certain animals, harvest and dry different types of fish, participate in specific traditions or ceremonies, or travel to ideal locales to renew social bonds with distant relatives were determined by living within this circle of time. Man and woman were not mere spectators within these culturally negotiated calendars rather they too played a vital and important role in its design and result.

Perimeter, area, volume were measured in very practical ways. A person constructing a shade house to provide comfortable, cool protection against the summer sun for instance, would "step off" a distance and count the steps that were taken to create the size of house to comfort ones family. The floor area of a shade house or teepee was determined adequate if its space served the function for which it was constructed.

Distances were described as a function of time and/or speed it took one to make the trip. A long journey was described by counting the number of sunsets, which elapsed during the travel. The light of the day was important; it was less difficult and safer to travel during daylight so sunsets provided an obvious stop to a day. Travelers would segment a trip into increments based on the time from sunrise to sunset. The speed or mode of travel (on foot or on horseback) influenced the "distance" traveled making the trip faster or longer.

Volume was considered by how much of something a vessel or container could hold. Weight of an item was always considered in relation to the person hefting or feeling; they made the determination if to them something was heavy or light. For cooking and food preparation, measurement again played an important function. Those doing the cooking used pinches, handfuls, and dashes of ingredients (such as crushed berries, salt, and other flavorings) as standard units. The experienced cook knew that a particular bowl or woven container when filled to a certain level would produce a specified amount to food. If the cook was preparing a meal for a greater number of people the container could have more added to it as the cook knew from previous experience how much more food would result.

Reconsidering what we had learned we see more deeply that the space between our western ways of understanding and that being expressed, clearly illuminated the life in numbers our cultural translators were sharing. For them, these were not figurative references at all but rather concrete examples of the matherizing (the active use of math) they were living! Indeed, each object did contain its own measurement along with a plethora of other information (temperature, pliability, density and weight, resiliency, etc.) that would make itself usefully available to one who had a specific need for such information in a relative context.

In discussing fractions, we learned that a recently harvested deer for instance could be shared equally even though different amounts of meat were provided to various families. Again, what dictated the importance was not an abstract measurement (one-half, one fourth, etc.) but how the use informed the understanding of the math that lived within. An Elder may be gifted a prime cut or his or her family an "equal" portion based on their stature within the community or the size of their family. Similarly, an equal amount (albeit lesser in weight or what part of the deer it came from) might be offered to someone younger or with fewer mouths to feed. When we stop to consider this perspective, consider how odd it must be to a child in this community to learn fractions exclusive of context as the teacher is teaching them about halves, fourths, eighths, or sixteenths! The obvious question for anyone attempting to make sense of the discussion must be, "One - fourth of what?"

In Conclusion – Implications for Scholars, Teachers, and Students

As previously stated, we see the article we wrote years ago in a much different light. Our 'noun dependent' reference of mathematics has become much more animated and better considered the verb Fabian suggested. Looking back using this lens, we realize there is a whole new story or meaning in that article. It is humbling to admit that it seems originally we heard or better stated, understood only a part of what Ute Elders were expressing. It was not until we ourselves became border crossers, willing tourists in the space between that a clearer understanding of how expansive mathematics in action truly is.

We suggest that scholars, teachers, and students can also make similar passage if they first consider how the reality they experience

NOTES

and the understandings they make are invariably derived in such in-between spaces. As we each engage the other, we become open to the realization that neither "we" or "they" really hold exclusive rights to who gets to define reality but rather all true understanding is an ongoing negotiation in this 'third space'.

We believe that instruction and curriculum emanating from this space between cannot be adequately simplified into a neat and tidy model. Rather each of us must discover for ourselves the life in numbers as participate in it.

References

Barta, J. and T. Shockey (2006). "The mathematical ways of an Aboriginal people: The Northern Ute." *Journal of Mathematics and Culture, 1*(1), 79-89.

Barta, J. (2009). Mathematics in the Milpa: Culturally relevant mathematics instruction in a Mayan village. *Teaching Children Mathematics, 16*(2), 3-9.

Bassanezi, R. (1994). Modelling as a teaching-learning strategy. *For the Learning of Mathematics: An International Journal of Mathematics Education, 14*(2), 31-35.

Bishop, A. J. (1991). *Mathematical enculturation*. Dordrecht, The Netherlands: Kluwer Academic Publishers.

Bjarnadottir, K. (2010). "Ethnomathematics at the margins of Europe: A Pagan calendar." *Journal of Mathematics and Culture, 5*(1), 21-42.

Cajete, G. A. (1999). *Igniting the sparkle:* An indigenous science education model. Durango, CO: Kivaki Press.

D'Ambrosio, U. (1985). Ethnomathematics and its place in the history and pedagogy of mathematics. *For the Learning of Mathematics, 5*(1), 44-48.

Gee, J. P. (1996). *Social linguistics and literacies: Ideology in discourses* (2nd ed.). London; Bristol, PA: Taylor & Francis.

Jaworski, B. (2008). Mathematics teacher educator learning and development. In B. Jaworski & T. Wood (Eds.), *The International Handbook of Mathematics Teacher Education Volume 4: The Mathematics Teacher Educator as a Developing Professional* (Vol. 4, pp. 1-13). Rotterdam: Sense Publishers.

Leontyev, A.N. (1979). The problem of activity in psychology (pp. 37-71). In J.V. Wertsch (Ed.). *The concept of activity in Soviet psychology*. Armonk, NY: Sharpe.

Lipka, J., Sharp, N., Adams, B., & Sharp, F. (2007). Creating a third space for authentic biculturalism: Examples from math in a cultural context. *Journal of American Indian Education, 46*(3), 94-115.

Lipka, J., & Mohatt, G. V. (1998) *Transforming the culture of schools: Yup'ik Eskimo examples*. Mahwah, NJ: Lawrence Erlbaum Associates.

Lipka, J., Shockey, T. L, & Adams, B. (2003). Bridging Yup'ik Ways of Measuring to Western Math: Exploring Ways of Connecting Everyday Knowledge to School Math. In D. H. Clements (Ed.), *Learning and Teaching Measurement* (pp. 181-194). Reston, VA: National Council of Teachers of Mathematics.

NCTM, (2000). *Principles and standards*. Reston, VA: National Council of Teachers of Mathematics. http://standards.nctm.org/document/chapter2/index.htm.

Perso, T. (2002). Closing the gap. *Australian Mathematics Teacher, 58*(2), 6-11.

Pike, K. L. (1967). *Language in relation to a unified theory of the structure of human behavior.* The Hague, Paris, France: Mouton.

Rosa, M., & Orey, D. (2013a). Ethnomodelling as a methodology for ethnomathematics. In G. A. Stillman, K. Gabriele, W. Blum & J. P. Brown (Eds.), *Teaching mathematical modelling: connecting to research and practice* (pp. 77-88). Dordrecht, The Netherlands: Springer.

Rosa, M., & Orey, D. (2013b). Ethnomodelling as a research lens in ethnomathematics and modelling. In G. A. Stillman, G. Kaiser, W. Blum & J. P. Brown (Eds.), *Teaching mathematical modelling: connecting to research and practice* (pp. 117-130). Dordrecht, The Netherlands: Springer.

Shockey, T. L. (2002). Etnomatemática de uma Classe Profissional: Cirurgiões Cardiovasculares. *BOLEMA, 15*(17), 1-19.

Shockey, T. L., & Pollington, C. (2005). Technology Tips: Curve Fit Challenge. *Mathematics Teacher, 99*(4), 288-291.

Shockey, T. L. (2006). Left Ventricle Reduction through an Ethnomathematics Lens. *For the Learning of Mathematics: An International Journal of Mathematics Education, 26*(1), 2-6.

Shockey, T., & Mitchell, J. (2006). An ethnomathematics approach toward understanding a Penobscot hemispherical lodge. *Horizontes, 24*(1), 69-76.

Solomon, Y.: (1998). Teaching mathematics: Ritual, principle and practice, *Journal of Philosophy of Education* 32, 377–387.

Strathern, A. (1993). *Landmarks*: reflections on anthropology. Kent, OH: Kent State University Press.

Whorf, B. L. (1956a). An American Indian model of the universe. In J. B. Carroll (Ed.), *Language, thought & reality*: selected writings of Benjamin Lee Whorf (pp. 57-64). Cambridge, MA: MIT Press.

Whorf, B. L. (1956b). The relation of habitual thought and behavior to language, *thought, and reality*: selected writings of Benjamin Lee Whorf (pp. 134-159). Cambridge, MA: The M.I.T. Press.

Whorf, B. L. (1956c). Science and linguistics. In J. B. Carroll (Ed.), *Language, thought & reality*: selected writings of Benjamin Lee Whorf (pp. 207-219). Cambridge, MA: MIT Press.

Whorf, B. L. (1956d). Linguistics as an exact science. In J. B. Carroll (Ed.), *Language, thought & reality:* selected writings of Benjamin Lee Whorf (pp. 220-232). Cambridge, MA: MIT Press.

Whorf, B. L. (1956e). Languages and logic. In J. B. Carroll (Ed.), *Language, thought & reality*: selected writings of Benjamin Lee Whorf (pp. 233-245). Cambridge, MA: MIT

Press.Vandendriessche, E. (2015). *String figures as Mathematics? An anthropological approach to string figure-making in oral tradition societies* (Vol. 36). New York: Springer.

Engaging in the Standards for Mathematical Practice by Investigating Vedic Mathematics

Dr. Bethany Noblitt
noblittb@nku.edu
Northern Kentucky University

Blaire Richter
blairerichter@cinci.rr.com

Volume 2
Community

Abstract

In September 2013, the Journal of Mathematics and Culture published a manuscript in which Vedic mathematics was used to make sense of a finger algorithm found in an elementary and middle grades teacher college textbook. The authors of this manuscript decided to take their investigations further by exploring more Vedic mathematics as well as how Vedic mathematics might encourage students to engage in the standards for mathematical practice found in the Common Core State Standards for Mathematics. This paper shares the authors' further investigation into both Vedic mathematics and how the standards for mathematics can be prompted by such explorations.

Keywords:

- Vedic Mathematics
- Common Core Standards for Mathematics

Introduction

In 2010, we began an investigation that would prove to be a very rewarding mathematical undertaking for both of us. We described our investigation in the September 2013 issue of the Journal of Mathematics and Culture in an article titled, “Using Vedic Mathematics to Make Sense of the Finger Algorithm.” In this article, we described a non-intuitive finger algorithm used to multiply two whole numbers between 6 and 10 and how it can be justified using Vedic Mathematics. The finger algorithm is found as a problem in the exercises in the text *Mathematics for elementary teachers: A contemporary approach* by Gary L. Musser, William F. Burger, and Blake E. Peterson (2004) and was discussed in my Mathematics for Middle Grades Teachers course. In the finger algorithm, you use one hand to represent a whole number between 6 and 10. To do this, typically, you would use two hands, the first hand having all five fingers raised and the second hand having between 1 and 5 fingers raised. In the finger algorithm, you represent a number between 6 and 10 by raising (on one hand) the number of fingers you would raise on the second hand as described above. Because the algorithm is used to find the product of two whole numbers, two hands will be used, each hand representing a factor in the product and each hand having between 1 and 5 fingers raised. To illustrate the algorithm, consider the product 6×7. On one hand, you will raise 1 finger (representing the 6) and on the other hand you will raise two fingers (representing the 7).

6×7

The total number of raised fingers (in this case 3) represents the number of tens in the product. The number of ones in the product is given by the product of the number of closed fingers on one hand with the number of closed fingers on the other hand (in this case, 4×3=12). To determine the product of 6×7, add the tens to the number of ones. In this case 6×7=3·10+12=42.

We set out to understand why this mysterious algorithm works and we used Vedic mathematics to do so. This began our love affair with Vedic mathematics and with the Nikhilam algorithm (a Vedic algorithm) in particular. The Nikhilam algorithm had a strong connection to the finger algorithm. In fact, we used the Nikhilam algorithm to justify why the finger algorithm is valid (Noblitt & Richter, 2013). Details of the Nikhilam algorithm appear later in this paper.

Vedic mathematics is a mathematics based on sutras from ancient Hindu texts known as the Vedas (Tirthaji, 1965). While Vedic mathematics focuses on doing tedious calculations efficiently and often mentally

NOTES

(Tirthaji, 1965), we looked closely at the mathematics behind many of the Vedic algorithms. By using Vedic mathematics to justify the finger algorithm, we had witnessed how Vedic mathematics could be used beyond the calculations and algorithms found in Tirthaji's 1965 text. It made us want to learn more about the Vedic algorithms. If they could be a tool used to justify the finger algorithm, what else could they be used for?

Around the same time our work in Vedic mathematics began, Bethany's state had just adopted the Common Core State Standards for Mathematics (CCSSM). With the CCSSM came a focus on standards for mathematical practice (SMPs). As we continued our study in Vedic mathematics, it became obvious that we were engaging in some of the SMPs. At the same time as we were engaging in the SMPs, we began to see even greater benefits to our Vedic mathematics explorations. We began to consider ways in which K-12 mathematics teachers could see the relevance Vedic mathematics could have in their classroom. While we believe that the Vedic algorithms present worthwhile mathematics in and of themselves, we also understand that the algorithms may not seem relevant to teachers because of their unfamiliarity. The SMPs provide relevance to Vedic mathematics. While we believe that students could engage in the SMPs in their classrooms by exploring Vedic mathematics, we also believe that in order for teachers to facilitate the SMPs in their classrooms, they must first experience them in their own mathematical work. An exploration of Vedic mathematics is a perfect means for teachers, both in-service and pre-service, to engage in particular SMPs. In this paper, we will present some of our Vedic mathematics explorations related to the Nikhilam algorithm and discuss particular SMPs that we engaged in during our work. It is our hope that teachers will see the value of both the Vedic mathematics and the SMPs.

In our previous Vedic work, we used the Nikhilam algorithm for determining the product of two single digit numbers justifying the finger algorithm (Noblitt & Richter, 2013). A natural extension of that work was to apply the Nikhilam algorithm to multi-digit numbers. We had applied the Nikhilam algorithm to another problem we had encountered. In order to take advantage of Vedic mathematics in future applications like we had the finger algorithm, we needed to explore more. We asked ourselves, "will the algorithm still work with multidigit numbers?" and "will the same justification work?" Below is an example of the Nikhilam algorithm for the product 89×95.

$$\begin{array}{cc} 89 & -11 \\ 95 & -5 \\ \hline 84 & 55 \end{array}$$

The -11 and -5 are the integers you must add to 100 (the base) to get the factors of 89 and 95, respectively. The 55 is the product of -11 and -5. This gives the number of "ones" in the product 89×95. The 84 is the difference 89+-5 or 95+-11. This is the number of hundreds in the product 89×95. Thus, the final product of 89×95 is 8,455. Because we had seen the same algorithm for a one-digit by a one-digit, we weren't surprised by the result, but nonetheless, it still seemed non-intuitive to us. Why does this work? We used the same justification as we used in our previous work but with a base of 100 (Noblitt & Richter, 2013). The base used for a particular problem is determined by the power of ten to which the two factors are closest. The reasoning behind the Nikhilam algorithm for 89×95 is found below.

Consider 89×95 as (100+(-11))×(100+(-5)).

$$\begin{aligned} 89\times95&=(100+(-11))\times(100+(-5)) \\ &=(100)(100)+(100)(-5)+(-11)(100)+(-11)(-5) \\ &=100(100-5-11)+(-11)(-5) \\ &=100(84)+55 \end{aligned}$$

Here we see the 84 hundreds and the 55 ones, or 8,455.

But, what if we were multiplying two numbers that were closer to 1000 than 100? Will the justification above still work? We wanted to generalize this proof. Tirthaji (1965) represented the product of the two numbers as $(x+a)(x+b)$, where x is the base and a and b are integers as $x(x+a+b)+ab$. But where did that come from? We looked at it like this:

$$(x+a)(x+b)=x^2+bx+ax+ab=x(x+a+b)+ab$$

Thus, the value of $(x+a+b)$ is the number of x's; that is, $(x+a+b)$ is the value in the x position in the base ten numeration system. It was our work justifying the algorithm that prompted our thinking about the standards for mathematical practice and how particular standards for mathematical practice were surfacing as we worked. In what follows, we present those standards, along with a discussion of the mathematics that prompted our engagement of the standards.

SMP 3: Construct Viable Arguments and Critique the Reasoning of Others

In Vedic mathematics, Tirthaji (1965) presents many computational algorithms that while efficient and interesting, are presented without much, if any, justification. While we were willing to use the algorithms and accept their validity in order to begin our learning, eventually we needed more. We wanted to understand why the algorithms worked and prove their generalizations. In other words, as discussed above, we critiqued Tirthaji's reasoning and tried to construct our own arguments for the Vedic algorithms. There are particular statements found in SMP 3 (National Governors Association Center for Best Practices & Council of Chief State School Officers, 2010) that we believe highlight how we engaged in SMP 3.

> *Mathematically proficient students understand and use stated assumptions, definitions and previously established results in constructing arguments.*

We took Tirthaji's assumptions, definitions and results and worked to make sense of them in order to understand the algorithms. As we did this, we started to question why they work. It was then that we used Tirthaji's assumptions and definitions to construct our own arguments in support of his results. Once we had established why the algorithm worked for a one-digit by one-digit product, we used that work to understand an argument not only for why the algorithm works in that case, but why it works in all cases where the base is a power of ten.

> *Students at all grades can listen or read the arguments of others, decide whether they make sense, and ask useful questions to clarify or improve the arguments. They are able to analyze situations by breaking them into cases.*

This is exactly what we did in our exploration. We read Tirthaji's arguments and sometimes they made sense to us. Sometimes they did not. When they didn't, we asked questions not only about his arguments, but also about other situations in which we might be able to apply his arguments. Some of our questions were: What if the numbers are multi-digit numbers? What if the base is any power of ten, not just 10 or 100? We also asked other questions that have not yet been discussed in this article: What about multiplying a two-digit number by a three-digit number? What if the factors are less than the base? What if one factor is greater than the base and one factor is less than the base? These questions

NOTES

gave rise to different cases when Tirthaji's algorithms were useful. For example, a quick look into the case of multiplying a two-digit number by a three-digit number yielded the realization that the algorithm works the same way with the same justification and is left to the reader to explore.

Another case that we considered emerged from the question, "what if both of the factors are greater than the base?" Consider the product 108×111.

108	8
111	11
119	88

If we proceed as previously discussed and follow the Nikhilam algorithm by finding the complements of the factors with respect to 100, we get 8 and 11. Then, after multiplication and subtraction, the 88 represents the number of ones in the product and the 119 represents the number of hundreds in the product, giving the final product of 11,988.

The previous case led naturally to the case of when one of the factors is greater than the base and the other is less than the base. What makes this case interesting is that negative numbers are involved differently from the above example. Consider the product 16×8. Again, proceeding as before, the complements from 10 will yield a +6 and a -2. If we do that, we get the following after applying the Nikhilam algorithm.

16	6
8	-2
14	-12

Interpreting this as before, we have -12 ones and 14 tens. That would suggest the product is 14(10)+-12 or 140-12=128. So, while the negative values are important to the algorithm, they do not affect the application of the algorithm. These cases of using the algorithm arose when we asked questions in order to make sense of the Nikhilam algorithm. We continued to ask questions, some of which pushed us to engage in other standards for mathematical practice. We engaged in SMP 7 as we explored our next question, "what happens when the two factors are not close to a power of ten?"

SMP 7: Look For and Make Use of Structure

> *Mathematically proficient students look closely to discern (a pattern or) structure...They also can step back for an overview and shift perspective. They can see complicated things ...as being composed of several objects.*

After exploring several different cases of the Nikhilam algorithm, we wondered if the algorithm was useful only when the two factors were close to a power of ten. Would the algorithm be helpful to determine a product such as 46×44? We did not doubt that if we used the algorithm and a base of 100 we would get a correct product. But, considering that 46 and 44 are both "far from" both ten and 100, it seemed that the computation would be, in Tirthaji's (1965) word, "cumbrous" (p. 17). So, how did Tirthaji address this? Interestingly, he described a way to use a working base that is not a power of ten to do the computation and then convert to a power of ten base (Tirthaji, 1965). Consider 46×44. Because both of these numbers are close to 50, Tirthaji (1965) suggested base 50. Proceeding as before, using base 50, we get the following.

46	-4
44	-6
40	24

Because we are using a working base of 50, we cannot interpret this as before. In other words, 40 does not represent the number of hundreds. Tirthaji (1965) provides two different ways to deal with the conversion from base 50 to base 100. First, think of 50 as 10×5 and apply this to the argument used earlier to make sense of the algorithm for 89×95.

$$46\times44=(50+(-4))\times(50+(-6))$$
$$=(50)(50)+(50)(-6)+(50)(-4)+(-4)(-6)$$
$$=50(50-6-4)+(-4)(-6)$$
$$=50(40)+24$$

As we can see, the 40 represents the number of 50s in the product. Rewriting 50 as 10×5, we get $(10\times5)(40)+24$. Thinking of this as $(10)(5\times40)+24$, we see (5×40) represents the number tens. This was a place where we were clearly making use of structure to make sense of the algorithm. Understanding and applying the associative property allowed us to more deeply understand the algorithm. Returning to the original notation of the algorithm, we get the following, which yields a representation of the product as 200 tens and 24 ones.

46	-4
44	-6
40	24
×5	
200	24

Thus, the product is 2,024. As we worked to understand the algorithm with a working base of 50, we noted that the conversion from the working base to the base that is a power of ten only affected the number of tens and not the number of ones. This is clear to see in the above computations.

Once we understood the above way of working with a different base, the working base of 50, we realized that Tirthaji (1965) had also presented us with a different perspective when the base is not a power of ten. In the above work, we considered 50 as10×5. The other perspective Tirthaji offered is presented below with a discussion of Tirthaji's changed perspective following.

46	-4
44	-6
40	24
÷2	
20	24

Knowing the product is 2,024, it should be clear that here, the 20 represent the number of hundreds and the 24 is still the number of ones. But, why divide by 2? To answer this question, we look at 46×44 as a product written in terms the base 50.

$$46\times44=(50+(-4))\times(50+(-6))$$
$$=(50)(50)+(50)(-6)+(50)(-4)+(-4)(-6)$$
$$=50(50-6-4)+(-4)(-6)$$
$$=50(40)+24$$

At this point, we had to realize that Tirthaji changed his perspective. In order to convert to a base that is a power of ten, he thought of 50 as $100\div2$ (instead of 10×5). In order to make sense of that, we realized that we need to get the base 100 by multiplying 50 by 2. But the question was how to do this without changing the value of 46×44.

$$50(40)+24=(50\times2\times1/2)(40)+24$$
$$=(50\times2)\times(1/2\times40)+24$$
$$=(50\times2)\times(40\div2)+24$$

This gave us the 100 that we wanted. Now we have

$$(50\times2)\times(40\div2)+24=100(20)+24,$$

which clearly shows the solution 2,024. We believe this part of our investigation also illustrates structure not only by again using

NOTES

the associative property, but also as viewing complicated things as being composed of several objects. On the surface, 46×44 is not a "complicated thing," but viewing the product as an application of the Nikhilam algorithm added some complexity when we asked ourselves about why the algorithm works. The work above shows how 46×44 can be viewed as a composition of a base (50), multiplication by 1 (×2÷2), and the associative property.

As our investigations in Vedic mathematics progressed, one phrase from the standards for mathematical practice stood out as capturing what we believe was Tirthaji's main objective. That phrase is "…look both for general methods and for shortcuts", found in SMP 8: Look for and express regularity in repeated reasoning (National Governors Association Center for Best Practices & Council of Chief State School Officers, 2010). We learned and used Tirthaji's shortcuts, but by engaging in the other SMPs, we were able to make sense of them. This included not only understanding why they work, but also which of the algorithms are more useful in particular situations. Tirthaji himself placed importance on knowing which algorithms were the most efficient for a particular situation (Tirthaji, 1965).

The exploration that we described in our previous Journal of Mathematics and Culture article fueled our passion to learn more about Vedic mathematics. That investigation only involved a limited view into the Nikhilam algorithm. But, that investigation led us to explore other uses for the Nikhilam algorithm, such as multiplying multidigit numbers and numbers that are not close to a power of ten. These investigations proved to be rich learning opportunities for us mathematically and pedagogically. One question (among others, I am sure) for future investigations in the Nikhilam algorithm is: What if the factors are numbers that are not close to the same base? Other questions for future investigations in Vedic Mathematics include (among others, I am sure): What about the other operations? Are there algorithms for other operations? What are they and how do they work? In this article, we hope that we have described how our learning has continued and how Vedic mathematics can be relevant to the K-12 classroom by providing opportunities for teachers and students to engage in the standards for mathematical practice.

References

Musser, G.L., Burger, W.F., Peterson, B.E. (2004). *Mathematics for Elementary Teachers: A contemporary approach.* 7 ed. John Wiley & Sons, Inc.

National Governors Association Center for Best Practices & Council of Chief State School Officers. (2010). *Common Core State Standards for Mathematics.* Washington, DC: Authors.

Noblitt, B. & Richter, B. (2013). Using Vedic mathematics to make sense of the finger algorithm. *Journal of Mathematics and Culture, 7*(1), 58-73.

Tirthaji, B.K. (1965). *Vedic Mathematics*. Delhi, India: Motilal Barnarsidass Publishers.

Relationships Between the Hopi Calendar and Measurement Concepts

Janet Sharp
Washburn University

Abstract

There is much to be relearned about mathematics by seeing it from a viewpoint of a particular society. This paper describes measurement concepts by illustrating how the ancient Hopi used them to actively mathematize their environment, building a calendar for celebrations and festivals necessary for their societal mission. The core mathematical concepts are: size, standard unit, zero-origin, tiling, continuousness, and part-whole. The Hopi calendar is presented in terms of the importance and development of the moon, as a counting unit. Ultimately, a 13th moon helps explain the Hopi calendar and further clarifies the mathematical concepts.

Introduction

Mathematics develops in response to humanity's efforts to better understand the environment. However, that humanistic intent is not always reflected in the mathematics of the classroom. Fostnot and Dolk (2002) posit that for years, mathematics has been taught as if it were a dead language. That is, something to be remembered for its own sake with little recognition of mathematics as a continually changing entity, known uniquely by each person and reflective of the times. Fortunately, researchers (e.g., Zaslavsky, 1973; Gerdes, 1998; Eglash, 1999) in the discipline of ethnomathematics are continually revealing mathematics through new eyes by reconnecting culture(s) to mathematics.

While preparing lessons about basic measurement concepts, we encountered such a journey. There is much to be relearned about measurement by seeing it within a viewpoint of a particular society. Understanding the construction of and the modification to the Hopi Festival Calendar is rich territory for illustrations of measurement content. I was not looking for something unique to "introduce" measurement. As Powell and Frankenstein (1997) warn, culture(s) must genuinely permeate mathematics lessons. Classroom discussions must not simply cite a few cultural connections, as an aside or a precursor that takes place before the "real" mathematics is presented. With this in mind, Barton's (2008) advice encouraged a deep study of the Hopi *system* of knowledge to try to draw parallels between their cultural practices and the mathematics (in this case, measurement) at hand.

Despite the initial expressed purpose of improving measurement instruction, I immensely broadened my personal mathematics knowledge. This new way of understanding measurement had a profound impact on the way I now understand and teach about measurement. It seemed important to share the information that was uncovered. What follows is a record of observations along the path I charted and the connections I have seen. This paper describes measurement concepts illustrating how the ancient Hopi used them to actively mathematize their environment and to build a calendar for celebrations and festivals necessary for their societal mission.

Ethnomathematics

In 1990, D'Ambrosio catapulted the field of mathematics forward by suggesting mathematics is a system of knowledge that should be understood within the context of social and cultural knowledge. In so doing, he questioned the prevailing paradigm that mathematics is static, antiseptic knowledge, devoid of cultural connections or influences. Knowing mathematics *includes* having awareness that it is a product of ongoing (and thus changing) efforts to explain and cope with an environment. Mathematics is created by and known by members of the society dealing with their environment, so is influenced by other knowledge from that culture as well as generations of knowledge produced by previous members. Understanding mathematics in this way is the field of study, ethnomathematics. This "programme ... looks into the generation, transmission, institutionalization and diffusion of knowledge with emphasis on the socio-cultural environment" (D'Ambrosio,

NOTES

1990, p. 369). More recently, D'Ambrosio (2006) suggested that ethnomathematics is more than recognizing mathematical ideas and practices of various cultural groups, it is deeper – it is also a comparison between several cultural groups.

Several ethnomathematics researchers (e.g., Powell & Frankenstein, 1997; Gerdes, 1998; Eglash, 1999; Barta & Shockey, 2006) continue to shed light on mathematics through careful study of cultural ramifications of mathematical ways of thinking. In some cases, researchers study familial cultural heritages, in other cases, they study other cultures. Some researchers physically immerse themselves, living in a culture different from their own, while others cull information from secondary and/or primary sources to reunite the mathematics with its cultural heritage. In all cases, the field of mathematics is enriched, as situations unique to various cultures are revealed to the global community through mathematical eyes in ways that enable researchers to study a multitude of views.

Eglash (1999) interviewed numerous African artisans and scientists as part of his efforts to deepen knowledge about fractals. He found systematic use of fractals in areas ranging from architecture to hairstyling across the diverse cultures of Africa. Zaslavsky (1973) documented the numeration concepts and procedures historically used among different African cultures. She describes a long process beginning with sifting through the literature of diverse disciplines ranging anywhere from African economics and anthroplogy to archaeology and linguistics. Then she followed the trail of her literary findings to personally interview African scholars, from many societies. Ortiz-Franco (1999) described the innovative ways various pre-Columbian cultures in the Americas created and used a positional numeration system not only to quantify their environment but to effectively mark time. Ancient Mayan astronomers predicted "solar and lunar eclipses to the day, hundreds of years in advance" (p. 221) and made these predictions hundreds of years before Galileo's birth. Ortiz-Franco pointedly challenges beliefs that ignore Chicano ancestors' contributions to mathematical knowledge such as beliefs that typically give credit of discovery of mathematics to civilizations, like those of Euclid and Aristotle, which actually emerged long after Mayan civilization. Barta and Shockey (2006) describe an historical account of the dynamic and personal ways the Northern Ute, a Native American people, use and know mathematics. Their work respectfully summarizes traditional Ute uses of a wide range of mathematical ideas and includes essential comments from tribal elders. Gerdes (1998) solicited example after example of unique ways of knowing geometry theorems among his secondary education students. They recollected numerous perspectives of the theorem, *diagonals of a rectangle are congruent* each related to specific cultures of their communities. After Moore (1988) studied Native North American petroglyphs, finding new ways to understand iteration, recursion and several other geometry ideas, he created meaningful classroom activities to illuminate these ideas holistically. As researchers endeavor to reconnect mathematics to the context of its socio-cultural environment, mathematics education has also benefited and grown.

Teaching and Learning

When viewing mathematics as indelibly linked to culture, teaching and learning mathematics takes on a different image than is often seen in school. Mathematics instruction becomes less of a chronicle of information (a static and unchanging set of rules and laws) and more of a dynamic problem solving investigation (something one *does* to handle an unexpected problem in an environment.) Changing instruction toward this view enriches learning, both because the mathematics is more tightly connected and because learners come to know mathematics while simultaneously honoring ways of knowing within the culture from which the problem originates. It is within this context that I study culture and mathematics.

As part of preparing for instruction, I find it not only satisfying, but enlightening to ground the lesson explicitly by making a meaningful cultural foundation integral with the mathematics topic(s). When learners layer mathematical ideas with knowledge about world cultures, their overall web of knowledge, including mathematics, is more dense. Richly connected knowledge is easier to recall, easier to use, deeper, and more long-lasting (NCTM, 2000). So, it is a natural extension to mathematics education to study mathematical ideas (like measurement) through cultural knowledge of members of the society who practice it (D'Ambrosio, 1990; Gerdes, 1998).

As a place where mathematics and education meet (Miarka, 2012), ethnomathematics can provide meaningful opportunities for studying mathematical ideas, including the focus of this paper: measurement. This paper explains how and why the inclusion of a thirteenth moon solved a societal "problem" for many Native North American tribes, including the ancient Hopi. Drawing parallels to the Hopi moons reveals the complexity of deceptively simple measurement concepts and weaves them together in a exceptional way.

Measurement

Learning and Teaching

Learning and teaching about measurement are intricate activities because it is composed of several related concepts, all of which must be applied and coordinated to appropriately determine the size of something (Lehrer, Jacobson, Kemeny, & Strom, 1999). Basic measurement concepts discussed in this paper are size, standard unit, zero-origin, tiling, continuity, and part-whole.

Basic Concepts

Before learning to measure, children must gain a basic awareness that objects have a size. Knowing that some things are somehow larger, deeper, or longer than other things is fundamental for building a knowledge base. Young children can order objects from smallest to largest whereas older children can compare objects to one specific object. The specific object of a size comparison might be a pencil, a footstep, or even recess. By mathematizing their classroom environment, measuring familiar things against familiar units such as these, students make connections between and among the relative sizes of elements of their immediate environment. After ordering and comparing, asking, "*How much* larger, deeper or longer?" piques children's interest in the inevitable follow-up task of communicating their ideas about size to others.

To explain measurements to people outside of their immediate environments, children build awareness that communicating how much larger, deeper or longer an object is requires some sort of *standard unit*. Instead of using a pencil, a footstep, or recess as their comparison unit, they develop conceptual understanding that standard units are things against which *everything* can be compared and about which *everyone* knows. Examples include inch, centimeter, second, year, degree, square yard, and cubic meter. Children are often interested in knowing that standards units like these exist beyond the classroom in every day life and that it is important for a culture or a society to agree on them.

NOTES

Basic Procedures

Sometimes, the typical measuring device does not properly align with the item being measured. Flexibility in using measuring devices depends partly on the learners' conceptual fluency with the *zero-indicator* on the device. Several concepts (like standard unit) are automatically embedded in most measuring devices, but the zero-indicator's importance is in showing where to start the measurement process. For example, when measuring recess, children need to recognize that even though the clock says 11:20 upon return from recess and it said 11:05 upon departure, they did not spend 11 hours and 20 minutes playing. An adjustment of 11 hours and 5 minutes is needed, making recess 15 minutes long. The zero-indicator was 11:05. At that exact moment, in fact the exact second, academics ended and recess began.

Joining Concepts and Procedures

The idea that the standard unit of a minute is used over and over again and that a new minute begins the exact moment the previous minute ends is an example of the concept of *tiling*. However, there are also moments when it is appropriate to pause the tiling when measuring in certain situations. Knowing when to pause the tiling is an important piece of the tiling concept. For instance, reconsider measuring the length of recess bearing in mind that it takes children 2 minutes to walk from the classroom to the playground. Those 2 minutes are not counted as part of the recess nor as part of academics.

Whenever the tiling process is being applied, the process is repeated *continuously* until reaching the end of the object being measured. Counting the number of iterations is the same as describing the size of the object under investigation. Of course, it is possible that the size of the object being measured may not coincide with a whole number of repeated standard units. So, understanding the notion of *part-whole* is important to fully appreciate the nuances of measurement and the ways that measuring devices can be applied. For instance, in the measurement of time, we have half hours, tenths of seconds, or quarters of years. Coordinating all these concepts to execute a procedure requires multiple learning experiences in varying environments. Investigating the calendar of an ancient agricultural culture, with attention to measuring time, reveals these concepts in a copious and rich way and brings new life to understanding them within a significant context.

The Ancient Hopi

The Hopi are one of the Pueblo peoples who, for centuries dating well before the time of Columbus, have made their home on the mesas in Northeastern Arizona in North America. They lived in the community but traveled out from their homes to farm their carefully nurtured desert fields.[1] Successfully growing crops in an inhospitable desert climate severely depended on a balance of moisture. Respecting rain was an integral part of Hopi culture and a cornerstone of Hopi history. This aspect of ancient Hopi history and knowledge was one of the mores passed down through oral tradition.

Celebrations

Such storytelling kept history alive during the Soyal celebration. Those who gathered for the sacred stories listened intently as Hopi historians used corporeal words to weave stories into the spiritual atmosphere. In particular, they repeatedly told the timeless story about Hopi emergence into the Americas and credited their long-lasting existence as a result of having been chosen as the caretakers of Earth. This solemn responsibility (that so depended on rainfall) impacted nearly all Hopi culture. Cultural

[1] Past tense is used throughout the remainder of the paper to base the discussion in ancient calendar-watching strategies. This is not meant to imply Hopi culture is not presently vibrant, nor that the ceremonies described in this paper are no longer practiced.

activities ranged from the material, planting corn seeds, to the spiritual, maintaining relationships with Katsinas, the spirit beings who guided them. The material activities relied on nourishment from Earth whereas the spiritual activities provided nourishment for the Earth. Naturally, it was of paramount importance for these activities to coincide with the rhythms and seasons of Earth. To further celebrate their role as nurturers of Earth, the Hopi constructed a framework of cultural festivals and celebrations to surround the hard work of tending for Earth. Certain agricultural activities corresponded with certain festivals and celebrations.

Forecasting celebrations. The responsibility to tend for Earth motivated accurate timing of these celebrations and festivals. Agricultural activities like planting and cultivating had to be carefully arranged, to coincide with the Earth's nourishments and needs. "... determination of the last freeze coincident with the vernal equinox is essential to crop survival" (Autrey and Autrey, 1981, p. 96). Planting beans or corn at the right time in the cycle, neither too soon nor too late was predicated upon appropriately forecasting the solar year (winter solstice to winter solstice). Embedding the Earth's rhythms in their cultural calendar reminded them of upcoming events. But, how did the Hopi elders schedule the community calendar? They matched celebrations and festivals with specific moons, each of which had a specific name. As those specific moons appeared, these ancient Hopi knew which ceremony and hence, which planting, tending, and harvesting activities had arrived. "Every ceremony is associated with a moon and every moon with a ceremony" (Ellis, 1975, p. 65). In other words, the moons measured time.

Measurement concepts of Moon time. The 29.5 day *standard unit* of the moon is itself a composite of phases. It moves from the first sliver of a moon, through waxing, on to a full moon, back through waning until just before the final sliver disappears. What exactly did the ancient Hopi need to measure with such a standard unit? To be sure, they needed to predict the Winter Solstice. But they could have done (and did) that by comparing their astronomical records with the position of the sun in the sky. When the sun's path is the shortest, it provides the least amount of daylight. The sun's shortest path also travels along its lowest path across the skyline, something visible even to the naked eye. The shortest and lowest path is the Winter Solstice, which marks the beginning of a new solar year. Certainly ancient Hopi could thusly predict this change over to a new year. And, this would have been enough if their cultural celebrations and festivals occurred only at the Winter Solstice. Because of the agricultural needs and responsibilities throughout the rest of the year, they needed to measure the solar year with some kind of smaller unit and they used the moon.

Appearance of that first sliver of moon, the crescent, marked the beginning of a new unit of moon time (Ellis, 1975). Determining placement of the *zero indicator*, or origin, is required in any measurement situation. In a Western[2] sense, this is akin to beginning the month of January at the stroke of midnight on December 31. Ancient Hopi began measuring moon time upon visual verification of that first sliver of new moon.

---

[2] The term "Western" is intended to imply a roughly European viewpoint and used throughout the manuscript to serve as a way to alert the reader if there is a need to translate a Hopi viewpoint into the cultural viewpoint of the Western author.

NOTES

There is a two-day lull (sometimes called a *new* moon) between the last sliver of one moon and the new sliver of the next moon. Did ancient Hopi worry about this lack of a moon? "The two preceding days in which no moon is visible are not counted; they fall between lunar periods" (Ellis, 1975, p. 64). So, the Hopi *standard unit* for measuring the lunar year was actually approximately 27.5 days, even though the Western description of a lunar month is 29.5 days. For the demarcation of time using the moon unit, the Hopi simply omitted those two days. That is, cultural events tended not to happen during a new moon.

Conceptualizing time. The way westerners would worry about these two days illuminates conflicting subtleties within the general concept of time from the Hopi point of view when compared to the Western point of view. The Hopi did not "count" moons or days. Counting, as in moving toward infinity, implies a view of measuring time as one of *linear passage*. Instead, the Hopi view of time was one of *constant cycling*. Each moon unit was associated with a recurring event rather than as one moon in a linear series of "next" events. "Our "length of time" is expressed by the Hopi not as a linear measurement, but as a relationship between two events" (Waters, 1969, p.104.) In effect, they did not tally time as much as they experienced it. During the two-day new moon, there was nothing to measure. That is, there was nothing to experience during that intermission. Since the western perception of time is linear, these ideas required me to stretch my thinking. But for the ancient Hopi, time was naturally a cyclic and symmetric affair. "The Hopi think of the moons of their calendar as being in a continuous succession which we can best picture as a moving chain encircling a wheel, ..." (Ellis, 1975, p. 69). They marked time not for calculating its *passage*, but more for properly aligning themselves with cyclic needs of the Earth. However, in so doing, they continuously tiled the *standard unit*, placing it end-to- end. But the calendar was not a line, it was a circle. They measured with the moon in order to connect two reappearing events, such as a moon and its agricultural celebration or festival. In so doing, they related the appropriate time during the solar year with the appropriate agricultural activity like planting or harvest.

Important events. There was a moon for Waiting (it is not yet time to plant corn, beans, and squash) and a moon for Planting[3]. There was even a moon, Sparrow-Hawk, designated as the time for taking account of celestial observations and guiding creation of the specific schedule for the upcoming year's community calendar (Secakuku, 1995). The (Autumn) Kelmuya season ceremony was always scheduled with Sparrow-Hawk Moon. In the Kelmuya ceremony, "singing is loud and deep and echoes the drumbeats, which are prayers for long and pleasant life for all mankind, moisture in the form of snow and rain, and a bountiful harvest of nourishing crops" (Secakuku, 1995, p. 6). Kelmuya prepared the Hopi for the next moon, Sacred Moon, which brought the ever-important Winter Solstice ceremony, Soyal.

Activities of the Soyal ceremony included the storytelling, noted earlier, that kept history alive. It also set in motion the first scheduled event in the yearly cycle of caretaking. During this ceremony, the Hopi persuaded the sun to begin to lengthen and raise its path across the sky, slowly relieving Earth of those short, dark, but necessary, winter days (James, 2000), and creating the Winter Solstice. As if further authentication of the importance of this ceremony were needed, for the first time since the preceding Summer Solstice ceremony, a Katsina returned from the spirit world to walk among the Hopi people. Any effort to understand ancient Hopi culture would be incomplete without studying the role and importance of the Katsina. That said, a westerner's efforts to really comprehend the

[3] Moon names were compiled from a variety of references: Sekakuku, 1995; Wright & Roat, 1965; Ellis, 1975; James, 2000; http://www.wwu.edu/skywise/indianmoons.html#Hopi

importance of the Katsina to Hopi culture will be incomplete due to the lifelong, dynamic nature of that relationship.

Katsinas. The depth of the relationship between Katsinas and the Hopi people must be understood as one where Katsinas are sacred "spirit beings to invoke substantial growth and maturity for all mankind" (Secakuku, 1995, p. 3). Since the Hopi were entrusted by the gods to provide watchful and careful nurturing of Earth, it was important they were given both spiritual support and earthly support (for instance, rain) to accomplish this directive. "Katsina dance events combine responsibility with pleasure because of the grave commission to provide moisture for the whole world" (Kealiinohomoku, 1989, p. 58).

For ceremonial purposes, selected Hopi people dressed like the spiritual Katsinas to symbolically represent the ethereal beings and to perform ritual singing, dancing and interactions with the people. Just as moons corresponded with ceremonies and agricultural activities corresponded with various ceremonies, Katsinas matched with specific ceremonies and dances. So, moons had not only to precisely match the Earth's rhythms, but also had to match the correct Katsina. Hosting the correct ceremony and thus interaction with the correct Katsina(s) signaled the correct agricultural activity for successful harvest and tending of the land. Correctly using moons to measure and plan these important events was imperative. Katsinas permeated ancient Hopi daily life in many ways. The Katsinas knew if the people had properly prepared for Soyal and upheld their promise to nurture Earth through another year (Secakuku, 1995). During a late winter moon, Purification Moon, Katsinas delivered a small bundle of bean sprouts to each family (Secakuku, 1995). These sprouts represented promises of the coming season. The importance of correctly timing ceremonies, so the appropriate Katsinas interacted with the Hopi people to nourish the correct agricultural activity, was the only way to ensure the Earth was properly tended.

With so much riding on the timing of the Soyal celebration, and with time being measured in moons, it was imperative and a somewhat delicate project to correctly match moons to the correct corresponding ceremony. Correctly timing Soyal, would mean the first sliver of Sacred Moon appeared *before* the days started to lengthen again, that is, before the turning point of the solar year, the shortest day, the Winter Solstice. Hopi Celestial Body-Watchers were expected to appropriately calculate this event beforehand. So, in effect, they had to be sure they knew when they were in the time of Sparrow-Hawk moon, when all subsequent times for ceremonies and festivals were planned (see Table 1).

NOTES

Table 1: Hopi Celebration Calendar

Moons	Celebrations/Events
Sparrow-Hawk Moon (This moon is immediately prior to the moon during which the Winter Solstice occurs)	• Second initiation period for youths • Youths became warriors (hawks symbolize this achievement) • Make new fire • Leads up to Winter Solstice ceremony
Sacred (or Respected) Moon (Moon during which Winter Solstice *must* occur)	• *Lengthy* ceremony, Soyal, First Katsinas appear *must occur during Winter Solstice*
Play Moon Joyful Moon	• Gaming and entertainment (weather is too cold to work outside) • Social dances are held • Secretly-planted beans & corn are brought to heated area, for sprouts
Purification Moon	• First initiation period for youths • More Katsinas appear to insure fertility of soil • Pre-planting ceremony • Present bean & corn sprouts • Plant more seeds, nurture in heated area • Young children receive Katsina dolls
Whispering Wind Moon	• Plant "early" corn, so Katsinas will have it for the harvest celebration (much later)
Windbreaks for Small Plants Moon	• Seeds are prepared for planting
Waiting Moon	• Planting of non-corn begins (e.g., melons, beans & pumpkins)
Planting Moon	• Corn planting begins • Young people are taught how to farm
Homedance Moon Moisture Moon	• Rains begin • Home-Going Ceremony (Niman) Katsinas travel "home" *occurs after Summer Solstice*
Joyful Moon	• Snake/Antelope or Flute ceremonies • Butterfly Dance
Big Feast Moon Moon of Plenty	• Smoke meeting - women's societies bless the harvest season • A few crops picked for feast
Harvest Moon	• Harvesting begins • Harvest Basket Dance
Corn Husking Moon *(If needed – some references do not include this moon)*	• Burden basket is carried

The Problem

Ancient Hopi elders needed to know exactly which moon was shining so the community calendar would be accurate. A successful harvest required things, like planting, to happen at appropriate times in the solar (or agricultural) year. And, planting had to occur during Planting Moon just like Winter Solstice had to occur during Sacred Moon. At first, it is tempting to think that keeping track of moons and reciting them in order would have been sufficient. However, when the ancient Celestial Body-Watchers compared the sun movement (solar year) to the moons, they found a small problem. The 365.25 day solar year (Winter Solstice to Winter Solstice) is not neatly divisible by the 29.5 day moon (new moon to new moon). There is a remainder of around 10 days: a solar year (365.25 days) is approximately 10 days longer than 12 iterations (354.9 days) of the moon. Simply counting out 12 moons would have resulted in incorrectly timed celebrations. Within a mere three years, an extra 30 days would occur, which encompassed an entire moon. We can see how quickly *tiling* the moon unit fell out of sync with the solar year and forced some sort of reconciliation of the *part-whole* concept. A solar year ends and begins partway through a thirteenth moon.

The Solution

So, approximately every 3 – 4 years, a 13th tiling of the moon could be fully completed before the height of the sun's path across the skyline alerted the Celestial Body-Watchers to conduct Sparrow-Hawk festivities. Sparrow-Hawk is considered the 1st moon of the cycle because this is the time when the year's festivities are announced. Sacred Moon, the moon bringing the Winter Solstice, is actually considered the 2nd moon in the cycle even though it's the first moon of the Solar Year. To correctly match the Winter Solstice ceremony to Sacred Moon means the ancient Hopi needed to figure out how to handle that part of a moon that caused a misalignment (every 3 – 4 years) between Sacred Moon and the Winter Solstice. This was compounded because they had no control over the *size* of their standard unit. Nor could they just pick up their measuring device and move the *zero-indicator* to the Winter Solstice and begin tiling again. They knew whatever decision they made to address the 10-day part of the 12th moon (or the 19.5 day part of the 13th moon) would have ramifications for the subsequent solar year.[4]

Connections to Measurement Concepts

The Hopi reapplied the *standard unit* over and over again, *tiling* it to measure time. They found a way to handle the *part-whole* problem: being unable to physically reset the *zero-indicator* to the Winter Solstice when the end of the 12th moon did not end in such a way that Sacred Moon would match up with the end of the solar year. This process fit well within the Hopi worldview of continuous symmetry within time. Just like they

[4] It should be noted that several sources (Ellis, 1975; Wright & Roat, 1965; Secakuku, 1995); McCluskey, 1981) indicate that 20th and 21st century Hopi do not measure exclusively with moons. Now, they recognize a 5 month time (July – November) when Katsinas are in the spirit world and a 7 month time (December – June), when Katsinas walk among them, which represents adaptation of the lunar month standard unit to the 12 month constructed calendar imposed upon them by the western world.

NOTES

anticipated the reiteration of cultural events during each solar year, they also counted on Katsinas to return to them each Winter Solstice, after having been in the spirit world since the ceremony following the previous Summer Solstice. They saw things like the recurring existence of a *longest* day and a *shortest* day as a promise that continuity of time would be preserved. Rather than time passing, it recurred.

Here, I came to appreciate the deft manner in which ancient Hopi demonstrated deep understanding of measurement concepts in dealing with their problem. They inserted another standard unit when confronted with the part-whole situation that was required and could accommodate the situation (approximately every 3rd or 4th year.) Viewing calendar measurement the way the ancient Hopi perceived it brings new life to my understanding of measurement concepts. Merging Hopi thinking with my understanding of measurement concepts wrapped cultural respect and appreciation around mathematical knowledge while also growing deeper mathematical knowledge from it.

Several mathematics education researchers believe teaching mathematics from an explicit ethnomathematics viewpoint may address the fact that many K-12 teachers are from different cultures than their students (Fuller, 1992; Gerdes, 1998; Malloy & Malloy, 1998). It is not atypical for a teacher to teach a room filled with students of many cultures, possibly none of whom represent the cultural group in which the teacher himself or herself claims membership. The study of ethnomathematics is dedicated to understanding mathematical concepts through the eyes of members of a social structure in which the mathematics is being implemented, invented, or studied. People's ways of knowing are cast in molds upon birth. These molds are influenced, in fact designed, by the culture into which the person is born. Knowing mathematics is no exception. Clearly, two is always two, circles are circles, and time is something we can measure. However, *ways of knowing* these mathematical ideas are not so antiseptic. While ethnomathematics does not suggest that in some cases two is three or that circles somehow look like squares in some cultures, how we think about and how we know about "two-ness" varies according to the culture in which we are members.

To teach mathematics within a context of such diversity, we ourselves must relearn mathematical practices from a variety of cultural groups. By connecting mathematics learning and teaching to cultural knowledge (our own and others'), we possess a richer tapestry of mathematics knowledge, hence an opportunity for better understanding and teaching of it.

The Resulting Mathematics Lesson

The goals of my lesson were for students to: (1) identify the measurement concepts embodied by the moon, (2) use a circular representation of a lunar calendar (moon ruler) to recognize the value and pitfalls of measuring years with moons, and (3) describe how the Hopi adjusted for these issues as evidenced by a statement acknowledging that 12 moons or 13 moons each year would put calendars out of sync with the planting seasons and (4) describe that the Hopi handled this by inserting a 13th moon only every 3 – 4 years. My lesson plan opened with a discussion question designed to focus their comments on a need to pinpoint a specific date during a year. I asked, "What if you lived in ancient times, when there were no paper calendars. How would you tell someone when your birthday was?" During the resulting discussion, I watched for opportunities to key the conversation toward using Winter Solstice as a landmark time, the need to measure specific moments during a year, using the moon as a standard unit, and the issues that would arise by using moon phases as the standard unit to measure time. After the discussion, where students recognized the plausibility of measuring a year using moons, I issued students a circular moon-ruler and a circular year chart and asked them to measure the year with a 12-moon cycle. After the students used the rulers they saw how quickly a 12-moon count ran out of sync with the seasons. Twelve iterations with the moon acting as a standard unit account for only 349° of the 360° in the circular year.

I then described how the Ancient Hopi had to know when the moon for Winter Solstice would occur and that 12 moons occurring in less than one year complicated timing for that event. The Ancient Hopi felt responsible for tending the earth and it was imperative for them to know when the Sparrow-Hawk moon would occur. It absolutely had to occur at the appropriate time. It had to be the moon immediately before the Respected Moon – the moon with Winter Solstice. Winter Solstice accompanied the ceremony that asked Katsinas for critical help in tending the earth. I also described the Celestial Body-Watcher, who was responsible for knowing when to insert the 13th moon, so that all of the year's festivals occurred at the proper time, with each of the corresponding moons. Students then read "The Problem" and "The Solution" before completing the Assessment. In the assessment, students calculated the first possible placement of Corn-Husking Moon, described how the measurement concepts interacted with the moon ruler, explained why the Hopi needed to measure years and explained how the Hopi Celestial Body-Watcher handled issues with the moon ruler.

Summary

As a place where mathematics and education meet, ethnomathematics can provide meaningful opportunities for studying mathematics ideas. The field of ethnomathematics can provide appropriate contexts for studying the teaching and learning of mathematics. It recognizes social activity and cultural membership as important components of mathematics knowledge. Inspiring diverse groups of children to build knowledge on their existing understandings seems more deftly handled when related social and cultural knowledge is taken into account. Studying and appreciating the thicket of connections between mathematics and culture (like measurement and the Hopi celebration calendar) allows teachers like me to deepen my knowledge. In the end, this grounded my efforts to create lessons that helped students develop connected knowledge about teaching, learning, and measurement concepts.

Endnote

The Hopi have a long history and a complex culture. Differences in knowledge due to the personal nature of storytelling are inherent even among different Hopi villages and across the three mesas. "No published material will ever meet the general approval of the Hopi people" (Secakuku, 1995, p. viii) and the Hopi people tend to be a private society. Some westerners have been allowed glimpses of enduring Hopi ideologies and beliefs (e.g., McCluskey, 1981; Carlson, 1986; Pacific Arts Video Publishing, 1990; Page & Page, 1994; Fewkes, 2001). The author acknowledges her decidedly and inescapable "western" viewpoint and offers a consequential apology to the Hopi people, who know that non-Hopi people cannot fully understand Hopi ways

NOTES

of knowing. She has treaded as lightly as possible, gathering only enough information to create this lesson's background. She requests their patience with her efforts to appreciate and respect their ways of knowing, as she studies the mathematics of some of their ancestors for the purpose of enhancing mathematics education.

References

Autrey, N. E. & Autrey, W. R. (1981). Zodiac Ridge.In Williams R.A. (Ed.), *Archaeoastronomy in the Americas* (pp. 81-99). College Park, MD: Ballena Press/Center for Archaeoastronomy,

Aveni, A. F. (1975). *Archaeoastronomy in Pre-Columbian America.* Austin, TX: University of Texas Press.

Barta, B. (2008). *Language of Mathematics: Telling mathematical tales*. New York: Springer Science.

Barta, J. & Shockey, T (2006). The Mathematical Ways of an Aboriginal People: The Northern Ute. *Journal of Mathematics and Culture, 1*(1), 79-89.

Bean, L. J. (1989). (Ed.) *Seasons of the Kachina*. Proceedings of the California State University, Hawyard conference on the western pueblos. Hayward, CA: California State University.

Carlson V. F. (1986). *Broken Pattern Sunlight and Shadows of Hopi History,* Happy Camp, CA: Naturegraph Publications.

D'Ambrosio, U. (1990). The History of Mathematics and Ethnomathematics: How a Native Culture Intervenes in the Process of Learning Science. *Impact of Science on Society, 160*, 369-377.

D'Ambrosio U. (2006). The Program Ethnomathematics: A Theoretical Basis of the Dynamics of Intra - Cultural Encounters. *Journal of Mathematics and Culture, 1*(1), 1-7.

Eglash, R. (1999) *African Fractals: Modern Computing and Indigenous Design*. Piscataway, NJ: Rutgers University Press.

Ellis, F. H. (1975). Pueblo Sun-Moon-Star Calendar. In Aveni, A. F. (Ed.), *Archaeoastronomy in Pre-Columbian America* (pp. 59-87). Austin, TX: University of Texas Press.

Fewkes, J. W. (2001). *Hopi Snake Ceremonies/Revised.* Albuquerque, NM: Avanyu Publications.

James, S. E. (2000). Some Aspects of the Aztec Religion in the Hopi Kachina Cult, *Journal of the Southwest, 42*(4), 897-926.

Kealiinohomoku, J. W. (1989). The Hopi Katsina dance event "doings". In Bean, L.J. (Ed.), *Seasons of the Kachina* (pp. 51 – 63). *Proceedings of the California State University, Hayward conference on the western pueblos*. Hayward, CA: California State University.

Lehrer, R., Jacobson, C., Kemeny, V., & Strom, D. (1999). Building on Children's Intuitions to Develop Mathematical Understanding of Space. In Fennema E. & Romberg T.A. (Eds.), *Mathematics Classrooms that Promote Understanding* (pp. 63-87). Mahwah, NJ: Lawrence Erlbaum Associates.

McCluskey, S. C. (1981). Transformation of the Hopi Calendar. In Williams, R. A. (Ed.), *Archaeoastronomy in the Americas* (pp. 173-182). College Park, MD: Ballena Press/Center for Archaeoastronomy.

Miarka, R. (2012). The Role of mathematics within Ethnomathematical Descriptions. *Journal of Mathematics and Culture, 6*(1), 296-307.

Ortiz-Franco, L. (1999). Chicanos have math in their blood: Pre-Columbian mathematics. In Short, I. & Pari, C. (Eds.), *Education is Politics: Critical Teaching Across Differences* (pp. 215 – 225). Portsmouth, NH: Heinemann.

Page, S. & Page, J. (1994). *Hopi*. New York: Abradale Press.

Powell, A.B., & Frankenstein, M. (Eds.) (1997). *Ethnomathematics: Challenging eurocentrism in mathematics education*. New York: State University of New York Press.

Secakuku, A. H. (1995). *Following the Sun & Moon: Hopi Kachina Tradition.* Flagstaff, AZ: Northland Publishing.

Waters, F. (1963). *Book of the Hopi*, New York: Penguin Books.

Waters, F. (1969). *Pumpkin Seed Point: Being Within the Hopi*. Athens, OH: First Swallow Press/Ohio University Press.

Williams, R. A. (Ed.) (1981). *Archaeoastronomy in the Americas*, College Park, MD: Ballena Press/ Center for Archaeoastronomy.

Wisconsin Educational Communications Board and Board of Regents UW System (1990). *Winds of Change: A Matter of Choice*. Beverly Hills, CA: Pacific Arts Video Publishing. ISBN: 1-56111-570-3.

Wright, B. & Roat, E. (1965). *This is a Hopi Kachina*. Flagstaff, AZ: Museum of Northern Arizona.

Zaslavsky, C. (1973). *Africa Counts*. Brooklyn, NY: Lawrence Hill Books.

Zaslavsky, C. (1996). *The Multicultural Classroom: Bringing in the World*. Portsmouth, NH: Heinemann.

Doing Mathematics with Roma: A Dialogue Between Dialogic Learning and Ethnomathematics

Javier Díez-Palomar
Department of Mathematics and Science Education
University of Barcelona
jdiezpalomar@ub.edu

Abstract

Roma have been excluded from education over the years. Their cultural heritage mediates their ways to learn and use mathematics within the classroom. Drawing on the dialogue between *Dialogic Learning* and *Ethnomathematics* open the possibility to incorporate Roma's voices to the learning process. Interactive groups become spaces for participants to share their own *cultural* strategies to solve mathematical tasks. Egalitarian dialogue, cultural intelligence, and equality of differences are *universal* principles that make possible for all participants in the classroom to draw on their own ways to do mathematics, as stated by the *Ethnomathematics* approach. The challenge for the near future is to identify these *universals* to design inclusive practices in mathematics education.

Keywords:

- Roma Studies
- Dialogic Learning
- Ethnomathematics
- Egalitarian Dialogue
- Cultural Intelligence
- Equality of Differences

Resumen

Las personas gitanas han sido tradicionalmente excluidas de la educación a lo largo de los años. Su herencia cultural media la manera en la que aprenden y utilizan las matemáticas en el aula. Partiendo del diálogo entre el *Aprendizaje Dialógico* y la *Etnomatemática*, se abre la posibilidad de incorporar las voces de las personas gitanas al proceso de aprendizaje. Los grupos interactivos se convierten en espacios donde los participantes comparten sus estrategias *culturales* para resolver tareas matemáticas. El diálogo igualitario, la inteligencia cultural y la igualdad de las diferencias son principios universales que hacen posible que todos los participantes en el aula usen sus propias formas de hacer matemáticas, tal y como establece el enfoque de la *Etnomatemática*. El reto de futuro es identificar estos *universales* para diseñar prácticas inclusivas en educación matemática.

Palabras clave:

- Estudios Gitanos
- Aprendizaje Dialógico
- Etnomatemática
- Diálogo Igualitario
- Inteligencia Cultural
- Igualdad de Diferencias

Introduction

Ten years ago I published the article *"Math is Everywhere:" Connecting Mathematics to Students' Lives* with my colleagues Ksenija Simic and Maura Varley. In this article, we described the experience of teaching and learning mathematics in an afterschool program addressed mainly to Latino children in an elementary school based in a southwestern city in the USA. That article was connected to my research work as *postdoctoral Fulbright Fellow* at The University of Arizona. There I had the opportunity to meet a great research team at CEMELA – *Center of the Mathematics Education of Latinos/as*. The work that I did gave me the opportunity to know and share the life experience of dozens of Latino families, and their struggle to bring real learning opportunities for their children. I was an observer (and participant actor as well) of the tensions that many Latino families (especially immigrants) experienced when their children asked them for help to solve mathematics homework, and their strategies did not match with the teachers' way to solve the tasks. I learned many resources new to me (as European visitor), and I helped to develop an inclusive curriculum of mathematics addressed to 3rd graders.

Drawing on the *funds of knowledge* (González, Moll & Amanti, 2006) approach, Ksenija, Maura and myself created an afterschool program addressed to Latino children. We developed memorable activities such as the one called "Cross the Border", in which the children calculated the amount of water (both in decimal and imperial system) that someone needs to carry on, in order to cross from Mexico into the United States through the border, and not to die during the journey into the desert in doing so; or "La Piñata", in which the children designed, bought the material, and build a *piñata* using geometric tools and knowledge; or "Mathematics through the professions of my neighbors", which drove us around the neighborhood to document with video cameras how different people use mathematics in their corresponding employments. This approach allowed us to make some connections between mathematics and children' everyday life in the *barrio*. But, more importantly, we were able to introduce an Ethnomathematics look to the mathematics present in their cultural context as members of a *borderland* community.

Back in Barcelona I had the opportunity to participate in the study with more resources and greater scientific impact funded by the European Commission within the VI Framework Program: the *INCLUD-ED: Strategies for Inclusion and Social Cohesion in Europe from Education* research project, lead by CREA (Community of Research on Excellence for All), from the University of Barcelona. The main aim of this project was to analyze educational strategies that contribute to overcome inequalities and promote social cohesion, and educational strategies that generate social exclusion, particularly focusing on vulnerable and marginalized groups. This project obtained a great social impact (Flecha & Soler, 2013). According to the goals set by Europe (*Horizon 2020*) and by the United Nations (*UN Sustainable Development Goals*), the researchers working within INCLUD-ED identify *Successful Educational Actions* (SEAs) that succeeded in reducing absenteeism and early school leaving in all the schools involved in the project (over 400 worldwide), increasing school performance (performance rates),

NOTES

and creating employment opportunities in locations of extreme poverty all over Europe (Flecha, 2014). Part of the success of INCLUD-ED is due to include the voices of members of the vulnerable groups through an egalitarian dialogue with the researchers. Using the communicative methodology, we were able to draw on participants' cultural backgrounds to deeply analyze situations within the school.

In this article I want to explain the case of Roma[1] people; I want to discuss how the Roma case illustrates a cultural view of mathematics, which can be *universal* as well. I will use Ethnomathematics and Dialogic Learning as theoretical lenses to think critically on this theme. In fact, this is a reflective article; it is not an empirical one. Roma people define their identity in cultural terms: being Roma means to share certain traditions and customs. Roma is a *nation* "without land," because for them their identity is not attached to the land, but to those traditions and customs shared by Roma people Worldwide. A common language (with some ethnic dialects), rituals, beliefs, religion, constitutes their *cultural heritage*, which is the basis for their identity as a *Pueblo*. They have survived over the centuries drawing on *oral traditions* passed from one generation to another (Vega Cortés, 1997). Roma still live in a situation of exclusion in Europe, due to the rejection that have experienced traditionally in our *white* and *western* societies. Roma have a culture with ancient roots, broadly present in Europe through the centuries.

Although more than six hundredth years ago a group of *Rromanò Thèm* people left the Punjab and Sinth regions in India, facing Europe until the Iberian peninsula, still nowadays they continue to be seen as *different* and they suffer rejection and persecution in Europe. Roma way of life (based on oral tradition, the respect towards elders, importance of trust word, honor, family, etc.) annoy Europeans, who still see them as "foreigners" and "invaders." (Unión Romaní, 2016). Roma people have been persecuted, enslaved (until the XIX century), castrated, and almost eliminated (according to Ian Hancock between 70% and 80% of the Roma population in Europe was eliminated during the II World War –Union Romaní, 2016–). However, it is also true that ways of peaceful coexistence and even integration (respecting the differences) have been founded. In recent years Spain and Spanish Roma have become international benchmarks for the rest of Europe, because of their serious work and their effort to create a social movement which is transforming Roma situation; nowadays Roma are becoming literate and many of them are developing academic careers at the University.

Ten years ago I discussed the case of Latino families living and experiencing mathematics in a southwestern town in the United States. Ksenija, Maura and I discussed the tensions that Latino children faced in that context, and we showed that it was possible to find sophisticated mathematics within an afterschool program designed with (rather than on) Latino people. In this article I pretend to discuss how Roma also makes essential contributions to the teaching and learning of mathematics, allowing us to broaden the Ethnomathematics approach. In order to do so, we need to create spaces free of stereotypes, where Roma culture could be appreciated as a source of mathematical knowledge. My hypothesis is that Dialogic Learning (DL) has the potential to further develop Ethnomathematics, because it opens the floor to find *universals* that can help us to develop high-quality curricula in mathematics, while being respectful of the diversity and idiosyncrasies of different cultural groups.

[1] Roma people prefer to call themselves "Roma" rather than "Gypsies," which is a non-Roma word to designate them.

Evolution of my Scholarship: Focusing on Vulnerable Groups and Roma Studies

There is very little scientific literature about Roma and education, especially regarding mathematics education and Roma culture (Chronaki, 2005; Stathopoulou and Kalabasis, 2006). This theme is also new from the perspective of the Ethnomathematics. There are no articles about the Ethnomathematics of Roma people.

According to the few studies that we have, Roma children use to be segregated from the mainstream in European schools. Fox and Vidra (2013) claim that Roma children are broadly segregated in the schools, both at the classroom level and at the school level in Bulgaria, Greece, Poland, and Romania. They claim that Roma students "typically receive a substandard education" (page 3). The reasons for segregation vary from a range of elements, including: school admission policies based on academic achievement rejecting [Roma] underachievers, population racism against Roma, alliances of teachers, parents and, eventually, school administrators as well, to remove "disruptive" Roma students from classrooms and schools. In Hungary, for instance, Roma students are systematically placed on schools for students with special needs (SENs) (Gurzó, 2014). Petrokou and Dimitrakopoulas (2002) report that some educational authorities claim (without any evidence) that "Roma families do not really want their children to attend school (…) the Roma are reluctant to integrate into society…" (New & Merry, 2012, page 630).

In the field of mathematics education, Stathopoulou and Kalabasis (2006) argue that Roma children failure in the school is due, mainly, to their culture as travellers, because Roma have a semi-nomadic way of life. The absence of "written" records in their culture provokes that Roma children "reject" the hegemonic discourse in the school, which is the use of elaborated codes (Bernstein, 1975). According to Stathopoulou and Kalabasis (2006), "Their socio-economic organization based upon family give children the opportunity to be taught by experienced members of the community without conceiving this process as teaching" (p. 152). Orality is in the center of their learning. Stathopoulou and Kalabasis (2006) as well as Chronaki (2005), highlight this aspect as essentially positive from the point of view of developing mathematical abilities such as mental calculation. According to Chronaki (2005) this is a clear advantage of Roma children towards non-Roma students. They are able to stand out using strategies such as piece mail, rounding, grouping, leading digit, compensation, break and bridge, etc. However, Stathopoulou and Kalabasis (2006) claim that the excess of using mental calculation induces children to have difficulties to develop mathematics understanding. According to them, "They had just made the right application, without taking into consideration the essence of the problem" (p. 154).

Chronaki (2005) does not agree with this view. For her, the oral tradition is not a problem, but a source of possibilities for Roma children. Chronaki (2005) describes and analyzes how two Roma girls use standard *home* strategies (such using fingers to solve arithmetic calculations), along with *school* strategies (defined by Chronaki as words, symbols and methods, formally talking). For her, what happens in the school is a process of transition from *home* towards school-related practices.

NOTES

This is consistent with previous studies in other cultural communities (Díez-Palomar and Civil, 2007; Gutiérrez, Baquedano-López, & Tejeda, 1999).

Nevertheless, this approach involves two main problems: (1) schools do not have spaces to incorporate different voices (those voices that are not aligned with the hegemonic discourse in the school); and (2) there is a clear negative stereotype concerning the cultural identity of Roma children. When I returned to Barcelona from Tucson in 2007, I participated within the research project INCLUD-ED (Flecha, 2014). There I had the great opportunity to work with people from the vulnerable groups, in the frame of Learning Communities (Gatt, Ojala & Soler, 2011; Garcia-Carrión, Girbés-Peco & Gómez-Zepeda, 2015; Ríos, Herrero, & Rodriguez, 2013). In this context, I met with Roma families, immigrants, people from low socio-economic, etc. All of them had a deep *informal* knowledge of mathematics. They taught me that our classic theories in mathematics education (such as the ones based on cognitive perspectives) have strong limitations to include other "ways to do mathematics."

But, Ethnomathematics has a conceptual and epistemological framework that makes the integration of *formal* and *informal* mathematics possible. Having in mind that my theoretical background is defined by the DL approach, which draws on the assumption that (a) everyone is a subject capable of speech and action (in Habermasianian' terms), and (b) our respective knowledge (whether or not formal) is always valid to the extend that it fits the criterion of validity claims, that is: it can be verified by anyone who wants to proof it within the classroom practice, then is it possible that a dialogue between DL and Ethnomathematics might move us [our work] beyond not only in terms of understanding how we teach mathematics inclusively, but also in terms of giving us guidelines to improve our teaching practices in the near future?

Pedagogical Focus: Creating Dialogic Spaces for Learning

Recent literature on education suggests that DL (Flecha, 2000) has contributed to providing an innovative vision of learning to help to overcoming traditional barriers. Based on Habermas, Freire, and others' contributions, Flecha theoretical proposal focuses on the idea of *egalitarian dialogue*, which is one of the seven principles characterizing the DL theory. According to Flecha, individuals use language (dialogue) as a tool for teaching and learning. Using language (as a communicative instrument, which may be in oral form, written, a gesture, a sign, etc.) individuals communicate ideas, concepts, knowledge, etc. Flecha defines different ways to use language depending on the intentions of the people involved in the communicative speech act. An individual can either justify a claim with an argument that can be validated by another individual within the dialogic process or s/he may try to impose his/her claim using his/her power position regarding the other participant(s) in the dialogue. This is what Habermas (1984) defines as power claims vs. validity claims. Drawing on the philosophy of language developed by Searle (1969), this is what characterizes and distinguishes the different speech acts.

Flecha assumes that all people have knowledge (*cultural intelligence*). This principle of the DL is common with the epistemological basis of the Ethnomathematics approach. The mathematical knowledge is contextual, situated (in Lave and Wenger's terms). However, the *cultural intelligence* as defined by Flecha is a universal concept: despite the differences among individuals, this concept means that all people has something to say when doing mathematics, everyone have their own (situated) mathematical knowledge, that can be used / shared in an egalitarian dialogue where individuals negotiate mathematical meanings embedded in the objects and representations used during the classroom practice. This position means expanding the scope and extent of Ethnomathematics as defined by D'Ambrosio (1985), Gerdes (1996) and others. In other words: the orchestration of knowledge is situated (contextualized), but such "knowledge" (the ability to know) is *universal*. Every human being is able to know mathematically because being a human being.

The dialogue between DL and Ethnomathematics has the potential to overcome the classic criticism against Ethnomathematics itself (Vithal & Skovsmose, 1997). The epistemological basis of the theoretical framework introduced by D'Ambrosio in late 1970s departs from three different aspects: *ethno* (that belongs to the culture of each human group), *mathema* (mathematical knowledge *per se*) and *techné* (which refer to forms of mathematical explanation). As defined by D'Ambrosio (1985): "we will call *Ethnomathematics* the mathematics which is practiced among identifiable cultural groups, such as national-tribal societies, labor groups, children of a certain age braked, professional classes, and so on" (page 45). This definition has been expanded by other authors like Gerdes (1996), who adds the idea of a *cultural system*. Similarly to DL, this approach emerges from the grassroots, as a way to incorporating the voices of people who traditionally are not taken into account within the community of *experts*. History demonstrates that all human groups have developed their own "mathematics," that is: their own way for counting and reasoning. Nevertheless, school mathematics (which uses what Bernstein would call *elaborated codes*) does not include forms considered as "non-accurate" by the international community of mathematicians. According to D'Ambrosio, this is because the mathematics that has been imposed in our society is the *western mathematics*, that is: the "heir" of Greeks' formalisms. But Ethnomathematics expands the scope incorporating "other mathematics." Some authors have argued that this broader view of mathematics suffers a clear political bias (Knijnik, 1998; Vithal & Skovsmose, 1997). Some of them even claim that Ethnomathematics is a form of postmodern mathematics (Dowling, 1993; Walkerdine, 1990). In several conversations I have had with Gelsa Knijnik over the years, I have always discussed the claim that Ethnomathematics is a form of post-modernism in the field of mathematics education. Throughout our discussions, we always end remembering that this theoretical approach has a *popular* sense *per se* (in Freirean terms). This fundamental meaning is shared with the DL. Ethnomathematics, similarly to DL, emerges from the individuals, includes their ways to do mathematics, and, for that reason, it offers a great platform to legitimize and re-value the different mathematics orchestrated by specific [cultural] human groups. This claim is especially true for those groups that traditionally have been excluded from the mainstream (the legitimated academic discourse), such peasants of the *Movimento Sem Terra* (MST) described by Gelsa, or Roma people, one of the vulnerable groups present in INCLUD-ED (Flecha, 2014).

"Re-incorporating the voices of vulnerable groups" has been a trend in recent decades, especially since social (or socio-cultural) approaches have colonized the field of education (and mathematics education as well). Concepts such as *funds of knowledge* (González, Moll & Amanti, 2006), *hybrid practices* (Gutiérrez, et al. 1999), *dialogic*

NOTES

talk (García-Carrión & Díez-Palomar, 2015; Díez-Palomar, & Cabré, 2015) allow us to update "old" instruments as the zone of proximal development (Vygotsky, 1980) or "scaffolding" (Wood, Bruner & Ross, 1976) to create more inclusive spaces where everyone can participate and share their ways of doing mathematics. Unlike Ethnomathematics, dialogic mathematics assumes that there are *universals* in the teaching and learning of mathematics; these *universals* must be the benchmark for thinking about successful strategies to teach mathematics in the classroom.

A clear example is *interactive groups* (IG). IG is a successful educational action that improves both students' performance and coexistence (Valls & Kyriakides, 2013). IG is a type of classroom organization based on heterogeneous grouping (especially in terms of ability). Students are split among three or four small groups. Each one must complete a task. All groups work simultaneously. After 15 or 20 minutes, the members of one group move to the next task, so at the end of the lesson, all students have completed all the tasks proposed. The groups are always facilitated by an adult (volunteer), who should never provide the answer to the task, but rather they should encourage the students to explain and justify with arguments (based on validity claims) their answers, to their peers in the group. The teacher is the one who corrects students' answers, lately. In these groups, many episodes occur, like the one in which a Roma boy is helping to do division his Senegalese fellow, who is a newcomer to Catalonia and never when to the school in Senegal. The task is to complete a worksheet with 2 digit, 3 digit divisions and so forth (see figure 1).

Figure 1. The Roma student and his peer from Senegal, working together to solve divisions

The Roma boy uses symbolic representations to explain the algorithm of distribution to his partner. He is drawing some puppets and then he distributes circles under each of them until he has no more circles left to distribute (see figure 2).

Figure 2. Roma' boy explaining how to divide using iconic representations

Thus, he is representing visually (through icons) divisor, dividend, quotient, and remainder. We can even appreciate that he is highlighting each of these components of the division in the sheet of paper. This iconic representation becomes a visual instrument that supports the Roma child argument. He uses it to explain the *formal* algorithm of division used in the Catalan schools, to his mate (see figure 3).

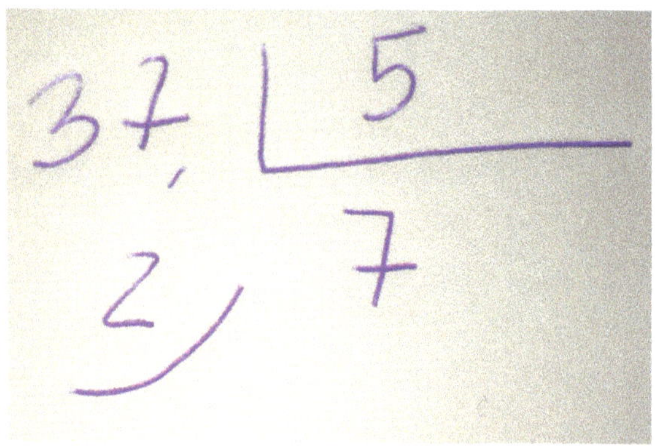

Figure 3. Formal algorithm of division used in Catalan schools

Drawing on an Ethnomathematics approach, we could claim that using this formal algorithm is part of the discursive social practice known as *western mathematics*. However, the Roma child is prompted to look for another strategy to explain how to divide to his peer from Senegal, who barely has being exposed to the basics in terms of

NOTES

literacy. "Explaining our answers to our peers using validity claims" is a requirement of working within IGs. Hence, the Roma child is "forced" to find a different way to explain division. He came with drawings (iconic representations) as a "teaching" instrument to represent visually every part of the formal algorithm. This is a non-standard algorithm equivalent to the one used in Catalan schools (as formal one), making possible to solve the division without knowing the symbolic representation. However, the reasoning embedded in both algorithms is the same in mathematical terms.

This case is not isolated. Stathopoulou and Kalabasis (2006) collected many cases of Roma using oral algorithms to solve problems. Their *classic* analysis of the school (classic in the structuralist sense of Bourdieu) leads them to criticize the school as a space impervious to the particularities of Roma culture: Roma children are marginalized and fail because neither their speech acts (based on *restricted codes*), nor their *habitus* (as a cultural group) correspond to those legitimated by the school. Being out of the hegemonic school culture is what explains why Roma are not able to succeed in the school (according to Stathopoulou and Kalabasis). But the focus on DL and IGs provides us the possibility to transform the context to turn up this situation: schools that use DL principles not only incorporate the voices of Roma students (and families as well) into the school discourse, but also transforms Roma identity presenting Roma culture as a culture of knowledge and learning. This helps to solve the first problem that I identified above (namely, that schools do not recognize the voice of Roma culture).

In addition, evidence from previous studies (Flecha & Soler, 2013) also contributes to overcoming the negative stereotypes towards Roma culture. Federico, a Roma colleague, and friend, who has participated in my research in the *Ramón y Cajal* Program on mathematics education of families of vulnerable groups traditionally excluded from the mainstream, reported that his mother taught him mathematics in order to "prevent that non-Roma people cheat on us". Federico's family, like many other Roma families, works on the street marked. When Federico was a child, he used to attend the street market to help. As in other famous cases of street vendors reported in the previous literature in mathematics education (Carraher, Carraher & Schliemann, 1985; Jurdak & Shahin, 1999; Saxe, 1988), Federico developed a number of mathematical abilities to do mental calculation, which allowed him to succeed in his academic career (and end up doing doctoral studies). Federico, like other Roma, had to face the contempt towards Roma culture and identity that exists in the hegemonic schools. The inclusion of his voice, similarly to the inclusion of the voices of people like him, contributes to overcoming the second problem that I identified earlier in this article. Our hypothesis for further discussion with the audience of this article is whether some of the DL principles (*egalitarian dialogue, cultural intelligence, and equality of differences*) may (or may not) help us to broaden the scope of Ethnomathematics to overcome its local and contextualized vision of mathematics practices, to enable us to identify *universals* that may lead to the development of successful educational actions.

Conclusion: Past, Present, and Future of the Ethnomathematics

In 2007, D'Ambrosio complained that the "resistance against Ethnomathematics may be the result of a damaging confusion of Ethnomathematics with ethnic-mathematics (...) which may lead to a folkloristic perception of Ethnomathematics" (p. ix). I think that this criticism of Ethnomathematics as being some sort of *folkloristic* perspective could be due to the fact that Ethnomathematics has been understood as an eminently contextualized (situated) theory of mathematics and its practice. The dialogue between DL and Ethnomathematics could be a way to overcome such criticism.

Contextualism is more and more questioned. Usually, it is introduced as a way to overcome cultural barriers towards the mathematics understanding (Boaler, 1993). But ultimately this approach did not give the expected result so far. We can identify what is different in cultural (ethnic) terms, regarding how different individuals use / apply mathematics.

However, in doing so does not mean that our practices as professionals in education will be inclusive (as D'Ambrosio complaints when he was a professor at SUNY). If this happens, the situation does not improve for the ones who use to be excluded from the regular educational system (like many Roma children). For this reason, many researchers worldwide suggest that the future trend for research is going to be to identify SEAs, that is: those *universals* which enable us to develop inclusive policies, actions and practices working for everyone, as well as respecting everyone's' cultural differences. After several decades studying and doing research, I tend to think that mathematics is a universal set of knowledge that manifest itself in situated practices. Concepts such as *hybrid practices* or *dialogic talk* are instruments to overcome the school resistance towards other types of *codes* that are not hegemonic.

In Ramon Flecha's words, we must find those active ingredients inherent to our human condition, and like happens with our bodies responding positively most of the times to a drug such penicillin (because everyone is made of the same material, regardless our beliefs, country of origin, socioeconomic status, skin color, gender, etc.), we need to find those *universal* ingredients in the educational field that make all individuals to respond positively (successfully) when learning mathematics. This is our challenge for the coming years.

References

Bernstein, B. (1975). *Class code and control: towards a theory of educational transmission.* London: Routledge and Kegan.

Boaler, J. (1993). The Role of Contexts in the Mathematics Classroom: Do they Make Mathematics More" Real"?. *For the learning of mathematics, 13*(2), 12-17.

Carraher, T. N., Carraher, D. W., & Schliemann, A. D. (1985). Mathematics in the streets and in schools. *British journal of developmental psychology, 3*(1), 21-29.

Chronaki, A. (2005). Learning about 'learning identities' in the school arithmetic practice: The experience of two young minority Gypsy girls in the Greek context of education. *European journal of psychology of education, 20*(1), 61-74.

D'Ambrosio, U. (1985). Ethnomathematics and its place in the history and pedagogy of mathematics. *For the learning of Mathematics, 5*(1), 44-48.

D'Ambrosio, U. (2007). Preface. In Favilli, F. (2007). Ethnomathematics and Mathematics Education. In *Proceedings of the 10th International Congress of Mathematics Education, Discussion Group 15: Ethnomathematics.* (pp. v-x).

NOTES

Díez-Palomar, J., & Cabré, J. (2015). Using dialogic talk to teach mathematics: The case of interactive groups. *ZDM, 47*(7), 1299-1312.

Díez-Palomar, J., & Civil, M. (2007). El impacto de la migración sobre el aprendizaje de las matemáticas en el contexto de una ciudad fronteriza en el Suroeste de los Estados Unidos. *Matematicalia: revista digital de divulgación matemática de la Real Sociedad Matemática Española, 3*(2), 1-9.

Dowling, P. (1993). Mathematics, theoretical "totems": a sociological language for educational practice. In Julie Ciryl, Angelis Desi (Ed.) *Political Dimensions of Mathematics Education 2: Curriculum reconstruction for society in transition* (pages 35-52). Johannesburg: Maskew Miller Longman.

Flecha, R. (2000). *Sharing Words. Theory and Practice of Dialogic Learning*. Lanham, M.D: Rowman & Littlefield.

Flecha, R. (2014). *Successful educational actions for inclusion and social cohesion in Europe*. Springer.

Flecha, R., & Soler, M. (2013). Turning difficulties into possibilities: Engaging Roma families and students in school through dialogic learning. *Cambridge Journal of Education, 43*(4), 451-465.

Fox, J., & Vidra, Z. (2013). Applying Tolerance Indicators: Roma School Segregation.ACCEPT PLURALISM research Project. 7th Framework Programme. European Commission.

García-Carrión, R., & Díez-Palomar, J. (2015). Learning communities: Pathways for educational success and social transformation through interactive groups in mathematics. *European Educational Research Journal, 14*(2), 151-166.

García-Carrión, R., Girbés-Peco, S., & Gómez-Zepeda, G. (2015). Promoting children's academic performance and social inclusion in marginalized settings. *World Education Research Yearbook 2015*, 30.

Gatt, S., Ojala, M., & Soler, M. (2011). Promoting social inclusion counting with everyone: Learning Communities and INCLUDED. *International Studies in Sociology of Education, 21*(1), 33-47.

Gerdes, P. (1996). Ethnomathematics and mathematics education. In *International handbook of mathematics education* (pp. 909-943). Springer Netherlands.

González, N., Moll, L. C., & Amanti, C. (Eds.). (2006). *Funds of knowledge: Theorizing practices in households, communities, and classrooms*. Routledge.

Gurzó, K. (2014). Does the SEN Category Discriminate Roma Children (only)?. *Procedia-Social and Behavioral Sciences, 116,* 5030-5035.

Gutiérrez, K. D., Baquedano-López, P., & Tejeda, C. (1999). Rethinking diversity: Hybridity and hybrid language practices in the third space. *Mind, culture, and activity, 6*(4), 286-303.

Habermas, J. (1984). *The Theory of Communicative Action,* Boston: Beacon.

Jurdak, M., & Shahin, I. (1999). An ethnographic study of the computational strategies of a group of young street vendors in Beirut. *Educational Studies in Mathematics, 40*(2), 155-172.

Knijnik, G. (1998). Ethnomathematics and political struggles. *ZDM, 30*(6), 188-194.

New, W., & Merry, M.S. (2012). Learning who they "really" are: From stigmatization to opportunities to learn in Greek Romani education. In Z. Bekerman, & T. Geisen (eds.), *International Handbook of Migration Minorities and Education* (pp. 623-639). Springer.

Petrokou, E., & Dimitrakopoulas, I. N. (2002). Racism and local authorities: The case of Roma in Nea Kios. *ANTIGONE*. Athens: RAXEN National Focal Point.

Ríos, O., Herrero, C., & Rodríguez, H. (2013). From Access to Education. The Revolutionary Transformation of Schools as Learning Communities. *International Review of Qualitative Research, 6*(2), 239-253.

Saxe, G. B. (1988). The mathematics of child street vendors. *Child Development*, 1415-1425.

Searle, J. R. (1969). *Speech acts: An essay in the philosophy of language* (Vol. 626). Cambridge university press.

Stathopoulou, C., & Kalabasis, F. (2006). Language and culture in Mathematics Education: Reflections on Observing a Romany Class in a Greek School. *Educational Studies in Mathematics, 64*, 231-238.

Unión Romaní (2016). Pueblo Gitano. Retrieved from http://www.unionromani.org/pueblo_es.htm

Valls, R., & Kyriakides, L. (2013). The power of Interactive Groups: how diversity of adults volunteering in classroom groups can promote inclusion and success for children of vulnerable minority ethnic populations. *Cambridge Journal of Education, 43*(1), 17-33.

Vega Cortés, A. (1997). Los gitanos en España. Retrieved from http://www.unionromani.org/histo.htm

Vithal, R., & Skovsmose, O. (1997). The end of innocence: A critique of'ethnomathematics'. *Educational Studies in Mathematics, 34*(2), 131-157.

Vygotsky, L. S. (1980). *Mind in society: The development of higher psychological processes.* Harvard University Press.

Walkerdine, V. (1990). Difference, cognition, and mathematics education. *For the learning of mathematics, 10*(3), 51-56.

Wood, D., Bruner, J. S., & Ross, G. (1976). The role of tutoring in problem solving. *Journal of child psychology and psychiatry, 17*(2), 89-100.

The Mathematical Ways of an Aboriginal People: The Northern Ute

Jim Barta
jbarta@cc.usu.edu
Utah State University
Elementary Education
Jones Education Bldg
Logan, UT 84322 USA

Tod Shockey
shockey@math.umaine.edu
University of Maine
Department of Mathematics & Statistics
238 Neville Hall
Orono, ME 04469 USA

Math is a part of our lifestyle too but we don't have it in the way that the White man has it, say... math like times tables, and addition or subtraction. We have those things also in our traditional ways of beading, our raising of our horses and cattle, or building fences, putting up teepees, etc. If you don't have your culture, if you don't have your language, your arts and crafts you are just another person out there in the world. If we have our beadwork, our songs and dances, if we believe in our old ways, our traditional ways then that is going to uplift us someday while the world fails around us. We will have something to fall back on. Right now we can still live in a White world; we are in a White world. My dad says we're considered White people because we speak their language; we go to work and pay bills. We are not like Indian people the way we used to be but we still have a part of our Indian ways and that belongs to us. We should hang on to that. It's good to have your culture and your language to fall back on. It gives you a good feeling to know it is a part of you and that is where you came from.

(Personal conversation with Fabian Jenks, Ute Tribal Elder, February 2001)

Ethnomathematics, the study of the relationship between mathematics and cultures represents a complex and multifaceted dynamic, which illustrates the culturally influenced uses of mathematics through its applications. D'Ambrosio's (1991) proposes that mathematics have been culturally derived since people first walked the Earth. The evolving use of mathematical ability as a component of human intelligence, helped human beings to not only survive but to succeed in their environment. The term Ethnomathematics is descriptive of both past and present histories of mathematical development and applications throughout the world related to and shaped by cultural influences across the millennia. (Zaslavsky, 1996).

Native American cultures are seldom recognized for their uses of mathematical processes yet it is through an examination of such applications in daily living practices that our understanding of the American Indian can be enhanced. This article depicts a Native American people, the Northern Ute and illustrates how mathematics was and is an integral aspect of Northern Ute culture highlighted in a variety of traditional and contemporary living practices. Ute elders, those with a deep understanding of their cultural heritage, beliefs, values, and traditions provided the insights and information upon which the article is based. The categories of counting, measuring, locating, designing, explaining, and playing (Bishop, 1994) will be used to frame how math has been used and explained in Ute culture. The history, richness, and complexities of the Ute culture are far too detailed to be easily described and certainly surpass what is known by the authors. Nevertheless, several important aspects of the Ute story will be shared.

A Brief History of the Northern Ute

The Ute successfully and continuously lived for centuries throughout the Rocky Mountain region in the western United States in what are now parts of Colorado and Utah. Traditionally the Ute were known as the Nuciu meaning "The People." (Lewis, 1994). The Ute language belongs to the Numic group of Uto-Aztecan languages shared by a number of other tribes living in the region. Historically, the Ute lived and traveled in small bands of extended family groups in numbers ranging from 20 to 100 persons. Twelve separate bands of Ute people existed, typically defined by the environment they inhabited. Occasionally these small bands of people would gather to trade, intermarry, participate in ceremonies or celebrations, or to conduct large hunts or foraging to obtain quantities of food for consumption or winter storage.

The people typically traversed the land following cyclical seasonal patterns where known foods such as seeds, nuts, berries, and roots, and medicinal plants could be gathered and harvested. Additionally, hunters supplemented food supplies with rabbits, birds and larger game such as deer, antelope, and bison. Fish were harvested in lakes and streams. The temporary homes of the Ute (wickiups) were cone-shaped and typically made of grasses and branches, the materials most readily available. In colder seasons when travel decreased and habitation sites became more permanent, teepees made of animal hides were constructed and warm blankets were woven from rabbit fur.

NOTES

In the early 1700's, the Ute's acquisition of horses from the Spanish explorers indelibly changed life for the people. Increased mobility meant more efficient travel, successful hunting, and raiding on enemies. The Ute became known for their horsemanship and fearsome exploits and appear to have benefited from the increased interactions of explorers through trading and bartering.

Beginning in 1847 the Ute traditional lifestyle changed dramatically with the arrival of the Mormon pioneers. The Mormon, who sought new lands where they could practice their religion unfettered, cherished the seemingly unclaimed land and set out to establish permanent towns and cities. Initially, relationships between the Ute and these new Mormon arrivals were peaceful. Eventually tensions arose over the contested resources; resources the Ute had managed and used for centuries. Violent conflicts between the pioneers and the Ute resulted. The Ute were eventually subjugated through military force and relocated to barren reservations set for them by the United States government.

Efforts were made to teach the Ute farming and ranching skills and were successful for some. Ute children were collected from their families and taken to boarding schools where they were taught to speak English and to learn skills seen necessary for integration into the dominating culture. This "educational" process disrupted and damaged the traditional practice of Ute culture and what a child would have normally learned within their family.

Today over 3000 people belong to the Northern Ute Nation, representing three different bands, live on or near the Uintah Basin reservation in northeastern Utah. Senior members or Elders of local reservation communities are those who most frequently practice traditional Ute culture. Spiritual significance is embodied in almost every aspect of daily living in traditional

Ute culture. Life and living is seen as being circular, relational, and reciprocal. The Ute reflect a humility and respect for life and the plants and animals with which they share the land. Efforts are being made to save and to renew cultural practices and awareness. Children are once again being taught the Ute language and the traditional ways of the people. The Ute today are a very proud people with a rich past and a challenging yet promising future.

Ute Mathematics

The following description of Ute math sets the starting point for further work that can and should occur by the current authors and others. Before this work, there does not appear to have been any description in the literature for teachers/researchers to access and use for elucidation of Ute math. The decision this article reflects, therefore, is to describe the Ute way of naming numbers, measuring things, and the like, rather than describing and analyzing the details of a Ute games, for example. Revealing the beautiful mathematical nuances that make the math of the Ute, their math, rather than another culture's mathematics, guided this ethnomathematical study.

In the Ute language, there is no one specific word for mathematics. Rather, mathematics is something you do as you live. Mathematics is witnessed through its many uses; in Ute culture it is a verb rather than a noun unlike the Eurocentric perspective. Some consider mathematics to be a gift from the Creator whose use allows for a successful life. The description of the traditional uses of mathematics of the Ute will begin with the act of counting.

Counting - nee nee aye (making a name)

The counting system of the Ute was based on groupings of ten that was used to quantify objects, people, and events encountered in daily living. It was not always necessary to name an exact count when describing a group of objects. Sometimes quantities were explained simply as "a lot, many, or a few."

When necessary, values ranging from one to endless (infinity) could be described. Zero meant, "having nothing or there is no more." Numbers too large to count were said to "just go on and on" or "those having no end." Virtually any quantity could be verbally described. Values through the tens and up to a few hundred were those most frequently explained because of their actual daily use. Very large numbers could be named but if there were no real use for them this was seldom necessary.

Hand gestures, similar to finger counting today were common and could accompany verbal counting or could be used in isolation. For values more than ten, the speaker would verbally indicate the value, such as tens or hundreds prior to gesturing. Usually the context of the conversation made such clarification unnecessary. Names for numbers one through nineteen and the 10s through 100 are listed:

#	Name
1	suwees
2	wyanee
3	bpaynee
4	whcheweenee
5	muhnugheenee
6	nahvayannee
7	nahvakahvuhnnee
8	wowh whcheweenee
9	suwah turgumsuweenee
10	turgumsuweenee
11	turgumsuweenee soocoos cheepeewahgudt
12	turgumsuweenee wykh cheepeewahgudt
13	turgumsuweenee bpaykwh cheepeewahgudt
14	turgumsuweenee whchook cheepeewahgudt
15	turgumsuweenee muhnuhgk cheepeewahgudt
16	turgumsuweenee nahvayouk cheepeewahgudt
17	turgumsuweenee nahvaykahvahouk cheepeewahgudt
18	turgumsuweenee wowh whchook cheepeewahgudt
19	turgumsuweenee suwah turgurmsok cheepeewahgudt
20	wahmsuweenee
30	paymsuweenee
40	wh chewee turgumsuweenee
50	muhnuhgkee turgumsuweenee
60	nahvah muhnugheenee
70	nahvahkahvah turgumsuweenee
80	wowh whchewee turgumsuweenee
90	suwah turgurmsuwe turgumsuweenee
100	sookoosmurh

NOTES

Eleven, twelve, etc. were literally translated as "one on top of ten," "two on top of ten", etc. 20, 30, were translated as "two counts of ten", "three counts of ten". Nine is one less than ten and the same pattern holds for other values immediately preceding groups of 10. Written numerals did not exist. Drawings of objects on hides, other objects, or rock faces also indicated quantities. Tally-like marks could be used to keep a count.

Ute elders described uses for the basic math operations of adding, (combining) subtracting (taking away), and dividing (sharing or splitting up). Repeatedly adding similar quantities when such amounts were counted took the place of multiplication. Division and fractional quantities were aspects of sharing items such as food and calculations were rather intuitive. One might get "a half" of the bounty and another might get another portion of what was left. Equal shares did not always mean everyone got the same amount but rather got what they and their family needed. "One-half" was described as "in the middle or center."

Measuring – tuehgyh (measuring)

Rather than creating a set of universal measurement standards Ute measurement was very contextualized and individual. A stick or a length of rawhide for instance possessed its own length that could then be used as a standard or comparison with another object. One Elder stated, "Everything was built with a measurement within itself." Parts of the body such as hands, length of an outstretched arm, paces, etc. were used and were specific to the person to which they belonged. In essence, each person carried with them their own standard units and thereby what one constructed was proportionately appropriate for the user. Hide shirts, buckskin dresses, moccasin footwear, and even tools and weapons were made with the body of the person acting as a measurement template.

Perimeter, area, volume were measured in very practical ways. A person constructing a shadehouse to provide comfortable, cool protection against the summer sun for instance, would "step off" a distance and count the steps that were taken to create the size of house to comfort ones family. The floor area of a shadehouse or teepee was determined adequate if its space served the function for which it was constructed.

Distances were described as a function of time and/or speed it took one to make the trip. A long journey was described by counting the number of sunsets, which elapsed during the travel. The light of the day was important; it was less difficult and safer to travel during daylight so sunsets provided an obvious stop to a day. Travelers would segment a trip into increments based on the time from sunrise to sunset. The speed of travel (on foot or on horseback) influenced the "distance" traveled making the trip faster or longer.

Volume was considered by how much of something a vessel or container could hold. Weight of an item was always considered in relation to the person hefting or feeling; they made the determination if to them something was heavy or light.

For cooking and food preparation, measurement again played an important function. Those doing the cooking used pinches, handfuls, and dashes of ingredients (such as crushed berries, salt, and other flavorings) as standard units. The experienced cook knew that a particular bowl or woven container when filled to a certain level would produce a specified amount to food. If the cook was preparing a meal for a greater number of people the container could have more added to it as the cook knew from previous experience how much more food would result.

The size of the container thereby functioned as a measurement device.

Designing – pur ur kut (how it is drawn or written down)

Designs carry messages and tell stories. A Ute beadworker for instance may use bead colors and the specific pattern of a flower or a geometric shape that has been passed down through their family for generations. Other people know their work and their identity in this way. The actual meaning of the pattern could be very personal and spiritual. Mathematical

calculations were intrinsically applied in beadwork from which the Ute are known. For instance, Ute beadwork involves balance and symmetry denoting a sense of beauty and harmony. The beadworker typically uses their sense of what looks good rather than making strict mathematical calculations to produce their work.

The applied mathematics functioned both as a guide to the construction of the object and as a nearly invisible byproduct. Beadwork was used to decorate clothing and items of daily use. Ute beadwork patterns are typically symmetrical as are those of most Native American populations. The balance inherent in beaded designs reflects spiritual understandings is thought to illustrate harmony, joy, and beauty; each of which are values respected for living a good life. The construction of a structure, tool, or beadwork design incorporated many geometrical and mathematical principles such as tessellations, transformational geometric concepts, or the use of a radius (rawhide strip) to draw a circle. A person who knew how to make an object implicitly used the mathematics necessary for construction in a most practical way.

Shapes were named by their characteristics. The shapes were not named simply to identify a geometry term but rather referred to the physical appearance of something real. The literal translation for the Ute word for "triangle" for instance is "having three corners." Hand gestures complemented verbal expression and could indicate length and width. The circle, known as "something round" was perhaps the most important geometric shape to the Ute. The circle

had significance because it was considered to have no starting or ending point. The circle illustrate the multiple circles or cycles a person observed everywhere around him or her such as the cycles of life and movement governing people, plants, and animals as well as those related to weather, the earth, moon, and the stars.

The circle also provided the most typical base shape for the traditional home. Mathematically for a given perimeter of a home, a circular shape provided the greatest amount of interior floor area. Building materials were collected from nature (sticks and branches) and the use of circular-shaped constructions reduced the impact on the environment because of the economical use of materials to achieve the greatest area for a fixed amount of resource and the amount of work required to gather those resources.

Locating – myh (finding)

Elders were respected for their knowledge of the land, its geography, geology, and resources. "Maps" existed in the memories of those that made a journey previously who would lead a group of travelers or describe verbally specific landmarks and distances

NOTES

expected. Occasionally, maps depicting routes were drawn on the ground or on hides using charcoal that one could then carry.

Cardinal directions were important and determined in relation to the rising or setting of the sun, the location of a mountain, river, lake or some other significant feature. Hand signals indicated directionality and appear to have been used consistently when describing a location to another. One could also situate their location by careful observation of the rising and location of the sun, moon, or stars which could then also be used to comprehend the time of year and events that could be expected (animals, plants, weather, etc.). Such insight and knowledge were paramount to continued survival. Pictographs and petroglyphs (drawings made on rock walls) helped communicate where people had previously hunted and camped. Certain signs or symbols were used to represent certain geographic or resource features. Spatial or topographical orientations such as right or left, above or below were very specifically described in relation where an object stood in relation to the person making the distinction. Again the context of the observation greatly influenced its description.

Explaining – myh kueh uhm (telling you something)

Mathematics is a form of communicating ideas and therefore is a form of explaining. The numbers four and seven were significant to the Ute and influenced cultural behaviors, construction, and design. The number four could indicate the cardinal directions or the number of times a prayer or ceremony was repeated or a song sung. An Elder described the importance of seven stating, "The seven are north, east, west, and south, and then we have the sky and the earth, and then the middle would be ourselves, our hearts, and that is the seven directions."

Time could be told by watching shadows and the directions in which they pointed. A stick could be placed in the ground or shadows created naturally could be observed. A shadow pointing west in the morning confirmed it was before noon. Exact time measurement was not considered necessary. The activities and actions of a person throughout the day were related to more natural rhythms such as when one was hungry or sleepy. However parts of the day or night could be explained and were referenced in relation to a description of where the sun appeared in the sky or by the color of the sky at dawn or dusk.

The Ute calendar was organized in relation to the moon and the various positions and shapes of it. Careful observation of the lunar cycle allowed those knowledgeable to know clearly what part of year it was and even more precisely phases of the night. A new moon indicated a new month was starting. This knowledge again was vital to Ute survival in knowing when berries would be right for picking, roots to be dug, plants to be gathered, animals to be hunted, and meat to be dried. There were four seasons described and within those times event representing the months could be described. The beginning and ending of a particular season was often related to the observation of a cyclical environmental or climatic event and may vary a bit from year to year. The naming of a month within a season noted, for instance, when eagles mated, when birds flew north or south, when plants budded, or when special foods could be gathered.

Wealth and prominence could be illustrated in the number of horses or cattle a Ute owned but were not merely determined by material possessions. A traditional Ute was considered wealthy if he or she possessed certain valued skills such as if one was a good beadworker or could dance or sing well. If you knew the traditional Ute ways, demonstrated humility in life, knew how to provide for your family, had a strong and healthy family with wise elders from which to learn, had fancy colors to design ones traditional clothing, you held a place of prominence.

Playing – keeyakyh (more than one person playing)

Playfulness was valued in Ute culture and a number of games existed. In some games, the use of mathematics was apparent, such as in those games where scores were kept. In others, the mathematics were embedded in the activities. Measuring, counting, probability were several of the mathematical aspects of the games. Invariably, scores were calculated to determine the winners.

There were games specifically for men, women, or mixed groups. The games could demand physical skills such as in running races, shooting arrows, or playing "Shinny", a game, which resembled field hockey of today. Other games were more cognitive such as those of the "Hand Game" or the "Stick Game." The "Hand Game" was and still is a popular guessing game played by two teams comprised of any number of men and women who gathered on either side of two long sticks or poles laid parallel to each other. Each team began the game with ten or twelve short sticks and several bones measuring two to three inches in length. One of the bones would have a piece of rawhide strip wrapped around it. The object of the game was to guess the hand in which one of the players on the opposing team was hiding the white bone (with no rawhide strip). When the opposing team guessed correctly, the losing team passed them one of their short sticks. The first team to lose all of their sticks lost the game and the bets that preceded the game. Items such as hides, clothing, personal items, and even horses were wagered.

The game of tu-roo-kweep was a stick game played on a two-to-three-foot square of buckskin using four playing sticks painted with combinations of colors and or markings, which represented certain numerical values which were determined by the players. Marks were drawn or painted on the buckskin along its edges. A flat rock was placed in the middle on which to bounce the playing sticks as players sat on opposite sides of the hide and took turns. Scores were determined in noting how the sticks landed and players moved their marker around the hide. The first one to mover their marker completely around the hide won the game.

Conclusion

This brief summary of traditional Ute uses of mathematics reflects its practical application by a people whose very existence relied on the effective use of mathematical concepts and principles. Ute mathematics defined the people through its many applications. The ingenuity and mathematical intelligence demonstrated portray a people with a rich heritage who used mathematics to enhance their lives and their culture. Today, with this understanding of this proud mathematical tradition, Ute children can know they too have an obligation to learn and to use mathematics as they honor Ute ways.

NOTES

References

Barta, J. (1999). Mathematics and beadwork. *Winds of Change: American Indian Science and Engineering Society, 14*(2), 36-41.

Bishop, A. (1994). *Mathematical enculturation: A cultural perspective on mathematics education.* Norwell, MA: Kluwer Academic Publishers.

D'Ambrosio, U. (1991). Ethnomathematics and its place in the history and pedagogy of mathematics. In M. Harris (Ed.), *Schools, Mathematics and Work* (pp. 15-25). Basingstoke: The Falmer Press.

Lewis, D. (1994). The Northern Ute. In M. Harris (Ed.), *Utah History Encyclopedia.* Salt Lake City, UT: The University of Utah Press.

Zaslavsky, C. (1996). *The multicultural math classroom: Bringing in the world.* Portsmouth, NH: Heinemann.

Note: Earlier versions of this paper appeared in *Pour la Science* and *Scientific American – Brasil*

Volume 2
Community

Juxtaposing Form, Function, and Social Symbolism: An Ethnomathematical Analysis of Indigenous Technologies in the Zulu Culture

Iman Chahine
ichahine@gsu.edu

Wanjira Kinuthia
wkinuthia@gsu.edu
Georgia State University
30 Pryor Street
Atlanta, GA

Abstract:

The purpose of this paper is to highlight the role of ethnomathematics as a model for translating and interpreting mathematical structures inherent in existing indigenous technologies used by South African indigenous people as means of survival and adaptation in a variety of environments. The present study focuses on the riches of two indigenous technologies as media for transmitting specialized mathematical knowledge produced and disseminated by the Zulu culture, namely beadwork and basketry. We argue that prospective work that investigates other indigenous technologies with underlying ethnomathematical ideas can be highly enlightening and reflect major breakthroughs in the field of indigenous mathematical knowledges.

Ethnomathematics and indigenous knowledge systems

As a new conceptual field and an emerging research program, ethnomathematics evolved from a rhetoric networking the interplay among mathematics, cultural anthropology, education, and politics. While too many definitions and associated perspectives prevail in the literature attempting to delineate ethnomathematics, two dominant conceptualizations are in use. D'Ambrosio (1985), who coined the term, insists on ethnomathematics as a rather broad conceptualization of mathematics and defines it as embracing "all culturally identifiable groups with their jargons, codes, symbols, myths, and even specific ways of reasoning and inferring" (p.17). Furthermore, D'Ambrosio (2001) explains, "Ethnomathematics is the mathematics practiced by cultural groups, such as urban and rural communities, groups of workers, professional classes, children in a given age group, indigenous societies, and so many other groups that are identified by the objectives and traditions common to these groups" (p. 1). In this perspective, ethnomathematics exists at the connection between mathematics, history, and cultural anthropology. Additionally, D'Ambrosio (1999) describes ethnomathematics as "a program in history and epistemology with an intrinsic pedagogical action. . . . [It] responds to a broader conception of mathematics, taking into account the cultural differences that have determined the cultural evolution of human mankind and political dimensions of mathematics" (p. 150). In this respect, ethnomathematics draws on the belief that humans, from prehistoric ages, have been accumulating knowledge to respond to their drives and needs. Such knowledge varies from region to region and from culture to culture and is transmitted through generations in serendipitous, informal ways.

In contrast, Ascher (1991) focuses more on the mathematics of indigenous societies and defines ethnomathematics as "the study of the mathematical ideas of non-literate people" (p. 125). Ascher (2002) indicates that there are 6000 cultures that have existed within the past 500 years and capitalizes on the riches that can be cultivated from studying cultural practices and heritage. Moreover, Ascher (2002) proposes that diverse cultures, such as the Inuit, Iroquois, and Navajo of North America; the Incas of South America; the Bushoong, Kpelle, and Tshokwe of Africa; and the Caroline Islanders, Malekula, Maori, Warlpiri, and Trobriand Islanders of Oceania, contribute to a laudable corpus of mathematical knowledge that evolved as a result of human praxis. Likewise, Nettle and Romaine (2000) discern that indigenous peoples comprise 4% of the world's population and speak 4,000 to 5,000 of the world's more than 6,000 languages.

One critical paradigm promoted by proponents of ethnomathematics is the belief that all peoples are capable of doing mathematics in their own unique and personal perspectives. Concomitant with such a framework is the view that ethnomathematics emerges from within individuals while interacting with their cultural and physical environments. Nunes, Schliemann and Carraher (1993) argue that ethnomathematics develops when there is a discrepancy between people's need for problem solving and the level of mathematics they have learned in school, that is, when people become involved in tasks requiring

NOTES

problem solving skills that are not learned in school. In this argument, ethnomathematics is believed to be closely tied to issues of access and equity (Anderson, 1997). In a similar vein, Mosimege and Onwu (2004) argue that ethnomathematics is an inherent component of indigenous knowledge systems, which the authors define as "an all-inclusive knowledge that covers technologies and practices that have been and are still used by indigenous and local people for existence, survival and adaptation in a variety of environments" (p. 2).

Indigenous Knowledge Systems (IKS) has been recently established by international world organizations as a top global priority for empowering indigenous and local communities in their quest for survival and development. Nonetheless, although these organizations recognize the strength of IKS in strengthening indigenous communities by preserving their cultural practices, little research to date has illuminated the contributions that IKS can have to mainstream knowledge and epistemologies.

An overarching goal of this paper is to epitomize the role that indigenous knowledges play in unearthing the contributions of non-Western indigenous cultures in the development of scientific and mathematical thinking. More specifically, our purpose is to examine mathematical ideas that thrive in existing indigenous technologies, which are still used by South African indigenous and local people to adapt and survive amidst social, economic, and contextual challenges. In this respect, we attempt an ethnomathematical analysis of indigenous technologies adopting Eglash, Bennett, O'Donnell, Jennings, and Cintorino's (2006) notion of *translation* as an inherent analysis technique to "describe the process of modeling indigenous systems with a 'Western' (i.e., mainstream, academic) mathematical representation" (p. 348) .

A major focus of this study was to investigate and *translate* the content and structure as well as mode of expression and dissemination of ethnomathematical ideas prevalent in two indigenous technologies which are designed and appropriated by the Zulu culture: beadwork and basketry. We conclude with a brief discussion of the importance of future research prospects in the area of indigenous knowledges and technologies.

Early research on ethnomathematics

Considering the wide consensus regarding the universality of mathematics, research on ethnomathematics and the cultural origin of mathematics emerged later than other ethnosciences (Prediger, 2002). In examining the link between ethnosciences and ethnomathematics, D'Ambrosio (2000) explains, "By ethnosciences I mean the corpora of knowledge established as systems of explanations and ways of doing accumulated through generations in distinct cultural environments. Particularly important for us is ethnomathematics-- knowledge derived from quantitative and qualitative practices such as counting, weighing and measuring, sorting and classifying" (p. 80). Furthermore, Rist and Dahdouh-Guebas (2006) define ethnosciences as "a scientific realm which aims to understand how humans—in spite of their fragmented and limited interactions with the world—are developing different forms of knowledge and beliefs" (p. 472). Likewise, ethnomathematics calls for systematically taking into account the flux of mathematical ideas that emerge as humans interact and make sense of their environment. To this end, studies on ethnomathematics challenge the evolutionist view that "indigenous societies had only simplistic technologies" (Eglash et al., 2006, p. 347).

The gist of early studies on ethnomathematics, which have been launched in the cause of justifying ethnomathematics as an internationally legitimate field of research, has been invoked in response to early writings on the nature of "primitive" thought. In his publication, *How Natives Think*, Lévy-Brühl (1966) mistakenly suggests a divided world of thought between literates and "non-literates," where he describes non-literate people as "pre-logical," primitive and mystical, thus incapable of analytic reasoning. He further explains that the minds of primitives do not function as those of the literate people. In explaining Lévy-Brühl's views on pre-logical thinking, Radford (2008) highlights major defects in Levy-Bruhl's theory as manifested in inconsistent characterizations as he attempted to draw explicit comparison between "primitive" and "non-primitive" mentalities. In a similar vein, Throop (2003) pinpoints internal tension in Lévy-Bruhl's writings particularly as he spoke of "individual" and "collective representations," as it was often difficult to differentiate between the two. According to Evans-Pritchard, Lévy-Bruhl (1952) attempted to distance himself from some of his earlier positions with regard to differences between "primitive" and "modern" mentalities as he states: "I admit that in my work (and it is here that 'I plead guilty') the savage is presented as more mystical and the civilized man as more rational than they in fact are. . . . Once this difference is recognized I have no objection to all that you say; that the savage is not so exclusively mystical, that the civilized man is not so consistently rational. Perhaps I have been wrong in insisting so strongly on these differences. I thought that the anthropological school had done enough to make the similarities evident" (p. 191).

An alternate proponent of Lévy-Brühl's thought are the views of Tylor (1958) in his book *Primitive Culture*, where he erroneously likened primitive cultures to "children" which can only progress and evolve through proper education. Another term ascribed to Tylor is his theory of "survivals." Tylor asserts that survivals are "processes, customs, and opinions, and so forth, which have been carried on by force of habit into a new state of society different from that in which they had their original home, and they thus remain as proofs and examples of an older condition of culture out of which a newer has been evolved" (p. 16).

Counter views to Tylor's and Lévy-Brühl's ideas fiercely challenge the traditional, linear path that stereotyped the evolution of human thought from "primitiveness" to civilization and propose rival views that acknowledge 'natural rationality' and 'organic reasoning' skills thriving in indigenous practices. In his book, *Mathematics as a Cultural System*, Wilder (1981) asserts that each culture has its own mathematics, which evolves and decays with that culture, and he further capitalizes on the role of cultural practices in appropriating human cognition. Wilder (1965) explains: "Mathematics, like other cultural entities, is what it is as a result of collective human effort directed along evolutionary and diffusionary lines. And what it becomes will not be determined by the discovery of "mathematical truth" now hidden from us, but by what mankind, via cultural paths makes it" (p. 298). Wilder (1981) further accentuates the role that cultural environment plays in inspiring creative mathematical thinking: "The individual mathematician cannot do otherwise than preserve his contact with the mathematical culture stream; he is not only limited by the state of its development and the tools which it has devised, but he must accommodate to those concepts which have reached a state where they are ready for synthesis" (p. 145).

By the same token, and in his seminal publication *The Oral Style* in 1925, French anthropologist Marcel Jousse (1886-1961)

NOTES

rejects the associated connotations of labeling 'oral' as primitive and 'literate' as complex and logical (as cited in Sienaert, 1990). Jousse states, "It is for this reason that all people who have not lost their natural spontaneity, who are not, as we literates are, disassociated - and that does not mean *primitive*, which is a term that must vanish from the vocabulary of every scientist" (p. 40). Jousse further insists on using the term "spontaneous" in lieu of primitive and explains that, within indigenous communities, thinking is rather spontaneous and rhythmic, marked by memorization as the primary manifestation of the human brain, that is rooted deeply in organic life. In a similar vein, and in his early works, Gerdes (1988) wrote of a "spontaneous, natural, informal, indigenous, folk, implicit, non- standard, and/or hidden (ethno) mathematics" (p. 138). Equivalently, D'Ambrosio (1982) spoke of spontaneous mathematics when he explains that each human being and each cultural group develops various mathematical ideas spontaneously. Ascher (1991) elaborates on the rich mathematical structures underlying indigenous knowledges and practices and asserts that "mathematical ideas are cultural expressions . . . and, moving beyond our own Western mathematics, a broader global history must acknowledge and include ideas of other cultures" (p. 196).

Extensive research that emphasizes the role of indigenous cultures in shaping mathematical thinking has been inspired from ethnographic literature (e.g., Carraher, Carraher, & Schliemann, 1985; Eglash, 1999; Gay & Cole, 1967; Gerdes, 1994, 1995, 2007; Harris, 1987; Jurdak & Shahin, 1999, 2001; Kyselka, 1987; Masingila, 1992; Millroy, 1992; Saxe, 1988; Zaslavsky, 1973). Such studies, and many others, reinforce the belief that mathematical thinking is rather culturally motivated, inherently bound by human experiences and, thereby, expressed in multiple and broadly diverse ways.

Ethnomathematics and orality

At its inception, the evolution of ethnomathematics as an epistemology can be perceived as a testimony demystifying the discrete divide between literacy and orality that tacitly underlies discourses in modern alphabetic cultures. In examining the historiography of the Andean knotted *khipu*, Brokaw (2003) explains: "An understanding of the *khipu* requires a deconstruction of this writing-(mnemonics)-orality opposition. Whether or not the khipu constitutes writing is not the issue I have pursued here. That is ultimately a semantic and, hence, political problem that may be solved by expanding—or not—the definition of writing." (p. 141)

While literacy requires the competency to assimilate information presented in written script, Ong (1982) defines orality, or orally-transmitted information systems, as signifying collective expressions of specialized knowledges that belong to a particular community of people. Ong further explains that extended thought in oral cultures is stored and retrieved from memory through mnemonic patterns called 'oral formulas.' The formulas organize information in carefully wrought clusters stored deep in the unconscious and preserved by oral cultures. Likewise, Devantine (2009) defines orality as "the restitution of memory transmitted through diverse expressions of voice or words of a culture" (p. 10).

> *Analogously, Nogueira (2008) defines "oral" tradition as a mode of expression that congregates knowledge, memories, values, and symbols generally configured in linguistic objects of non-literary or aesthetic-literary nature, objects with or without consignment in written testimonies, accomplished vocally and recognizable collectively and during consecutive generations*

> *in an anatomy built by the laws of traditionality (anonymity, persistence, variation). (p. 164)*

As such, a vital function of oral tradition is its thrust of creating communal identity and a sense of amity by virtue of sharing the same past experiences. From an ethnomathematical standpoint, the vehicles by which a body of knowledge is preserved and transmitted are inextricably determined by formats fashioned and enacted solely by the community itself. As D'Ambrosio (2000) states,

> *Mathematics, as a human endeavor, is the same. This is one focal point of the research program known as ethnomathematics, which deals with the generation, the intellectual and social organization and the diffusion of different ways, styles and modes (tics) of explanation, understanding, learning, coping with and probing beyond (mathema) the immediate natural and socio-cultural environment (ethno). (p. 90)*

Ethnomathematics and indigenous technologies

As a mediator of people's existential experiences (Cuthbertson, Socha, & Potter, 2004), technology is a significant part of everyday life in any society as it plays a key role in the development and maintenance of cultures. Rogers (2003) defines technology as "a design for instrumental action that reduces the uncertainty in the cause-effect relationship involved in achieving a desired outcome. Technology often has two components: the tool that embodies the technology and the information base for the tool" (p. 13). The idea of technology is to make life easier for its users, and, in whichever form technology is manifested, it is inherently free in its development and in how it is operated. While there are all sorts of technological developments that continue to shape society and how we live (Cuthbertson et al., 2004), technology acts as a tool of humanity that influences social values and practices and thus becomes a part of the culture.

Research on the ethnomathematics of different cultures has unearthed numerous indigenous technologies that inherently embody mathematical structures, such as number systems, folk games and puzzles, kinship relations, divination systems, and symmetric strip decorations (Chahine, 2011). Bogoshi, Naidoo, and Webb (1987) cite one of the earliest indigenous technologies that is assumed to represent a manifestation of human mathematical activity: "a small piece of the fibula of a baboon, marked with 29 clearly defined notches discovered in the early 1970s during an excavation of Border Cave in the Lebombo Mountains between South Africa and Swaziland and initially dated to approximately 35,000 BC" (p. 294). It is believed that the bone closely mimics the calendar sticks that are used up to the present day by many cultural tribes in Namibia and Bostwana (Gerdes, 2000). Amara (1987) defines indigenous technology as the use of traditional methods and locally available materials to produce essential goods for sale or domestic consumption. In many cases, such endeavour requires the design and construction of devices to facilitate the production process. Amara also provides a variety of examples of indigenous technologies that include production; processing and preservation of local food stuffs; production of chemicals, such as salt;

NOTES

metal work; charcoal production; preparation of herbal medicines; production of local brews; trap making; and arts and crafts (e.g., weaving, carving, tie dying, pottery).

In this paper, we define indigenous technologies as embodiments comprising information codes that reflect potential sources of implicit and explicit narratives and modes of strategic thought most valued by the culture and which are communally accepted as emulations of the people's cultural heritage and social identity. The significance of examining those indigenous technologies from an ethnomathematical perspective stems from their inherent ability to unfold the cultural and sociopersonal identity of the South African cultures. Of major interest are two vital indigenous technologies that thrive in South African context and that provide information about the cultural heritage and social norms of the culture: beadwork and basketry.

Rogers (2003) asks why indigenous knowledge systems are often ignored or denigrated by change agents who introduce innovations. Furthermore, Rogers believes that this attitude often leads to fallacies and views that the old and new are not compatible. While to dichotomize traditional and modern technologies is less relevant for our purposes in this paper, we argue that indigenous technologies are in fact off-scale vis-á vis the old/new dichotomy in that such technologies have different attributes in design and in purpose. In addition, we emphasize that the motivation behind exploring those indigenous technologies is not to present such technologies as pedagogical illustrations of activities that can be incorporated in the classroom. Therefore, we exercised caution to avoid falling in the trap of reducing the argument to a mere exposition of "microscopic" dissection of these technologies in isolation of the sociocultural and political circumstances that instigated its emergence (Pais, 2011). In this article, we argue that beadwork and basketry are quintessential illustrations of existing indigenous technologies that thrive in the Zululand and which represent an intimate disclosure of highly specialized knowledges that are owned and appropriated by its practitioners.

Methodology

Context: Durban, South Africa

We conducted our investigation in Durban, the largest city in the South African province of KwaZulu-Natal (KZN) and the third largest city in the country. Durban is a coastal city located on Natal Bay of the Indian Ocean, with warm climate and sandy beaches (around 97 km of coastline). eThekwini Municipality is the local government body responsible for governing and managing Durban. The eThekwini Municipal Area (EMA) is 2297 km in size (1.4% of the province of KZN). Durban has a population of around 3.5 million people of whom 68% are Black African; 20% Asian; 9% White; and 3% Colored. Being the busiest port in South Africa, Durban attracts tourists from different parts of the world and is famous for the traditional crafts and art work that is perfected by the indigenous communities and exhibited by the street sellers along the coastline (Ethekwini Municipality: online reference).

In the Zulu culture of South Africa, orally-transmitted and tactile-based indigenous technologies play a major role in the production, dissemination, and preservation of a rich body of information that is concealed and estranged from a foreign viewer and which is inextricably linked to everyday life of the people. D'Ambrosio (2000) explained that cultural ethos encloses mechanisms generated by the culture to guard against violations and breaching of codes from outsiders. The only way to decode the syntax of this information system and retrieve meaning is to penetrate the Zulu culture's systems of social organization, symbolism, belief, and ideology. Furthermore, Eglash et al. (2006) touched upon ethical issues concerning the "ownership" and accessibility concerns with IKS and stated that "for the moment, we can only say that we maintain it is possible to make respectful, ethical use of those materials, given a dialogue with community representatives" (p. 360).

Procedure

The first author visited Durban in South Africa to learn about the culture and to examine firsthand the intrinsic cultural artifacts, including beadwork and basketry, that thrive on the streets of the coast of Port Natal in Durban. We used 3 methods for collecting data: (a) informal interviews with bead workers and basket makers; (b) collection of artifacts including beadwork items, baskets, and pictures; and (c) researcher's introspection during field observations. Some of the sites visited while in Durban included the William Campbell Museum of African Art and Furniture; the Phansi Museum; the Shakaland Cultural Village; and the beachfront bead and basket street sellers.

While walking on the beachfront in Durban, the researcher had the opportunity to converse with some Zulu women and men, who exhibited their beadwork and basket ware on the pavement under the shaded huts. The women, who were so proud of their artistic work, were keen on explaining the variety of formats and techniques they used in the bead-woven bracelets, necklaces, picture frames, belts, bookmarkers, and head bands. The myriad techniques such as netting, wrapping, fringing, braiding, and weaving introduced with many vivid designs and bright colors embody artistic expressions of social narratives and cultural discourses. Geometric structures and concepts prevalent in the beadwork designs, such as symmetries and transformations, unraveled cultural codes (i.e., conventions or patterns) that convey expressions of thought from the Zulu ancient traditions and that hold the key to intrinsic knowledge systems.

Method and analysis

An ethnographic approach using participant observation and artifact collection was employed over a period of one week. The researcher, posing as a customer, collected artifacts from street sellers on the coast of Durban. The researcher also asked permission to take photographs of collectible pieces of bead ornaments and baskets when visiting museums and art collections. A total of 54 pictures was collected, 18 of which were examined and analyzed for further reflection.

We employed the object- and photo-voice (Wang, 2006) method in our analysis of indigenous technologies inherent in the artifacts collected. We used object- and photo-voice technique as an action research strategy, where objects and their photos were read as narratives through which experiences of indigenous Zulu communities are textualized and translated using an ethnomathematical perspective. As Burnett (1995) emphasizes in relation to photography, "photographs encourage the viewer to displace the conventional parameters of language use in order to create a new context for the articulation of meaning. A radically different conceptual framework may be initiated that does not, in and of itself, permit the easy use of either the verbal or the written for descriptive or interpretive purposes" (p. 44).

In analyzing the photos and objects, we performed a feat of reconstruction, calling upon the sensory experiences of the designers and the contextual and social narratives evoked by multiple layers of textures and

NOTES

designs. In some cases, we used photographs as a kind of artefactual notation to elicit and stand for the original object.

Ethnomathematical Analysis of Indigenous Technologies

Beadwork technologies: Semantics and syntax

In the realm of oral traditions, the technology of bead working represents a form of an intercultural visual language that textualizes geometric patterns into ideas which are communicated and exchanged vividly and meaningfully (see Figure 1). For the Zulu people, this orally-based, naturalistic system of visual representations emblematize a mode of information dissemination, an embodiment of thought and belief system drawn from lived experiences of ancestors across many generations. Barta and Silverman (1997) argue that: "Mathematical application and thinking are embedded holistically in the beading process" (online reference).

Figure 1. The earliest beadwork example in the Killie Campbell Africana Collection dates from about 1927.

Concomitantly, our field work observations and informal discussions with the Zulu natives revealed the primacy of these indigenous technology systems in the sociocultural fabric of the community. Imbibed with metaphoric connotations in the meaningful portrayal, preservation, and transmittal of complex and insightful solutions, such technologies are created and appropriated by the Zulu culture in response to a spectrum of emerging quotidian problems. Narratives associated with various bead ornamentation and designs can range from everyday endeavors, such as making embroidered and beaded covers for dishes to keep the flies away, to more complex issues that address fertility (e.g. fertility dolls), individual identity, and the struggle for survival (Sienaert, Cowper-Lewis, & Bell, 1994).

Narratives attached to beadwork: The Zulu bead messages

The most vivid illustration of an indigenous technology created by the Zulu culture for the purpose of communication and dissemination of information is the widespread practice of *Ibheqe* or Zulu bead letters (see Figure 2). As an ingenious medium of information expression and exchange that is neither oral nor written per se, the Zulu bead messaging technology attaches narratives to an assortment of color coded geometric patterns.

There are varied and interesting associations of narratives and bead patterns that flourish in South Africa. Zulu bead geometric bricolage mimics in different ways personal messages and narratives and captures metaphors that transcend time and embody information in a visuo- perceptual form. As such, Zulu bead patterns act as a concretized information system with inherent algorithmic and geometric processes that create and transmit pieces of information in a more specialized and highly contextualized way. Nonetheless, the dominant view that prevails concerning traditional bead ornament is their intrinsic association with tribal courtship and marriage (Schoeman, 1968). "The object—[Zulu bead message]—is more than a mnemonic. It seems to present itself as a puzzle and a challenge: why is it there? What explanation can be given for its presence" (Barber, 2005, p. 269). Interpreting and decoding bead messages for meaning are complicated by the wealth of proverbs rich in linguistic logic and metaphors. The different structural analyses of a single motif in a beadwork pattern reveals a complex hybrid of intertwined multidimensional layering of information which, if appropriately decoded, makes explicit the meaning embedded in the motif. The first layer embodies the physical and material texture of the pattern as different materials might be incorporated in the design. In addition to glass beads which are most commonly used in South Africa today, other kinds of beads are also used. These are locally made from wood, shells, animal teeth, seeds, metal rings, and clay (see Figure 2).

Figure 2. A magician's headrest.

The choice of the material is deliberate, as it signifies the tone, depth, and hierarchy in power relations and social statuses. The second layer consists of an assortment of discursive visual patterns or structural units, such as dots, lines, triangles, symmetries, dimensions, scales, angles, and proportions, adding vivid rhythmic texture to the meaning. The third layer comprises the colors, which act as signifiers of visual metaphors such that each color evinces dichotomous positive and negative connotations related to cultural position and personal and social accomplishments. For example, Figure 3 shows an *iNcwadi* Zulu love letter that represents a message woven in beads by Zulu maidens to be given to their lovers as a symbol of love and affection.

NOTES

Figure 3. A Zulu love letter.

Women bead makers we informally interviewed indicated the importance of color combinations and coding techniques in conveying messages and narratives. For example, a linear grouping of blue, white, and black is a favorable combination and subsequently signifies a positive message (see Figure 4).

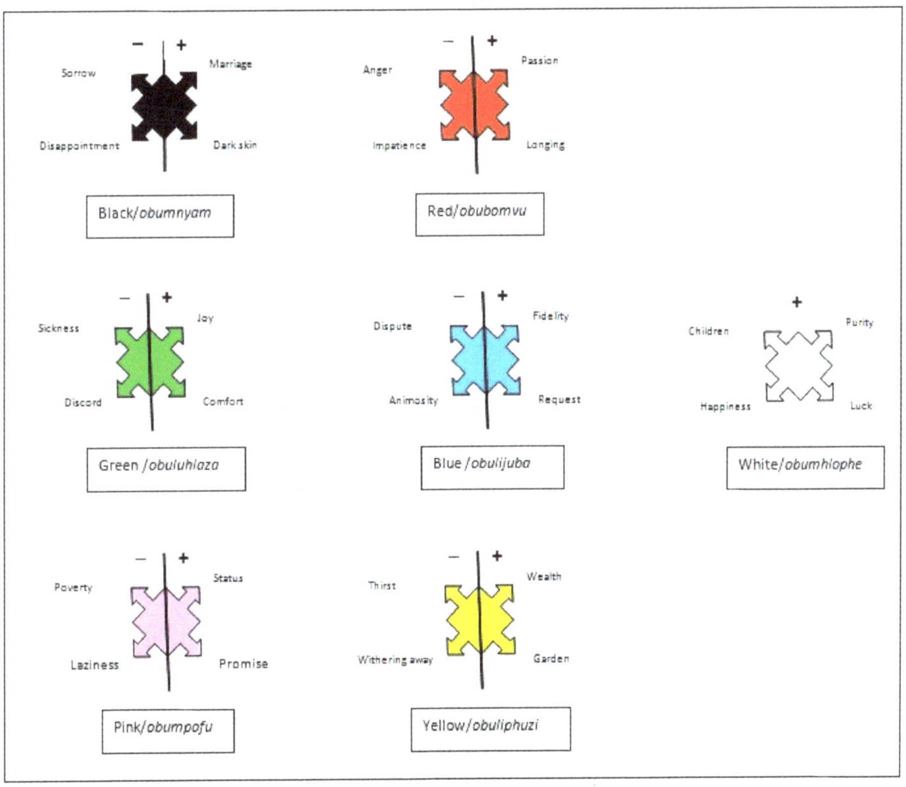

Figure 4. The association of meaning to basic color codes (The figure is produced by the first author).

On the other hand, a grouping of red, pink, and yellow is not favorable and thus carries a negative connotation. In addition to color codes, the order and sequence of the color combinations is primal. For example, in interpreting the symmetrical arrangement of colors, Schoeman (1968) explains that a vertically symmetrical arrangement of red; yellow; blue (at the center) then yellow; and red may convey the message shown in Figure 5.

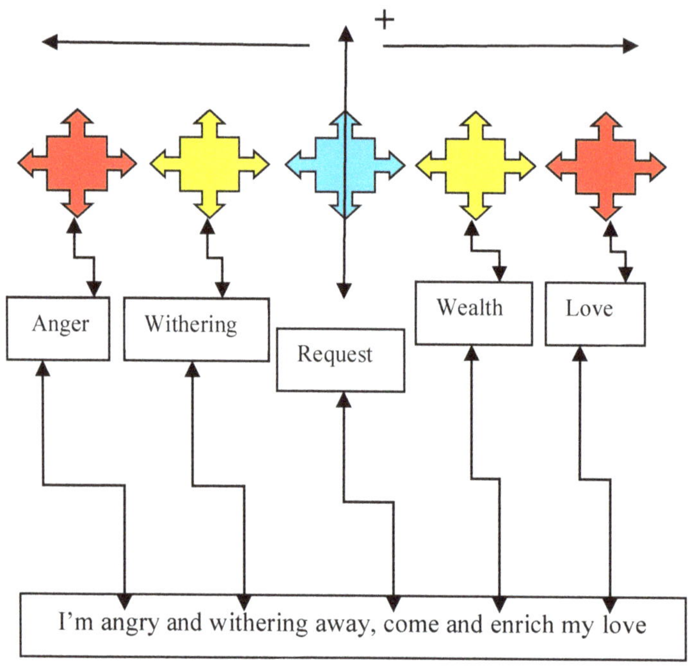

Figure 5. Decoding bead messages from a combination of color codes and symmetrical structures (The figure is produced by the first author).

Thus, the synergy of texture, geometric patterns, and color combinations provides information which is visually synchronized and transmitted to the viewer. The recurrence and nesting of motifs constitute a fluid visual narrative that represents momentous disclosures communicating a story, a social encounter, or a religious event.

In terms of functionality, this visually-based information system is inherently diverse and serves many purposes. Be it as an ornament, a jewelry piece (see Figure 6), a textual diction (e.g., a Zulu letter), or a commemorative object, such functionalities provide various venues for exploring different narratives by virtue of its association with the context.

NOTES

Figure 6. Beadwork designs on bracelets and head bands.

In terms of patterns, a common geometrically tractable motif that thrives in most of the Zulu beadwork designs is the 'stepped diamond.' Constructed from the triangle as a "basic Zulu design unit," the diamond represents a signature image template whereby bead workers experiment and play with symmetrical patterns.

Papini (1994) argues that the stepped diamond shape emanates from a common divinity design known as *vesica piscis*. The vesica, which literally means "fish bladder," is created by the intersection of 2 circles of equal diameters with the perimeter of each passing through the center point of the other (see Figure 7).

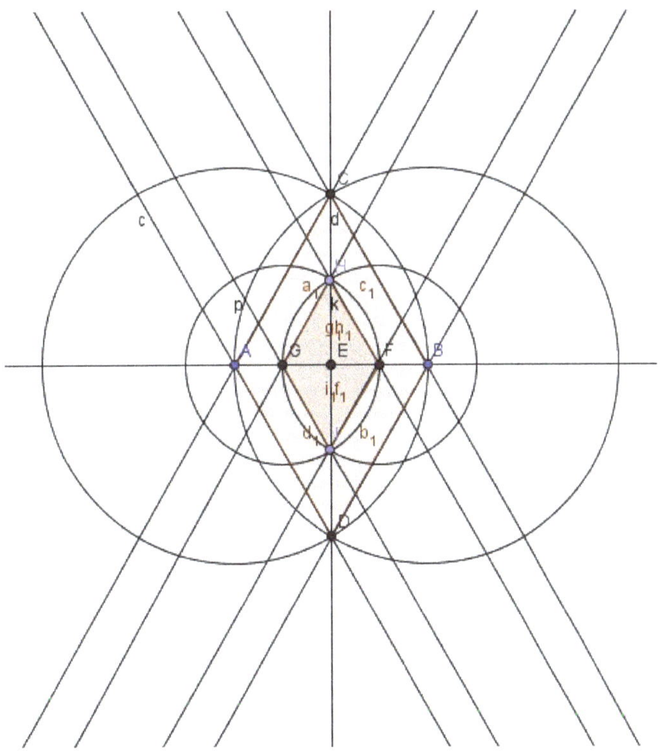

Figure 7. Geometric construction of the vesica piscis underlying the structure of a stepped diamond.

Thus the triangular shape of a stepped diamond structure as enunciated by Papini (1999) is iconographic of reproductive fertility emblematized in "geometric expression of two enacting entities reconciling relationship, being the third factor symbolizing the principle of harmony" (p. 165).

Basketry

As one of the most ancient handicrafts, baskets have acquired a special dimension of meaning in the social fabric of the Zulu culture. In major rural areas on the outskirts of the city, basket makers at work are a familiar sight. With intricate designs and maximized functionality, hand-woven baskets from the KwaZulu Natal area continually illuminate the struggle between the artisanship and machine production (see Figure 8).

Figure 8. Displays of traditional South African baskets at the Phansi Museum in Durban.

As with bead work, structural analysis of the motifs accentuated in basketwork reveals a complex array of parameters that determine collectively the characteristics of the final product. Different basket designs reflect various constructional solutions synchronized to meet certain criteria, such as the material used in making the basket, specialty and purpose of the basket, and the size of the person who will be carrying the basket (i.e., adult or child.).

For the Zulu culture, basketry technology requires thinking in terms of form, functionality, and the process of handling the vegetal materials. The materials, which are confined to the local vegetation of the Zululand, determine the kind of construction and consequently influence the shape of the baskets. To avoid trial-and-error while the design, which may wear down the quality of the materials used, the weaver performs basketry techniques with extreme deliberation and decisiveness. While in the act of weaving, the basket maker falls in a state of relentless contemplation and placid

NOTES

communication with the materials. Most basketry materials are kept the maximum length that they occur naturally in order to avoid starting of new materials which, in turn, takes time, interrupts the rhythm of the design, interferes with the uniformity of the repeated units, and weakens the structure. In response to the delicate nature of the elements, the process of bending and folding gently but firmly follows a methodical configuration that adheres to certain specifications in the dimensions of the piece, to the flexibility of the design, and to the strength of the structure. The basket is composed of systematically repeated modular units or repeated sequences of units (see Figure 9).

Figure 9. A small round flat bowl with rotational order.

From a stacking of these units and through arrangement of colors and textures infinite patterns can be produced. The question of manipulating the dimensions and forming space is intrinsically meaningful to the nature of this technology. The basket begins its existence as a flat surface and progressively takes a three dimensional shape, a kind of hemisphere in many variations. To get the desired dimensions, the basket maker resorts to successive subtractions, reducing the size of the materials used without interrupting the iterative patterns and the structural organization of the elements. Within the changing number of units, the geometric pattern is constructed in synchrony with the changing number of modular units, which is more complex than incorporating geometric patterns where the number of units per row is consistently fixed (see Figure 10).

Figure 10. A thickly woven basket with spatial depth.

In the making of the baskets, which may take many months, the weaver has to make a series of heuristic-based computational decisions to ensure optimal precision and accuracy in the execution of the design. For example, we envision the basket maker while in the process of constructing the basket shown in Figure 9 as caught between increasing the units at a greater rate to create an enlargement of the interior of the basket concurrently proceeding at a lesser rate in consonance with the development of the motif. Notwithstanding the fact that the geometric patterns depicted on the body of the basket employed bilateral symmetry along the vertical and horizontal axes, which consequently imposes restrictions on the possibilities of manipulating the structure, the basket maker remarkably succeeds in overcoming the problem of preserving regularity in geometric designs while continuously changing the number of modular units. By the same token, Gerdes (1998) capitalized on the specialized knowledge that the *sipatsi* (hand bags) and *titja* (coiled baskets) weavers in Mozambique and their expertise in inventing various strip pattern designs that demonstrate the seven possible symmetry groups.

Contrary to beadwork, the hands are the basic tools and machinery required in basket making technology; hence, no device or implements are necessary to hold the material together. Nonetheless, a fine degree of equilibrium, tension and loosening without false straining are optimized. Like beadwork, Zulu basketry calls for marked orderliness during execution and an optimal skill to manipulate the materials so as to follow configurations that allow the elements to support each other under tension whereby all are held in place by virtue of their relative positions to one another. As such, different materials induce different designs and a different array of expression. Because this technology does not allow corrections or adjustment, geometric patterns which result so naturally from the process are favored because of their predictability. When asked about how to remedy if a problem in the structure has been encountered, one basket maker mentioned that even before the work begins, the picture of the finished piece stands before the weaver's eyes, clear in its complete form and in its details. Thereby, errors are practically minimized.

Looking at myriad designs of Zulu basket technology does not provide a full understanding of the functionality of the different genres of designed baskets. One must experience their commodious usage in the everyday life of the Zulu people. From an outsider perspective, it is congenial to decide whether the shapes, colors, textures and designs of the baskets as decorative pieces are appealing. What is most interesting is the all-inclusiveness of the form where the unity of mass and proportion, the weaving and binding, the materials and techniques all combine in one smooth flow. When closely observing the regularity and orderliness of the structures as well as the proportions and logical organization of the motifs and patterns, the viewer cannot but discern the skill and expertise of the maker and the time and dedication incurred to produce such a technological artifact that is widely usable, inherently useful, and intensely expressive.

Closing remarks

In this article, we address the ethnomathematical structure underlying two indigenous technologies that undergird the daily practices of Zulu culture. Measuring, counting, timing, devising and structuring patterns, bending and folding according to afforded dimensions, and decision making are all examples of activity-situated ethnomathematical ideas that thrive in the design and execution of indigenous technologies prevalent among crafts people in the Zululand. We contend hitherto that such technologies embody an intensification of subjective and highly specialized indigenous knowledges and skills that are methodically systematized yet orally- based. The patterned beadwork designs and the 3-dimensional basketwork structures reflect geometric expressiveness endlessly synchronized with symmetrical forms and colors that attest not only to the proficiency of the Zulu communities but also to the riches of their ancestral heritage and collective cultural persona.

At the present and as most of oral traditions are at risk of being fungible by a thirst for material novelty and enthusiasm for the new, exploring the wealth of existing indigenous technologies as well as preserving their edifices would be a worthwhile endeavor for indigenous cultures sustainable local and global wellbeing.

References

Amara, J. (1987). Indigenous technology of Sierra Leone and the science education of girls. *International Journal of Science Education, 9*, 317-324.

Anderson, S. E. (1997). Worldmath curriculum: Fighting Eurocentrism in mathematics. In A. Powell & M. Frankestein (Eds.), *Ethnomathematics: Challenging Eurocentrism in mathematics education* (pp.291-306). Albany, New York: State University Press.

Ascher, M. (1991). *Ethnomathematics: A multicultural view of mathematical ideas*. Pacific Grove: Brooks/Cole.

Ascher, M. (2002). *Mathematics elsewhere: An exploration of ideas across cultures*. UK: Princeton University Press.

Barber, K. (2005). Text and performance in Africa. *Oral Tradition, 20*(2), 264-277.

Barta, J., & Silverman, R. (1997). Mathematics and beadwork: Notes from a conversation with a Sioux artist. Retrieved from http://www.ethnomath.org/resources/ISGEm/092.htm

Bogoshi, J., Naidoo, K., &Webb, J. (1987). The oldest mathematical artefact. *The mathematical Gazette, 71*, 294-302.

Brokaw, G. (2003). The poetics of khipu historiography. *Latin American Research Review, 38*(3), 111-147.

Burnett, R. (1995). *Cultures of vision: Images, media, and the imaginary*. Bloomington: Indiana University Press.

Carraher, T. N., Carraher, D., & Schliemann, A. D. (1985). Mathematics in the streets and in schools. *British Journal of Developmental Psychology, 3*, 21-29.

Chahine, I.C. (2011). Beyond Eurocentrism: Situating ethnomathematics within the history of mathematics narrative. *International Journal for Studies in Mathematics Education, 4*(2), 35-48.

Cuthbertson, B., Socha, T. L., & Potter, T. G. (2004). The double-edged sword: Critical reflections on traditional and modern technology in outdoor education. *Journal of Adventure Education & Outdoor Learning, 4*(2), 133-144.

D'Ambrosio, U. (1982). *Mathematics for rich and for poor countries*. Paramaribo: CARIMATH.

D'Ambrosio, U. (1985) . Ethnomathematics and its place in the history and pedagogy of mathematics. *For the learning of mathematics, 5*, 44-48.

D'Ambrosio, U. (1999). Literacy, matheracy, technocracy: A trivium for today. *Mathematical thinking & Learning, 1*(2), 131-153.

D'Ambrosio, U. (2000). A historiographical proposal for non-western mathematics. In H. Selin (Ed.), *Mathematics across cultures: The history of non-western mathematics* (pp.79-92). Dordrecht: Kluwer Academic Publishers.

D'Ambrosio, U. (2001). *Ethnomathematics: Link between traditions and modernity*. Rotterdam, Taipi: Sense Publishers.

Devantine, F. (2009). Written tradition, oral tradition, oral literature, fiuriture. *Shima: The International Journal of Research into Island Cultures, 3*(2), 10-14.

Eglash, R., Bennett, A., O'Donnell, A., Jennings, S., & Cintorino, M. (2006). Culturally situated design tools: Ethnocomputing from field site to classroom. *American Anthropologist, 108*(2), 347-362.

Ethekwini Municipality (n.d.). The official site of the city of Durban. Retrieved from http://www.durban.gov.za/durban/

Gay, J., & Cole, M. (1967). *The new mathematics and an old culture: A study of learning among the Kpelle of Liberia.* New York: Holt, Rinehart, and Winston.

Gerdes, P. (1988). On possible uses of traditional Angolan sand drawings in the mathematics classroom. *Educational Studies in Mathematics, 19*, 3-22.

Gerdes, P. (1994). *Sona geometry: Reflections on the tradition of sand drawings in Africa south of the Equator*. Maputo, Mozambique: Ethnomathematics Research Project.

Gerdes, P. (1995). *Women and geometry in Southern Africa: Suggestions for further research*. Maputo, Mozambique: Ethnomathematics Research Project.

Gerdes, P. (1998). *Women, art, and geometry in Southern Africa*. Asmara, Eritrea: African World Press.

Gerdes, P. (2000). On mathematical ideas in cultural traditions of central and Southern Africa. In H. Selin (Ed.), *Mathematics across cultures: The history of non-western mathematics* (pp. 313-344). Dordrecht, The Netherlands: Kluwer Academic Publishers.

Gerdes, P. (2007). *African basketry*. Maputo: Center for Mozambican Studies and Ethnoscience.

Harris, M. (1987). An example of traditional women's work as a mathematics resource. *For the Learning of Mathematics, 7*(3), 26-28.

Jurdak, M., & Shahin, I. C. (1999). An ethnographic study of the computational strategies of a group of young street vendors in Beirut. *Educational Studies in Mathematics, 40*, 155-172.

Jurdak, M., & Shahin, I. C. (2001). Problem solving activity in the workplace and the school: The case of constructing solids. *Educational Studies in Mathematics, 47*, 297-315.

Kyselka, W. (1987). *An ocean in mind*. Honolulu: University of Hawaii Press.

Lévy-Bruhl, L. (1966). *How natives think*. New York: Washington Square Press.

Lévy-Bruhl, L. (1952). A letter to E. E. Evans-Pritchard. *The British Journal of Sociology, 3*(2), 117-123.

Luria, A. R., & Vygotsky, L. S. (1998). *Ape primitive man and child: Essays in the history of behavior*. Boca Raton, FL: CRC Press.

Masingila, J. O. (1992). Mathematics practice and apprenticeship in carpet laying: Suggestions for mathematics education (Doctoral dissertation, Indiana University, 1882). *Dissertation Abstracts International, 53*, 1833A.

Millroy, W. (1992). An ethnographic study of the mathematical ideas of a group of carpenters. *Journal for Research in Mathematics Education, Monograph 5*.

Mosimege, M., & Onwu, G. (2004). Indigenous knowledge systems and science education. *Journal of the Southern African Association for Research in Mathematics, Science and Technology Education, 8*, 1-12.

NOTES

Nettle, D., & Romaine, S. (2000). *Vanishing voices: The extinction of the world's languages*. London, United Kingdom: Oxford University Press.

Nogueira, C. (2003). Oral tradition: A definition. *Oral Tradition, 18*(2), 164-165.

Nunes, T., Schliemann, A. D., & Carraher, D. W. (1993). *Street mathematics and school mathematics*. London, United Kingdom: Cambridge University Press.

Ong, W. J. (1982). *Orality and literacy: The technologizing of the word*. New York: Methuen

Papini, R. (1994). The stepped diamond of mid-century Zulu bead fabric: Aesthetic assimilation beyond the second Izwekufa. In E. Seinaert, M. Cowper-Lewis, & N. Bell (Eds.), *Oral tradition and its transmission* (pp. 27-46). Durban, South Africa: Center for Oral Studies.

Papini, R. (1999). Izazi on the Eleusine Mother: Uncommon sources for precolonial religious history in KwaZulu-Natal. *Voices: A Journal for Oral Studies, 2*, 156-193.

Pais, A. (2011). Criticisms and contradictions of ethnomathematics. *Educational Studies in Mathematics, 76*, 209-230.

Prediger, S. (2002). Consensus and coherence in mathematics: How can they be explained in a culturalistic view? Retrieved from http://people.exeter.ac.uk/PErnest/pome16/ consensus.htm

Radford, L. (2008). Culture and cognition: Towards an anthropology of mathematical thinking. In L. D. English (Ed.), *Handbook of international research in mathematics education* (2nd ed., pp. 439-464). New York: Routledge.

Rogers, E. (2003). *Diffusion of Innovations*. New York: Free Press.

Rist, S.; & Dahdouh-Guebas, F. (2006). Ethnosciences—A step towards the integration of scientific and indigenous forms of knowledge in the management of natural resources for the future. *Environment, Development and Sustainability, 8*, 467-493.

Saxe, G. B. (1988). The mathematics of child street vendors. *Child Development, 59*, 415-425

Schoeman, H. S. (1968). A preliminary report on traditional beadwork in the Mkhwanazi area of the Mtunzini district, Zululand. *African Studies, 27*(2), 57-81.

Sienaert, E. (1990). Marcel Jousse: The oral style and the anthropology of gesture. *Oral Traditions, 5*(1), 91-106.

Sienaert, E., Cowper-Lewis, M., & Bell, N. (Eds.). (1994). *Oral tradition and its transmission.* Durban, South Africa: Center for Oral Studies.

Throop, C. J. (2003). Minding experience: An exploration of the concept of "experience" in the early French anthropology of Durkheim, Lévy-Bruhl, and Lévi-Strauss. *Journal of the History of the Behavioral Sciences, 39*(4), 365–382.

Tylor, E. B. (1958). *Primitive culture*. New York: Harper.

Wang, C. C. (2006). Youth participation in photovoice as a strategy for community change. *The Journal of Community Practice, 14*(1), 147-161.

Wilder, R. (1981). *Mathematics as a cultural system*. Oxford: Pergamon Press.

Wilder, R.L. (1965). *The foundations of mathematics*. New York: John Wiley.

Zaslavsky, C. (1973). *Africa counts: Number and pattern in African culture*. Boston: Prindle, Weber & Schmidt.

Pacific Ethnomathematics: The Richness of Environment and Practice

Nicholas J. Goetzfridt
Professor of Library Science and Micronesian Studies
University of Guam
671-735-2300
ngoetzfr@gmail.com

Abstract

Contrasting geographic features that are evident between atolls and islands in Micronesia and in large Pacific land masses, particularly those of Papua New Guinea, have historically compelled indigenous peoples to apply distinctive linear systems related to distance, place, and ancestral origins. In the Caroline Islands of Micronesia – particularly on the islands of Puluwat and Satawal – traditional land finding techniques have applied linear approaches to the rising and setting of stars as they relate to specific sea routes between home islands and target islands. While linear concepts used in the culturally nuanced, earth grounded sand drawings in the Melanesian nation of Vanuatu are unique in their own right, linear based mnemonic systems for the recollection of ancestral origins and prestige granting cultural knowledge at sea in the Caroline Island and on the ground in Melanesia have interesting similarities. A greater academic concentration on the rich ethnomathematics of the Pacific could yield a trove of revelations in comparative cultural practices.

Atoll Linearity and Distance Calculations in the Caroline Islands

Numerical, spatial, and linear concepts in Melanesia, Polynesia, and Micronesia vary in accordance with distinctive physical environments and the social and cultural histories embodied in these Western assigned boundaries of the Pacific. Counting systems, agricultural calendars, and methods of conveying currency values for exchange are, in a comparative sense, dominant in the scattered ethnomathematical references to Melanesia where Papua New Guinea is predominantly represented. And ethnomathematical concepts and practices related to space, distance, and linearity generally form the environmental structure of ethnomathematics in Micronesia and Polynesia.

For Oceania, it is of course the environment of islands in a seemingly endless ocean upon which environments are not only known but perceptually understood in the context of cultures, change, and in the acculturating forces that colonial governments have wrought over the centuries. Smaller atolls, for example, hundreds of miles from larger land areas, develop linearity, space, and horizon-based concepts for measuring distance or direction and for more immediate measurement ideas. The navigators of the central Caroline atolls of Satawal and Polowat represent not only the past, but also a fairly recent renaissance of indigenous pride over the reality that their ancestors did indeed control their voyages by numerous environmental means including aligning distances with stars and sectionally dividing a journey between two points into several segments using a midway, out-of-sight island that was either real or imaginary.

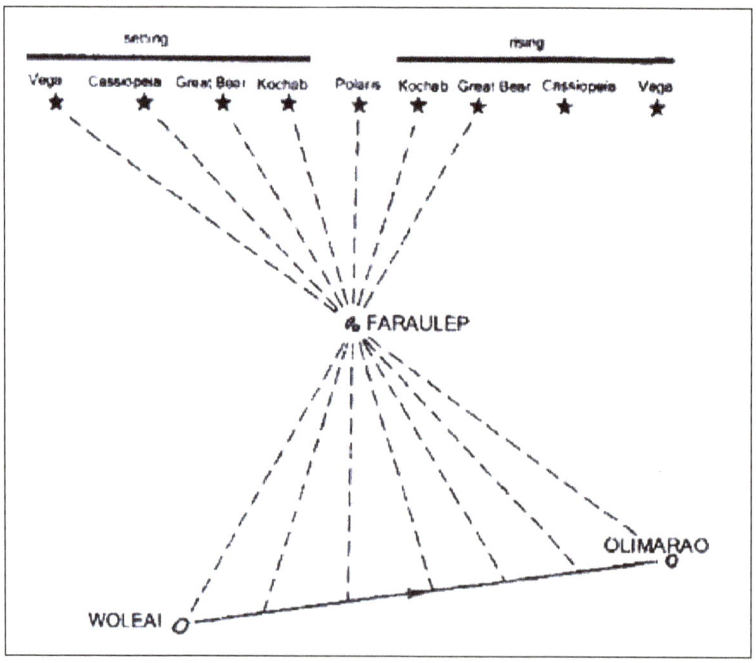

Figure 1, from Alkire 1970, page 53.

NOTES

This technique, known as *etak*, provides the means by which the navigator can tell where he is – essentially which segment (and thus distance traveled) – during a specific voyaging route. This is but one mechanism of orientation in a three-pronged task for leaving and finding land: orientating oneself in relation to a home and a target island, keeping on course through dead reckoning, and achieving landfall. Like the star compass described briefly below, the *etak* technique for understanding one's progression qualifies as an example or product of a cognitive mapping process that does not follow the standards of western maps of longitude and latitude but rather mnemonic sources that contribute to the ultimate achievement of landfall. More than anything else, it is a method of organizing information in ways that are primarily concerned with non-instrumental dead reckoning on the way to a target. In the case of *etak*, the canoe conceptually remains stationary as the stars, particularly those used to mark the location of islands through the places of their rising and setting, remain fixed. (Gladwin [1970] uses the near stationary mountains that a passenger on a train sees as a comparative example.) One out-of-sight island is chosen for each sea route and it serves as a fixed marker for that route. It is ideally located around fifty miles off to the side of the canoe and about midway between the home island and the target island. The star bearings for the reference island are known before departure as are other navigational star positions that the reference island will pass under as this island "moves" during the voyage. Each time the island "moves" under a new, previously known navigational star, this marks one completed *etak* or segment in the voyage. According to Hutchins (1983), these stars bearings "radiate out from the navigator" as they are applied to the reference island and the segmentation of the journey thus contextualizing the practice as something more than simple line orientation in space.

So, if a particular sea route has four such *etak*, the completion of the first *etak* would mean that one quarter of the voyage has been completed. If there were six *etak* to a particular sea route (as in the illustration above), then one-sixth of the voyage would have been completed. Estimation however of actual distance traveled (as in a western idea of miles or kilometers) is relative to a subjective understanding of wind conditions and the passage of time marked by the movement of stars. (Gladwin [1970] also compares this subjective understanding to a man walking in a familiar field at night and approximating his position through an innate familiarity with the field.) The *etak* segments form uneven triangles which are potentially made more uneven, or rather longer, at the start and end of the voyage if the reference island is relatively near to the canoe.

The system of *etak* is fixed and each *etak* system is attached to a specific sea route. Early European attempts at understanding the *etak* system mistook the reference island for a refuge island. However, in order for a navigator to know where the *etak* reference island is in relation to a specific route, the navigator would also have to know where all of the islands around him are, any one of which could provide a refuge island if sea conditions required it. (Not surprisingly and probably because of the patriarchal nature of voyaging knowledge, women as navigators are not found in European historical references which rarely attempted to interpret these systems.) It is also possible that an imaginary reference island could be employed so long as the same mechanism of the movement of the island under each established navigational star for a specific sea route is conceptually allowed to proceed. Because these *etak* segments are of temporal duration as the voyage proceeds, the system also allows for an *etak* of bird segments (designated for an *etak* near the completion of the voyage where land based birds would likely appear as land drew nearer) or an *etak* of "sighting" typically for an *etak* within which a home island or a target island is actually in sight.

The *etak* system is derived from a larger and more generally applied Carolinian "star compass" (also sometimes referred to as a "compass rose") whose thirty-two uneven compass points based on the rising and setting of specific stars (the Carolinian prefix *tan* is applied to names for rising points and *tubul* for setting points) are traditionally depicted in a circle with pieces of coral representing the stars, eight coconut leaves just inside the ring of coral to represent swell directions, and several banana fibers stretched across the face of the compass to emphasize reciprocal star courses. (See figure 2.) Given the east- west orientation of the Carolinian island chain across Micronesia, reference to a primary star for a particular sea route is always made in relation to its rising in the east. While most depictions of this star compass by western writers is typically circular, very early twentieth century German reports on the subject noted that Carolinian navigators used a quadrangular structure with Alkire (1970) noting that navigators from Woleai Atoll always depicted the star compass in the quadrangular. The four corners actually aided in the memorization of star and constellation point order along the compass' perimeter (Alkire 1970, 41-43). Finney (1998) also notes that because most scholars consider the quadrangular form to have been original, the eventual presence of the magnetic compass with its thirty-two points may have transformed the quadrangular compass into a circular one.

In applications of the star compass specifically to the *etak* approach, travel time and the actual distance that exists between a home island and a target island are of very secondary importance considering the variables of sea voyaging. However, being able to know the distance that one has traveled in order to fix a position in the context of a sea route is of primary concern. Measuring distance equates to determining one's location. These rising and setting points are applied to specific voyaging routes, depending upon the target island to be reached. The spacing is "uneven" in terms of the standard expectation that there be even spaces and points on a western mechanical compass since the coral pieces represent the actual rising and setting of individual stars – none of which are obligated to meet this standard.

NOTES

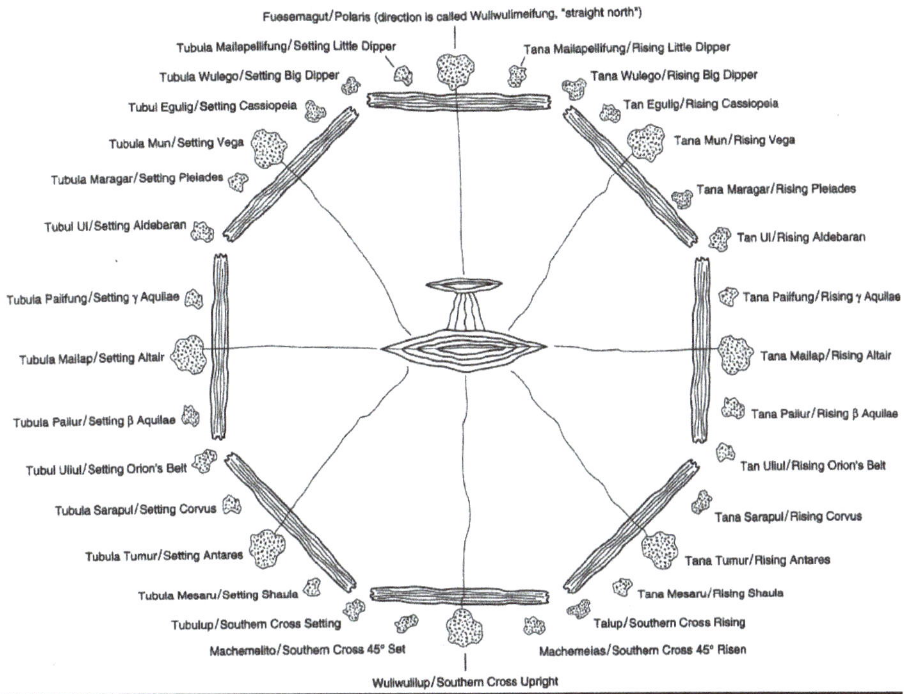

Figure 2, from: Finney, Ben. (1998). Nautical cartography and traditional navigation in Oceania. In: J. B. Harley and D. Woodward (eds.), *The history of cartography*, vol.2, book 3: *Cartography in the traditional African, American, Arctic, Australian, and Pacific Societies* (462). Chicago and London: University of Chicago Press.

Atoll Mnemonics

Indigenous principles of mnemonics are integral to the learning, recall, and recovery of components of linear based sequences – whether those sequences involve *etak*-like stellar systems of Puluwat to reach specific atolls using star driven courses or the recollections of story-based ancestral names and contexts. As is the case in much of Pacific scholarship, colonial and imperialistic forces have generally shaped the geographical structures of analysis so that a comparative study of, say, the patterns of Yapese formal seating arrangements based on protocols of authority in Micronesia with those found in Fijian formal settings, would require a significant break with the North Pacific's American dominated scholarship with the South Pacific's British based analyses and all of the scholarly antecedents from each respective colonial power. Fortunately recent history and anthropological conferences have helped weaken this colonially induced paradigm.

That is what makes Per Hage's 1978 comparative study of mnemonic systems unique in its analysis of techniques used by the navigators of Puluwat to organize their non-instrumental navigational knowledge as well as prestige granting cultural information and the mnemonic techniques the Iatmul people of Papua New Guinea use to implement their ancestral recall system by employing a name cycle with "phonetically and semantically interacting pairs" and clusters as is discussed in the next section. Although the ideas of cultural recollection are similar between the geographically distinctive societies on Polowat atoll in Micronesia and the Iatmul people on the far greater land mass of Papua New Guinea – this recollection solidifying prestige and power on Puluwat while supporting land claims for the Iatmul – the conceptual diagrams that they produce, not to mention their functionality, are distinct from each other.

The Puluwat people have used points composed of real islands or reefs and imaginary "ghost islands" and fantastic, nonexistent creatures that are joined by lines representing star courses between these points. This mnemonic system may be a structure embedded in a larger image, such as a trigger fish that encompasses a large sea area of several island destinations or in the movements of a mythical hero or a legendary navigator (or a fish that represents that navigator) along a series of courses. Or five trigger fish, in triangular shape, could be linked together via their dorsal fins, each encompassing the unique elements of these points and lines all the way from the central Caroline Islands to Guam and the Mariana Islands hundreds of miles to the north. This mnemonic device allows for navigators to use a set of locations that are each associated with an "image of an item to be remembered." Each item is then "recalled by mentally traveling from place to place and observing each item stored there." (Hage 1978: 87.)

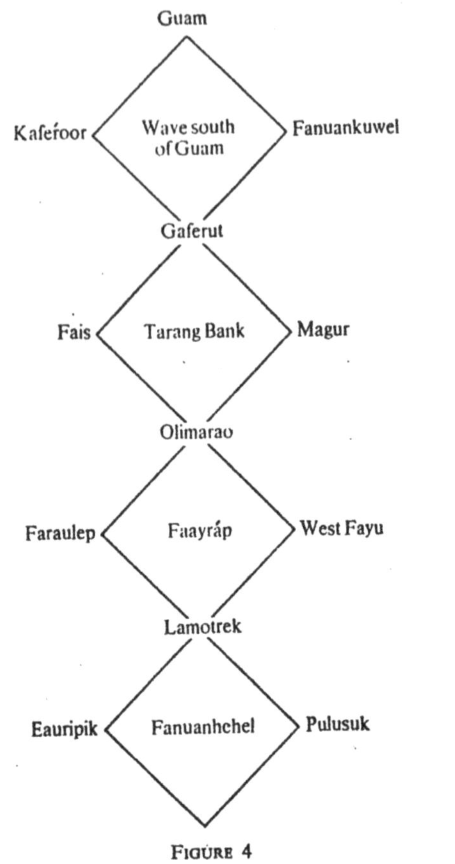

Figure 3, from Riesenberg, page 30

Here, for example, is Saul Riesenberg's rendering (1972) of a Polowat mnemonic scheme using the metaphor of a giant trigger fish envisioned as a diamond – the head, tail, dorsal fin, and the ventral fin representing four different geographical features. (See figure 3.) A fifth feature to the trigger fish is the backbone composed of geographic phenomena, real or imagined which function as centering agents. None of these geographic places, either real or imagined, lay precisely on these diamond formations. But they are approximations of geographic markers which

Volume 2
Community

NOTES

in the case of these four linked diamonds or rather, trigger fish, would lead a navigator from Polowat in the central Caroline Islands to Guam in the Mariana Islands.

It is important to note that this scheme and others like it also serve important mnemonic functions for recalling important cultural information such as myths, chants, spells, and so forth. While metaphors like this strengthen the recollection of star courses, they also enable navigators to participate in an esoteric language common to their class and to increase their status through demonstrations of cultural recall that these systems enable. Think of it perhaps as individual markers located in a high rise building that, upon reaching, remind the climber of a set of distinct facts or processes for an entire organization. Or a medical student who uses acronyms to remember a series of muscles. Each letter reminds the student of an item from a set of letters that ultimately represents this ordered sequence of muscles. (Hage 1978: 85.)

Another example is the parrot fish metaphor in the Reef Hole Probing system. Initially meant to enable a navigator to find an entrance through the reef into an island's lagoon, the system is suddenly enlarged when the navigator pokes a stick into a deep hole of that reef, scaring the parrot fish off to the reef hole of another island. When the navigator reaches that island and probes that reef hole with a stick, the parrot fish takes off again to the next island, following the established star course for reaching each island. The navigator reaches that island, probes again for the parrot fish and the parrot fish swims again to the next island and so on until the navigator is back at his starting point where he finally catches the fish. Each island point in this sequel of star courses and perhaps at points in between where real or imaginary creatures - such a two headed whale or a spirit in flames - can appear, prompt the mnemonic recollection of a traditional story or other cultural information that strengthens the navigator's status in his society. Such systems can also include more uncommonly known remote places for the sake of finding the solitude that enables the outlines of the system or systems to remain sharp in the navigator's mind without interference from others.

Linearity and Mnemonic Distinctions between Atolls and Larger Land Masses

Mnemonic systems of the Iatmul people of Papua New Guinea employ name clusters representing "classes of objects conceived as totemic ancestors" (Hage 1978: 88) rather than the fixed location sequences that the Polowatese people use to store and retrieve cultural information. In either case, the mnemonic effect appropriate to their cultural, social, and survival needs is achieved. Recall of Iatmul names is facilitated through story narrative in the same way that the mnemonic structures of the Polowat system is designed to be remembered as a unit. According to Hage, "the names are classes of objects conceived as totemic ancestors. Each name refers to a detail or segment of a myth and groups of names are corporately owned and individually assigned . . . . Itamul names are patterned in some way, just as the Puluwatese locations are patterned in some way. They are grouped into homogeneous sets – "name cycles" – and within each set they are grouped into phonetically and semantically interacting pairs so that each name is a cue for the other through sound and association." As is the case with Pulwatese knowledge of different linear structured destinations to locations of cultural information, recall name clusters achieve the same goal. Those learned in either technique possess various components of cultural knowledge in the case of the Puluwatese and name clusters in the case of the Iatmul – both in large quantities, with a "learned" Itamul person possessing "from ten to twenty thousand names including those of many other clans in addition to his own" (Hage 1978: 88).

For the Iatmul people, legitimate possession of these names becomes dependent upon one social group's capacity to associate mythology or more appropriately, stories, in greater detail than another group. Almost all of the names, which are arranged in pairs (and "each pair generally resemble each other much as the word Tweedledum resembles the word Tweedledee" (Bateson, 1958: 222-223) are compounds composed of four to six syllables. The mnemonic relationship between the names and the myths or stories "appear to be reciprocal in nature," Hage writes. "The names may be more easily remembered because they are embedded in a narrative. The stories may be more easily remembered because each detail or segment is cued by a particular name or pair of names. This type of reciprocal relation also characterizes the star courses and narrative content in the Puluwatese system . . . . The entire cognitive structure is designed to be easily memorable – as a unit." (Hage 1978: 89.) The legitimate possession of these names by a social group is eventually confirmed however only through the demonstration of its knowledge of related mythological references in a public debate during which methods of cuing aid one's cognitive performance. (See Bateson [1958], pages 125-131 for a description of this ceremonial debate.) On Puluwat, such cuing occurs on "learning mats" upon which shells are placed to help in the memorization and recall of star positions. "In both cases," Hage observed, "erudition . . .[which] relies exclusively on memory is generalizeable": Puluwatese navigators may achieve higher status and office and Iatmul "learned men," assert themselves, according to Bateson (1958), "not only in the debates which concern totemic names but also in those on every subject from initiation to land tenure" (Hage 1978: 91). Bateson also points out that "the people are very proud not only of the number of their totemic ancestors and their esoteric exploits in the origins of the world but even of the "straightness" of their song cycles" (Bateson 1936: 128).

Impact of Atoll and Island versus Large Land Environments on Information Recall Systems

Environments and indigenous responses to them reflect forms of ethnomathematical concepts that have often gone unrecognized as such. Nowhere is this perhaps more apparent than when one considers the differences between communal practices in Papua New Guinea, marked by tribally distinguished counting systems for agricultural and inter-tribal exchanges and the comparatively minor role of such systems in the Micronesian islands of Chuuk where the heavens and distinct boundaries of the sky and a unique organization dominate ethnomathematical-like thinking. These two distinct cultural perspectives make considerable sense when one thinks of the massive land of Papua New Guinea with its hills, mountains, valleys, and extensive concentrations of unique peoples as contrasted with the atolls of Chuuk which were traditionally created by the deep, underwater 'Bearer of Islands' (*Nikowupwuupw Fénú*) or simply the 'Bearer' or 'Nurturer' who heaped stones upon her back and arms until they broke the ocean surface, creating Chuuk. Another interpretation from German anthropologists from the 1920s has The Bearer as a great eel upon whose breast the earth rested (Bollig 1927, 12).

NOTES

But up and beyond the sea spans *Fachchamw* – the 'Under Brow' just under the arch of heaven where the gods of pre-Christian Chuuk resided, the most prominent being the 'Great Spirit' with an assembly of gods and spirits such as 'Good Determining' (*Óórorofich*) who is the brother of 'Father of Determining'. Other godly residents of the 'Under Brow' included 'Mother Age Worn' (*Inemes*) – sister of the 'Great Spirit', 'Lady Wishing' (*Neyáneyán*), 'Rains Father' (the god of weather), along with 'Finisher' (*Ä remei*), 'Gossiper' or 'Babbler' and 'Lord of Plover Birds' (*soukuling*). 'Middle of Heaven' (*Nuukeyinen*) or simply 'Middle' served as the Great Spirit's intervener in human affairs who itself was the originator of various manifestations who served specific functions in these interventions. Goodenough notes that Suuta – one of his informants - associated *Fááyiro* (meaning the 'under-arc' or 'under-bow') with a meeting house in 1964. "The implication seems to be," Goodenough writes, "that people are assembled under the aegis of the arc of heaven and also under the aegis of the *Fááyiro* school of *itang*" which is linked to "an outrigger stanchion as a support that serves as the 'maintainer' or 'repairer' of the land" (Goodenough 318-319). (*Itang* were political priests holding substantial power who possessed special bodies of knowledge and the rhetorical skills associated with that knowledge.)

This vision of the heavens and thus the inherent attachment of distances to that envisioning also appear in the atolls of Kiribati where the evening sky is likened to a vast roof supported by imaginary rafters – three on the eastern slope of the roof and three to the western correspondence. (See Figure 4.)

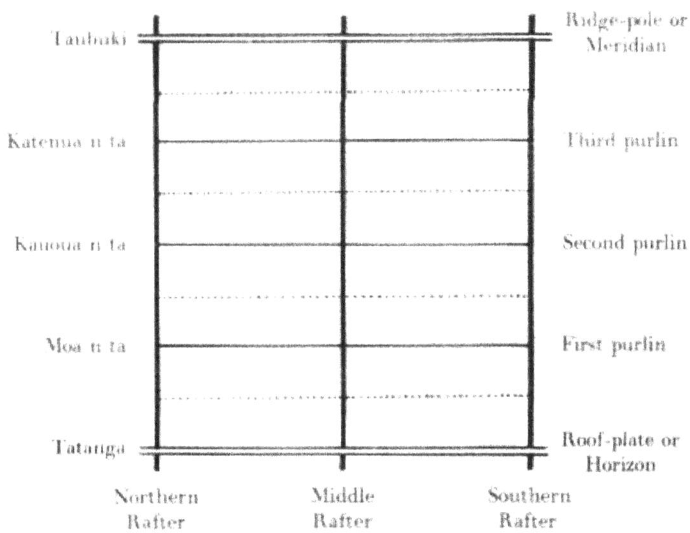

Figure 4, Roof of Voyaging. From Grimble, page 198.

Drawing from information provided to him by some of the approximately twenty persons on Kiribati still versed in traditional stellar knowledge in the 1920s and 1930s, Arthur Grimble describes the I-Kiribati conceptualization of the night sky as being a vast roof – a vast 'roof of voyaging' to be specific - in which the eastern horizon functions as the 'roof-plate of east' (*te tatanga ni mainiku)*, the western horizon as the 'roof-plate of the west' (*te tatanga ni maeao*), and the meridian (*te tabuki*) serving as the ridge-pole for this heavenly roof. The I-Kiribati appear to be more specific than the Chuukese about how this sky is actually held up using imaginary rafters – three on its eastern slope and three on the corresponding western slope with a middle pair of rafters representing the I- Kiribati celestial equator that is about 8 degrees south of the standard Western equator. The apex of this middle set of rafters is found at the point at which the star Rigel (beta Orionis) crosses the meridian. The point at which the Pleiades crosses the meridian represents the apex of the northern rafters. Antares provides the meeting point for the southern rafters at 18 degrees south of the I-Kiribati equator.

Across these rafters, three series of cross-beams, equally spaced across each slope of the roof, are imagined. Grimble likens them to "steps of a ladder up the sky" of the east (Grimble 216). While the western sky is structured exactly the same, this portion of the sky is envisioned as having numbered purlins leading down from the meridian to the horizon. It is over this tremendous framework that the stars provide a continuously shifting thatch under which the observer can calculate the altitude of a star with the imaginary rafters enabling the individual to also measure the star's declination. Indigenous names for various points on each purlin across which specific stars move are also used in the system. Before being allowed to identify specific stars upon this imaginary but intricately structured roof, Grimble's informant was required to memorize the exact names for at least 178 stars and constellations – sometimes using or weaving original tales about the stars as "persons or objects seen during the voyage of some fictitious character" (Grimble 218).

These tales were often transformed into songs whose lyrics functioned as storage containers and apparently enabled ancient voyages by Carolinians to Guam to resume in 1788, at least a century after Carolinians began to steer clear of Guam because of a perceived brutality of Spanish colonialists and then again in 1805 after they had been convinced by a Chamorro administrator (who traveled to Woleai atoll to reestablish connections) that the 1788 fleet that was lost at sea perished because of a storm and not at the hands of the Spanish or the Chamorros. The songs were definitely used by voyagers from Polowat and Satawal to Guam and the Northern Mariana Islands in the 1960s, 70s, and 80s after a voyaging hiatus of about 90 years.

In contrast, the piecemeal references to ethnomathematical practices found in the ethnographic and anthropological literature of Melanesia (and dominated by references to Papua New Guinea that range from the early 1900 until the present) underline the centrality of geographic context that becomes more apparent when contrasting Micronesian and Polynesian focuses on the heavens and the horizon. Alan J. Bishop's research (1979) on the visual and spatial abilities of first-year students at the University of Papua New Guinea, for example, demonstrated that not only did these students fail to use roads in drawing a routing map between two points, they lacked indigenous words for "line," "horizontal," "vertical," and "above." Ten years later, Bishop was offered words by peoples of the Papua New Guinea Highlands for the steepness of slopes but he still failed to find indigenous words that could be used to describe the idea of something being "horizontal."

In one study of indigenous mathematical, conceptual qualities of Papua New Guinea children – one of the few subjects in Pacific ethnomathematical literature that is responsive to other scholarly efforts – these children were found to have difficulty in

NOTES

"conceptualizing spatial relations in terms of Euclidean and projective relations involving awareness of straight lines, rectangles, metric relations and proportions, unless the circumstances of their material existence demand such conceptualizations." (Hallpike: 1979, 318.) One could perhaps say that this observation is indicative of Eurocentric expectations not only in what one anticipates seeing and not seeing but setting the standards for what to look for in the first place. But in this case, the research was set in the context of needing to discover ways to help students enrolled in the western modeled schools of Papua New Guinea.

It is nevertheless obvious that the physical environment of these areas of the Pacific – the natural priority that residents of far flung atolls and islands have placed upon the sea and the horizon in contrast to the priority that peoples of Papua New Guinea have placed upon inland environments of a very large island – have a tremendous, natural impact upon the mathematical and geometric systems that these cultures have developed over the centuries. And certainly the importance of centuries of use and change or sustainability is clearly evident both in the complex network and independence of the numerous indigenous counting systems throughout Papua New Guinea, many of which do not rely on a base ten foundation and in the methods of navigation in Micronesia which today contribute toward building a sense of cultural pride for an indigenous seafaring history.

Melanesian Linear Concepts and Cultural Practices

While linear concepts geared toward the horizon can of course differ among Micronesians, lines and linear concepts attached to the earth and things of the earth are at the same time as traditionally bound in Melanesian societies as are traditional hierarchical concepts of the sky or the heavens in Micronesian and Polynesian societies.

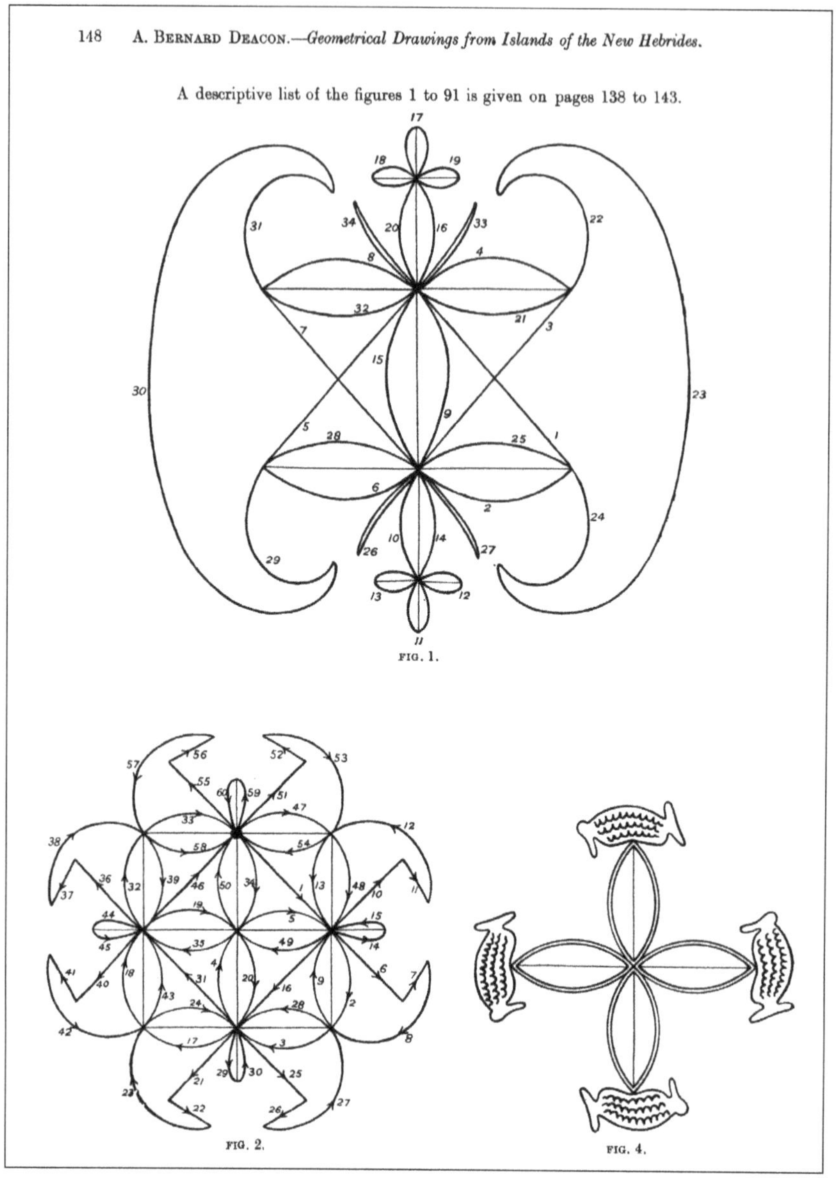

148 A. BERNARD DEACON.—*Geometrical Drawings from Islands of the New Hebrides.*

A descriptive list of the figures 1 to 91 is given on pages 138 to 143.

FIG. 1.

FIG. 2.

FIG. 4.

Figure 6, from Deacon and Wedgwood, page 148.

This may be most evident in the sand drawings of the Malakula people of Vanuatu (see Figure 6) whose tracings are perhaps most dramatically represented (and from Western eyes, you could perhaps charge exotically as well) by the drawing that the female ghost Temes Savsap half erases and expects each dead soul, based upon his or her fluency in culture, to correctly reconstruct this half before being allowed to pass on to the land of the dead. If the dead fails in this task, he or she is automatically consumed by Temes Savsap. (See Figure 7.)

NOTES

Temes Savsap was eventually vanquished – at least in the Seniang District of Malakula - by a deceased man who woke up, asked for his bow and arrow, and implored mourners to check to see if the rock of Temes Savsap had fallen into the sea the following morning - thus proving his successful killing of this demonic guardian of the land of the dead – before falling dead again. The rock was found fallen into the ocean, subsequently leaving the design associated with Temes Savsap without function.

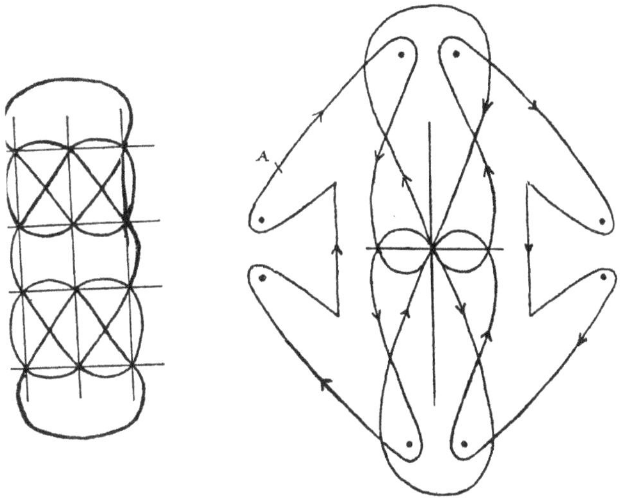

(a) Nahal. (b) Nevet hor Iambi.

ɜ. 37.—The geometrical figures *Nahal* (from Seniang) and *Nevet hor Iambi* (from Lambumbu), both of which are connected with the journeyings of the soul after death.

Figure 7, from Deacon, page 555.

The geometric drawings that A. Bernard Deacon studied in the 1930s in Vanuatu are characterized by a central purpose which Deacon stressed was to usually “execute the designs perfectly, smoothly, and continuously” without stopping at any point in the drawing’s composition and with no line ever being traversed twice. The strong symmetry of these designs also include those that are naturalistic in nature which according to Camilla Wedgewood’s study (1934) of these drawings and Deacon’s interpretation are also more maze-like in an effort to add elements of realism to the depictions of animals and plants.

One cannot simply generalize, however, that because Micronesia’s geographic situation places vast oceanic distances between atoll and islands, a vast cosmology of structure must therefore be birthed by the vastness of the heavens. A grounding of a cosmology and social identity within the landscape of an island - Tanna in Vanuatu – is a good example of how the drawing of distinctions between the linear concept formulations among atoll inhabitants and the more earth focused concepts of larger islands like Papua New Guinea can also become blurred. Although there are some contextual distinctions to be drawn between the anthropological attachment of the rhizomatic model of topographic inscription to Oceanic conceptualizations and practices as opposed to the arborescent or tree-like Western models, the horizontal root stem system that links up and crosses over the territorial creations of primordial ancestors in Tanna have been contrasted and compared with the Dreaming tracks of Australia’s Aboriginal peoples. (Rumsey 2001.) Although the people of Tanna have not historically been subjected to the dispossession of large tracks of land as the Aboriginal peoples of Australia have, their historical resistance of colonialism and Christianization has led to a renewed emphasis on Tannese *"Kastom"* upon which rests the indigenous ethnographic research and application of chants, traditional stories and narratives in order to apply customary links – and thus identity - to the land and subsequently "marking out the truly settled people of *kastom* in relation to it" (Rumsey 2001: 39).

On the other hand, while there is no particular order of precedence among Aboriginal origin places, a place called *Kwavar* (now underwater in the Torres Strait) has served as a "single origin place for all the Moorehead people." After survivors of the flood that inundated Kwavar moved to a second place – *Tjuari* – "they moved on to a number of other places, sometimes in linear series but often also in successive phases of proliferation out from common nodes" (Rumsey 2001: 28). Efforts to link ancestral origins and tracks to the land have also been central to the politics of Aboriginality in Australia. Distinct forms of topographic inscriptions are involved in both environments while the emergence and the movement of these ancestors across the landscape are traditionally nonhierarchical and are thus more reflective of the rhizomatic model of topographic inscription than of the arborescence model.

Deleuze and Guattari (1987) have stressed distinctions between rhizomatic and arborescent models as they relate to topographic inscriptions of native peoples. The former is associated with "nomadic socioterritorial regimes" – which characterizes these Melanesian methods of ancestral tracings and subsequent tribal claims to lands. The hierarchical-based arborescent system is associated with a "centralized, state-based" regime (Rumsey 2001: 20) (i.e. "the West") which may appear to be unrepresentative of what we are describing here. "The tree" in this sense provides a centering system within which units and subunits lead upward toward higher levels of influence to the extent that only higher levels possess higher information as growth occurs along "preestablished paths" (Rumsey 2001: 22). Deleuze and Guattari observe that "the rhizome is reducible neither to the One nor the multiple. It is not the One that becomes Two or even directly three, four, five, etc. It is not a multiple derived from the One, or to whch One is added . . . .It has neither beginning nor end, but always a middle from which it grows and which it overspills" (Rumsey 2001: 22). Rumsey argues, however, that in forcing this binary abstraction of the arborescent model of thought representing "the West" and the rhizomatic model representing "the East" (i.e. Oceania), they themselves fall victim to the arborescent mode of thought.

The Morehead River system of totemic landscape or a "system of places" (Rumsey 2001: 27) in Papua New Guinea has an elaborate manner of precedence among origin places. Although there are two tree like indigenous metaphors evoked by this precedence (the Black Palm and the *wasur* - a type of fig tree), neither assumes the Western tree model but are rather models based, respectively, on a "lineal segmentation from a unity" and, for the *wasur*, a "scattering or fragmentation from a starting point or center" through which "center places" can be "reconstituted by people moving back to them from the places to which they had dispersed" (Rumsey 2001: 28). (See Figure 8.) No matter how many different strands occur from the universal origin place of

NOTES

Kwavar for all Morehead people, "each episode of differentiation replicates the last," thus enabling things to "remain in their originary state" (Rumsey 2001: 28). This lineal "system of places" that targets ancestral origins and their emanations through tracks of land as specified and controlled by traditional narratives and chants, mirrors in a sense Puluwatese mnemonic schemes such as the trigger fish track upon oceanic spaces discussed earlier. As Ayres emphasizes, a claim to land on this "system of places" is made "not so much on the site of the story-place itself, but on the secret knowledge about it" (Ayes 1983: 122, 162).

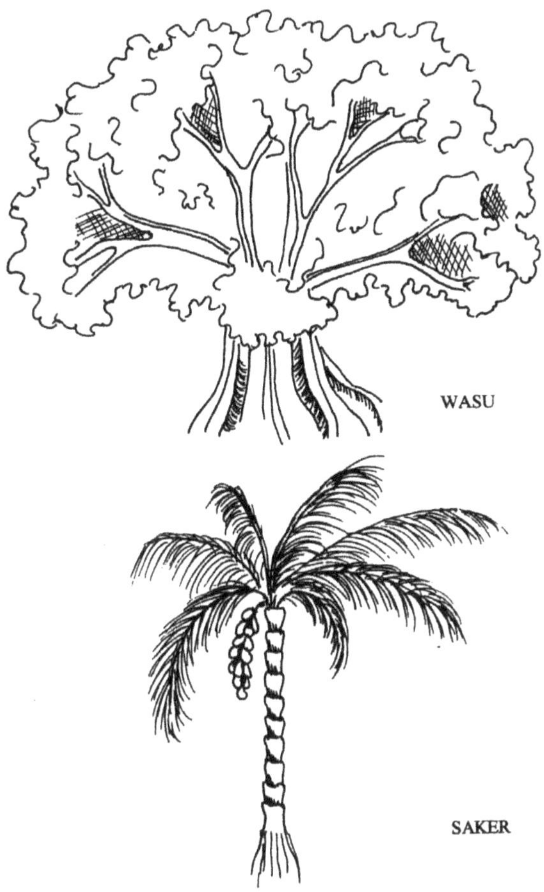

Figure 1-3. Two types of origin trees from the Morehead River area of South Coast New Guinea. (From Ayres 1983: 101; redrawn for this publication by Doerte Sueberkrueb. Reproduced with the permission of Mary Ayers)

Figure 8, from Rumsey, page 30.

In both the Moorhead and Tanna examples there are also precise differentiations of social space in land challenges associated with mythical recall that do not replicate the rhizomatic model. In the localization of power and knowledge, it is also interesting to note that Westerners in these Pacific contexts are considered to be not of this place and are therefore nomads of the rhizomatic sense because they are not the indigenous peoples of the land.

Primordial ancestors among the Iatmul people of the middle Sepik, Papua New Guinea originated from a single pit who then moved about, creating various linear paths to specific places and at unique events that each clan claims ownership to. Totemic songs for these places, along with mnemonic devices such as the use of a twine cord with a knot for each named place among the people of Central and Western Iatmul, provide communal connections to this land-based cosmology within which place abstraction is more pronounced than in the forms of Australian Aboriginal topographic inscription. Nevertheless social identity and these cosmologies are indelibly linked.

Ideas of distance and that distance's linear and horizontal foundations and the cultural complexities that form the marrow of cosmologies in landscape or in the heavens, are the products of their environments to which Pacific ethnomathematical concepts respond to. This is why generalized references to "the Pacific," or worse, to "the South Pacific" have, in the course of historical Western influence, served to diminish these very intricate similarities and differences that have been represented in this paper by Carolinian *etak*, star compass, linear mnemonic and I-Kiribati "roof of voyaging" systems and by Melanesian linear information recall systems of the Iatmul, Malakula, and Tannese peoples. It is interesting to consider what cognitive similarities could be drawn, for example, between the Puluwatese linear systems of land-finding and mythical recall for societal standing and the Morehead cosmology of an ancestral origin and subsequent ancestral tracks over the land and how the Deleuze and Guattari models might bring them closer together in understanding or pull them further apart – if Pacific ethnomathematics is given more concentrated attention.

One can legitimately hope, therefore, that if Oceania gradually receives more attention in the field of Ethnomathematics, these environmental distinctions and the indigenous histories that they pronounce will receive their legitimate place in the study of the many culturally nuanced ways in which Pacific peoples have responded to the unique worlds around themselves.

NOTES

References

Alkire, W. H. 1970. Systems of measurement on Woleai Atoll, Caroline Islands. *Anthropos, 65*, 1-73.

Ayres, M. 1983. This side, that side: Locality and exogamous group definition in the Moorhead area, southwestern Papua. Ph.D. diss., University of Chicago.

Bateson, G. 1936. *Naven: A survey of the problems suggested by a composite picture of the culture of a New Guinea tribe drawn from three points of view*. Cambridge: Cambridge University Press.

Bishop, A. J. 1988. *Mathematical enculturation: a cultural perspective on mathematics education.* Dordrecht: Kluwer Academic Publishers.

Bishop, A. J. 1979. Visualising and mathematics in a pre-technological culture. *Educational Studies in Mathematics, 10*(2), 135-146.

Bolling, P. L. (1927). Die Bewohner der Truk-Inseln. *Anthropos Ethnologische Bibliothek, 3*(1), 12.

Deacon, A. B. (1970). *Malekula: A vanishing people in the New Hebrides,* C. H. Wedgwood, ed. Oosterhout, The Netherlands: Anthropological Publications.

Deacon, A. B. and C. H. Wedgwood. (1934). Geometrical drawing from Malekula and other islands of the New Hebrides. *Journal of the Royal Anthropological Institute of Great Britain and Ireland, 64*, 129-175.

Deleuze, G. and F. Guattari. (1987). A *thousand plateaus: capitalism and schizophrenia.* Minneapolis: University of Minnesota Press.

Finney, B. (1998). Nautical cartography and traditional navigation in Oceania. In: J. B. Harley and D. Woodward (eds.), The history of Cartography, vol.2, book 3: *Cartography in the traditional African, American, Arctic, Australian, and Pacific Societies* (443-492). Chicago and London: University of Chicago Press.

Goodenough, W. H. (2002). *Under heaven's brow: pre-Christian religious tradition in Chuuk.* Philadelphia: American Philosophical Society.

Grimble, A. (1931). Gilbertese astronomy and astronomical observations. *Journal of the Polynesian Society, 40*(160), 197-224.

Hage, Per. (1978). Speculations on Puluwatese mnemonic structure. *Oceania 49*(2), 81- 95.

Hallpike, C. R. (1979). *The foundations of primitive thought*. Oxford: Clarendon Press.

Hutchins, Edwin. (1980). Understanding Micronesian navigation. In: D. Gentner and A. L. Stevens (eds.), *Mental models* (191-225). Hillsdale, N.J.: L. Erlbaum Associates.

Riesenberg, Saul H. (1972). The organization of navigational knowledge on Puluwat. *Journal of the Polynesian Society*, 81, 19-56.

Rumsey, A. (2001). Tracks, traces, and links to land in Aboriginal Australia, New Guinea, and Beyond. In A. Rumsey & J. Weiner (eds.), *Emplaced myth: space, narrative, and knowledge in Aboriginal Australia and Papua New Guinea* (19-42). Honolulu: University of Hawai'i Press.

Volume 2
Community

The Incan Abacus: A Curious Counting Device

Molly Leonard
maleonard1@stthomas.edu

Cheri Shakiban
c9shakiban@stthomas.edu

University of St. Thomas
2115 Summit Ave.
St. Paul, MN

Abstract:

Little is known about the ancient calculating device of the Incans la yupana. It was a tablet upon which stones, grains, or beans were placed and manipulated to perform calculations. This paper critically examines the little information available on la yupana and proposes a new interpretation, taking into consideration both mathematical and cultural factors. The conclusion reached is that la yupana is an abacus that directly correlates with the quipus, performs addition, subtraction, multiplication, and division in base ten, can be used to decompose numbers, and reflects linguistic principles of the Aymara language. A better understanding of the Incan abacus could lead to further insights about the civilization.

Introduction

La yupana

La yupana[1] was the ancient calculating device of the Incas. In the indigenous Andean language Quechua "*yupar*" means "to count." There are numerous historical accounts that mention the existence of such a calculating device, but the way they worked is left very unclear. What we do know for sure is that the Incas used some sort of tablet upon which stones or beans were placed. These markers were moved around in order to perform the various calculations needed to make the *quipu* records.[2] Both the *quipus* and *la yupana* are shown in the image to the right. The *quipucamoyoc* is holding the *quipus* while the abacus is located in the lower left corner (see Figure 1). It would not have been very feasible to tie and untie the knots on the *quipus* itself, hence, the versatility the abacus provides would have been essential. Stones or other markers were used because they could easily be moved and relocated. Once the operation was performed on the abacus the final computation could then be tied permanently into the *quipus*. It is easily seen that some sort of calculating device was used, but much more difficult to identify exactly how *la yupana* was used.

Figure 1. From Day, page 35

The most important piece of information in solving this riddle is the sketch referred to previously by Guaman Poma de Ayala, a chronicler of the 16th century. While a picture is worth a thousand words, I really wish that Poma de Ayala had left 1000 words describing exactly how *la yupana* worked. All that is available is this single picture and a few brief written accounts. That's not to say that we are unable to know anything of interest about this counting mechanism- it will just require more work and diligence on our part. Much can be inferred about this counting device while examining it within a comprehensive overview of the Incan way of life. This essay will introduce a few possible interpretations of this Incan calculating device, and determine the best one (or combination thereof). Because we are analyzing a math

[1] *La yupana* and "abacus" will be used interchangeably throughout the essay

[2] The *quipus* is an artifact of the Incan civilization that is directly related to *la yupana*. It is a knot-tying record composed of pendant cords hanging from a main cord. The number and types of knots on each pendant cord determine the numbers that are represented. These numbers denote all the tracked quantities of the empire from potato inventory to taxes paid to the Sapa Inca.

NOTES

tool from a very distinct time period and culture, all the theories should be criticized accordingly. The two most important factors to consider are: (a) the ease of manipulating the numbers and making calculations and (b) the consistency with Incan culture and their perception of numbers. The most cogent solution is the one that makes the most sense mathematically as well as culturally.

Historical Accounts

The first attempted solution was the work published in 1931 by Henry Wassén. His article "The Ancient Peruvian Abacus" appeared in Nordenskiold (1931), a collection of comparative ethnographical studies. His solution is based on two written accounts of Incan arithmetic. The first of these accounts is the manuscript accompanying Poma de Ayala's sketch of the abacus. Ayala writes:

> *"Haciendan Chasquicoc, the main treasurer, says that [Tahuantisuyo Runa Quipoc Incap, the main accountant] had great ability. In order to determine his skill, the Inca sent him to count and adjust Indian accounts in this region. . . His skill was greater than it would have been with pen and paper. [The counters] count with tablets. They number 100,000 and 10,000 and 100 and 10 until arriving at 1. They count everything that goes on in the kingdom: the parties, the Sundays, the months, the years, and these eight treasurers and accountants were in every city, village, and Indian town of the kingdom. They counted in this way: starting with one, two and three- 1 – 2 – 3 – 4 – 5 – 6 – 7 – 8 – 9 – 10 – 20 – 30 – 40 – 50 – 60 – 70 – 80 – 90 – 100 – 1000 – 1000 – 100000 – 1000000 – 10000000 – infinity."*[3]

The second reference used is that of Jose de Acosta taken from *"La Historia Natural y Moral de Las Indias"* originally published in 1586:

> *"To see them use another kind of* quipu *with maize kernels is a perfect joy. In order to effect a very difficult computation for which an able calculator would require pen and ink for the various methods of calculation these Indians make use of their kernels. They place one here, three somewhere else and eight I do not know where. They move one kernel here and three there and the fact is that they are able to complete their computation without making the smallest mistake. As a matter of fact, they are better at calculating what each one is due to pay or give than we should be with pen and ink. Whether this is not ingenious and whether these people are wild animals let those judge who will! What I consider as certain is that in what they undertake to do they are superior to us."*[4]

From these sources, Wassén deduces that (1) the operator was very skilled (2) he was commanded by the Sapa Inca to perform these calculations throughout the kingdom (3) he used wool and *quinoa*[5] and could compute more cleverly than those with pen and paper. Many words from the indigenous Andean language of Aymara allude to these types of computation. Excerpts taken from Bertonio's Spanish-aymara dictionary "Vocabulario de la Lengua Aymara" are evidence of this. The Quechua words *calana, aponocatha, iranocatha, saraatha*, and *velinocatha* all mean "to count with little rocks."[6] *Cchaara* is a "rock count to show what is due" and *hanka* is "what has been paid."[7] Other

[3] Ayala quoted in Burns, page 60, my translation

[4] Acosta, translation

[5] A very important grain and staple for the Incans

[6] Bertonio, Page 139, my translation

[7] Bertonio, Page 367, my translation

evidence suggests that these operations were endemic to Ecuador and performed by other indigenous groups as well:

> *"The archives reduced down to wooden, rock, or clay deposits with many compartments in which rocks of all sizes, colors, and angles were placed; they were excellent lapidaries. With the diverse combinations of the rocks, they perpetuated the events and created accounts of everything."*[8]

Wassén's Solution

The synthesis of this information led Wassén to the following solution. In this interpretation *la yupana* serves as a coordinate system with the vertical columns based on the decimal system and the numerical values of the horizontal rows "based on the important part that the number 5, i.e. a hand, plays among many people, and evidently played in the development of the decimal system in Peru."[9] Addition would be a simple enough computation. If you start with three stones in Ae for example and would like to add three more, you would add two more stones to Ae (until that square is full), replace those stones by one stone in Be and the last stone would be placed in Ae. Wassén argues that any summation can henceforth be directly related to the *quipus*. The problem with this idea is that if this were the case, the only holes needed in the abacus would be column A and one of the holes in each of the boxes of column B. Such an abacus could sufficiently hold all values. Wassén concedes that this is the case, but that columns C and D would be necessary for values larger than 222,220. It is a very improbable conjecture. (Especially since the Incas would rarely need to make such large calculations- the Ayala manuscript even implies that on the computing boards they numerated from one to 100,000; even if the Incas did work with larger numbers, it is even more unlikely that they would change the whole structure and functionality of their computing device in order to accommodate for the larger numbers.)

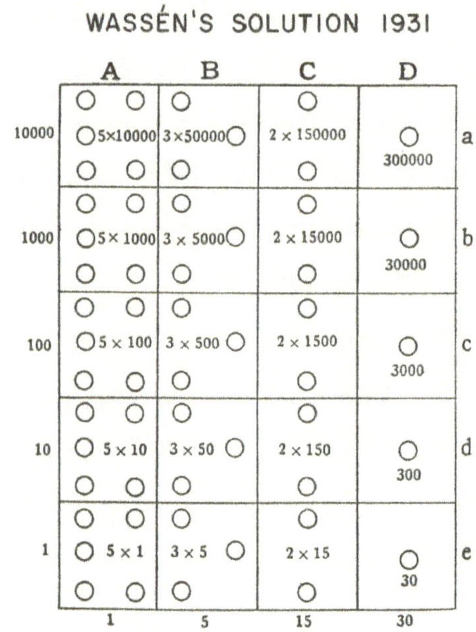

Figure 2, from Day, page 35.

Wassén's theory is indeed representative of a possible way to value and arrange the rocks on *la yupana*, but is far from likely. This type of number manipulation is neither derived easily nor is very natural or intuitive. Even with a strong background in mathematics, one would find it difficult to conceive and effectively utilize such a method, with so

---

8 Wassen in Nordenskiold

9 Wassen in Nordenskiold, page 200

NOTES

many unnecessary holes and place values. There is a definite gap in logic and the cultural significance of such a board structure is just as underdeveloped. Wassén is not able to support his claims. "We find that the solitary hollows in row D, counting from the bottom, denote the values of 30, 300, 3000 and so on. The importance of these squares in chronological computation by for instance setting down ten years as equaling 3650 days is no doubt self-evident."[10] I would argue that making such an unfounded jump is the only thing that is self-evident here.

Symbolic Values

Another possibility that has surfaced but has come up short from fully explaining the enigma, is that proposed by JA Mason and Diego Pareja. Pareja (1986) notes that many pre- Hispanic civilizations have made use of symbolic notation to represent different numerical values. The Babylonians used vertical and horizontal hooks and wedges, the Chinese used vertical and horizontal lines, and the Mayans were able to express all mathematical values with only three symbols: a dot, a bar and a shell-like symbol (representing zero). Applying the same logic, Pareja proposes that the Incas employed two symbols to denote values: a white circle representing "one" and a black circle representing "five." The distinction would be made by using stones of two different colors. Every number could be represented by at most five stones and an empty spot in the abacus would be understood as a zero. This explanation is clearly lacking because such a system has no apparent correlation between stone values and the differing number of holes in each square of the abacus. Considering that each square has a different number of holes (one, two, three, or five), blank squares represent zero, and each stone has a different face value (either one or five), the values of each *yupana* square could range from zero to 25. This lack of uniformity doesn't seem very plausible.

Pasquale's Solution

More recently in 2001, an Italian engineer came up with a solution that is uniform and does indeed take into consideration and utilizes the fact that the tablet squares have various numbers of holes; the problem is that this solution is not in conformity with the cultural norms and characteristics of the Incan culture and civilization. Nicolino De Pasquale proposes that *la yupana* actually calculated in base 40. The stone in the square with one hole would have value one, the two stones in the square with two holes would each have value two, the three stones in the square with three holes would each have value three, and finally the five stones in the square with five holes would each have value five. This yields a total of $1+2(2)+3(3)+5(5) = 39$. The values one through 39 could all be represented by placing the corresponding number of stones in the first row; once this row was filled, one stone would be placed in the second row, indicating 40, and the process would continue. While this method is indeed more structured and logical than the previous proposals, it likewise lacks depth of content. De Pasquale claims "It took me about 40 minutes to solve the riddle. I am not an expert on pre-Columbian civilizations. I simply decoded a 16th century drawing from a book on mathematical enigmas I received as a Christmas present." Just because something seems plausible from a mathematical perspective, does not mean that it is meaningful with regards to the cultural context. No matter how brilliant a solution might seem, it should not be accepted definitively before a rigorous analysis and understanding of the context from which the problem stems; this cannot possibly be done in 40 minutes.

10 Wassen in Nordenskiold , page 201

Alternate Interpretation

It is very easy to criticize hypotheses and accentuate shortcomings, but it is a much more difficult feat to propose alternatives. Considering *la yupana* in light of *The Social Life of Numbers: A Quechua Ontology of Numbers and Philosophy of Arithmetic* by Gary Urton (1997), some interesting possibilities emerge. Originally, I considered turning the abacus on its side so that the row of ones would be the first row. This perspective lends itself to a very unique interpretation. The five ones could be correlated to the hand, each representing a finger, with value one. Once these five holes were filled, you would have successfully completed a "hand." This interpretation has potential because it applies the Incan notion of completion. Incas were very aware of the "complete" state of numbers and strived to make numbers more complete. Complete numbers (such as paired or even numbers) were always valued over incomplete numbers (such as odd numbers). Correspondingly, numbers were never thought of as solitary entities but rather as members of some larger communal entity.[11]

Once all the stones in the first row were filled, you would be left with five fingers and you could substitute that for a stone in the second row, representing a hand and being worth 5. This has a linguistic basis. In Quechua "the fingers represent an important set of tools not only for counting but also for naming and organizing ordinal relations."[12] The squares in the second row each have two holes, corresponding with two hands. Once again, the notion of pairing two hands together is a very important one for the Incans and could have even given rise to the Incan use of the decimal system. Two hands together represent 5+5, or 10, the base of the decimal system. The pairing together of two hands in Incan culture is known as the unity of "*khalluntine*." Once each square had two stones, that square would be complete and would represent all ten fingers of one person. Perhaps when all the five squares of the second row were filled, it could represent a complete group of five people (perhaps a family unit?). A full second row would value 50 and could be substituted by a single stone in the third row with value 50. Arriving at a complete grouping of 50 has significance for the Incans as well. For example, when measuring things, it is always more desirable to arrive at the 50 mark. Consider this Quechua saying for example: "The needle of the balance is still not showing 50 kilos; add a sombrero full of potatoes more."[13]

While this method might not be the most straightforward in terms of calculating values, having a method grounded in cultural tendencies might actually make calculations easier. For example, even though the values of the stones (1, 5, 50, and 750) are row-dependent, perhaps they can be used with ease if they represent a cultural manifestation (example: a finger, a hand, a person, a family, etc.) The previous proposals all make sense logically but are devoid of meaning because they have no social or cultural ties. These interpretations would be readily accepted

[11] In the indigenous language Quechua, the numbers one through five related to a family unit. When counting corn for example, the number one referred to a mother and the numbers that followed were all thought of as her offspring. chuqllu (mom), apaña (offspring), iskay apaña (second offspring), kinsa apaña (third offspring), etc. Taken from Urton, *The Social Life of Numbers*, 86.

[12] Ibid., page 73

[13] Ibid., page 151

NOTES

in a western society that always perceives numbers in an abstract realm, but this is the underlying assumption that contradicts the very way the Incans viewed numbers. For the Incans, "numbers are conceptualized in terms of social- especially family and kinship- roles and relations."[14] Numbers are not simply symbolic representations but rather integral components of culture and societal relations; without these ties there can be no counting device in the first place.

Glynn's Solution

William Burns Glynn definitely took these cultural characteristics into consideration when forming his ideas about this topic. Glynn was a textile engineer from England who had an incredible knack for both mathematics and Quechua. He spent a considerable amount of time working in Peru and was more than familiar with the Incan civilization. His work entitled "*La Tabla de Calculo de los Incas*" published in 1981 presents a very convincing argument about the method of use of the Incan abacus. Glynn starts his analysis with the same quotation from de Ayala stated at length above. From this quotation various basic, but nonetheless fundamental, conclusions are drawn: (1) the decimal system is used, (2) words are assigned to the different numbers, (3) a calculating tablet is indeed used, and (4) stones or other counters were used to represent the numerical values.

Glynn sets up the abacus so that the long side of the abacus with the five-hole squares are closest to the operator. (See Figure 3). The rightmost column is reserved for the ones digits, followed by the tens, hundreds, thousands, and ten-thousands in order going from right to left, similar to the place notation system used in modern western mathematics. Each column has a total of 11 holes with the top hole being "*la memoria*"- the "memory." The introduction of the "memory" hole is rather revolutionary because it allows for the decimal system to be in place even though each column has 11 holes instead of ten. When counting, the holes from each row are filled (from bottom to top) and once all ten holes are full, they are removed from the board and one stone is placed in the memory hole. The stone can then easily be moved into the second column because it represents the number ten. Having this memory placeholder helps eliminate human error because it temporarily holds the value while the stones are being removed. The operator doesn't need to store that information in his own memory and the operation becomes more mechanic. Once the stones are removed, the memory stone can simply be moved into the next column to the left. This set-up allows for addition, subtraction, multiplication, and division.

Figure 3. from Cahill

Another breakthrough is that Glynn's analysis is the only one to find a purpose behind the hole structure. Each column is composed of squares that break down into one, two, three, and five holes. These numbers represent the infamous Fibonacci Sequence. Utilizing this sequence greatly reduces the number of steps needed to perform a multiplication or division. For example, if you were to multiply 150 x 70, instead of adding 150 to itself 70 times, these Fibonacci numbers could be used to simplify the problem into only two additions. This

[14] Ibid., page 13

trick is used in the interpretation of *la yupana* that follow; a more detailed explanation can be found in the section on multiplication.

Improved Abacus Methods

While the Burns proposal is the most probable, there are a few discrepancies and further commentary is needed. The first main incongruity deals with the way the abacus itself is situated. Burns argues that the abacus should not be situated as it is in the picture, but rather with the long side (of squares of five holes) towards the *quipucamayoc*. He uses the *Margarita Philosophica* by Gregorius Reisch as the basis for this claim. This 16th century encyclopedia has European origins and was used particularly in Germany. In this work, there is a picture of a counting device, situated so that the long side is closest to the operator. (See Figure 9). Burns writes that "the rules of Industrial Engineering in so far as they refer to the proper way to situate a work, suggest that the longest side should be placed by the operator. Likewise, the squares with more circles should be closer to the operator to avoid unnecessarily large movements."[15] While this might be the case in many European industrial settings, it is not likely that the Incans positioned the abacus in that way. This line of thought is clearly influenced by the western emphasis on efficiency as well as the way numbers are written in the western world: with the ones, tens, hundreds, etc. moving from right to left. While this correspondence with western practices doesn't necessarily disprove the theory, it disregards the Incan practices that lend themselves to a more probable positioning of the abacus. The Incas kept track of numbers on the *quipus* with the ones, tens, hundreds, etc. starting from the bottom of the cord and working upwards. It only makes sense to model the abacus the same exact way. This allows the numbers to be taken directly (and very easily) from (and transcribed onto) the *quipu* cords. There would be a direct correlation between the abacus and the *quipus*. This also coincides with what Acosta writes about the Incan mode of writing: "Their mode of writing wasn't in a straight line, but rather from top to bottom."[16] It seems only logical that the abacus be used in the very manner it appears in the drawing of De Ayala- and more exact evidence would be required in order to rotate the image as Burns so suggests. The remaining part of this essay will use the position of the abacus and row labels as shown in Figure 5 below.

[15] Burns, page 60, my translation

[16] Acosta, page 292, author translation

NOTES

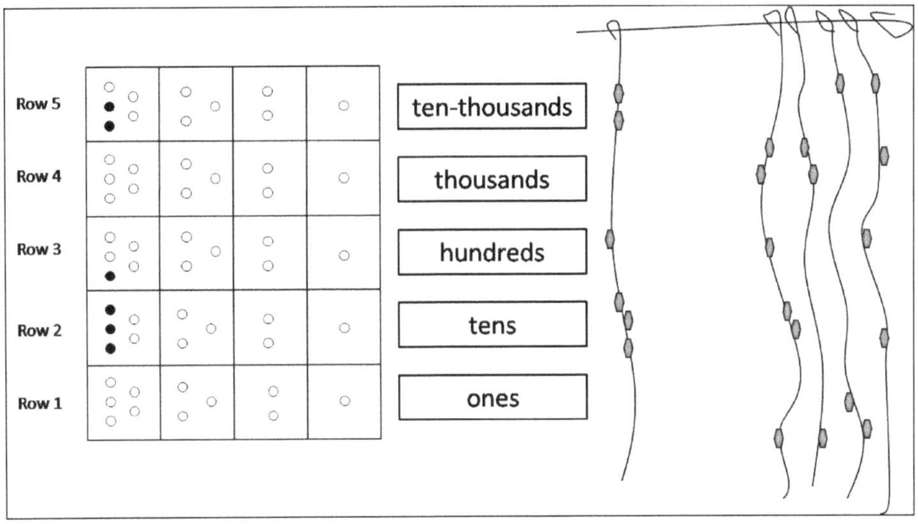

Figure 4 The abacus when vertical , corresponds directly with how quipus are read and used. The abacus and the first cord of the quipu are displaying the number 20, 130.

There is also a better way to consider the decomposition technique used in the multiplication and division problems. Burns proposes that the decomposition can easily be seen as a matter of looking at the stones and separating them into groupings of one, two, three, and five. Instead of resorting to a visualization of the stones, the abacus itself can be used to more exactly and effectively determine the decomposition of any number into its corresponding factors. (The purpose of this will be made clear in the following examples.) The stones could be placed on the left side and translated to the right until the squares completely filled. (This process is also very consistent with the cultural aim of "completion" discussed earlier). For example, when displaying the number 47 on the abacus (see Figure 6), stones would be moved to the right, with three stones remaining in the in the second square and the last stone completing the last square. The seven stones in the ones row already complete the first square with five holes and the remaining stone would fill the square of two holes. Once the number is displayed on the board this way, it is very obvious to see that the number can be broken down into 30, 10 (the three and one stones in the second row, because in the tens row, take on the values 1 x 10 = 30 and 1 x 10 = 10 respectively), 5, and 2. The abacus structure lends itself to immediately finding the decomposition of any number; this is more effective than any visual technique.

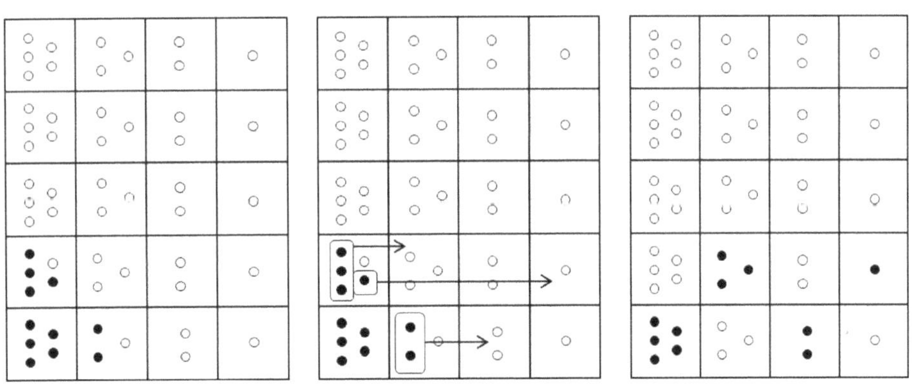

Figure 5 Showing that the abacus can be used to easily demonstrate that 47 = 30 + 10 + 5 + 2

One last addition that must be made deals with the effectiveness of the "memory hole." The stone placed in the memory hole could remain there until all the remaining stones were added to the abacus. In this way, the person performing the calculation could place all stones on the abacus in one fluid motion, eliminating possible error. The final step would be to move the stone from the memory hole into a hole from the row above. Saving this step for the end would indeed serve the purpose of a "memory holder," freeing up the operator to continue the calculation and the stone addition without any interruptions of motion or train of thought.

Learning from previous methods and taking into consideration Glynn's work, we are able to devise an interpretation of the abacus that proves most logical considering the information available. The following operations will be executed using the author's original interpretation of the counting device. They follow the basic structure of the operations outlined in *La Tabla de Calculo* de los Incas, but take into consideration the improvements suggested above and are simplified considerably (especially in the division problem) to the benefit of the operator of the abacus.

Addition

Addition (as executed in Figure 7) is the most straightforward of all calculations, and probably one of the most common operations. One of the summands is placed on *la yupana* while the other summand is placed outside *la yupana*, to the right of the corresponding rows (A). The stones outside the abacus in the ones row are added to *la yupana* first (B), using the memory stone as needed. Whenever the total number of stones in any given row reaches ten (hence the 5- square, 3-square, and 2-square are all full) then those stones can be removed and one stone placed in the single hole (the memory hole) on the right (C). The rest of the stones can then be added to the now-empty holes of that same row (D). After all the stones have been added, the stone from the memory hole can be shifted up to occupy a space in the next row up (E) and the process continues by adding the summand from that next row. All the place values are computed in the same fashion, always moving from the bottom to top. Once all the summand stones on the right side of the abacus have been worked into the calculation, the number remaining on *la yupana* is the answer, in our case 516.

NOTES

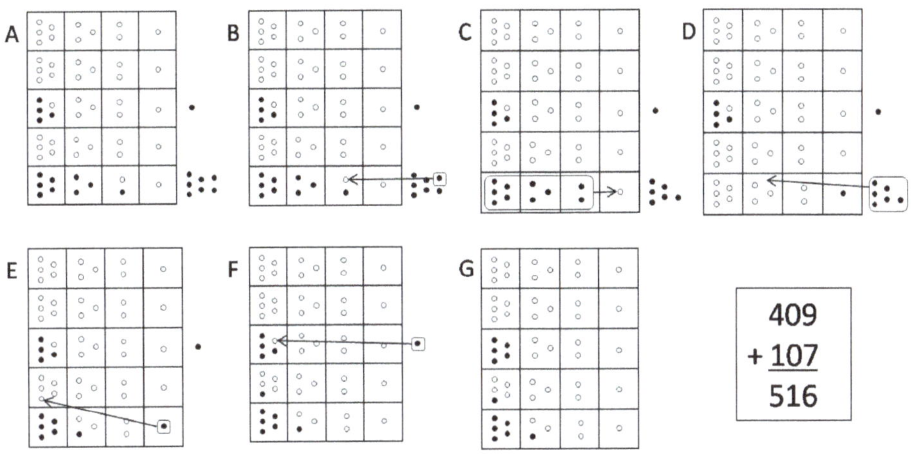

Figure 6

Subtraction

Subtraction is a similar computation, but the steps are done in reverse. (Refer to Figure 8 to follow along with the written explanation of 715-260). Subtraction would have played just as an important role as addition in order to rectify imbalances and keep track of the accounting transactions. In this case, the larger number is placed on the abacus and the number being subtracted is placed outside the abacus (A), to the right of the appropriate rows. (Refer to Figure 7). This time the calculation is performed from top to bottom. The stones are removed from the abacus and placed to the left of the tablet (B-C). Stones are removed until the number of stones on both sides of the abacus match. If there are not enough stones, then a stone from a superior row (the row directly above) is moved to the memory space of the row just below it (D). That memory stone can be converted into ten stones which occupy the holes in that row (E) and the stone removal can continue (F). Stones are transferred and removed until the number of stones displayed to the right of the board and the number of stones removed (represented by the stones that are found to the left the board) are the same (G). At this point, you have a visual confirmation that the calculation is completed; the number that remains on the abacus is the result of the subtraction, this time being 455, see Figure 8.

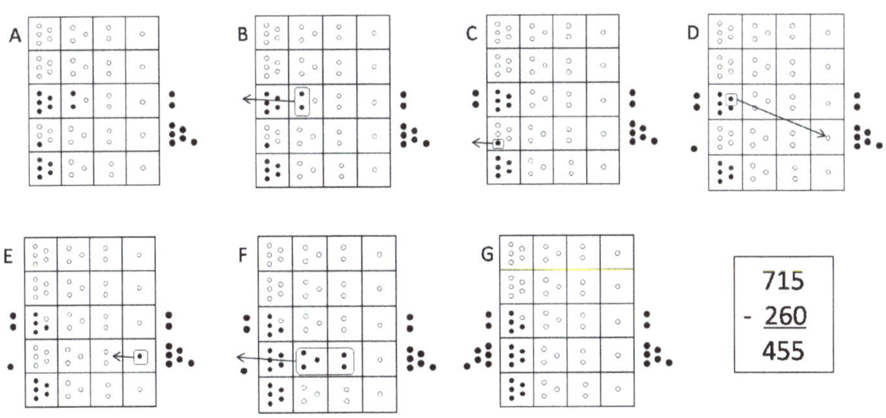

Figure 7

Multiplication

Using the abacus to multiply is a bit more involved, but the genius of the operation lies in the fact that it makes use of the hole structure of the abacus (i.e. the squares having 1, 2, 3, and 5 holes). We notice that the presence of the Fibonacci Sequence is not only an interesting observation, but that it has practical significance in this interpretation of the abacus. The Fibonacci sequence is constructed by summing the two previous numbers in the sequence. This aids in multiplication because it is no longer necessary to add the same number together as many times as the value of the multiplier; we can take advantage of the progression developed by adding together partial products.

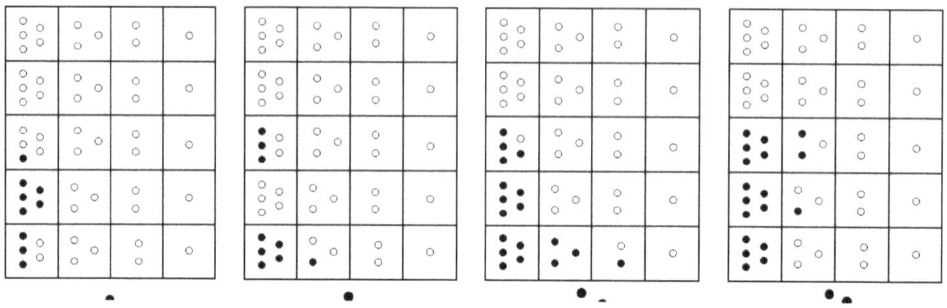

Figure 9, the preparation tables for the multiplication operation, representing 153 x 1 = 153, 153 x 2 = 306, 153 x 3 = 459, 153 x 5 = 765.

Multiplication requires many preliminary steps before the calculation itself is carried out. Perhaps this explains the fact that there were always as least a minimum of two *quipucamayocs*. Each *quipucamayoc* could help perform the preliminary steps on his own tablet. Let's look at the example 153 x 47. In order to prepare, the following information would be collected about the greater of the multipliers: 153 x 1 = 153, 153 x 2 = 306, 153 x 3 = 459, and 153 x 5 = 765. These calculations could be very easily resolved. 153 times 1 is obvious: 153. 153 times 2 is just that number added together twice. 153 times 3 is the two previous results added together and so forth. Each of the results could easily be displayed on a separate tablet (like those shown in Figure 9).

The next important observation is that the second multiplier can always be broken down into a sum of numbers that are easier to work with. Every multiplier can be decomposed into a sum of Fibonacci numbers, including various powers of ten. For example, the number 47 could be thought of as (30+10+5+2). To review, after the counters are placed on the tablet, they can very easily be moved around in order to determine the factors that can be summed together to arrive at the desired result. In accordance with the Incan principle of completion, any counters that are in a square but not able to complete it, are translated to the square to the right, until they are able to completely fill the square. Once all counters are located in complete squares, the factors of the decomposition are simply the value of each square. (Refer back to Figure 6 for clarification).

NOTES

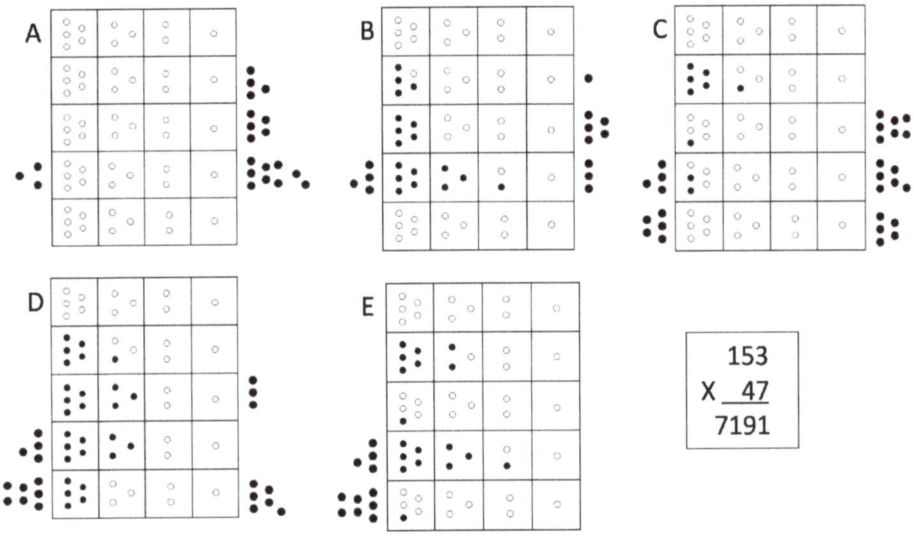

Figure 10

After the second multiplier is broken down into ones, twos, threes, and fives, the problem can be solved through repeated addition. Instead of adding 153 forty-seven times, we can simply add together four number (the number of additions matches the number of terms in the decomposition); the problem becomes greatly simplified and efficient. (See Figure 10). Recalling the decomposition already established previously (47 = 30 + 10 + 5 + 2) we start with the largest factor: 30. We find the number that corresponds with the partial product of the third and multiply it by a power of ten. (153 x 3 = 459 x 10 = 4590). This number is placed to the right of the abacus, and three stones are placed to the left of the second row (A) to represent the factor that has already been taken into consideration. The next step is to take the number that corresponds with the factor 10. (153 x 1 = 153 x 10 = 1530). 1530 stones are placed to the right of the abacus while one stone is placed in the tens row to the left of the abacus (B). The summation is performed. Next, the number corresponding to the factor five is added into the equation. (153 x 5 =) 765 is added to the 4590 + 1530 already found on the *quipus*. Five stones are then placed to the left of the first row (C). Finally 278 is added to the abacus and two stones are placed to the left of the last row to denote the addition of the last factor (D). At this point, you are left with the number 47 on the left side of la yupana confirming the fact that the entire number was factored into the calculation. The number that remains on the abacus is the product of the multiplication (E). In this case, that number is 7191 (4590 + 1530 + 765 + 306 = 7191).

Division

Division is perhaps the most difficult operation, but is more easily understood with the aid of a counting device like the Incan abacus. Just like with multiplication, the first part of the process is to make a table of multiples of the divisor. Let's use the same tables that were prepared for the multiplication problem (Figure 6) and look at the problem 7191/153. The dividend 7191 is placed on the abacus. (Refer to Figure 11). Working top to bottom (like in the case with subtraction), we look at the value in the highest row- in this case there is a 7 in the thousands row. We must refer back to the preparation tables in order to find something that is equal to or less than that value. None of the values are less than 7, in fact the smallest value shown is 153 x 1 = 153. We see right away that we must move down a row but using the hundreds row as our starting place we still only have the value 71 which is not sufficient. Moving down again we stop at the tens row and see that if the "decimal point" is placed on this row then we are looking at the number 709. We can now go back again to the preparation tables to find the highest value represented that does not surpass 709. The last value (765) is too large and we settle for the third tern- 459. Three stones are placed on the right side of the abacus (in- line with the tens row) and 459 stones are removed from the abacus and placed on the left side (A). The removal of stones follows the same rules as a traditional subtraction problem. In some cases, a stone from a higher row will have to be converted into ten stones of a lower row (B). After the stones on the right side and the stones on the left side perfectly match the respective term of the preparation table (C matches the third *yupana* of the preparation *yupanas* of Figure 6), the subtracted stones on the left side of *la yupana* can be removed in order to simplify the computation. The number left on the abacus is 260 (we are still using the tens position as our point of reference). Are any of the values from Figure 8 260 or less? The first factor of 153 is and so it can be subtracted as well. Again, one stone (representing that we are using the first factor) is placed to the right of the abacus while 153 is subtracted from the abacus and placed to the left (D-E). The number that remains is 107 and the subtracted stones on the left side of *la yupana* can be cleared away (F). 107 is smaller than all the factors of 153 so we must move our reference point to the ones position. From that vantage point we are left with 1071 and can subtract the largest factor, 153 x 5 = 765. Five stones are placed on the right side of the abacus (in-line with the first row) and 765 must be subtracted (G). After subtracting the single stone in the first row, that row is vacated and so the stone in the thousands-row must move down a row and be converted into 10 one-hundreds. We are now able to subtract the remaining stones, by moving them to the left of the abacus and adding a five to the right of the ones row (H). Once the stones are subtracted and match the number of stones as represented in the preliminary table of values (of the last table, correlated with the 5-factor) they can be removed from the work space (I). It is now readily seen that the 306 stones left on the abacus correspond with the second factor of 153, 153 x 2 = 306. Therefore the last factor of two is placed on the right side of the abacus in-line with the ones row and the 306 stones are removed to the left side (J).

Once the abacus is empty, the quotient remains on the right side of *la yupana* (K). In our example, the answer is 47 because we had placed three stones in the tens row, one stone in the tens row, five in the ones row and finally two more in the ones row. (30 + 10 + 5 + 2 = 47). This answer is expected from the results of the previous multiplication problem but this example shows how each step of the operation is computed. Additional abacus movements are needed in between some of the steps shown in Figure 11 but the main steps are all illustrated (some of the steps that involve changing a stone from a higher row into ten stones in the lower row via the memory hole are taken as understood because they are shown in previous examples). While this operation does seem rather complicated, it is simply a repeated subtraction problem just as multiplication is a repeated addition problem. Once the addition and subtraction methods are understood and the basic structure of the multiplication and division problems, any computation can be executed rather simply and straightforwardly.

NOTES

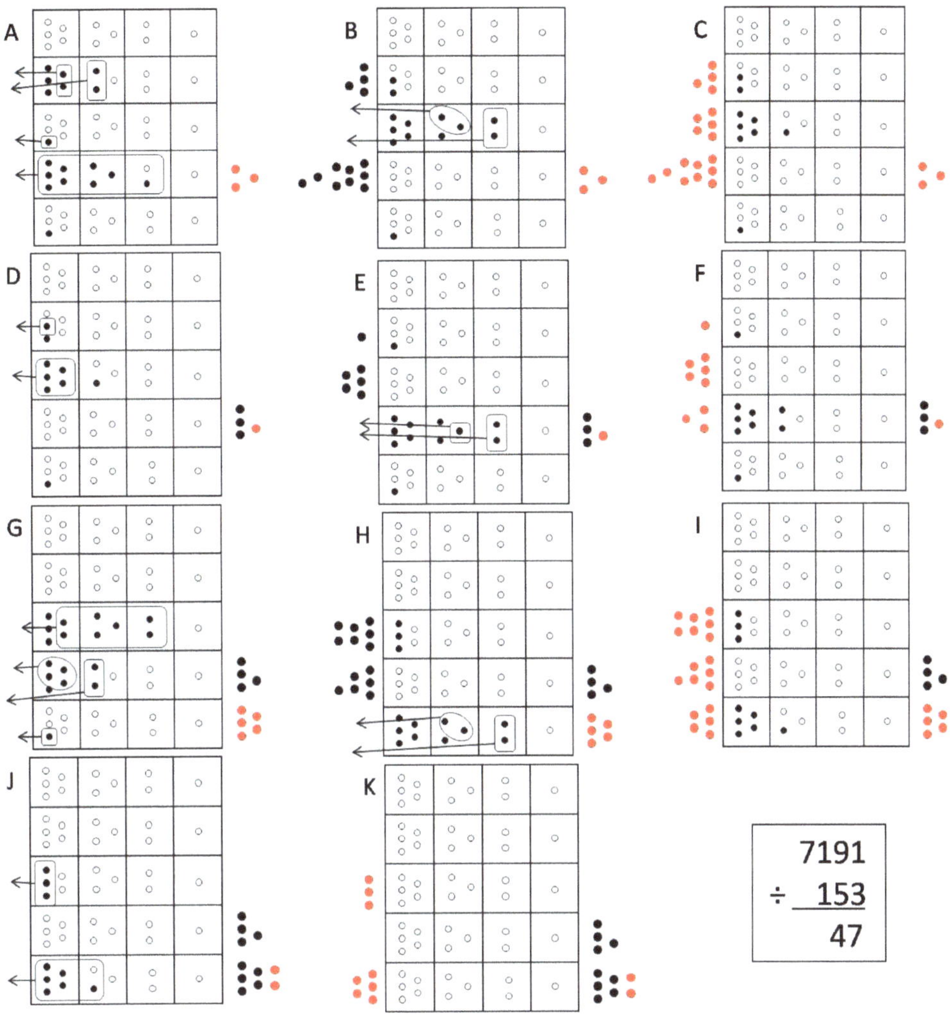

Figure11

Linguistic Link

Because *la yupana* is a cultural artifact as well as a counting device, it should have a cultural explanation as well as a mathematical one. The Burns interpretation is more than sufficient to show that everyday calculations can effectively be carried out, but there must be cultural evidence that supports the same interpretation as well. The Andean-linguistic expert Guido Alfredo Pilares Casas provides the missing link. In his article entitled "*Los Sistemas Numericos Del Quechua y el Aymara*," Casas discovers a connection between *la yupana* and the structure of the ancient Andean language of Aymara. This language was endemic to the very region in which the abacus was found, and Gerald Taylor suggests that Guaman Poma de Ayala the chronicler and creator of *la yupana* image itself spoke some form of Aymara that is now extinct. Casas considers from a linguistic perspective the distribution of holes into groups of five, three, two, and one. The answer relates nicely to the quinario-decimal structure of the Aymara language. While in the basic decimal system, each number builds off a base of ten, twenty, thirty, etc., in the quinario-decimal system each number builds off multiples of five as well as multiples of ten.

In this way the word five is denoted in Aymara by "*qallqu*." The three stones in the next square can all be derived from this value of five. Six is known as "*maqallqu*" ("*ma*" + "*qallqu*" = 1 + 5), seven as "*paqallqu*" ("*pa*" + "*qallqu*" = 2 + 5), and eight as "*kimsaqallqu*" ("*kimsa*" + "*qallqu*" = 3 + 5). The next two stones are grouped together and can be derived from the value of ten. Nine is known as "*llatunka*" in Aymara, meaning "ten minus one." Ten is simply "*tunka*." Once these stones are all combined and placed in the memory hole, there is another word in the Aymara language that means ten. "*Matunka*" which literally means "one ten." The numbers in the Aymara language coincide with the structure of the Incan abacus perfectly. This would make the instrument all the more accessible to the Incan people; this very same interpretation of *la yupana* is still used today and is part of the curriculum in the rural, bilingual schools of Peru.

Conclusion

While the above interpretation of the Incan abacus is the most accurate and comprehensive to-date, the lack of primary resources on the topic renders it impossible to know exactly how each minor movement was made. While the overall computation can be determined based on the mathematical and cultural clues, some of the smaller steps as presented could actually be performed differently. For example, in the division operation, the stones wouldn't necessarily have to be cleared away after each repeated subtraction. Also, the addition and subtraction could be computed in another order (it wouldn't matter whether the movement was from top to bottom or bottom to top). Likewise, since multiplication is a repeated addition problem, it makes no difference the order in which the additions are executed. Other possibilities include switching the side of the abacus that the stones are placed or even using additional *yupanas* on either side of the main *yupana* in order to store the stones. While these distinctions appear arbitrary, they are not completely so because the operations chosen in this analysis were those which were of greatest ease-of-use and clarity. While distinctions can be made, it is important that unity is remained throughout.

La yupana was a very important instrument in the Incan empire and not nearly enough information has been gathered about it. From the few primary and secondary sources that exist, we can reach some basic conclusions about how this instrument was used. Burns made significant progress and was the first to give an adequate explanation of the abacus, but was missing some key pieces of the puzzle. After comparing his conclusions with additional cultural information, we were able to enhance our understanding. *La yupana* was most likely used in the same way as it is seen in the Ayala's drawing; this renders a direct correlation between *la yupana* and el *quipus*. The tablet itself and its structure can be very easily used to factor any number. Another very pertinent observation is that the structure of the counting board directly correlates with the structure of and is reflected in the Aymara language itself.

It is also very fascinating to point out that such an interpretation is similar to that of the Chinese abacus. The Chinese abacus has beads of values 1 and 5. The numbers 6, 7, 8, and 9 are established by adding "ones" to the "five" bead. This abacus operates in a quinario-decimal system just like *la yupana*. These two counting instruments were developed independently of one another and the similarities found between the two only make the argument and interpretation of *la yupana* proposed in this essay all the more credible. Even though we are far from

NOTES

exhausting this discussion and decoding the complete mystery behind the Incan abacus, we are putting the stones in place for building a solid foundation and a better understanding of the mathematical practices of the Incas.

References

Acosta, J. (1940). *Historia Natural y Moral de las Indias*. Mexico: Fondo de Cultura Economica.

Ascher, M., & Ascher R. (1981). *Mathematics of the Incas: Code of the Quipu*. Mineola: Dover Publications.

Bertonio, L. (1984). *Vocabulario de la Lengua Aymara*. Cochabamba: Centro de Estudios de la Realidad Económica y Social.

Cahill, H. (2006). *Margarita Philosophica*. Retrieved July 15, 2008, from ISS: Information Services and Systems. Web site: http://www.kcl.ac.uk/depsta/iss/library/speccoll/bomarch/bomapril06.html

Casas, G. (2005). *Los Sistemas Numericos del Quechua y el Aymara*. Lima: Direccion Nacional de Educacion Bilingue.

Cobo, B. (1979). *History of the Inca Empire: An Account of the Indians' Customs and their Origin Together with a Treatise on Inca Legends, History, and Social Institutions*. Austin: University of Texas Press.

Day, C. (1967). *Quipus and Witches' Knots: The Role of the Knot in Primitive and Ancient Cultures*. Lawrence: University of Kansas Press.

Garcilaso de la Vega. (1989). *Royal Commentaries of the Incas and General History of Peru*. Austin: University of Texas Press.

Glynn, W. B. (1981). La Table de Calculo de los Incas. *Boletin de Lima*, 11, 1-15.

Locke, L. L. (1923). *The Ancient Quipu or Peruvian Knot Record*. New York: The American Museum of Natural History.

Locke, L. L. (1923). *A Peruvian Quipu*. New York: Museum of the American Indian, Heye Foundation.

Locke, L. L. (1932). The Ancient Peruvian Abacus. *Scripta Mathematica*, 1, 36-43.

Mora, L. C., Valero N., & Arias C. *La yupana como Herramienta Pedagogica*. Universidad Pedagogica Nacional.

Nordenskiold, E (Ed.). (1931) *Origin of the Indian Civilizations in South America*. Goteborg: Elanders Boktryckeri Aktiebolag.

Pareja, D. (1986). *Instrumentos Prehispanicos de Calculo: el Quipu y la yupana. Integracion. 4*(1), 37-56.

Poma de Ayala, G. F. (1980). *Nueva Coronica y Buen Gobierno*. Caracas: Biblioteca Ayacucho.

Urton, G. (1997). *The Social Life of Numbers: A Quechua Ontology of Numbers and philosophy of Arithmetic.* Austin: University of Texas Press.

Urton, G. (2003). *Signs of the Inka Khipu*. Austin: University of Texas Press.

Mathematics of the Hñähñu: the Otomies

Thomas E. Gilsdorf
Department of Mathematics
University of North Dakota
101 Cornell Street, Mail Stop 8376
Grand Forks, ND 58202-8376
thomas.gilsdorf@und.nodak.edu

Abstract:

English: This article is about the mathematics of the cultural group known as the Otomies, of present day central Mexico. In particular, I discuss the Otomi number system and a comparison of that system with Aztec counting, Otomi art and decoration, mathematical symbols that appear in some Mesoamerican codices, and the Otomi calendar.

Español: Este artículo es del tema de matemáticas culturales de los Otomíes, de la región central de lo que hoy en día es México. En particular, describo el sistema numérico de los Otomíes y una comparación con el sistema numérico de los Aztecas, el arte de los Otomíes, los símbolos matemáticos que aparecen en unos códices mesoamericanos, y el calendario otomí.

1. Who are the Hñähñu?

The cultural group we are going to consider is generally referred to as the Otomies (pronounced "oh- toh- MEES"). The name Hñähñu ("hñ" is pronounced like the "ny" in "canyon", preceded by a nasalized "h") is one of the names by which people of this culture refer to themselves. In general, people of Hñähñu culture use a variant of this word when referring to themselves with other members of the same culture, but use the word *Otomi* in communications in Spanish or with people from outside the culture. The term *Otomi* has been used in a derogatory way since before the arrival of the Spanish. Nevertheless, there is some consensus among anthropologists who study *Hñähñu* culture that it is better to use the word *Otomi*, presumably in a positive way, so as to work toward creating a more positive image of the people who are often referred to with this word. A discussion of the meanings of the terms *Hñähñu* and *Otomi*, and the negative ways in which the Otomies have been portrayed over the years can be found in (Wright, 2005; p. 19) and (Wright, 1997). I will use the term *Otomi*, and hopefully this work will serve as a positive description of the people, as viewed through an analysis of their number system, weaving and embroidery, and calendrics- the main themes of this article.

The Otomies are an indigenous group of what is now central Mexico. The modern-day territory in which significant numbers of Otomies live and practice some form of their traditional culture includes large parts of the Mexican states of Hidalgo and Mexico, with smaller areas in the states of Puebla, Guanajuato, Michoacán and Veracruz. There is also a significant number of Otomies in present day Mexico City (the Federal District). Over the last several centuries the Otomies have migrated because of issues such as conflicts with the Aztecs[1] and the arrival of the Spanish in 1521.

The Otomies represent one of the largest and oldest indigenous populations of what is now Mexico. According to Lastra (2006; pg. 25), the number of speakers of Otomi (including those who also speak Spanish) as of 2000 was approximately 292,000. This gives a rough estimate of the number of people who practice Otomi culture to some extent. Wright (1997; pg 439 – 441), shows that there is significant evidence to conclude that the Otomies represent one of the oldest cultural groups of what is now central Mexico. Moreover, the Otomi language is part of the Otomangue family (Lastra, 2006 or Soustelle, 1993 [1937]), which is neither Yuto Aztecan nor Mayan. Thus, the Otomies represent a very old and very large cultural group, and one that is distinct from the more well- known Aztec and Mayan cultures.

A consideration that is pervasive in the study of Otomi culture is the interaction between the Otomies and the Aztecs. The history of Aztec– Otomi interaction and conflict is long and complicated[2], and one consequence of that history is that it is often difficult to distinguish between cultural aspects of these two groups. Understanding the mathematics of the Otomies requires understanding some basic facts about the mathematics of the Aztecs, and we will encounter this issue several times. Our first topic is the Otomi number system.

[1] See pages 3- 4 and Note 2, pg. 294 of (Smith, 2003) regarding the use of the word "Aztec" versus "Mexica". Though both terms are widely used, I have chosen to use "Aztec" here.

[2] See (Soustelle, 1993 [1937]), pages 463 – 467

NOTES

2. Numeration.

In this section, I will give a list of Otomi number terms and indicate the mathematical constructions that go with it. What follows is a comparison with Aztec number words, and conclusions that come from that comparison. The Otomi number words in Table 1 below come from the dialect of Otomi spoken in the Mezquital Valley of the state of Hidalgo. I have taken the information for these number words, including the linguistic notation that appears, from (Barriga Puente, 1998; p. 229), originally referenced in (Merrifield, 1968). See also (Acuña, 1990) and (Thomas, 1897). A comparison of Otomi dialects in Barriga Puente (linguistic category 49) shows that the Mezquital Valley dialect has had less influence from Spanish than most of the other Otomi dialects.

Table 1. Otomi number words.

Number	Otomi expression
1	*ˀna*
2	*yoho*
3	*hñu*
4	*goho*
5	*kïtˀa*
6	*ˀrato*
7	*yoto*
8	*hñato*
9	*gïto*
10	*ˀrætˀa*
11	*ˀrætˀa ma- ˀra*
12	*ˀrætˀa ma- yoho*
20	*ˀnate*
39	*ˀnate ma- ˀrætˀa ma- gïto*
63	*hñu ˀrate ma- hñu*
85	*goho ˀrate ma- kïtˀa*
100	*ˀna nthebe*
1000	*ˀna ˀmo*

For the numbers from one to five, the only connection between terms that appears is the possibility that *goho* for four might represent twice two. On the other hand, starting at the term for six, there begins to be a pattern. Notice that the word for six, *ˀrato*, can be split into the parts *ˀra* and *to*. The part *ˀra* is similar to *ˀna* from the number one, but it is a good idea to look further down the list before concluding anything. The word for seven, *yoto*, splits into *yo* and *to*, and the word for eight, *hñato* splits into *hña* and *to*. From these examples, we start to see part of the word for two in the word for seven, part of the word for three appears in the word for eight, and part of the word for four appears in the word for nine. The word for ten no longer follows the pattern, however the word for eleven appears to represent the construction of "ten plus one". We can now conclude that the *ˀra* in the term for six is probably "one", and that *to* means "to add to five". Hence, six is "one plus five", seven is "two plus five", eight is "three plus five", and nine is "four plus five".

Let us now look again at the expression for eleven, *ˀrætˀ*a *ma- ˀra*. It is easy to deduce that "11 = 10 + 1", and similarly, "12 = 10 + 2". It turns out that this pattern continues until twenty, when a new word appears for that value. Table 2 shows the constructions of the values shown in Table 1, with an added number, 2488.

Volume 2
Community

Table 2. Otomi number expressions and structures.

Number	Otomi phrase	Construction
1	*ˀna*	
2	*yoho*	
3	*hñu*	
4	*goho*	
5	*kïtˀa*	
6	*ˀrato*	1 + 5
7	*yoto*	2 + 5
8	*hñato*	3 + 5
9	*gïto*	4 + 5
10	*ˀrætˀa*	
11	*ˀrætˀa ma- ˀra*	10 + 1
12	*ˀrætˀa ma- yoho*	10 + 2
20	*ˀnate*	
39	*ˀnate ma- ˀrætˀa ma- gïto*	20 + 10 + 4 + 5 = 20 + 10 + 9
63	*hñu ˀrate ma- hñu*	3 x 20 + 3
85	*goho ˀrate ma- kïtˀa*	4 x 20 + 5
100	*ˀna nthebe*	1 x 100
1000	*ˀna ˀmo*	1 x 1000
2488	*yoho ˀmo ne- goho nthebe ne-goho ˀrate ma- hñato*	2 x 1000 + 4 x 100 + 4 x 20 + 3 +5= 2 x 1000 + 4 x 100 + 4 x 20 + 8

These constructions of Otomi numbers have been known for some time. Specific details of this counting system can be found in (Barriga Puente, 1998; pg. 229, Table 49ch). Nevertheless, we can ask ourselves here, "What is the counting base for the Otomi number system?" Indeed, there are values based on five (for numbers 6 to 9), while other values are constructed using ten and twenty. So, our conclusion is that Otomi counting is not done by a single base but rather by a combination of bases. In particular, the Otomi system would be called a 5 - 10 - 20 system; see (Closs, 1986; pg. 3). The Otomi system follows the counting patterns of several Mesoamerican cultural groups. For example, the Aztecs use[3] a similar 5 - 10 - 20 system; see for example, (Payne and Closs, 1986; pg. 214 – 218). The Mayans use a 10 - 20 system according to (Closs, 1986; pg. 292 – 293). Examples of counting systems of other Mesoamerican groups can be found in (Barriga Puente, 1998).

I will now compare the Otomi system with Aztec counting system. Because the structure of (counting) words is a

[3] I use the present tense in describing numerical expressions in Otomi, Náhuatl, and Mayan as a reminder that there are large numbers of people who still speak the languages of those groups.

NOTES

linguistic consideration, the Aztec words are described in terms of their language, Náhuatl (pronounced "NAH- wah- tl"). Here is a list of Náhuatl numbers words, as described in (Payne and Closs, 1986; pg. 214 – 218).

Table 3. Náhuatl number words.

Number	Náhuatl phrase	Construction
1	*ce*	
2	*ome*	
3	*yei*	
4	*nahui*	
5	*macuilli*	
6	*chicuace*	1 + 5
7	*chicome*	2 + 5
8	*chicuei*	3 + 5
9	*chiconahui*	4 + 5
10	*matlactli*	
11	*matlactli once*	10 + 1
12	*matlactli omome*	10 + 2
13	*matlactli omei*	10 + 3
15	*caxtolli*	
16	*caxtolli once*	15 + 1
17	*caxtolli omome*	15 + 2
20	*cempoalli*	
30	*cempoalli ommatlactli*	20 + 10
37	*cempoalli oncaxtolli omome*	20 + 15 + 2 = 20 + 17
60	*eipoalli*	3 x 20
100	*macuilpoalli*	5 x 20
400	*tzontli*	

The first similarity between the Otomi and Aztec number systems we encounter is that the structures for the expressions for six to nine are the same. This is worth looking into deeper.

It is generally understood that the Aztecs are a Yuto- Aztecan speaking group that migrated to present day Mexico City from somewhere north of that location. As indicated in the introduction, the Otomies are a very old Mesoamerican culture, and were already established in central Mexico when the Aztecs arrived. Therefore, it makes sense to examine maps of linguistic groups. In (Valiñas Coalla, 2000), there are several tables and maps of 16th Century linguistic information regarding number words of Northern Yuto-Aztecan and of Southern Yuto- Aztecan. The geographical area he studies includes the present day southwestern United States and northwestern Mexico. One such group from the northern part of this geographic area is the Opata ("oh PAH tah"). The following Opata number expressions are taken from (Valiñas Coalla, 2000). The expressions for one to ten are: 1- *seni*, 2- *gode*, 3- *vaide*, 4- *nago*, 5- *marizi*, 6- *bussani*, 7- *seni- bussani*, 8- *go nago*, 9- *kimakoi*, 10- *makoi*. Now, notice that the word for six has no connection with the word for five. The expression for seven is: "7 = 1 + 6". For eight we have: "8 = 2 x 4", and nine is "9 = 10 – 1". A look through the other Yuto- Aztecan dialects from the northern part of the region (roughly the part that would be in the present day southwest of the U.S.), shows that most such dialects follow a pattern similar to the Opata. Meanwhile, checking the dialects that are further south, we see that most of those Yuto- Aztecan dialects have the same structures for the numbers six through nine as do cultural groups from central (present day) Mexico, including the Otomies. Several cultural groups of Yuto-Aztecan speakers migrated toward central Mexico over the years before the rise of the Aztecs (Smith, 2003; p. 37). It appears that groups that migrated closer to central Mexico adopted number expressions similar to those of groups such as the Otomies. Based on these observations, we can conclude that the structures of the Aztec number expressions for six to nine represents an influence by existing cultural groups of central Mexico, including the Otomies. This conclusion is supported by Doris Bartholemew (see Bartholemew, 2000), in which she concludes that the Aztec expressions for six to nine were directly influenced by the Otomies. The difference between her analysis and ours is that hers is based on anthropological considerations, while ours is based on considering mathematical structures.

3. Weaving and embroidery.

The Otomies have traditionally been highly regarded for their skills at textile arts, and I would like to examine the mathematics of Otomi weaving and embroidery. Some general observations about weaving and embroidery will motivate the mathematical considerations. I will focus on the Otomies, though many of my statements hold for traditional textile artists in general. Also, I will tend to refer to Otomi textile artists as female because most are women, though there are male weavers and embroiderers.

First, to become a traditional Otomi weaver is a life long process. A common situation is one in which there are several generations involved in creating woven or embroidered projects. That is, there is often a gathering that includes girls, mothers, aunts, grandmothers, and so on. Second, Otomi weavers and embroiderers must keep track of many counts of threads, and must make precise measurements. They must know the entire design from memory. The impressive part of the weaving or embroidery process is that the artist typically is not using diagrams for the patterns, nor is that person using a ruler to measure distances[4]. A third consideration is that the products finished by the artists (as is the case in many traditional contexts) often have important cultural significance. Finally, it is important to mention that many kinds of weaving and embroidery designs are very

[4] From discussions with Otomi artists and with Prof. Richard Ramsey (personal communication), some Otomi artists now calculate thread counts on paper, and beginning artists sometimes work from a previously finished design. Nevertheless, most experienced textile artists still create their work completely from memory.

NOTES

time consuming to make. It is not unusual for traditional textile artists to take several weeks to make a work of their art.

We start with weaving, a textile art that is common in Otomi culture. Weaving is a practice that dates back to pre Columbian time, though some European techniques have been introduced to it over the years. The mathematical aspects of weaving involve a lot of thread counting and keeping track of those thread counts. Conversations I had with traditional Otomi weavers in Tolimán (in the Mexican state of Querétero) in 2001 revealed that in the past Otomi weavers kept track of all thread counts by memory. Jacques Soustelle (1993) [1937] published one of the most complete anthropological studies of Otomi culture and in it he describes his observations of traditional Otomi weavers and embroiderers. On page 94 he states, "La base es la cuenta de hilos. . . . La memoria de los tejedores Otomies es tan perfecta, sus cuentas de hilos tan exactas, que . . . , queda perfectamente simétrico." This translates to "[the art of weaving] is based on thread counts. . . . . The memory of the Otomi weavers is so perfect, their thread counts are so exact, that . . . , [the final product] ends up perfectly symmetric." The description by Soustelle indicates that what the Otomi artists are doing does in fact require mathematical processes of organizing information (via the thread counts) and knowledge of concepts of symmetry.

Embroidery is a related textile art for which Otomi artists also typically have substantial skill. In the case of embroidery there is usually more complicated patterns, often including several figures that have properties of symmetry.

Let us look at some specific examples of Otomi embroidery. Figure 4 shows a cloth from an Otomi market in Ixmiquilpan, in the Mexican state of Hidalgo.

Figure 1. An Otomi cloth. Photograph by the author.

If we look at the pattern of the bird images that appear in the figure, we can see that they are symmetric about a vertical line that could be drawn down the middle of the figure. Also, they are horizontally symmetric, if we draw an imaginary horizontal line through the middle of the figure.

In fact, the person who embroidered this piece understands the concepts of vertical and horizontal symmetry. Evidence of this comes from observations by anthropologists. I quote (Soustelle, 1993) [1937]) again, p. 94, ". . ., se hace generalmente el motivo de manera que quede situado justo en medio del sarape, . . . , es necesario . . . hacerlo en dos etapas, una mitad sobre el borde derecho de la banda, la otra sobre el izquierdo." This translates to ". . . ., in general the pattern is made in such a way as to be situated in the middle of the shawl, . . . , and it is necessary . . . to make it in two stages, half on the right hand border and the other half on the left side." Thus, the artist creates half of the figure, then completes the other half. In order to make the figures symmetric, the artist must create the second half of the figure in the proper orientation. This means the artist must understand a *priori* how the symmetric pattern must look.

As for measurements, a close look at Otomi textile work shows very precise spacing, such as the sizes of figures, spaces between figures, and so on. Nevertheless, the artists create their works without the use of diagrams or rulers. The precise measurements are made from memory and from a lot of practice in the embroidery process. Figures 2 and 3 show other Otomi textile examples. I leave it to the reader to decide which of the symmetry properties (described for example in (Ascher, 1991; Chapter 6) these examples possess.

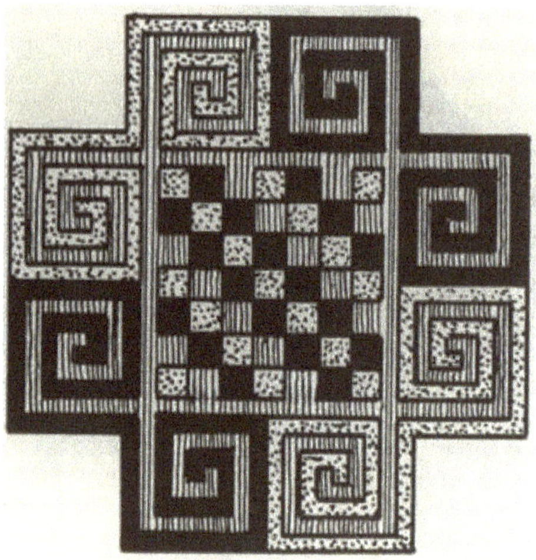

Figure 2. Otomi pattern from a shoulder cape. From Irmgard Weitlaner- Johnson, *Mexican Indian Folk Designs, 252 Motifs from Textiles.* Dover Publications, 1976, Figure 64, p. 24.

Figure 3. Otomi pattern from a shoulder cape. From Irmgard Weitlaner- Johnson, *Mexican Indian Folk Designs, 252 Motifs from Textiles.* Dover Publications, 1976, Figure 66, p. 25.

4. Otomi mathematics in the Codices.

Part of the long and complicated history of interaction between the Aztecs and Otomies involves the fact that much of Otomi territory was eventually controlled by the Aztecs which led to ongoing conflict between the two cultures. The Aztecs collected tributes from groups they controlled and descriptions of such tributes appear in various codices relevant to Aztec history. Several such

NOTES

codices contain mathematical information, and we will see that this information tells us something about Otomi mathematics.

First, a description of the mathematical symbols that appear in such codices is shown in Figure 4 below. These symbols appear most often in context of tributes that were given to the Aztecs from people in territory they controlled, and later, in the context of tributes given to local Spanish officials.

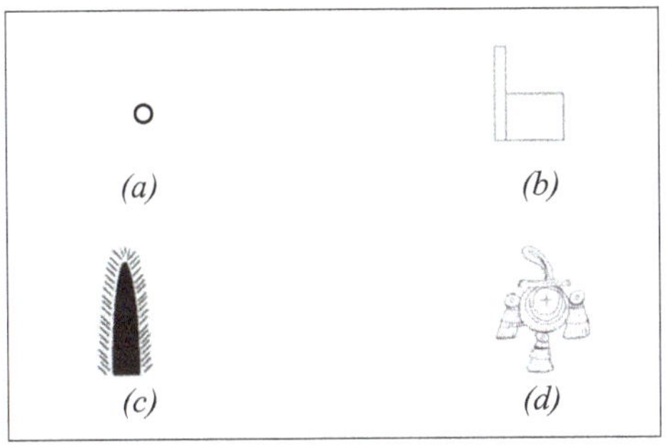

Figure 4. ***Mathematical symbols from Mesoamerican codices*****: (a): The symbol for 1 (units), (b): The symbol for 20, (c): The symbol for 400, (d): The symbol for 8000. These figures were made by the author, using pages 42 – 56 of (Ross , 1978) as examples.**

Figure 5 shows a folio (page) from the *Mendoza Codex* in which some mathematical symbols appear. This codex was written by a person of Aztec culture, shortly after the Spanish conquest of Mexico[5] (see Ross, 1978; p. 11).

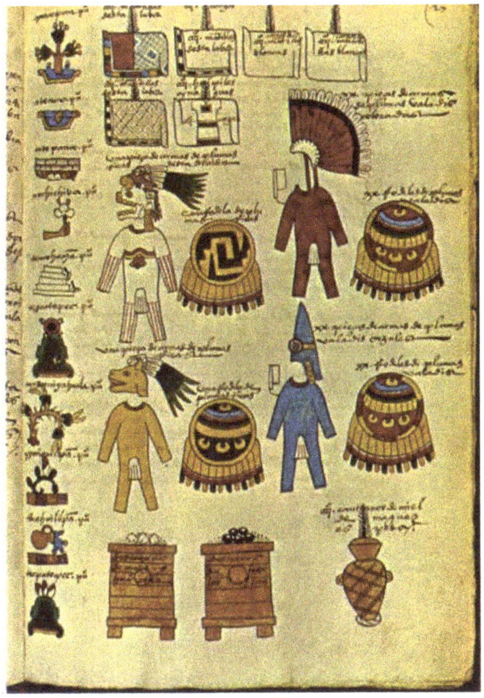

Figure 5. A folio from the Mendoza Codex. To the left are the glyphs for Mixquiahuala and Ixmiquilpan. This image is from: *Mendoza Codex*, (1978), with Comments by Kurt Ross, Figure (F), p. 46.

Several folios of the codex describe tributes the Aztecs collected from towns and locations they controlled. There is handwriting in several places. The handwriting was done by a Spanish person who gave explanations for the images. Observe the symbols for 20 and 400 in several places. Along the left side of the folio there are ten drawings with writing above each one. These drawings are glyphs that identify which towns and locations had delivered tributes that are described on that folio. The third and fourth glyphs from the bottom are those of Ixmiquilpan and Mixquiahuala[6], respectively. These two locations are in the Mezquital Valley, and have been considered part of Otomi territory since before the *Codex Mendoza* was written. Thus, in this folio some of the tributes were delivered by Otomies to the Aztecs. It is

[5] From 1519 to 1521.

[6] These glyphs are known to be those of Ixmiquilpan and Mixquiahuala, however if you look up the work by Ross and magnify those images, it is possible to read the words "Ixmiquilpan" and "Mixquiahuala" directly.

reasonable to conclude that the Otomies that were involved with that delivery must have understood the mathematical symbols that were used to represent the quantities.

Below I will describe another codex that we could use as a comparison, but first, some more details about the folio from the *Mendoza Codex* are in order. Notice the images near the top of the folio. They are square- looking images with a symbol of 400 above each one. In fact, these images represent textiles that were given to the Aztecs as part of the tributes they collected. According to (Ross, 1978: p. 37), the icons with colored patterns in them represent such things as mantles and tunics that have some pattern to them. Ross explains (ibid, p. 46) that such textiles were "richly embroidered." We can interpret "richly embroidered" to mean that the textiles contain several patterns. As we know from the discussion of Otomi art, this means the textiles likely included properties of various kinds of symmetry, and took a long time to make. The blank icons represent larger mantles without special patterns. There are a total of 6 x 400 = 2400 total textiles that were delivered as tributes by the ten locations to the left. The images below the textiles represent war suits (also called war dresses) and shields. On page 37 of (Ross, 1978), he tells us that tributes of mantles were collected every six months and the war suits every year.

Two such images have the symbol for 20 connected to them, so more than 40 such suits were delivered to the Aztecs. From the previous section we know that creating these textile products is a complicated process that includes mathematical concepts and thinking. Also, making more than 2420 mantles, tunics, war dresses, and shields represents a lot of work by many people! The fact that the Aztecs collected them shows that they had an appreciation of the work involved in making them, and probably held the textile artists in high regard.

Figure 6 below shows a folio from a fragmented codex known as the *Mixquiahuala Codex*. The most recent and complete description of the *Mixquiahuala Codex* that I know of is (Hermann Lejarazu, 2001).

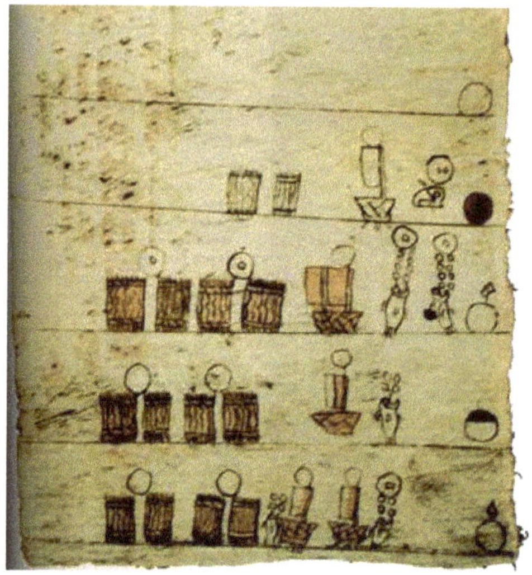

Figure 6. The Humbolt 7 Fragment, from the Mixquiahuala Codex.

On page 96 of Hermann Lejarazu's article he explains that the page known as the *Humbolt 7 Fragment* that appears in Figure 6 was written in about 1571. Above each drawing there we can see either some small circles, one or more flags, or a larger circle that may have a small circle drawn inside it. The larger circles with (in some cases without) smaller circles within them represent the amount in tomines (the local currency used at the time) that the local Spanish authority paid for the quantities of goods delivered. The figures immediately above each drawing is represented by either the symbol for one, or

NOTES

flags for twenty. As an example, the last row of figures represents (from left to right) four bundles of firewood, four fish, 20 baskets of tortillas, another 20 baskets of tortillas and four more fish. The symbol at the end of that line describes an approximate date of delivery in the form of a representation of the moon. What distinguishes this fragment from the folio of the *Mendoza Codex* of Figure 5 is that the *Humbolt 7 Fragment* represents an Otomi-Spanish communication. We can see that the pre- Hispanic symbols were still in use at time of the writing of this fragment. This gives us more confidence to conclude that the mathematical symbols for one, twenty, 400 and 8000 were known to and understood by people of Otomi culture. There are more examples of the use of these mathematical symbols by people of Otomi culture, such as in the *Osuna Codex*; see (Valle, 2001).

5. The Otomi calendar.

The Otomies used a calendar system that consisted of a 365 solar calendar and a 260 day ritual calendar. This structure is similar to many Mesoamerican calendars; see (Broda de Casas, 1969). Our discussion of the Otomi calendar here can be modified for consideration of some other Mesoamerican cultures such as the Aztecs or Mayans. First I will describe the basic structure of the Otomi calendar and later consider the mathematical aspects of Leap Year, and the implications of this question to interpreting the mathematics of the Otomi.

The primary source for information on the Otomi calendar comes from the *Huichapan* (pronounced "wee- CHAH- pahn") *Codex*. The *Huichapan Codex* was probably written sometime shortly after 1632, according to (Caso, 1967; p. 211). It is a post Conquest document that is written in Otomi with Spanish translations written near the Otomi text. Many anthropologists have studied it and have agreed that it was originally written by a person of Otomi culture.

The structure[7] of the Otomi solar calendar is that of 18 cycles of 20 days each for a total of 360 days, followed by 5 extra days added at the end for a total of 365. The ritual calendar is calculated as 13x20 = 260 days. The Otomi expression for a given year[8] appears as a drawing that consists of one to thirteen small circles placed near a glyph that represents one of twenty day names. Figure 7 below shows two such dates from the *Huichapan Codex*.

[7] The structure of the Otomi calendar follows a pattern similar to calendars used by many Mesoamerican cultural groups. There are unresolved questions about how the18 x 20 + 5 structure made its way across Mesoamerican cultures. Because the Otomies are a very old cultural group, further information about how they created their calendar system might lead to more insights into this issue.
[8] The Otomi description of years is also similar to that of other Mesoamerican calendars.

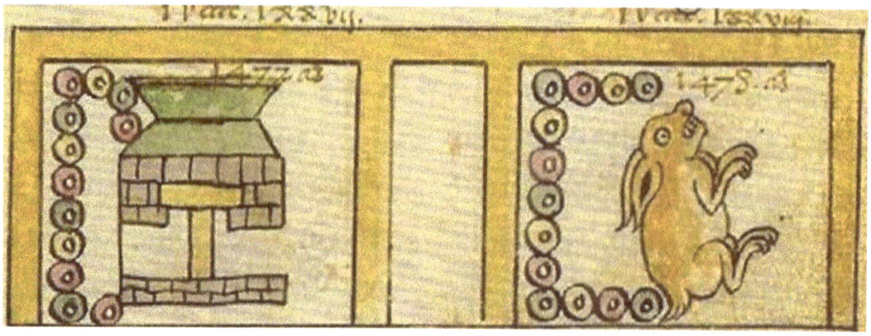

Figure 7. Two Otomi calendar dates from the Huichapan Codex.

From left to right, the two dates shown above are: 12 *House* and 13 *Rabbit*. We will see shortly that the explanation for using numerical values of one to thirteen and twenty day names is that by counting cycles of 365 days in this manner, it is easy to predict when the 260 day ritual calendar and 365 day solar calendar coincide. That coincidence occurs when the least common multiple (lcm) of both calendar cycles agree, which is:

$$lcm(365, 260) = lcm(73x5, 13x20) = 13x73x20 = 18{,}980 \text{ days} = 365x52 = 260x73.$$

Thus, every 52 solar years the two calendars coincide.

Now let us look at the description of the Otomi years. Table 4 shows the list of all twenty day names used by the Otomies in their calendar.

Table 4. The twenty day names of the Otomi calendar. Translated from (Soustelle, 1993 [1937]; p. 519).

Day	Name	Day	Name
1	Knife	11	Monkey
2	Wind	12	Yellow herb
3	House	13	Cane (sugar)
4	Lizard	14	Bite
5	Snake	15	Eagle
6	Death	16	Turkey
7	Deer	17	Earth movement
8	Rabbit	18	Flint
9	Water	19	Rain
10	Dog	20	Flower

NOTES

In descriptions of Otomi years, such as in the *Huichapan Codex*, there are only four day names that are ever used to describe years: *House, Rabbit, Cane*, and *Flint*. Why only four? The explanation is that for 365 days, we have 365 = 360 + 5 = 18x20 + 5, so after five cycles of 20, there are five more days that are named. Because 20/5 = 4, it leaves only four choices for the name of the day at which the new year begins. Observe that this way of using four day names to denote years and up to thirteen numerical symbols means that the solar and ritual calendars will coincide when the first pattern repeats; that is, when the thirteen number symbols have cycled four times for a total of 52 solar years.

Now it is time to consider the question of Leap Year in the Otomi calendar. Because of the proximity of the writing of this codex and the declaration in Europe of the Gregorian calendar (by Pope Gregory VIII in 1582), there was likely a strong interest in whether calendars of indigenous groups of the Americas had made adjustments to their calendars to account for what Western calendars refer to as Leap Year: The addition of an extra day every four years, with minor adjustments. In the case of the Otomi calendar, the author of the *Huichapan Codex* states in Otomi on folio 13: (Lastra y Bartholomew, 2001; p. 43):

> *"N˜uccãdaandaqhueya. Edettatemahiãntitemaquüttamapa. ccclxv yquüttzi edato oras. ãnãbeattegui. Nãh˜u oras. Emãh˜equi e oras."* This statement is followed by written comments in Spanish: *"Cada año [consta de] dieciocho veintanas más cinco días [= 365] ccclxv . . se añaden seis horas [del] reloj . . ., se llaman horas iguales horas."*

The above translates to an explanation of a calculation of Leap Year by adding six hours to each year, the total over four years giving 6x4 = 24 hours. This would seem to imply a calculation of Leap Year in the Otomi calendar. Although there seems to be some logic in this description, from a cultural-mathematical point of view, we are led to doubt it[9]. The first problem is that there is no evidence that the Otomi divided their days into smaller units ("hours"). Although this is certainly possible (e.g., by observing shadows on a fixed object), the Otomies would not have made a calculation of 6 x 4 = 24 in this context. Indeed, we have seen that the Otomies use a 5- 10- 20 number system in which 24 would not be a natural choice.

The fact that only four names are used to describe years also tells us something about the possibility of adding a day to calendar after every four years, as would be done to include Leap Year as it is known in Western terms. In fact, adding one more day would break the pattern of using only four day names to describe years. For example, if at the end of a year 7 *Cane* there would be a Leap Year calculation, the following year would then be named 8 *Rain* because of the addition of one day. However, as far as we know the Otomies never made this change in their descriptions of years.

Our final conclusion of the 6x4 = 24 statement that appears in the *Huichapan Codex* is that the Otomies did not measure Leap Year this way. The question then becomes: Did the Otomies account for Leap Year in some other way? We must be careful not to conclude that just because a culture does not make a Western style calculation, it does not imply that the culture is ignorant of the relevant mathematics. It is widely known that many Mesoamerican cultures were very knowledgeable of astronomy and cycles of the sun, the moon, Venus, and other astronomical phenomena such as equinoxes and solstices. Moreover, the Otomi calendar was very closely connected to agricultural cycles, as described for example in (Albores, 2006). Hence, in order to maintain their

[9] Jacques Soustelle (1993 [1937]; p. 520) raises similar doubts in his discussion of the Otomi calendar. Our observations here are mainly on the mathematical implications of making such a calculation.

calendar as an accurate tool for agriculture, the Otomies must have adjusted it from time to time. How the adjustments were made is not known. This question is addressed to some extent in (Broda de Casas, 1969), and Beatriz Albores describes a possibility that some Mesoamerican cultures related to the Otomies may have used the cycle of Venus to adjust their solar calendar; see (Albores, 2006).

Acknowledgements:

I would like to acknowledge research support for this paper as part of a Fulbright García Robles Scholarship at the Instituto Tecnológico Autónomo de México (ITAM), Mexico City, during the 2006- 2007 school year. I am also grateful to the University of North Dakota for Developmental Leave support during that time as well. Finally, I would like to thank the people who participate in the annual Coloquio Otopame meetings in Mexico, and those who participate in the Otopame seminar at the National University of Mexico (UNAM) in the Instituto de Investigaciones Antropológicas (Institute of Anthropological Studies). The many discussions regarding Otomi culture have been very helpful to me in preparing this paper.

References

Acuña, René. 1990 *Arte breve de la lengua Otomi y vocabulario trilingüe, Español- Náhuatl- Otomi, de Alonso Urbano, 1605*. Reprinted by the National University of Mexico (UNAM), Mexico City.

Albores, Beatriz. 2006 *Los graniceros y el tiempo cósmico en la región que ocupó el Matlanzinco. In Estudios de Cultura Otopame*, vol. 5. Mexico City: Instituto de Investigaciones Antropológicas, UNAM, 71 - 117.

Ascher, Marcia. 1991 *Ethnomathematics: A Multicultural View of Mathematical Ideas*. Pacific Grove: Brooks/Cole.

Barriga Puente, Francisco. 1998 *Los Sistemas de Numeración Indoamericanos; un enfoque areotipológico*. Mexico City: UNAM.

Bartholomew, Doris. 2000 *Intercambio lingüístico entre Otomi y Náhuatl*. In *Estudios de Cultura Otopame*, vol. 2, Y. Lastra and N. Quezada, Eds. Mexico City: UNAM, 189 - 201.

Broda de Casas, Johanna. 1969 *The Mexican calendar as compared to other Mesoamerican systems*. Acta Ethnologica et Linguistica, Series Americana, vol. 4, Wien: Institut für Völkerkunde der Universität Wien.

Caso, Alfonso. 1967 *Los Calendarios Prehispánicos*. Mexico City: UNAM, Institutode Investigaciones Históricas.

Closs, Michael P., Ed. 1986 *Native American Mathematics*. Austin: University of Texas Press.

Gilsdorf, Thomas E. 2008 *Etnomatemáticas de los Otomíes*. In *Estudios de Cultura Otopame*. Vol. 6, Mexico City: UNAM, 165 - 181.

NOTES

Hermann Lejarazu, Manuel. 2001 *Códices tributaries de Mezquiahuala.* In *Códices del Estado de Hidalgo, State of Hidalgo Codices.* Laura Elena Sotelo Santos, Victor Manuel Ballesteros García, Evaristo Luvián Torres, Eds. Pachuca: UAEH, 88 – 99.

Lastra, Yolanda. 2006 *Los Otomies, su Lengua y su Historia.* Mexico City: Instituto de Investigaciones Antropológicas, UNAM.

Lastra, Yolanda, y Bartholomew, Doris, Eds. 2001 *Códice de Huichapan. Paleografía y traducción* (of notes by Lawrence Ecker). Mexico City: Institutode Investigaciones Antropológicas, UNAM.

Merrifield, William R. 1968 *Number names in four languages of Mexico.* In Grammars for Number Names, H. Brandt Corstius, Ed. Dordrecht: D. Reídle Publishing Company.

Ross, Kurt. 1978 *Codex Mendoza, Aztec Manuscript.* Miller Graphics Liber S. A., CH- Fribourg.

Smith, Michael. 2003 *The Aztecs, Second Edition.* Malden: Blackwell Publishers.

Soustelle, Jacques. 1993 *La famille Otomi – pame du Mexique Central*, Travaux en Mémoires de l'Institut d'Ethnologie, Paris, 1937. Translation into Spanish: *Centro de Estudios Mexicanos y Centroamericanos*, Fondo de Cultura Económica, 1993.

Thomas, Cyrus. 1897 *Numeral systems of Mexico and Central America.* In Nineteenth Annual Report, Bureau of American Ethnology. Washington, DC.

Valiñas Coalla, Leopoldo. 2000 *Lo que la lingüística yutoazteca podría aportar en la reconstrucción histórica del Norte de México.* In *Nómada y Sedentario en el Norte de México, Homenaje a Beatriz Branniff.* Marie- Areti Hers, José Luís Mirafuentes, María de los Dolores Soto, Miguel Vallebueno, Eds. Mexico City: UNAM, 175 – 205.

Valle, Perla. 2001 *Códice Osuna.* In *Códices del Estado de Hidalgo, State of Hidalgo Codices.* Laura Elena Sotelo Santos, Victor Manuel Ballesteros García, Evaristo Luvián Torres, Eds. Pachuca: UAEH, 56 – 63.

Weitlaner- Johnson, Irmgard. 1976 *Mexican Indian Folk Designs, 252 Motifs from Textiles.* Mineola: Dover Publications.

Wright, David. 1997 *Manuscritos Otomies del Virreinato.* In *Códices y Documentos sobre México, Segundo Simposio, Volumen II*, Rueda Mithers, Salvador, Vega Sosa, Constanza, and Martínez Barca, Rodrigo, Eds., Mexico City: InstitutoNacional de Antropología e Historia (INAH), 437 – 462.

Wright, David. 2005 *Hñahñu, Ñuhu, Ñhato, Ñuhmu.* In Arqueología Mexicana, vol. 13, No. 73, May - June, 19.

A Etnomatemática na Comunidade Piscatória de Câmara de Lobos em Portugal

The Ethnomathematics of a Fishing Community at Câmara de Lobos in Portugal

Pedro Palhares
CIEC, Instituto de Educação, Universidade do Minho, Portugal
palhares2307@gmail.com

Filipe Sousa
CIEC, Instituto de Educação, Universidade do Minho, Portugal
filipe.fcm@gmail.com

Abstract

The mathematics education research community has been discussing the issue of the connections between mathematics and culture. This is the main concern underneath this work, as it tries to reveal everyday life mathematics of a fishing community in Câmara de Lobos, Madeira, Portugal. The study followed a qualitative approach based on an ethnographic model. In order to collect data the researcher remained during long periods of time in the field, using participant observation, interviews, photographic and video registers and document collection. Data analysis was of an interpretative nature. We focus here on the mathematical notions of ratio and fraction and how they are used within the community, especially by one caulker living there.

Resumo

A comunidade de Investigação em Educação Matemática tem vindo a discutir as conexões entre a matemática e a cultura. Esta é a preocupação por detrás deste trabalho, que tenta revelar as práticas matemáticas do quotidiano de uma comunidade piscatória em Câmara de Lobos, Madeira, Portugal. Este estudo seguiu uma linha qualitativa baseada num modelo etnográfico. De forma a coligir dados, o investigador permaneceu durante longos períodos de tempo no local, usando observação participante, entrevistas, registos fotográficos e vídeo e recolha de documentos. A análise de dados seguiu uma linha interpretativa. Neste artigo, acabamos por focar nas noções matemáticas de fracção e razão e como são usadas na comunidade, em especial pelo calafate desta comunidade piscatória.

Introdução

O conhecimento matemático é algo que o Homem utiliza ao longo dos tempos, de uma forma mais ou menos rudimentar, mais ou menos profunda. Desde as mais simples contagens dos animais dos seus rebanhos aos complexos cálculos utilizados na construção de uma estação espacial, passando pela utilização de aspetos de simetria em produtos artesanais que o Homem recorre à sua competência matemática.

Neste artigo, procurar-se-á mostrar as competências matemáticas que a comunidade piscatória de Câmara de Lobos utiliza no seu quotidiano, e com base nos propósitos da etnomatemática, reflectir sobre a sua utilização em contexto escolar.

Etnomatemática

Em muitos países, a matemática é vista como algo criado e desenvolvido na Europa, ou de forma mais geral, na civilização dita ocidental. Mas através de uma enorme variedade de estudos feitos, verificou-se que além dessa matemática, subsiste uma matemática indígena (Gerdes, 1999), mas também uma matemática praticada por grupos culturais, como por exemplo, o cálculo mental da comunidade cigana no norte de Portugal (Cadeia, Palhares e Sarmento, 2008), do movimento dos sem-terra (Knijnik, 2008) e de laborais específicos como pescadores, calafates, serralheiros, carpinteiros, latoeiros, tanoeiros, em muitos casos com expressão nos artefactos decorrentes da atividade específica (Palhares, 2010). Essa matemática, embora não seja aprendida nas escolas, mas no ambiente da família ou da comunidade ou no do trabalho (D'Ambrósio, 2002) proporciona conhecimentos matemáticos com utilização prática no quotidiano.

Para D' Ambrósio (2006), o termo etnomatemática significa a arte ou técnica de explicar, de entender, compreender a realidade, num contexto cultural próprio. Neste trabalho, a etnomatemática enquadra-se na compreensão da realidade do quotidiano da comunidade piscatória de Câmara de Lobos, na procura de conhecimentos e processos matemáticos presentes em minorias sociais, mas também assumindo uma utilização futura desses conhecimentos e processos na educação matemática (D'Ambrosio, 2002; 2006; 2007).

Resumidamente, a etnomatemática assume um carácter investigativo da matemática presente em minorias sociais, como por exemplo, a comunidade piscatória de Câmara de Lobos e, também, um carácter educacional que pretende servir-se desse conhecimento matemático informal para o relacionar com a matemática escolar (D'Ambrósio, 1993; 1998; 2002; Monteiro, 2004; Nunes, 2010; Oliveira, 2004). O carácter investigativo, enquadra-se na recolha de informação directamente dos vários elementos da comunidade piscatória, de práticas matemáticas do quotidiano [(etno) matemáticas[1]] utilizadas nas suas actividades profissionais. O carácter educativo enquadra-se na análise dessas (etno)matemáticas e na verificação da presença destas nos Programas de Matemática do Ensino Básico.

[1] O termo (Etno)matemática é assumido aqui como a prática matemática desenvolvida e usada numa determinada cultura ou grupo social, sem que seja utilizada de uma forma *educacional*, por exemplo, em contexto de sala de aula.

NOTES

Na Comunidade Piscatória de Câmara de Lobos, o sucesso escolar dos alunos é bastante insatisfatório, triste panorama, em que se inclui a educação matemática. O abandono escolar é uma preocupação. Muitos jovens abandonam a escola para se dedicarem a outras atividades que são normalmente ligadas ao mar. Assim, esta investigação baseia-se nos pressupostos da etnomatemática para tentar conhecer a matemática utilizada na Comunidade Piscatória de Câmara de Lobos, questionando sobre a sua possível utilização para melhorar o ensino da matemática e o sucesso escolar dos alunos desta comunidade. Tal como Bishop (2010), entendemos que urge construir um currículo matemático balanceado, que respeite e legitime as práticas matemáticas locais dentro do contexto escolar, sendo este um primeiro passo nesse sentido da nossa parte.

Metodologia

Neste estudo, desenvolve-se um trabalho de natureza qualitativa. A investigação qualitativa representa um processo sério, rigoroso, carregado de dúvidas e inseguranças (Garcia, 1992). A sua característica fundamental é a flexibilidade, a capacidade de adaptar-se a cada momento e circunstância em função da troca que se produz na realidade que se está indagando (Gomez, Flores & Jinenez, 1996).

A investigação qualitativa valida a utilização de variados desenhos de investigação. Neste trabalho, o recurso metodológico utilizado foi a etnografia, porque propõe-se apreender a vida, tal qual ela é quotidianamente conduzida, simbolizada e interpretada pelos actores sociais nos seus contextos de acção (Sarmento, 2003). Optou-se pela descrição densa, pelo facto de se enquadrar com a abordagem etnográfica. Mas também porque, a tarefa do etnógrafo é descrever as práticas ou concepções dentro do contexto da outra cultura (Barton, 2004).

De um modo particular, utilizou-se a observação participante, as entrevistas não estruturadas e a análise de documentos principalmente os oficiais ligados à construção de barcos. Por vezes, quando em grupo, as respostas de um entrevistado podem influenciar as respostas dos restantes, tornando a informação contorcida, não correspondendo fielmente à realidade. Para evitar o enviesamento dos dados, efectuaram-se entrevistas em todos os momentos do trabalho de campo, sendo aplicadas aleatoriamente ao calafate e pescadores, em momentos e entrevistados distintos. Em conjugação com as entrevistas, utilizou-se a observação participante e a análise de documentos, de forma a complementar as restantes técnicas de recolha de dados. Também se efetuaram registos fotográficos, áudio e vídeo.

Câmara de Lobos

A Madeira é um arquipélago, habitado desde a sua descoberta pelos navegadores portugueses no início do século XV. É composto por quatro ilhas, duas das quais desabitadas. A ilha da Madeira é a mais populosa, com cerca de 250 mil habitantes. A principal atividade económica é o turismo. No entanto, a pesca é uma parte importante da ocupação das pessoas. Existem na ilha, três zonas nas quais a pesca é a atividade principal, sendo uma delas a comunidade piscatória de Câmara de Lobos. Esta pequena vila tem cerca de 30 mil habitantes, dos quais dois terços são analfabetos. A comunidade de pescadores vive na periferia da vila, perto do mar, numa pequena superfície rochosa e tem desenvolvido um estilo de vida independente. Fechada em si própria, tem o seu próprio dialeto, o seu código não-escrito e sentido de honra, que inclui uma enorme solidariedade interna quando ocorre qualquer tragédia (Brandão, 2004; Fernandes, 2004). A figura 1 mostra alguns pescadores de Câmara de Lobos em meados do século XX.

Figura 1 - Pescadores de Câmara de Lobos em meados do séc XX (www.ilhas.org)

Esta pequena comunidade é de certo modo depreciada pelo restante sociedade, por se refugiar no seu mundo, constituindo uma colónia pouco aberta ao convívio e usos dos restantes habitantes da localidade (Lamas, 1956). No entanto, adquiriu uma identidade e uma cultura de tal forma distinta e vincada que consegue por vezes sobrepor-se à restante sociedade, tornando-se o ponto de referência de todas as curiosidades e observações. Aos pescadores são atribuídos quatro epítetos depreciativos: *pesquito* (pescador); *chernoca* (pequeno cherne); *xavelha* (nome de embarcação) e *tangerino* (de origem africana) (Lamas, 1956).

A vida de pescador é difícil como é evidenciado por um deles:

> *É muite sacrifície. Muites dia sem tomar banhe, os dentes começam a apodrecer, depois começa a se sofrer do estômague, comeres em cima de comeres, nãm há higiéne, o sône também é 3; 4; 5 dias quase sem dormuir (...) Iste é tude ruim*[2].

Eles sentem que têm sido explorados. Em tempos monárquicos, centenas de anos atrás, houve um imposto especial sobre os seus rendimentos. Atualmente, não é especialmente tributado, mas é de senso comum que o proprietário do barco e o comerciante, enriquecem e somente o pescador continua pobre (Brandão, 2004).

Em Câmara de Lobos, o peixe capturado é, predominantemente, o peixe-espada preto. O seu nome científico *é Aphanopus carbo*, que significa, *sem pés aparentes*. Na verdade, este peixe não tem barbatanas ventrais ou membros posteriores. Popularmente é denominado por *peixe-espada preto* pela sua forma alongada e fina e pela sua cor escura. É um peixe que mede cerca de 1 metro de comprimento e 10 centímetros de largura na parte abdominal. Possui pele negra e viscosa, cabeça alongada e feroz com os seus dentes aguçados e enormes e olhos esbugalhados. É de grande importância na alimentação dos madeirenses, a ponto de lhe chamarem "o trezentos e sessenta e cinco" (Lamas, 1956) por ser alimento de pobres e remediados durante os 365 dias do ano.

Atualmente, e à semelhança de há muitos anos, "a venda do peixe ao público, depois da lota na praia, é feita por negociantes, nos mercados das mais importantes localidades da ilha e por vendedores ambulantes" (Lamas, 1956). Há até mesmo um dia especial para comemorar. É a festa do peixe-espada preto. Esta festividade acontece anualmente durante

[2] Efetuou-se transcrições fonéticas, para caraterizar a especificidade da linguagem na comunidade piscatória de Câmara de Lobos.

NOTES

a última semana de Julho. É uma tradição com cerca de 200 anos de existência, o que é para os pescadores motivo de muito orgulhoso.

Os Pescadores e as suas etnomatemáticas

Uma das coisas que os pescadores têm de aprender é como dividir o dinheiro. Após a venda do peixe, ocorre o seguinte processo. Eles duplicam o número de pescadores do barco e dividem o valor total do peixe vendido por este número. Cada pescador recebe uma parte. O restante é dividido em dois. Uma das metades, o proprietário do barco utiliza para pagar todas as despesas relativas à faina, a outra metade ele guarda. Pela época natalícia, todo o dinheiro guardado é distribuído por todos os que participaram nas fainas ao longo do ano. Vejamos um exemplo:

Se a embarcação for constituída por 11 pescadores, a quantia recebida é dividida equitativamente por 22. Dessas 22 partes, é entregue a cada um dos pescadores uma parte, ou seja, os 11 pescadores arrecadam 50% do total recebido na lota, que corresponde a 11 partes. A restante metade é dividida em duas partes iguais. Uma das partes (50%) destina-se ao pagamento de despesas de alimentação e manutenção da embarcação, como por exemplo, o gasóleo, os seguros, as licenças, a pintura, o material de pesca, etc. A outra metade (50%) é guardada cumulativamente nas várias fainas pelo armador da embarcação, que será distribuída em partes iguais por todos os tripulantes na época de Natal. É o 13.º mês ou subsídio de Natal dos pescadores, que é denominado de *paga*. Neste caso, os pescadores não calculam a parte dos 50% que o armador guarda, pois determinam 50% do somatório das quantias que vão recebendo em cada faina. Por exemplo:

> *O Sr. A perguntou-lhe se tinha rendido bem. O pescador respondeu que tinham pescado 4000 espadas. O Sr. A. com cara de admiração respondeu:*
>
> *- Bem bom! Já dá para ganhar perto de 500 euros. Entretanto os jovens ausentaram-se e o Sr. A. contou:*
>
> *- Tá vende, tiveram no mar 12 dias e deu pra ganhar 500 euros. E depois ainda recebem a pague que é 50% do que se ganhe.*

Apresenta-se um exemplo concreto da divisão do dinheiro pelos pescadores. Supondo- se que determinada faina rendeu 8000 euros e que a embarcação tem 5 pescadores, estes fazem os cálculos seguintes:

- 8000 euros ÷ 10 pescadores = 800 euros

Cada pescador recebe de imediato 800 euros. Os 5 pescadores recebem 50% do total facturado. Os outros 50% que correspondem a 4000 euros, são divididos em duas partes iguais.

- 4000 euros ÷ 2 = 2000 euros
- 2000 euros – Despesas
- 2000 euros – Paga
- 2000 euros ÷ 5 pescadores = 400 euros

Dessa maneira, cada pescador, relativamente a esta faina, tem direito a 800 euros de imediato, mais 400 euros em Dezembro.

O armador fica com o dinheiro não entregue aos pescadores, pelo que metade desse dinheiro é para este pagar todas as despesas necessárias, a outra metade será para juntar ao longo do ano até Dezembro, mês em que divide o total acumulado equitativamente pelos pescadores.

No entanto, os restantes pescadores calculam de forma diferente. Considerando como rendimentos mensais por pescador, os constantes na tabela 1, o pescador adiciona todos os rendimentos mensais e calcula metade dessa soma, ficando a saber o valor da *paga*. O quadro 1 mostra o rendimento mensal de cada pescador em euros.

Quadro 1: Rendimento mensal de cada pescador em euros

Jan	Fev	Mar	Abr	Mai	Jun	Jul	Ago	Set	Out	Nov	Dez
350	420	640	480	760	890	320	200	830	600	1500	900

Por exemplo,

- 350 + 420 + 640 + 480 + 760 + 890 + 320 + 200 + 830 + 600 + 1500 + 900 = 7890
- 7890 ÷ 2 = 3945 - Valor da *paga*
- 7890 + 3945 = 11,835

O rendimento anual de cada um dos pescadores seria 11,835 euros.

Outra prática que está presente entre eles é a conversão entre unidades de medida, que é efetuada num processo de aproximação, suficientemente adequada na maioria dos casos. A braça, é uma unidade de medida utilizada pelos pescadores de Câmara de Lobos para medir a profundidade a que o peixe se encontra no mar e que nesta comunidade é apelidada por *braçada*. A braça é assumida por estes pescadores como sendo a medida de comprimento entre as extremidades dos dedos médios das mãos, quando os braços estão abertos e na posição perpendicular em relação ao tronco.

> *Sr. J.: Com a colaboraçãm deles vocês ficam a saber come é a espade, qual é a fundure (...) a fundure é o quê? Mil metres nãm é? Mil...mil e quinhentes metres...Com mil metres já se mate espade...*
>
> *Sr. A.: Com seiscentes braçades (...)*
>
> *Sr. J.: Ume braçade dá um metre (...) um metre e tal (...) (exemplificou abrindo totalmente os dois braços) (...) um metre e meie (...) equivale a seiscentes braçades (...)*
>
> *Investigador: Um metro e meio são seiscentas braçadas?*
>
> *Sr. J.: Ume braçade é um metre e meie (...)*
>
> *Sr. A.: Um metre e meie... ume braçade... Ume braçade é um metre e meie (...) na sonde (...) O mínime é seiscentes braçades, novecentes, mil braçades (...)*

Um dos intervenientes não é pescador (Sr. J.), revelando alguma confusão no seu discurso. O outro (Sr. A.) é pescador e evidencia os seus conhecimentos relativo à medida, afirmando que uma braçada é 1,5 metros e que o peixe-espada preto é pescado entre as seiscentas e as mil braçadas.

O Calafate

Na comunidade piscatória há um calafate, o Sr. J., que tem vindo a construir barcos de pesca. Ultimamente construiu uma caravela, navio utilizado nos descobrimentos portugueses, para uma exposição em Lisboa. A entrada de Portugal na União Europeia introduziu mudanças no seu trabalho, pois agora todas as embarcações são construídas respeitando copiosamente os projetos

NOTES

que lhe são enviados. Neste artigo, serão analisados os procedimentos utilizados na construção de barcos sem recurso a projetos, ou seja, utilizando apenas os conhecimentos na construção de barcos que foram sendo transmitidos verbalmente de geração em geração e interiorizados com a prática do quotidiano. Estas técnicas de construção de barcos, não se socorrem da consulta de qualquer documento elaborado por engenheiros navais, onde se podem consultar medidas, cálculos, plano geométrico e memória descritiva da embarcação. Na construção de barcos os calafates servem-se principalmente de conhecimentos geométricos, proporcionalidade direta e o cálculo mental utilizado de forma aproximada.

Uma fase importante na construção de um barco é a colocação da principal caverna. O Sr. J. alerta que a principal caverna tem que ser colocada rigorosa e obrigatoriamente no ponto médio da quilha e não no ponto médio da embarcação como alguns calafates erradamente fazem. Por exemplo, a entrevista com o Sr. J. revela esse a aspecto:

> *(...) Sr. J.: Há pessoues que ai vezes... há barques que fiquem defeituoses, muite largues à frente. É porque há pessoues que agarram na caverne, na principal caverne do meie e puxam um bocadinhe mais pá frente e fique um bocadinhe mais abolade. Tem que ser sempre centrade ao meie. Nãm é medir... nãm é medir o barque no comprimente e depois centrar a caverne ao meie... é medir a quila e entãm dar ao meie da quila, aí é que dá certe. Nãm é medir o barque e dizer: aqueí é que fique o meie. Com a fite por aqueí abaixe, põe o prume que aqueí fique o meie. Nãm, nade disse. A quila é que vale. A quila é éste parte de baixe (...).*

A fase seguinte destina-se a dar forma ao barco com a ajuda de ripas e a partir daí fazer moldes para as cavernas seguintes. Estas têm de ser colocadas de forma a que se aproximem o mais possível de uma reflexão relativamente à quilha, o eixo de reflexão. A figura 2 mostra a simetria de reflexão em relação à quilha do barco.

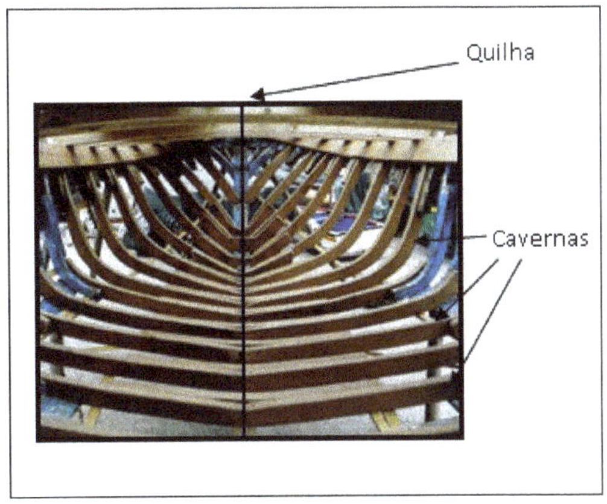

Figura 2: Simetria de reflexão relativamente à quilha

O que se pretende é que o ponto médio das extremidades superiores das cavernas *coincida* com a quilha. Mas há uma outra variável que influencia a *perfeita* reflexão das cavernas, que é a abertura que cada uma delas tem junto à quilha. A amplitude do ângulo formado pela quilha e a caverna do lado direito do barco, tem de ser igual à amplitude do ângulo formado pela quilha e a caverna do lado esquerdo do barco. A figura 3 mostra um barco em construção.

Figura 3: Barco em construção

O calafate parece ter em mãos um problema de difícil resolução uma vez que não utiliza transferidor e o esquadro não se adequa porque as cavernas não são fixadas à quilha na perpendicular. O calafate socorre-se de um instrumento denominado suta (figura 4) que lhe resolve o problema evitando as amplitudes.

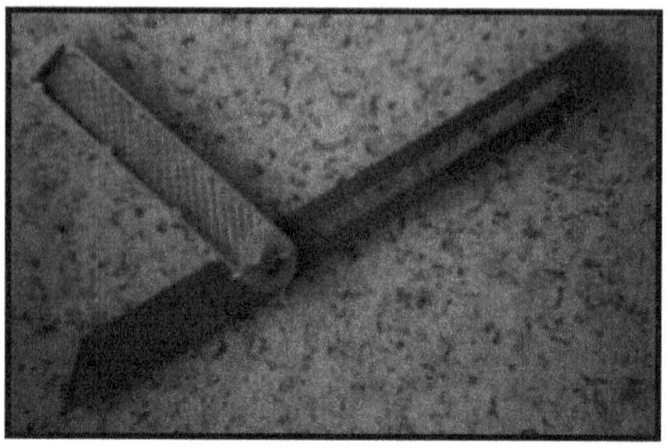

Figura 4: Suta, instrumento utilizado pelos calafates em Câmara de Lobos

Com a ajuda da suta, o calafate marca numa das cavernas a abertura necessária, fixa a abertura da suta de forma que as suas peças móveis não se desloquem e atribui à outra caverna a mesma abertura que aplicou na caverna simétrica.

Tanto para os calafates da Comunidade Piscatória de Câmara de Lobos como para os carpinteiros navais de Abaetetuba, a suta é um dos instrumentos mais interessantes para medir, comparar e indicar parâmetros de medidas de ângulos, tanto para a elaboração de peças como para a própria estruturação do barco (Lucena, 2002). Em Abaetetuba, no estado do Pará, no Brasil, existe um pequeno estaleiro de construção de barcos, onde os mestres artesãos executam o seu trabalho de forma ainda rudimentar e utilizando técnicas e conhecimentos que passaram de pais para filhos e assim sucessivamente. Em 2002, Lucena desenvolveu uma investigação com enquadramento na etnomatemática, que se centrou nas práticas e conhecimentos matemáticos dos carpinteiros navais usados na construção das embarcações.

Na comunidade piscatória de Câmara de Lobos, as principais fases de construção de um barco são análogas às que os artesãos utilizam em Abaetetuba. Isso também se verifica no que diz respeito à relação entre as três dimensões do barco como o comprimento, a boca e o pontal (Lucena, 2002). Por exemplo, um dos pescadores afirma que:

> *É porque é consoante o comprimente da embarcaçãm. A gente tem de (...) de (...) calcular sempre consoante o comprimente da embarcaçãm. O comprimente do barque é que mande a alture e a largure, tá a perceber? O comprimente mande tude, quase. Mande a largure ou boca e a alture, o pontal...que se chame o pontal.*

A *boca* é a maior distância perpendicular às bordas da embarcação, que coincide com o ponto médio do comprimento da quilha. A medição da *boca* é realizada exactamente no local da principal caverna. É aí que a embarcação tem a sua maior largura. A figura 5 mostra um esquema para calcular a *boca* de um barco.

Volume 2 Community

NOTES

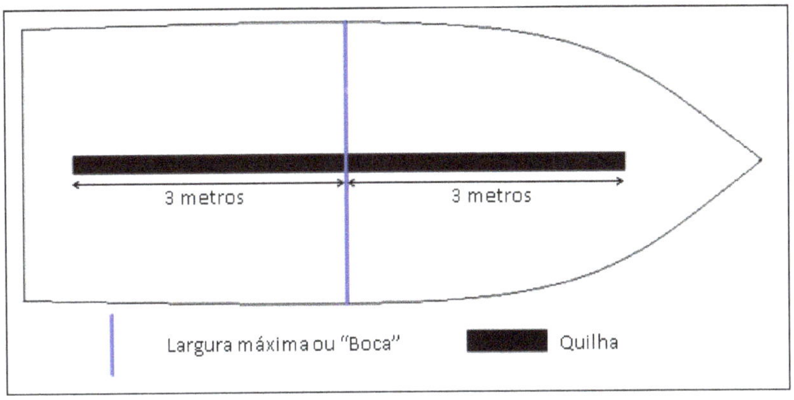

Figura 5: Esquema para calcular a *boca* de um barco

A altura do barco é medida a partir da parte superior da quilha até ao convés. Esta medida deve também efetuar-se no local da principal caverna. A altura é a medida de comprimento de um segmento de reta que é perpendicular ao segmento de reta, que representa a largura, e à quilha. As três dimensões acima descritas, o comprimento do barco, a largura e a altura, são os parâmetros mais relevantes na construção deste artefacto. A figura 6 ilustra a boca e o pontal, bem como os locais onde se devem efetuar as respectivas medidas na embarcação.

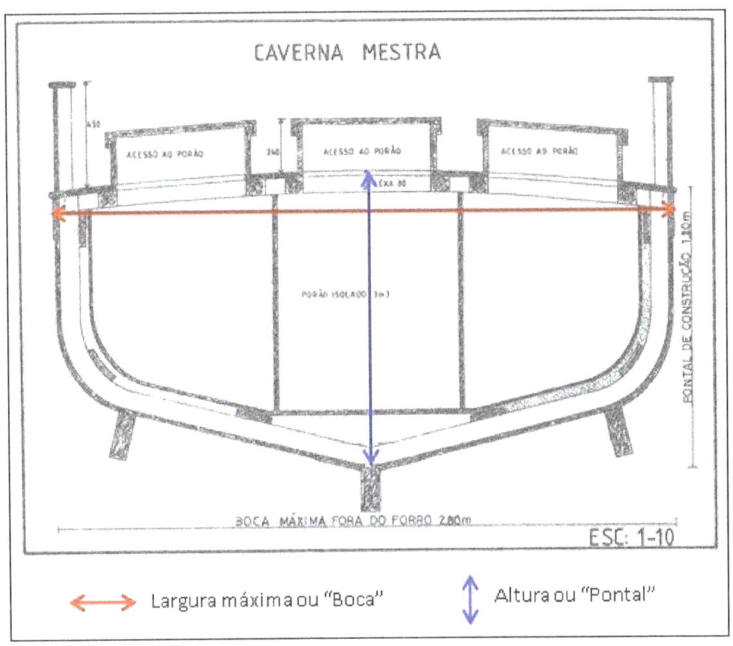

Figura 6 – Largura ou boca e Altura ou pontal de uma embarcação

No entanto estas dimensões não são estanques entre si. Existe uma íntima ligação que as relaciona do ponto de vista matemático. O Sr. J. explica que para um dado comprimento x, a largura ou *boca* tem de ser y e a *altura* ou *pontal* tem de ser z, com possíveis variações, mas mínimas.

Para o calafate a medida da largura deve ser cerca de um terço da medida de comprimento, pois afirma que:

> *Portante... é mais ou menes (...) Portante, nove... nãm, é mais um bocadinhe de um terce... é. É mais*

ou menes dentre de um terce, um bocadinhe mais, um bocadinhe menes... sãm aqueles barques feites de cabece, tá a perceber?

Quando calcula mentalmente as dimensões das várias partes do barco, o calafate termina a discussão, concluindo que a largura é um pouco mais do que um terço da medida de comprimento. O diálogo abaixo mostra um trecho da entrevista entre o Sr. J. e o investigador.

Sr. J.: Imagine um barque de (...) 12 metres. O barque tem 12 metres (...) ele tem de ter no máxime, no máxime (...) nãm desculpa (...) tem de ter (...) 12 metres de comprimente, ele tem de ter no mínime 4 metres e veinte centímetres de larque, no mínimo. Quatre metre e veinte de larque. E o pontal, ele tem de ter um metre e sessenta.

Investigador: Quatro metros e vinte de largo, para 12 de comprimento?

Sr. J.: Para 12 de comprimente. Mínimo.

Investigador: Pode ter mais (...)

Sr. J.: Pode ir até aos 4 metres e meie, também fique bem, pode dar mais 30 centímetres, 15 pra cade lade, fique bem.

Investigador: Por exemplo 4 metros já não pode ser...

Sr. J.: EEEhh (...) 4 metres nãm, ele fique muite estreite, fique um canude, chama-se um canude. (...) Nãm é precise desenhes. Aquile nãm foi precise desenhe e tá um barque bem armade acolá. Se você reparar bem ele tem consoante a largura, consoante o comprimente (...) tá a perceber?

Quanto à altura, também se aproxima de um valor proporcional, mas devemos lembrar que nenhuma das medidas é exata, pois são calculadas mentalmente por aproximação e estimativa.

Na tabela abaixo, podemos ver os valores para nove barcos, as medidas de cada dimensão, e as relações entre algumas delas. O quadro 2 mostra os valores das medidas utilizadas em nove barcos.

NOTES

Quadro 2: Valores das medidas utilizadas em nove barcos

Medidas (m)	Barco1 (Pré-1986)	Barco2 (Pré-1986)	Barco3 (Pré-1986)	Barco4 (Pré-1986)	Barco5 (Pré-1986)	Barco6 (Pós-1986)	Barco7 (Pós-1986)	Barco8 (Pós-1986)	Barco9 Caravela
Comprimento	6.3	9	10	12	20	7	8.9	9	22
Boca	2.2	3.2	3.8	4.2	6.9	2.8	3.6	3.7	7
Pontal	0.8	1.1	1.3	1.7	2.4	1.1	1.5	1.6	2.6
Comprimento /Boca	2.9	2.8	2.6	2.9	2.9	2.5	2.5	2.4	3.1
Comprimento /Pontal	7.8	8.2	7.7	7.1	8.3	6.4	6	5.7	8.5
Boca/Pontal	2.8	2.9	2.9	2.5	2.9	2.5	2.4	2.3	2.7

Após 1986, com a utilização de escalas, o quociente das razões *Comprimento/Boca* e *Boca/Pontal* é aproximadamente 2,5. A razão *Comprimento/Pontal* deve ser próxima de 6. Antes de 1986, estas razões são ligeiramente superiores (para segurança das embarcações), com as razões *Comprimento/Boca* e *Boca/Pontal* a aproximarem-se de 3 e a razão *Comprimento/Pontal* aproximadamente 8.

Na tabela 2 pode verificar-se que o trabalho executado pelo calafate após 1986 (barcos 6; 7 e 8, com utilização de escalas) demonstra mais rigor, relativamente às construções de 1 a 5. O barco 9 é um conceito totalmente diferente. Foi um pedido especial para a comemoração da chegada de Cristóvão Colombo ao continente americano, pelo que foi construída uma réplica de um dos barcos de sua frota. Este barco tem a estrutura apropriada a travessias de oceanos e possibilidade de ser carregado com um número considerável de toneladas de mercadorias. Os outros barcos, além der serem mais pequenos, são embarcações pesqueiras e com as características apropriadas à arte de pesca.

Após 1986, o calafate tem de considerar as escalas dos projetos. Embora tenha dificuldade em explicar o que é uma escala, consegue fazer muito bem equivalências entre medidas. Consegue fazer com rigor a medida no desenho e, consultando a escala, fazer corretamente os cálculos e indicar a medida real no barco. É provável que subsista aqui uma dificuldade discursiva, contudo existe uma utilidade prática de escala. A conversa abaixo revela esse fato:

> *Investigador: Então sem escala não conseguia fazer nada?*
>
> *Calafate: Nãm, nãm. É difícil, é difícil porque eu nãm sei qual é as medides da escale. Porque há uns (arquitectos) que fazem este desenho num tipe de escala 1: 20 e este noutre tipe de escale 1: 25, tá a perceber... é por isse (...) sem a escala o senhor nãm consegue só através de metre medir, agore com a escala (...)*

O calafate mostrou em várias ocasiões que sabia o conceito de escala. Numa das situações, faz alguns cálculos para explicar ao investigador.

> *Investigador – Então o que para si significa este 1?*
>
> *Calafate: O 1? O 1 pronte, quer dizer (...) 1 ponte veinte, 1 ponte queinze, 1 ponte dez, tem que ter sempre o 1 atrás nãm sei porquê...*
>
> *Investigador – E o 20 o que é que significa?*

Calafate: O 20 significa, portante na escale 1:20 o barque diminui, se for na escale 1:10 o plane geométrique já é maior, tá a perceber? (...) se reparar bem um metre no barque equivale aquei no metre a ceinque centímetres. Investigador: E 1 centímetro a quanto equivale?

Calafate: Um centímetre...pronte a pessoue tem que se ir lá de cabece buscar ehh (...) 1 centímetre equivale... porque ceinque equivale a um metre... um equivale a veinte centímetres (...)

Apesar de sua capacidade de calcular mentalmente alguns valores, o calafate não se limita apenas ao cálculo mental na construção dos barcos, pois usa uma régua de conversão para o trabalho de maior responsabilidade. É com este instrumento que o calafate confirma os cálculos que faz mentalmente. É de tal forma rigoroso que uma vez detectou um erro num projeto e corrigiu-o.

Numa das visitas a Câmara de Lobos, o investigador, em conversa com o Sr. J., pediu- lhe que exemplificasse e explicasse, os procedimentos e cálculos matemáticos utilizados no momento de construção de um barco. Num dos armários da sua pequena oficina remexeu uma quantidade de papéis dos quais escolheu um projecto de um dos barcos que tinha construído. Abriu-o em cima da sua bancada de trabalho, pegou nas suas réguas com as escalas, uma esferográfica e começou a medir e a explicar:

(...) vames lá ver. Por éste linhe fore, por éste linhe fore (...) espere, deixe-me pôr (...) (Abriu o projeto totalmente colocando-o sobre o balcão) (...) Este risque é que mande... dá 8 metres e novente, come você tá a ver (...) portante a escale equivale, 1 metre, (...) 2 metres, (...) 3 metres, (...) 4 metres, (...) 5 metres, (....) 6 metres. Acaba aquei. E depois vames aquei novamente, 7 metres, (...) 8 metres, (...) e aquei quer dizer 9 metres, neste risque, só que (...) portante (...) eeh, eeh, a rode de proua quase nãm conte, vinhe pr'aquei, dave 8 metres e novente. Dá 8 metres e novente, tá a perceber?

Numa das suas explicações algo curioso aconteceu. Quando o Sr. J. calculava a medida real da largura de um barco, partindo do desenho e utilizando a escala, o calafate conclui que o projecto não está com as medidas correctas, ou seja, as medidas da memória descritiva não coincidem com as medidas do desenho, fazendo os cálculos com a escala indicada no mesmo desenho. Na primeira vez que o Sr. J. fez os cálculos, achou que tinha sido erro de medida. Para esclarecer essa dúvida, o calafate mediu novamente no desenho e registou. Fez de novo os cálculos, mas igualmente não eram coincidentes com a memória descritiva. Uma vez retificadas as medidas, julgou ter errado nos cálculos, visto terem sido feitos de cabeça. Calculou uma vez mais as dimensões pretendidas, verificou se a escala a utilizar nos cálculos era a indicada no projecto e insistentemente os resultados obtidos não estavam de acordo com os do documento já aprovado pela Inspecção de Navios e Segurança Marítima. O Sr. J. deu um passo atrás, olhou fixamente o projecto amarelecido e afirmou "Tá dande errade (…) não dá certe, enganaram-se (…) sãm desenhes que pronte, é precise olhar (…) porque os desenhadores também se enganem, tá a perceber?". O Sr. J. explica o sucedido da seguinte forma:

NOTES

Sr. J.: Agore conte aquei asseim no meie do barque. Escale 1: 20, nãm é? Escale 1:20. 3 metres e 60. (Pausa) Tá dande errade. Portante, iste é o centre do barque (...) portante 1 metre, 2 metres, 3 metres, e 10, e 20, e 30, e 40 (...) nãm pode ser (...) 50, 60 (...) entãm é éste linha (...) mesme asseim pronte é pra você ver qu'eles às vezes nos projectos até enganam-se. Se for esta linha dá 3 metres e 55 (...) mesme asseim não dá certe, enganaram-se. (...) O barque neste momente tem trêis mestres e quarenta e dois e meie... neste momente a boca máxima tem... e aquei marca trêis metres e sessenta. (...) Investigador : Pois, ou se enganaram a fazer o desenho ou se enganaram a escrever a medida aqui. (...)

Sr. J.: Nãm eles enganaram-se foi aquei (no desenhe) porque eu lembro-me que fiz o barque com 1 metre e sessenta e passou na prova. E come você vê o desenhe tá aprovade.

Visto que o barco passou na prova de estabilidade e o projecto foi aprovado, o investigador, com algumas dúvidas, questionou o calafate sobre a forma ou estratégia utilizada na construção do barco. Que medidas ter em conta? As medidas do desenho, feito com todo o rigor de um desenhador profissional? As medidas da memória descritiva? Ou talvez enviar o desenho à entidade competente para que seja rectificado?

O Sr. J. já tinha construído este barco, por isso a resposta foi prontamente pronunciada. O calafate ignorou as medidas do desenho e utilizou as medidas da memória descritiva, fazendo todos os cálculos necessários com a escala indicada no projecto, em todas as peças de todas as fases de construção da embarcação. Este episódio revela que as práticas matemáticas e os conhecimentos do quotidiano do calafate, são vincados, e por vezes, sobrepõem-se às diretrizes de entidades superiores. Mostra ainda que o calafate possui conhecimentos matemáticos sobre escalas, medida e proporcionalidade direta, que lhe permite afirmar com toda a convicção que os "engenheiros também se enganam", ou seja, o calafate não tem qualquer dúvida que o seu raciocínio e os seus cálculos estão corretos.

Conclusões

Constatou-se que no quotidiano desta comunidade piscatória, é realmente utilizada matemática que, devido ao abandono escolar precoce ou ao baixo nível de escolaridade, por vezes, vai além da matemática escolar que aprenderam. É o caso da utilização de fracções para calcular salários, do raciocínio proporcional ou da conversão entre unidades de medida para determinar se a profundidade é suficiente para capturar o peixe-espada preto.

Na construção de barcos, o calafate utiliza conhecimentos matemáticos que nalguns casos são invisíveis aos olhos do construtor de barcos ou não são vistos como matemática. O ponto médio é algo que o calafate determina sempre que pretende calcular o centro de um barco. Fracções, razões e proporções são utilizadas em várias etapas da construção. A simetria de reflexão é imprescindível, como condição para a estabilidade do barco.

Atualmente o calafate tem que trabalhar com desenhos à escala e com um nível diferente de precisão. Por vezes, tem dificuldade em exteriorizar o que é uma razão ou outros conceitos matemáticos são, mas é capaz de trabalhar com eles e até mesmo de corrigir nos projetos, eventuais erros humanos.

Como se pode verificar na Comunidade Piscatória de Câmara de Lobos existem práticas matemáticas que são usadas nas necessidades do quotidiano. Pelo menos parte dessas práticas não resultam de conhecimentos matemáticos adquiridos nas escolas, uma vez que a grande maioria dos investigados não tinha o primeiro ciclo completo (primeiros quatro anos de escolaridade).

Pode ainda verificar-se que grande parte da matemática detectada e descrita, faz parte dos programas de Matemática do 2.º ciclo do ensino básico (5.º e 6.º anos de escolaridade). Nenhum dos investigados frequentou este ciclo de ensino, à semelhança da esmagadora maioria dos indivíduos da Comunidade Piscatória de Câmara de Lobos. Assim, pode-se concluir que estes adquiriram conhecimentos e competências matemáticas fora da instituição Escola. Possuem (Etno) matemáticas obtidas na lide do quotidiano que lhes proporciona sucesso nas suas actividades profissionais e que ultrapassam os saberes proporcionados pela escola que frequentaram.

Referências bibliográficas

Barton, B. (2004). Dando sentido à etnomatemática: etnomatemática fazendo sentido. In: J. P. M. Ribeiro; M. C. S. Domite; R. Ferreira (Org.) *Etnomatemática: papel, valor e significado* (pp. 39-74). São Paulo: Zouk.

Bishop, A. (2010). Directions and possibilities for research on mathematics and culture, in relation to mathematics education: a personal view. In Márcia M. F. Pinto and Teresinha F. Kawasaki (Eds.). *Proceedings of the 34th conference of the International Group for the Psychology of Mathematics Education, v. 1,* (338-349). Belo Horizonte, Brazil: PME.

Cadeia, C., Palhares, P. e Sarmento, M. (2008). Cálculo mental na comunidade cigana. Em Pedro Palhares (Coord.) *Etnomatemática: um olhar sobre a diversidade cultural e aaprendizagem matemática* (67-103). V. N. Famalicão: Húmus.

D'Ambrósio, U. (1993). Etnomatemática: um programa. *A Educação Matemática em Revista* (pp. 5-11). V1, nº 1.

D'Ambrósio, U. (1998). *Etnomatemática: Arte ou técnica de explicar e conhecer.* 5ª Edição. São Paulo: Editora Ática.

D'Ambrósio, U. (2002). *Etnomatemática: o elo entre as tradições e a modernidade.* 2ª edição. Belo Horizonte. Editora Autêntica.

D'Ambrósio, U. (2006). The Program Ethnomathematics: A theoretical basis of dynamics of intracultural encounters. *The Journal of Mathematics and Culture, v. 1*(1).

D'Ambrosio, U. (2007). The Program Ethnomathematics and the challenges of globalization. *CIRCUMSCRIBERE, International Journal for the History of Science*, vol.1, 74-82.

Fernandes, João Carvalho. (2004). *Madeira: A Pérola do Atlântico*. Retirado em 9 de Julho de 2006 de http://www.ilhadamadeira.weblog.com.pt

Garcia, Marcelo Carlos (1992). Dar sentido a los datos: la combinación de perspectivas cualitativa e cuantitativa en el análisis de entrevistas. *In Marcelo Garcia Carlos. La investigatión sobre la formatión del professorado. Métodos de Investigatión y Análisis de Datos*. Argentina: Editorial Cincel.

NOTES

Gerdes, P. (1999) *Geometry from Africa: Mathematical and Educational Explorations*. Washington, DC: The Mathematical Association of America.

Gómes, G. R.; Flores, J. G.; Jiménez, E. G. (1996) *Metodología de la investigación cualitativa*. Granada: Ediciones Aljibe.

Knijnik, G. (2008). Educação matemática e diversidade cultural: matemática camponesa na luta pela terra. Em Pedro Palhares (coord.) *Etnomatemática: um olhar sobre a diversidade cultural e aaprendizagem matemática* (131-156). V. N. Famalicão: Húmus.

Lamas, M. (1956). Arquipélago da madeira, Maravilha Atlântica. In: Câmara Municipal de Câmara de Lobos. *Site Oficial da Câmara Municipal de Câmara de Lobos*. Retirado em 17 de Maio de 2005 de www.camaradelobos.org

Lucena, I.C.R. (2002). *Carpinteiros Navais de Abaetetuba: Etnomatemática navega pelos rios da Amazónia*. Natal: Universidade Federal do Rio Grande do Norte.

Monteiro, A. (2004). A Etnomatemática em Cenários de Escolarização: alguns elementos de reflexão. In: G. Knijnik; F. Wanderer; C. J. Oliveira (Org.). *Etnomatemática: Currículo e Formação de Professores* (pp. 432-446). Santa Cruz do Sul: UNISC.

Monteiro, A.; Orey, D. C.; Domite, M. C. S. (2004) Etnomatemática: papel, valor e significado. In: J. P. M. Ribeiro; M. C. S. Domite; R. Ferreira (Org.). *Etnomatemática: papel, valor e significado* (pp. 13-37). São Paulo: Zouk.

Nunes, T. (2010). Continuities and discontinuities between informal and scientific mathematical thinking: insights for education. In Márcia M. F. Pinto and Teresinha F. Kawasaki (Eds.). *Proceedings of the 34th conference of the International Group for the Psychology of Mathematics Education, v. 1*, (328-332). Belo Horizonte, Brazil: PME.

Oliveira, H. D. L. (2004). A Etnomatemática em Cenários de Escolarização: alguns elementos de reflexão. In: G. Knijnik; F. Wanderer; C. J. Oliveira (Org.). *Etnomatemática: Currículo e Formação de Professores* (pp. 305-322). Santa Cruz do Sul: UNISC.

Palhares, P. (2010). Studying artifacts in order to find out people's ways of thinking mathematically. In Márcia M. F. Pinto and Teresinha F. Kawasaki (Eds.). *Proceedings of the 34th conference of the International Group for the Psychology of Mathematics Education, v. 1*, (323-327). Belo Horizonte, Brazil: PME.

Palhares, P; Gomes, A. (2006). A Formação em Matemática para Professores do 1.º Ciclo – Em que Bases nos Podemos Apoiar? In: Pedro Palhares; Alexandra Gomes (Coord.). *MAT1C – desafios para um novo rumo*. Braga: IEC, Universidade do Minho.

Sarmento, Manuel J. (2003). O Estudo de Caso Etnográfico em Educação. In: N. Zago; M. P. Carvalho; R. A. T. Vilela (Org.). *Itinerários de Pesquisa: Perspectivas qualitativas em Sociologia da Educação* (pp. 137-179). Rio de Janeiro: DP&A Editora.

Culture *That* Counts

Volume 3 ◆ Teaching

Volume 3
Teaching

Cultural and Institutional inequalities: The case of mathematics education in Flemish schools

Karen François
karen.francois@vub.ac.be
Center for Logic and Philosophy of Science
Free University Brussels (VUB)

Brendan Larvor
phlqbpl@herts.ac.uk
University of Hertfordshire, Hatfield, UK

Abstract

This paper has three parts. The first sketches the evolving notions of mathematical and statistical literacy. The second presents the twin research programmes of Ethnomathematics and Critical Mathematics Education. The third part considers education in mathematical and statistical literacy in Flanders secondary schools. Attainment outcomes in Flemish schools show marked differences when analysed by native language and, separately, for immigration status, even when controlled for social and economic status. We claim that these data demonstrate the significance of two levels of Bishop's five-level analysis and indicate the practical and political importance of research into Ethnomathematics and Critical Mathematics Education.

Keywords:

- Mathematics education
- Human rights
- Social justice
- Ethnomathematics
- Critical mathematics education
- Flanders

Introduction

The main purpose of this article is to present a study of attainments in mathematics and statistics of 15-year-old pupils in Flanders secondary schools –in some ways a typical Western school context – and to show that unequal outcomes can indicate a problem that probably is connected to the cultural and societal background of the learner. Although pupils from distinct social groups have the same institutional opportunities, as we will explain later, they do not result in equal outcomes across those groups. Before elaborating on the specific case we will first present the theoretical frame of our analysis in which we will discuss the relation between education, human rights and social justice.

The method used for our purpose is literature review of both theoretical work and empirical research. The first literature review concerns theoretical investigations on the relation of education and human rights, the social and cultural context of mathematics, mathematical practices and mathematics education, and finally on the concepts of social justice within the field of mathematics education. A second literature review concentrated on empirical research on mathematics and statistics outcomes of 15 year old pupils as analysed by the Programme for International Student Assessment (PISA) (OECD, 2004, 2005). For the interpretation of mathematics and statistics outcomes of 15-year-old pupils in a Flanders secondary schools, we used the analysis of empirical data at a national level (De Meyer, Pauly & Van de Poele, 2004; De Meyer, Warlop & Van Camp, 2013). In the following section we will start by sketching the notions of mathematical and statistical literacy and the relation with human rights.

Mathematical and Statistical Literacy, and Human Rights

The content of these two literacy-concepts has evolved in the decades since they were first introduced. The difference between mathematics and statistics is beyond the purpose of this paper, and in any case their literacy-concepts have evolved in similar ways. We include both concepts in the analysis is because the empirical data as collected by PISA (OECD, 2004, 2005) assessed the outcomes of mathematics and of statistics under the umbrella of 'mathematical literacy'.

Although the initial notion of statistical literacy was exclusively related to the technical dimension of statistics (Haack, 1979), in the nineties it broadened in an attempt to include the kinds of statistical skills that people need in everyday life (Evans, 1992). As an example of this shift, we refer to Wallman's (1993) statement in her Presidential Address to the American Statistical Association:

> *Statistical literacy is the ability to understand and critically evaluate statistical results that permeate our daily lives–coupled with the ability to appreciate the contributions that statistical thinking can make in public and private, professional and personal decisions (1993, p. 1).*

In this broad conception of statistical literacy, people should be able to understand and critically evaluate everyday statistical information. That will benefit every person in professional life, in personal life and as a critical citizen. We can observe the same reasoning concerning mathematical literacy.

NOTES

Ten years after Wallmann's declaration (1993), the Organisation for Economic Co-operation and Development offered this definition (OECD 2004):

> *Mathematical literacy is an individual's capacity to identify and understand the role that mathematics plays in the world, to make well-founded judgments and to use and engage with mathematics in ways that meet the needs of that individual's life as a constructive, concerned and reflective citizen (OECD 2004, p. 37).*

This is very similar to the descriptions of statistical literacy. Again, one can observe the claim that mathematical literacy benefits the individual. It is a handy tool for a rich, a critical and a constructive role in contemporary society. Our society is increasingly defined by mathematical discourses. Each individual thus has an interest in being able to understand this mathematized world, if only to criticise it.

The Universal Declaration of Human Rights asserts a right to education, and stipulates "education is key to social and economic development [...] achieved by providing education for all. (UNESCO, 1948, art. 26). It is, presumably, through the activity of mathematically and statistically literate citizens that mathematics and statistics education can hope to contribute to the social and political goals expressed in the UN and UNESCO (1948) declarations.

Both of these UN and UNESCO declarations are compatible with the view that mathematics and statistics are value-free, culturally neutral technical tools. Indeed, this view may be one source of confidence that education in mathematics and statistics can support social and political progress. However, in recent decades researchers have explored the suggestion that mathematics education is socially embedded, and comes laden with culture and values. This complicates the optimistic picture presented in the quoted declarations. It may mean that pursuit of the progressive goals of these declarations may require more than simply teaching the same curriculum to everyone in the same way. It is to these complications that we now turn.

The entanglement of social justice and mathematics and statistics education

Since the middle of the nineteen-eighties it has been generally accepted among educational researchers that mathematics and statistics education is socially embedded and this in several layers (Gutiérrez & Boero, 2006; Lerman, 2006; Gates, 2006). Alan Bishop (1985) developed a scheme for the analysis of the social structure of this educational process. He distinguishes five significant levels of analysis, going from a macro perspective (culture) to a micro perspective (the individual). The first of these is the cultural level, which shows how the development of mathematical ideas is embedded in culture. For example, the ideal of pure mathematics emerged in a German academic context, while at the same time in France it was considered valuable for mathematics to be closely associated with engineering and natural science (Hacking 2014; Ferreirós 2016). The second, the societal level, investigates the influences of institutions other than schools and colleges with stakes in mathematics education. Some of them are formally concerned with education, e.g. the ministry of education, but many are not e.g. businesses that require mathematically literate employees but do not train them in mathematics. The third level is the institutional level. Research at this level looks for the influences within school systems to attain the targets of the mathematics curriculum. Topics at this research level are e.g. the influence of school culture or the instruction language of the learning process. With the fourth level we enter the classroom. Most important research at the pedagogical level concerns the didactics of mathematics education. Finally Bishop (1985) points to the individual level as a research domain of the

Volume 3 Teaching

sociological study of mathematics education. The focus at this level is on the learner from a social perspective.

Research on the fourth (pedagogical) and the fifth (individual) levels is mostly practiced by Social Psychology and Cognitive Science. We will not go into detail about these research fields because they are in general already better known. Such research is mainly presented at conferences such as Psychology of Mathematics Education (PME) and the International Congress on Mathematical Education (ICME). Research topics in this area concern the social motivation for studying mathematics, the fear of mathematics, the fear of success (preferring not to succeed academically in order to be socially successful and accepted by adolescent peers), teachers' perception of pupils, learning styles of pupils, self-concept of the learner, social cognition, social interaction in the learning process, etc. (Coessens, François, & Van Bendegem, 2012).

Research on the first three (socio-cultural) levels is mainly done within the research field of Ethnomathematics and Critical Mathematics Education, mostly focussing on the cultural, the societal and the institutional level. Gerdes' (2001) definition of Ethnomathematics refers to it as 'the study of mathematical ideas and activities as embedded in their cultural context'. Ethnomathematics originated in a post-colonial context, where an imported Western mathematics curriculum stands in contrast to local or indigenous mathematical practices. The research programme on Ethnomathematics has its roots in São Paulo, Brazil. Ubiratan D'Ambrosio—a mathematician and philosopher of mathematics education—is considered the intellectual father of Ethnomathematics. Since then it has spread all over the world. The notion of Ethnomathematics evolved from an exotic meaning of the concept—as the mathematical practices of non-literate people—to the general concept of mathematical practices of all people, including academic Western mathematics (Gerdes, 2001; François & Van Kerkhove, 2010). From this more general perspective mathematics is seen as a human practice, which emerges and develops within a socio-cultural context. The development, transmission and distribution of mathematical knowledge, including academic Western mathematics, is a dynamic process, embedded in time and culture. The challenge for the ethnomethodological approach is not just to pay more attention to other traditions and cultures, but also, importantly, to take 'the other' as given, rather than pressing their practices into fictitious categories (e.g. exotic, primitive, indigenous) that stand in opposition to our `regular' or `normal' Western practices (Said, 1995).

Even though Ethnomathematics is a critical research program and a critical practice regarding mathematics education, in literature it is still distinguished from so-called Critical Mathematics Education (Vithal & Skovsmose, 1997). Critical Mathematics Education originated within a Western high-tech society (Denmark), and it defines a number of suppression types and stereotyping, for example suppression based on class and gender (Atweh, Forgasz & Nebres, 2001; Burton, 2003). Burton (2003) and Sriraman (2008) provide an interesting overview of research from the perspectives of critical theory, feminist theory and social justice. Vithal & Skovsmose (1997) insist on the differences between these research

NOTES

programmes. Increasingly, however, there is research lying 'between' the two, e.g. research on the stereotyping of specific groups in non-Western societies. The gender issue which originates from the research field of Critical Mathematics Education—and thus from a Western context—is an intriguing example. There is growing interest in gender issues in non-Western contexts. Meanwhile, Australia (Vale, 2010), the United States (Paek, 2010) and Iceland (Steinthorsdottir, Dadisman, Robertson & Steinthorsdottir, 2010) have shifted their focus to the 'boy-problem'. Looking at the overview of international trends of gender differences in mathematics achievements, Ma (2010) firstly concludes that gender differences are small in magnitude and that they are limited to a small number of countries. Secondly there is no longer an exclusively male predomination in mathematics. Although there are more gender differences in favour of boys than girls, there is a growing number of countries revealing significant differences in favour of girls (Blömeke & Kaiser, 2010).

The program of Critical Mathematics Education and (the extended notion of) Ethnomathematics agree that pupils should have equal access to education and to mathematical and statistical literacy. This notion of mathematics for everyone is an expression of pedagogic optimism and egalitarianism.[1] Specifically; this approach takes parity of outcomes as the criterion of equality, rather than equality of opportunity (Hirtt, Nicaise & De Zutter, 2007; Roemer, 1998). We will adopt this criterion in our analysis.

In the following section we present a case of mathematics and statistics outcomes of 15-year-old pupils in a Western school context. We selected a specific case to show that unequal outcomes can indicate a problem that probably is connected to the cultural and societal background of the learner. Although pupils have the same institutional opportunities (Bishop's level 3), they do not result in equal outcomes.

A practical implication: the case of the Dutch-speaking community of Belgium

In this case we consider mathematics and statistics education in Flanders secondary schools, based on the results of the Programme for International Student Assessment (PISA) international comparative research.[2] PISA conducts its survey on a three-year cycle, starting in 2000. In each round, one domain is taken as the main subject. In this study, we consider the empirical data from the PISA 2003 and 2012 surveys because their focus was on mathematical literacy (including statistical items). The principal aim of the OECD/PISA assessment is to look 'at young people's ability to use their knowledge and skills in order to meet real-life challenges rather than how well they had mastered a specific school curriculum' (OECD 2005, p. 9). In other words, PISA assumes the same broad notion of mathematical literacy that we encountered above.

Looking at the 2003 results, we make two observations. First, the results between European countries are very different—even when those counties are near neighbours with similar educational systems. Second, even within one national context the results can indicate a big gap between high and low achievers. In the following paragraph we will pursue the second observation by examining the example of the Dutch-speaking community of Belgium. In this example, a high average for mathematical literacy hides significant differences within the country. Table 1 presents the average scores of secondary school pupils for mathematical literacy.

[1] For alternatives to this mood, see Gardiner (2004) and Simeonov (2016).

[2] We will from now one speak in terms of mathematics education since statistics is part of the mathematics curriculum and the results reflect both mathematical and statistical literacy.

The Dutch-speaking community of Belgium (Vlaanderen) (553) is, together with Hong Kong (550) (although there is no statistical significance) at the top of the international ranking. The Dutch-speaking community of Belgium is well ahead of its neighbours e.g. The Netherlands (538), the German-speaking community of Belgium (515), France (511) Germany (503) and the French-speaking community of Belgium (498). The table shows the figures for Belgium together with the scores of the Dutch-speaking community of Belgium (Vlaanderen), the French-speaking community of Belgium (Franse Gem.) and the German-speaking community of Belgium (Duitst. Gem.). Since Belgium is a federal state with three language communities (the Flemish, the Walloon and the very small German community) and educational matters are under the control of these language communities, national figures on mathematics education are subdivided into the averages of the language communities of Belgium. They have authority to decide on their own educational system and structures, and they develop the curriculum that will be enforced by the parliament of the respective language community.

This is the main reason why language community represents Belgian national figures.

The figures from 2012 show almost exactly the same pattern. The figure for the Dutch-speaking community of Belgium decreased slightly from 553 to 531. The first six positions are all taken up by Asian countries (with scores ranking from 631 to 538). The following five countries are ranked with a score of 536 to 523 (without statistical significance); the Dutch-speaking community of Belgium is one of them with a score of 531.

The Dutch-speaking community of Belgium is still ahead of his neighbours e.g. The Netherlands (523), the German-speaking community of Belgium (511), France (495) Germany (514) and the French-speaking

Table 1 Average scores on mathematical literacy for secondary education learners. PISA-2003 results (Adapted from De Meyer, Pauly & van de Poele, 2004, p.5)

Countries - mean - standard deviation

Wiskundige geletterdheid		
Landen	Gem.	St. Fout
Vlaanderen	553	(2.1)
Hongkong	550	(4.5)
Finland	544	(1.9)
Korea	542	(3.2)
Nederland	538	(3.1)
Liechtenstein	536	(4.1)
Japan	534	(4.0)
Canada	532	(1.8)
België	529	(2.3)
Macao-China	527	(2.9)
Zwitserland	527	(3.4)
Australië	524	(2.1)
N-Zeeland	523	(2.3)
Tsjech. Rep.	516	(3.5)
IJsland	515	(1.4)
Duitst. Gem.	515	(3.0)
Denemarken	514	(2.7)
Frankrijk	511	(2.5)
Zweden	509	(2.6)
Oostenrijk	506	(3.3)
Duitsland	503	(3.3)
Ierland	503	(2.4)
Slow. Rep.	498	(3.3)
Franse Gem.	498	(4.3)
Noorwegen	495	(2.4)
Luxemburg	493	(1.0)
Polen	490	(2.5)
Hongarije	490	(2.8)
Spanje	485	(2.4)
Letland	483	(3.7)
Veren. Staten	483	(2.9)
Russ. Fed.	468	(4.2)
Portugal	466	(3.4)
Italië	466	(3.1)
Griekenland	445	(3.9)
Servië	437	(3.8)
Turkije	423	(6.7)
Uruguay	422	(3.3)
Thailand	417	(3.0)
Mexico	385	(3.6)
Indonesië	360	(3.9)
Tunesië	359	(2.5)
Brazilië	356	(4.8)

Volume 3 Teaching

community of Belgium (493). The figure for Belgium (the average of the three language communities) is 515. The Dutch-speaking community of Belgium is still top of the European countries. Six Asian countries dominate the international top (four Chinese provinces plus Singapore and Korea).

Without going into the advantages and disadvantages of international comparative research (such as PISA or TIMMS), we will use the figures to demonstrate that the success (of some countries) is not the full story. Behind the Flemish success story there are hidden differences and gaps. By presenting these figures in more detail, we will show how these differences and gaps are related to cultural background, language and socio-economic status. From this observation we will argue for the relevance of Ethnomathematics and Critical Mathematics Education as research programmes that investigate the relation between mathematics education and culture, language and socio-economic status. To this end, we present figures that describe the relation between pupils' origins and their scores for mathematics (Figure 1) and the differences between native and non-native speaking pupils in the different countries of the PISA 2003 research (Figure 2).

In Figure 1, the variable 'origin of pupils' divides into three categories:

1. native students (students and parents are born in the land of test taking): red line
2. the first-generation students (students but not parents are born in the land of test taking): blue dot
3. immigrants (not born in the land of test taking): green dot

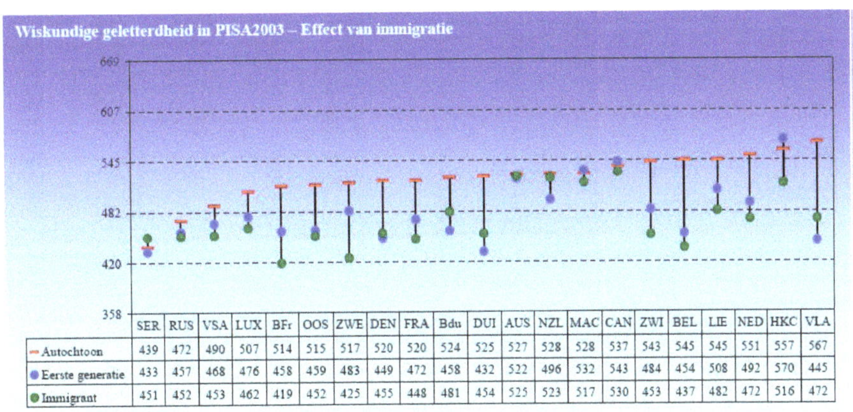

	SER	RUS	VSA	LUX	BFr	OOS	ZWE	DEN	FRA	Bdu	DUI	AUS	NZL	MAC	CAN	ZWI	BEL	LIE	NED	HKC	VLA
Autochtoon	439	472	490	507	514	515	517	520	520	524	525	527	528	528	537	543	545	545	551	557	567
Eerste generatie	433	457	468	476	458	459	483	449	472	458	432	522	496	532	543	484	454	508	492	570	445
Immigrant	451	452	453	462	419	452	425	455	448	481	454	525	523	517	530	453	437	482	472	516	472

Figure 1 Mathematics performance and impact of immigration

Note. Adapted from De Meyer, Pauly & Van de Poele, 2004, p. 32.

Note. The abbreviations stand for the following countries: Serbia (SER), Russian Federation (RUS), United States (VSA), Luxembourg (LUX), French Community (BFr), Austria (OOS), Sweden (ZWE), Denmark (DEN), France (FRA), German-speaking Community (Bdu), Germany (DUI), Australia (AUS), New Zealand (NZL), Macao-China (MAC), Canada (CAN), Switzerland (ZWI), Belgium (BEL), Liechtenstein (LIE), Netherlands (NED), Hong Kong China (HKC) and Flanders (VLA).

Note. The three groups of pupils are: Autochtoon = Dutch-speaking native pupil (red line), Eerste generatie = First generation (blue dot), Immigrant = Immigrant (green dot).

Looking at the results one can observe that the Dutch-speaking community of Belgium is at the top for the degree of inequality between natives and first-generation students. The difference is, on average score 122 points. Note that e.g. the difference between boys and girls (not in this figure) is 15 score points in favour of boys. The smaller difference between the immigrants (compared with the first generation students) is explained by a group of children who live in the Netherlands (and are therefore registered as immigrants).

These families have a high socio-economic status (SES) index because they are children from wealthier families (Hirtt, Nicaise, De Zutter 2007, p. 35).

The figures from 2012 show a similar pattern. The Dutch-speaking community of Belgium is still at the top for the degree of inequality between natives and first-generation students.

Based on the 2012 report, an additional analysis was added to control for the SES of pupils. After control for SES variables, the gap decreases from 97 points to 86, which means that SES can explain a small part of the observed differences. A considerable difference between natives and immigrants remains.

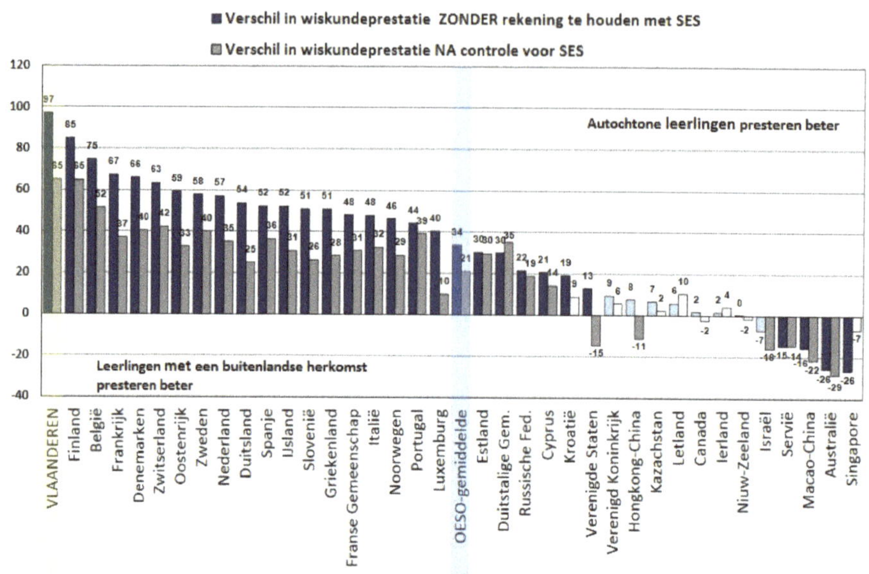

Figure 3 PISA-2012 Mathematical Literacy of 15–years old pupils at Flemish schools. Difference in score between Dutch native pupils and pupils with migration background (before and after control for Socio-economic status).
Note. Adapted from De Meyer, Warlop & Van Camp, 2013, p. 110.

As we demonstrate in Figure 2, the language at home also plays a major role in mathematics performance.

NOTES

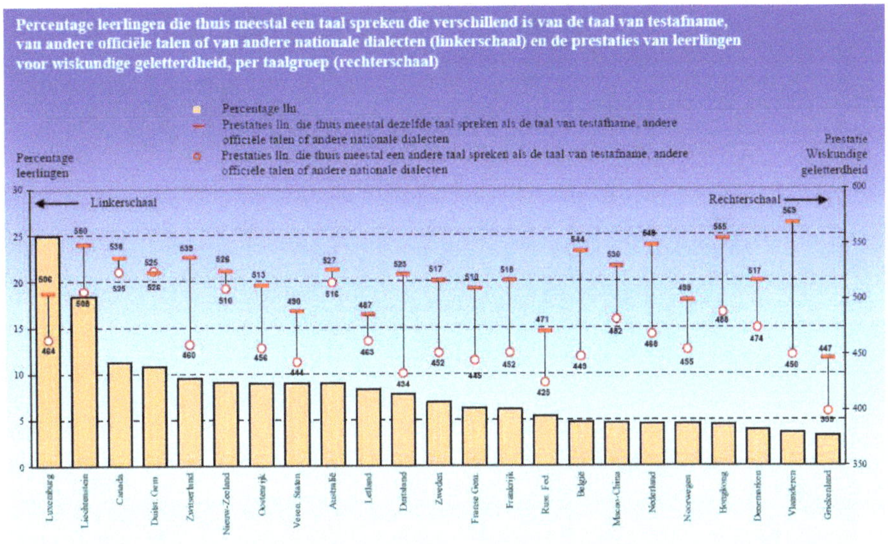

Figure 2 PISA-2003 achievements for mathematical literacy. Difference between native speakers and non-native speakers

Note. Adapted from De Meyer, Pauly & Van de Poele, 2004, p. 33.
Note. Histogram = % of non-native pupils; red line = Mathematics performance of native pupils; dot = Mathematics performance of non-native pupils.

The histogram represents the percentage of non-native speakers. The boxplot (vertical lines) represents the differences between the achievements of native (red line) and non-native (red circle) speaking pupils. The figures indicate the score of mathematical literacy. In the case of the Dutch-speaking community of Belgium (Vlaanderen), the histogram indicates that the Dutch-speaking community of Belgium has an average of 3.3% of non-native speaking pupils. The score on mathematical literacy for native speaking pupils is 569 (red line), for non-native speaking pupils 450 (red circle) which makes a giant gap of 119 between native and non-native speaking pupils.

The figures from 2012 show the same pattern. In 2012, the proportion of non-native pupils increased from 3,3% in 2003 to 7% in 2012 and the gap between native and non-native speaking pupils is 118 points. Again, for the 2012 figures, a SES control was added to the report. After control for SES variables the gap decreases from 118 points to 78, which means that SES can explain part of the observed differences although a considerable difference between native and non-native speaking pupils persists.

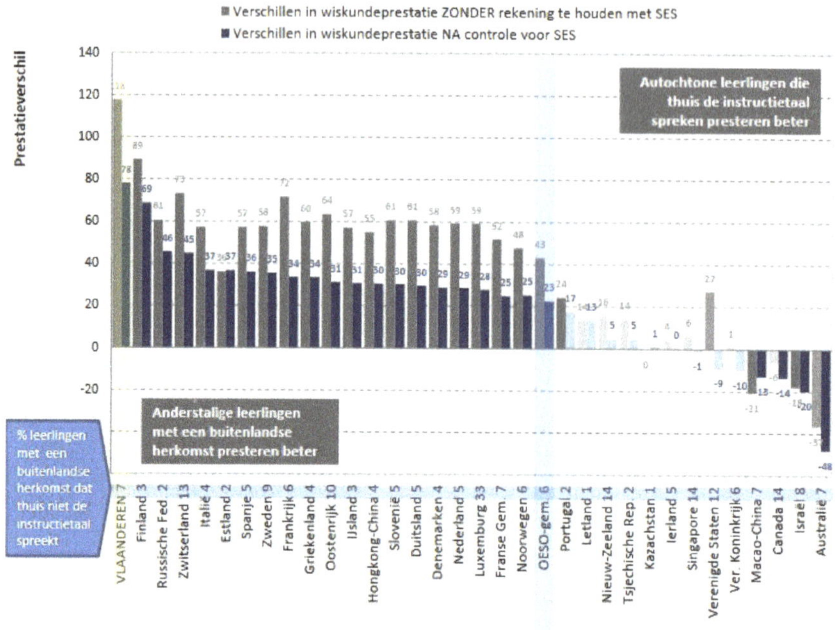

Figure 4 PISA-2012 Mathematical Literacy of 15–years old pupils at Flemish schools. Difference in score between Dutch native pupils and non-native pupils ((before and after control for Socio-economic status).

Note. Adapted from De Meyer, Warlop & Van Camp, 2013, p. 115.

Note. The figure after the country indicates the % of non-native pupils.

Our analysis of PISA figures and the conclusion that the educational system of the Dutch-speaking community of Belgium reflects social inequality is confirmed by the recent UNICEF report on how inequality affects children in high-income countries (UNICEF, 2016). The report ranks 41 EU and OECD countries according to inequality of income, educational achievement, self-reported health and life satisfaction. Denmark is at the top of the table with the lowest inequality among children. Israel ranked lowest across all domains. Belgium follows at the ranking of the lowest for the domain of educational achievement.

The educational system of the Dutch-speaking community of Belgium confirms social inequality. This inequality already starts at primary education and it increases in secondary and higher education (Groenez, Van den Brande & Nicaise, 2003; Hirtt, Nicaise, De Zutter, 2007). The Dutch-speaking community of Belgium can be called the champion in mathematical literacy but at the same time the champion in inequality.

Discussion

In this paper we focused on the interrelation of mathematics and statistics education and social justice. We noted that the aspirational documents of transnational bodies (UN &

NOTES

UNESCO) link mathematics and statistics education to human rights, and to the ability to participate as full citizens. In the theoretical part of the paper we first argued that mathematics and statistics and indeed mathematics and statistics education are socially embedded. With Bishop (1985) we argued that this field is and can be studied from different socio-cultural levels. Second, we presented two research domains: Ethnomathematics and Critical Mathematics Education. Both research domains deal with the same concern, namely, mathematical and statistical literacy for all and equality of educational outcomes, and both offer resources for understanding phenomena at Bishop's levels one to three.

In the empirical section, we presented the case of Flanders secondary mathematics education. We selected a social perspective from the *cultural* level (the immigrant status of pupils) and one from the *institutional* level (pupils' native languages). As the results show we can see an inequality based on these socio-cultural variables, even when controlled for SES. This corroborates the analytic value of two of Bishop's five levels of analysis, since measurable effects are discernible at those two. That these effects occur at the cultural and institutional levels indicates the relevance of the research into mathematics education at precisely those levels carried out within the Critical Mathematics Education and Ethnomathematics. This is a significant result because it is easy to multiply distinctions, propose taxonomies a priori and invent analytical grids. Such schemes (including Bishop's five-level model) require empirical corroboration to show that their categories correspond to discernible phenomena in reality. We conclude that even those educational systems with the highest scores have to evaluate their systems on the basis of equality. The research programmes of Ethnomathematics and Critical Mathematics Education can be of great value for this equality assessment and offer good practices to close the gap between high and low achievers keeping in mind that mathematical and statistical literacy is a basic right for *all* pupils.

Volume 3 Teaching

References

Atweh, B., H. Forgasz, H. & B. Nebres (Eds.), (2001). *Sociocultural research on mathematics education: An international perspective*. Mahwah, NJ: Lawrence Erlbaum Associates.

Bishop, A.J. (1985). The social psychology of mathematics education. In L. Streefland (Ed.), *Proceedings of the 9th PME International Conference, Vol. 2* (pp. 1–13). NA. Illinois: Northern Illinois University.

Blömeke, S. & G. Kaiser (2010). Mathematics teacher education and gender effects. In J.H. Forgasz, J.R. Becker, K.-H. Lee & O.B. Steinthorsdottir (Eds.), *International perspectives on gender and mathematics education* (pp. 263–283). Charlotte, NC: Information Age Publishing.

Burton, L. (Ed.), (2003). *Which way social justice in mathematics education?* Santa Barbara, CA: Praeger Publishers.

Coessens, K.; François, K. & Van Bendegem, J.P. (2012). Mirror neuron, mirror neuron in the brain, who's the cleverest of them al" From the Attraction of Psychology to the Discovery of the Social. In P. Smeyers & M. Depaepe (Eds.), *Educational research. The attraction of psychology* (pp. 91–104). Dordrecht: Springer.

D'Ambrosio, U. (2007a). Peace, social justice and Ethnomathematics. *The Montana Mathematics Enthusiast.* Monograph 1: 25–34.

D'Ambrosio, U. (2007b). Political issues in mathematics education. *The Montana Mathematics Enthusiast*. Monograph 3: 51–56.

De Meyer, I., J. Pauly & L. Van de Poele (2004). *Learning for tomorrows' problems. The first results of PISA 2003, OECD/PISA*, Gent: Ministerie van de Vlaamse Gemeenschap, Departement Onderwijs, Universiteit Gent, Vakgroep Onderwijskunde.

De Meyer, I., Warlop, N; & Van Camp, S. (2013). *Wiskundige geletterdheid bij 15–jarigen. Vlaamse resultaten van PISA2012 (Mathematical Literacy of 15–years old puplis. Flemish results of PISA2012).* Gent: Universiteit Gent, Vakgroep Onderwijskunde.

Evans, J. (1992). Mathematics for adults: Community research and 'barefoot statisticians'. In M. Nickson & S. Lerman (Eds.), *The social context of mathematics education: Theory and practice* (pp. 202–216). London: Southbank Press.

Ferreirós, J. (2016). Purity as a Value in the German-Speaking Area In B. Larvor (Ed.), *Mathematical cultures* (pp. 215–234). Switzerland: Springer.

Forgasz, J.H., J.R. Becker, K.-H. Lee & O.B. Steinthorsdottir (Eds.), (2010). *International perspectives on gender and mathematics education.* Charlotte, NC: Information Age Publishing.

François, K. & Van Bendegem, J.P. (Eds.), (2007). P*hilosophical dimensions in mathematics education.* New York: Springer. 240 pp.

François, K. & Van Kerkhove, B. (2010). Ethnomathematics and the philosophy of mathematics (education). In B. Löwe, & T. Müller (Eds.), *Philosophy of mathematics: Sociological aspects and mathematical practice.* Series: Texts in philosophy (pp. 121–154). London: College Publications.

Gardiner, A. (2004). *What is mathematical literacy?* Paper presented at the ICME 10, Copenhagen.

Gates, P. (2006). The place of equity and social justice in the history of PME. In A. Gutiérrez & P. Boero (Eds.), *Handbook of research on the psychology of mathematics education. Past, present and future* (pp. 367–402). Rotterdam: Sense Publishers.

Gerdes, P. (2001). Ethnomathematics as a new research field. Illustrations by studies of mathematical ideas in African history. In J.J. Saldaña (Ed.), *Science and Cultural Diversity. Filling a Gap in the History of Science* (pp. 11–36). México: Cuadernos de Quipu.

Gutiérrez, A. & P. Boero (Eds.), (2006). *Handbook of research on the psychology of mathematics education. Past, present and future.* Rotterdam: Sense Publishers.

Groenez, S., Van den Brande, I. & Nicaise, I. (2003). *Figures on social inequality in Flanders education.* LOA-rapport nr. 10. Leuven: Vlaamse minister van Onderwijs en Vorming. [In Dutch].

Haack, D. (1979). *Statistical literacy: A guide to interpretation.* North Scituate: Duxbury Press.

Hacking, I. (2014). *Why is there philosophy of mathematics at all?* Cambridge: Cambridge University Press.

Hirtt, N., Nicaise, I. & De Zutter, D. (2007). *The school of inequality.* Berchem: EPO. [In Dutch].

NOTES

Lerman, S. (2006). Socio-cultural research in PME. In A. Gutiérrez & P. Boero (Eds.), *Handbook of research on the psychology of mathematics education. Past, present and future* (pp. 347–366). Rotterdam: Sense Publishers.

Ma, X. (2010). Gender differences in mathematics achievement. Evidence from regional and international student assessments. In J.H. Forgasz, J.R. Becker, K.-H. Lee & O.B. Steinthorsdottir (Eds.), *International perspectives on gender and mathematics education* (pp. 225–248). Charlotte, NC: Information Age Publishing.

Paek, P.L. (2010). Factors contributing to gender differences in mathematics performance of United States high school students. In J.H. Forgasz, J.R. Becker, K. -H. Lee & O.B. Steinthorsdottir (Eds.), *International perspectives on gender and mathematics education* (pp. 203–224). Charlotte, NC: Information Age Publishing.

UNESCO. (1948). *Universal declaration of human rights*. Available at http://www.tashkent.unesco.org/en/education/ [25 April 2016].

UNICEF (2016). *Fairness for children. A league table of inequality in child well-being in rich countries*. Innocenti Report Card 13,Italy: UNICEF Office of Research.

OECD. (2004). *Learning for tomorrow's world: First results from PISA 2003*. Paris: OECD.

OECD. (2005). *PISA 2003* Technical report, Paris: OECD.

Pinxten, R. & François, K. (2011). Politics in an Indian canyon? Some thoughts on the implications of ethnomathematics. *Educational Studies in Mathematics, 78*(2), 261–273.

Roemer, J. (1998). *Equality of Opportunity*. Cambridge: Harvard University Press.

Said, E.W. (1995). *Orientalism. Western conceptions of the Orient*. London/New York: Penguin (orig. 1979).

Simeonov, E. (2016). Is Mathematics an Issue of General Education? In B. Larvor (Ed.), *Mathematical cultures* (pp. 439–460). Switzerland: Springer.

Sriraman, B. (Ed.), (2008). *International perspectives on social justice in mathematics education.* Charlotte, NC: Information Age Publishing.

Steinthorsdottir, O.B., K. Dadisman, D.L. Robertson & K. Steinthorsdottir (2010). Mathematics achievement in Icelandic playschools. In J.H. Forgasz, J.R. Becker, K. -H. Lee & O.B. Steinthorsdottir (Eds.), *International perspectives on gender and mathematics education* (pp. 249–261). Charlotte, NC: Information Age Publishing.

Vale, C. (2010). Gender mainstreaming. Maintaining attention on gender equality. In J.H. Forgasz, J.R. Becker, K. -H. Lee & O.B. Steinthorsdottir (Eds.), *International perspectives on gender and mathematics education* (pp. 111–143). Charlotte, NC: Information Age Publishing.

Vithal, R. & O. Skovsmose (1997). The end of innocence: A critique of Ethnomathematics. *Educational Studies in Mathematics. 34*(2): 131–157.

Wallman, K. (1993). Enhancing statistical literacy: Enriching our society. *Journal of the American Statistical Association, 88*(421): 1–8.

Volume 3
Teaching

Mathematics, Statistics, and (Jewish) Culture: Reflections on Connections

Lawrence Mark Lesser
Lesser@utep.edu

Amy E. Wagler
awagler2@utep.edu

Dept. of Mathematical Sciences
The University of Texas at El Paso
500 W. University Avenue
El Paso, Texas 79968-0514 USA

Abstract

The first author reflects (using first-person pronouns) upon the publication of Lesser (2006) in the debut issue of *Journal of Mathematics and Culture*, including how the subsequent ten years has not only made him aware of even more connections between mathematics and Jewish culture, but has also moved him more generally to incorporate more awareness of culture and equity into his writing, teaching, and outreach. He has also conducted subsequent scholarly exploration (e.g., with the second author of the current paper) into matters of context, language, diversity, and culture, and the results of a randomized experiment are reported here.

Keywords:

- Probability
- Experiment
- Judaism
- Mexican
- Culturally Relevant Mathematics

Background

It is a great pleasure and privilege to reflect upon a decade of my journey since publishing Lesser (2006), my first paper delving into a culture to make mathematical connections for pedagogical purposes. The paper involved connecting the domains of Judaism and mathematics at Houston's Emery High School, which I subsequently sought and received permission from head of school Stuart Dow (2006) to name. For many reasons, there was much excitement and creativity in teaching at a Jewish high school only in its second and third years. After deciding to integrate Judaism into Emery's mathematics classrooms, I aimed to stay within what I knew was covered in the students' Judaics curriculum. There was diversity within this Jewish school and so I aimed to be perceived (p. 11 of the 2006 paper) as teaching *about* religion, rather than teaching religion, per se. This helped me write the paper emphasizing cultural aspects of Judaism, because even religious Judaism has distinctive "culture" (e.g., Benor, 2012).

While Emery had integrated Judaics with subjects such as history and English, there was no precedent for doing so with mathematics. Indeed, this absence is a trend in the literature, where Shargel (2012, p. 38) notes, "The three aforementioned studies of interdisciplinary learning described curricula that linked subjects within the humanities like Jewish literature and history. None included math or science within an interdisciplinary framework." Shargel's case study of the Keshet school included a rare exception in which a math lesson used the validity of non-Euclidean geometry as a vehicle to understand there can be different ways of understanding Torah, and thus sustain conflicting viewpoints.

When I mention that I wrote about connections between mathematics and Judaism, the most common response is "oh, you mean *gematria* [using numerical equivalents of the letters in words; e.g., Blech 1991]". Lesser (2006), of course, does include a section on gematria, but also includes many other topics for classroom use, including: quotations about math in traditional Jewish sources, mathematical firsts, counting (permutations, marking time, etc.), infinity, pi, mathematical modelling, use of geometry and logic, and connections to Jewish text/customs/holidays/games. As satisfying as it was to identify and share connections between Judaism and mathematics, it was no less meaningful during the process to encounter parallels in other cultures, such as an African parallel (Zaslavsky, 1973) to a taboo on counting people.

Connections I have encountered since the paper include: (1) connecting spherical trigonometry to the matter (relevant to a space of worship) of identifying the direction to Jerusalem (e.g., Levin, 2002; Schechter, 2007), (2) identifying "culturally relevant hypothesis tests" such as whether the mean number of seeds in a pomegranate is 613 (see Zivotofsky, 2008), (3) noting when studying statistics that comparative experiments in written history are as old as the first chapter of the book of Daniel (Dodgson, 2006; Neuhauser & Diaz, 2004), and (4) noting connections to probability concepts (e.g., Rabinovitch, 1973).

Regarding the latter, there are well-known examples of lotteries that appear in Jewish history and holidays, such as Purim – an Akkadian word meaning "lots," referring to Haman's lottery to choose the date to kill the Jews (Esther 9:24-26). Other

NOTES

examples include the Kohen Gadol casting lots (Leviticus 16:5-10) on Yom Kippur (i.e., "a day like Purim") to determine which of two identical goats would be offered as the scapegoat, the division by Joshua (Joshua 18:10) of the Promised Land among the tribes, and the finding of the culprit in the tale of Jonah (Jonah 1:1-7). It was believed that divine intervention accompanied the lots when they were cast fairly (Responsa Havot Ya'ir, section 61, quoted in jewishvirtuallibrary.org) and there are indeed numerous instances (Yoma 37a, Yoma 39a, Jerusalem Talmud Yoma IV 1) of precautions Jews took to ensure that procedures for drawing lots were free of human bias or potential for manipulation, long before the modern-day advent of state lotteries (which I have explored; e.g., Lesser, 2013a) having accounting firms conduct independent monitoring. Those precautions included making the lots identical in size and material, having the urn's opening be no bigger than necessary for drawing lots, and shaking the urn before the drawing.

A decade after the *JMC* paper, it still seems to be the case that "there does not appear to be a single book or bibliography specifically and comprehensively on Jewish mathematics from which a classroom teacher could readily teach." (Lesser, 2006, p. 9). (An Internet search reveals, however, many "Christian mathematics textbooks" marketed to Christian schools and homeschooling parents.) I would one day love to see (if not write) such a book in the style of the book by Barta, Eglash and Barkley (2014). My paper still appears to be the only published paper on Jewish mathematics that aims for comprehensiveness (i.e., not limited to one particular mathematical topic) and a pedagogical audience (i.e., not limited to scholars). If readers are aware of others, I'd be very grateful to hear about them! Because of the specialization of the niche, perhaps I should not be surprised that Katsap and Silverman (2008, 2016) appear to be the only authors so far who have cited my 2006 paper. But even if I cannot now claim the paper has had high impact on the field, I can confidently claim it has had a major impact on me, as explained in the next section.

The Paper's Impact on My Journey

While I wrote the piece without explicitly positioning myself (e.g., D'Ambrosio et al., 2013) as a Jew (it is certainly possible a non-Jew could have acquired much knowledge about Judaism), publishing in *JMC* planted the seed to position myself in later pieces, most notably in a reflective essay about being a newcomer to the language of a domain, which I initially published in *NASGEm News* (Lesser, 2013c) and then published a version (with an added section on privilege) in the refereed journal of the NCTM affiliate group TODOS: Mathematics for ALL (Lesser, 2015a).

Writing the *JMC* piece gave me the confidence and desire to give more presentations about culturally relevant mathematics. In addition to national presentations (e.g., Lesser, 2013e), I gave various local presentations, including: adult education classes at community-wide (*Yom Limmud* or *Tikkun Leil Shavuot*) events, noncredit classes for Jewish high school students, and talks at individual congregations (from Modern Orthodox to post-denominational) and dayschools. I also was moved to write more about culturally relevant mathematics, which I did with Jewish culture (pi day in Lesser, 2013b), Hispanic culture (*toma todo* in Lesser, 2010a; *la lotería* in Lesser 2013f), and Hawaiian culture (Lesser, 2016b). Inspired by Ron Eglash's suite of culturally situated design tools to teach mathematics through students' culture, I also wrote about connecting mathematics to guitars (Robertson & Lesser, 2013) and to song (Lesser, 2014b, Lesser 2015b, Lesser, An, & Tillman, 2016) and to poetry (Lesser, 2013, 2017). I also looked to include historical, cultural, or etymological connections from varied cultures in my curriculum writing, such as my university's Bhutanese architecture (Lesser, 2008b).

I also allowed connections in Judaism to inspire my overall educational work, such as adapting and extending a reading from the Reform Jewish prayerbook into a statement of intention and welcome for the first day of any class (Lesser, 2010d, 2014c). I also discovered the work of Scottsdale (AZ) Community College's Christopher Benton in making mathematical connections within Judaism. One of many things I learned from Benton (2009-2010) is that interest in optimum class size dates back to the Talmud (B. Baba Bartha 21a): "The number of pupils to be assigned to each teacher is 25. If there are 50, we appoint two teachers. If there are 40, we appoint as assistant at the expense of the town." This was particularly fascinating to me because I had just finished a piece on average class size (Lesser, 2010c).

Because of the great potential of ethnomathematics to support the goals of social justice (D'Ambrosio, 2007), it is in retrospect not surprising that it was around the time of writing my *JMC* paper that I also began writing about social justice in the mathematics classroom, especially in the statistics classroom. While there was already a literature on "teaching mathematics for social justice", and despite the major role of statistics in identifying instances and patterns of inequity, Lesser (2007) appears to be the first juried article comprehensively connecting "social justice teaching" to "statistics education", and yielded followup webinars for K-12 (http://magazine.amstat.org/videos/education_webinars/UsingSocialJusticeExamples.wmv) and college (https://www.causeweb.org/cause/webinar/2007-07/) teachers and was discussed as part of a 2016-17 "A Call for Collection Action" national webinar series on equity sponsored by NCTM with seven other organizations. The Lesser (2010c) paper was expanded into a book chapter (Lesser & Marsan, in press). I also reviewed a book on equity for *JMC* (Lesser 2013g). A particular equity subfocus we've developed (e.g., Lesser, 2011; Lesser, Wagler, & Salazar, 2016; Lesser & Winsor, 2009) since the 2006 *JMC* paper is researching equitable teaching for statistics students whose strongest language is not English, and this entails both cultural and linguistic aspects, such as how certain mathematical contexts (even as simple as flipping a coin; see Lesser, Wagler, & Salazar, 2016) can be laden with culture and how being bilingual can be a great resource in the mathematics classroom. Another subfocus has been supporting diversity (Lesser, 2010b) and confronting harmful stereotypes (Lesser, 2014a, 2016a, 2016c).

Subsequent Randomized Experiment

As a way to help ethnomathematics gain broader attention, I found myself interested in applying rigorous quantitative methods to assess the effect of culture in a mathematics activity to increase student interest and learning. Most ethnomathematics scholarship seems best described as qualitative or descriptive and examples that are quantitative either use no randomization (e.g., quasiexperimental such as Achor, Imoko, & Uloko, 2009, or Ogunkunle & George, 2015) or use it with very small sample sizes (e.g., Sankaran, Sampath, & Sivaswamy, 2009). One of the few examples of randomized experiments in ethnomathematics research is Kisker, Lipka, Adams, Rickard, Andrew-Ihrke, Yanez, and Millard (2012). So this research arguably fills a major gap in the literature as well as helps establish whether

NOTES

there may be an important role for cultural contexts to facilitate student interest and learning, as well as whether it matters if the cultural context is a familiar one.

I did an exploratory study at a moderately large doctoral/research university located in a large city in the southwestern United States by the México border. Over 75% of the student population (and of the county) is Hispanic and about 7% of the Hispanic student population are Mexican nationals who commute across the border to take courses, and so awareness of Mexican culture is high. At the same time, the percentage of the city's residents that are Jewish is about 1% (roughly half the national average) and there is no reason to believe it is any higher in the university's population. Therefore, I expected this population can to be more familiar with one culture (Mexican) than another (Jewish), giving the experiment a nice design contrast.

This study took place within two sections of the aforementioned university's statistics literacy course taught by the same instructor. The sections enrolled 79 students as described in Table 1:

Table 1: Demographics of sections where experiment conducted

gender	85% female, 15% male
year in school	15% first-year, 32% sophomore, 43% junior, 10% senior
major	91% future teachers; 9% majors outside College of Education

The experiment (which the IRB exempted from review) occurred within 30 minutes of each section's November 29, 2010 meeting, with all 65 attending students participating. Section results are combined because the instructor was the same, the day of the week was the same, class times were similar (11:30 versus 1:30), and student demographics were similar.

The instructor told students they would be doing a worksheet individually and anonymously as ungraded classwork for the instructor to assess how effective a worksheet with a game of chance is in motivating and helping students explore and solidify probability concepts. While students had not previously participated in explicitly cultural activities in this course, they had been exposed to probability-based games, including SKUNK (Brutlag, 1994), Pass the Pigs (Hildreth & Green, 2016), and the game show "Deal or No Deal" (Baker, Bittner, Makrigeorgis, Johnson, & Haefner, 2010).

Three versions of the worksheet (see Appendices) were shuffled so each student (within each course section) was randomly assigned to a version. Each version's game involves rolling or spinning a symmetric object, an activity which facilitates rich conceptual explorations (e.g., Ely & Cohen, 2010) and which has been connected to many cultures (McCoy, Buckner, & Munley, 2007). All three versions were designed to be identical in mathematical structure (i.e., equiprobable sample space where some outcomes involve gain and some involve loss) and difficulty, differing only in cultural context (see Table 2). In particular, Versions 2 and 3 used games from Mexican and Jewish culture (e.g., Krause, 2000), while Version 1 was a "control" game I invented to have similar structure but no real culture. We note, however, that the instrument in the experiment referred to familiarity with the game rather than of the probability device used in the game. While this distinction seems unlikely to have affected Versions 2 or 3, it may have affected Version 1 if respondents felt that they were familiar with a die, but not with the particular manufactured game that was presented.

Table 2: Treatments in randomized experiment

Appendix	Culture	Probability device in game	*n*
A	N/A	6-sided die (number cube)	23
B	Mexican	6-sided top (*pirinola; topa*)	21
C	Jewish	4-sided top (*dreidel*)	21

All three versions of the worksheet packet (two double-sided pages) began with the same generic coversheet, which said, "I am trying out a probability learning game/ activity. You don't need to put your name on this – I just need you to answer honestly and work independently so I can learn as much as I can about how this activity works for [name of course] students. I (and future students) thank you for your cooperation! [name of instructor]". The coversheet was followed by: 1) the description of a game (noting any cultural origins), 2) a description of the physical object (number cube or spinning top) that gives the game its element of chance and how it could be built, 3) rules of the game and initial distribution of chips, 4) a series of content questions structured to be as identical as possible across versions, and 5) statements [see items 7-10 in any Appendix] for which respondents indicated level of agreement or disagreement regarding the game's familiarity, interest, understanding and use in teaching.

The content questions [see items 1-6 in any Appendix] were adapted and extended from a lesson described anecdotally in Lesser (2010a), building on a comment in Lesser (2006). Sources such as Krause (2000) have typical questions about comparing experimental frequencies to the (equally likely) theoretical frequencies, but this worksheet goes beyond by, for example, including expected value of the second player's first turn (Lesser, 2006, 2010a) and thus allowing for the assessment of whether there is an advantage in going first.

Results

The second author (a statistician) conducted the data analysis knowing only the general idea of the study, but blind to treatment group and the first author's expected results. A Fisher's exact test yielded no evidence of a difference among the proportions of content questions correct for the three groups (*p-value* = 0.3158). This implies that there is no difference in knowledge with regard to context. Since the intent of the activity was not to teach the concept, but to assess how context affects the ability to transfer knowledge already gained from instruction, this supports the claim that the three groups were uniform with regard to concept knowledge before the activity.

Following the test for equality among the proportions, the second author compared correlation matrices among the three groups to assess any differences in association among

NOTES

the items assessing familiarity, interest and helpfulness in understanding due to context. This analysis implies that the covariance structures for the Toma Todo and Dreidel groups are different from each other but neither one substantively differs from the number cube structure. This result warrants further consideration of these covariance structures. (For the familiarity ratings, the Toma Todo group had 65% of responses at the extremes (response levels 1 and 6), while the Dreidel group had only 40% at those extremes.) In order to investigate these covariance structures further, the second author conducted individual tests of the correlations with multiplicity controlled at an overall error rate of 5%. For the Toma Todo group, there is a positive correlation between the context of Toma Todo helping understanding and believing that they will use this in teaching in the future ($r = 0.63$, $n = 20$, *p-value* = 0.030). Similarly, for the Dreidel group, the correlation between the context helping in understanding and having more interest was also positive ($r = 0.73$, $n = 20$, *p-value* <0.001). Since these correlation tests are protected by first assessing overall differences in the covariance matrix, it is also reasonable to assess the pointwise observed significance levels for statistical significance. For the Toma Todo group, all of the correlations among familiarity, interest and helpfulness are statistically significant with pointwise *p-values* smaller than 0.06, with the exception of the correlation between teaching and familiarity. In contrast, for the Dreidel group, fewer of these ratings are correlated (only those for familiarity and interest and understanding with both teaching and interest). This provides evidence that for a context that is more familiar among a subset of students, the relationships among these ratings of context are more dependent. Another way to say this is that the level of familiarity with the context may affect the dependency among these ratings for interest, teaching, and helpfulness and that, overall, higher familiarity ratings may increase interest and helpfulness in understanding.

Closing Remarks and Future Directions

Because the analysis of the above experiment was done during the spring 2016 semester, I have not yet had the chance for it to inform the planning of my future courses, but the results give me no reason not to continue what I have already been recently doing in the classroom. While arguments might be made for giving students only the version they most connect with (to increase their motivation and ability to make content connections), I believe it is important to also expose them to examples of less familiar cultures not only for the overall humanizing benefits of ethnomathematics, but also to help them more clearly verify that they have connected to the underlying mathematics by being able to make transfer across cultures. It might be too repetitious to give them multiple versions of the entire worksheet, though, so if I used worksheets in the future, the initial worksheet could have full detail and the ensuing variations could be scaled down or streamlined to what is needed to demonstrate transfer to the new cultural context, including making observations such as that the game of dreidel favors the first player, but Toma Todo does not (Lesser, 2010a). Also, for non-research classroom use, I would likely remove items 7-11 from the end of the worksheet.

In my spring 2016 class, I brought in a *pirinola* and discussed the Mexican game *toma todo* (e.g., as in Lesser, 2010a) and my students were surprised and delighted that I knew about the game, cared to bring it in and make content connections to it, and that I even asked them some questions (e.g., "Do you view the game as a family game or a drinking game, or can it be either?" their answer: either!) so that I could learn more about it. Then I also talked about how many other cultures have some game with a spinning top and then I placed a different top [a *dreidel*] under the document camera and asked if they recognized it (some did). This happened to be a day footage was obtained from the class for university publicity purposes, so you can see this moment during 1:24-1:32 of the video embedded at http://news.utep.edu/mathemusician-named-piper-professor/.

This example of games with a spinning top may offer professors less familiar with ethnomathematics a way to visualize how one type of artifact can be explored across cultures in a way that not only makes the underlying mathematics more memorable and interesting, but also communicates or reinforces the implicit life lesson that diversity is a strength and that cultures have differences but generally even stronger similarities. An additional though less distinctive feature of this example is that it offers a low barrier to entry in the sense that it is a concrete lesson that need not consume more than one class meeting and its materials (the tops) are not expensive or difficult to obtain.

While my more quantitatively-focused students will respect that the above experiment uses quantitative methods, I know this will not guarantee an increase in their appreciation of the qualitative aspects and richness of ethnomathematics itself. Towards that end, I sometimes try to expand their horizons by gently exposing them to the inherently qualitative or cultural aspects of the "value-free, objective" mathematics and statistics they study. For example, when I teach my master's level research methods classes for secondary mathematics teachers, we discuss how Huberty (2000) shows how a quantitative study typically involves many qualitative judgments, how Romagnano (2001) shows how even quantitative mathematics assessments may not be as "objective" as they would like to believe, and how Best (2002) argues that statistics are "socially constructed". I also share how students from other cultures and countries solve even basic mathematics problems with different-looking algorithms that are just as valid (e.g., Moschkovich, 2013) and that we will not be fully effective teachers or practitioners of the mathematical sciences if we think diversity or culture can be ignored (Celedón-Pattichis, Musanti, & Marshall, 2010; Lesser, 2010c).

While nowhere near as well-known or used as the term *ethnomathematics*, it is worth noting that there is some literature on *ethnostatistics*. Gephart (1988) articulates three categories spanned by the term: (1) ethnographic studies of groups that produce statistics, (2) testing the technical and operational assumptions involved in the production of statistics, and (3) the use of statistics as a rhetorical device. I plan more exploration on how culture has a natural synergy with the emphasis statistics already places on context (e.g., Moore, 1988; Pfannkuch, 2011).

Earlier in this paper, I mentioned how meaningful it would be to see curriculum fully developed on Jewish cultural mathematics that could be used as a textbook or supplement for one or more courses. It would be fascinating to conduct a case study on how students (e.g., in a pluralistic Jewish high school) responded to using such curricula. And in general, there is much room to explore the nature and magnitude of the benefits when students are taught mathematics that integrates ethnomathematics – both in the case where students already have connection to the culture and in the case where they do not. The randomized experiment reported for the first time in the present paper suggests a template for isolating the effect of a particular culture, possibly adapted to have the activity actually introduce new content and having both pre- and post- assessment of both understanding of and attitude towards mathematics.

NOTES

References

Achor, E. E., Imoko, B. I., & Uloko, E. S. (2009). Effect of ethnomathematics teaching approach on senior secondary students' achievement and retention in locus. *Educational Research and Review, 4*(8), 385-390.

Ascher, M., & Ascher, R. (1981). *Code of the quipu: A study in media, mathematics, and culture.* Ann Arbor, MI: The University of Michigan Press, 1981.

Baker, A., Bittner, T., Makrigeorgis, C., Johnson, G., & Haefner, J. (2010). Teaching prospect theory with the 'Deal or No Deal' game show. *Teaching Statistics, 32*(3), 81-87.

Barta, J., Eglash, R., & Barkley, C. (2014). *Math is a verb: Activities and lessons from cultures around the world.* Reston, VA: National Council of Teachers of Mathematics.

Benor, S. B. (2012). *Becoming frum: How newcomers learn the language and culture of Orthodox Judaism.* New Brunswick, NJ: Rutgers University Press.

Benton, C. J. (2009-2010). Some personal reflections on educational practices found in rabbinic literature. *The Maqom Journal for Studies in Rabbinic Literature, 17.* http://www.maqom.com/journal/paper31.pdf

Best, J. (2002). People count: The social construction of statistics. Talk given at Augsburg College, Minneapolis, MN. http://www.statlit.org/PDF/2002BestAugsburg.pdf

Blech, B. (1991). *The secrets of Hebrew words.* Northvale, NJ: Jason Aronson, Inc.

Brutlag, D. (1994). Choice and chance in life: The game of 'Skunk'. *Mathematics Teaching in the Middle School, 1*(1), 28-33.

Celedón-Pattichis, S., Musanti, S. I., & Marshall, M. E. (2010). Bilingual elementary teachers' reflections on using students' native language and culture to teach mathematics. In M. Q. Foote (Ed.). (2010). *Mathematics teaching and learning in K-12: Equity and professional development.* (pp. 7-24). New York: Palgrave Macmillan.

D'Ambrosio, B., Frankenstein, M., Gutiérrez, R., Kastberg, S., Martin, D. B., Moschkovich, J., Taylor, E., & Barnes, D. (2013). Positioning oneself in mathematics education research. *Journal for Research in Mathematics Education, 44*(1), 11-22.

D'Ambrosio, U. (2007). Peace, social justice and ethnomathematics. *The Montana Mathematics Enthusiast,* Monograph 1 (International Perspectives on Social Justice in Mathematics Education), pp. 25-34.

Dodgson, S. J. (2006). The evolution of clinical trials. *The Journal of the European Medical Writers Association, 15*(1), 20-21.

Dow, S. (2006, May 21). Personal [e-mail] communication with the first author.

Ely, R. E., & Cohen, J. S. (2010). Put the right spin on student work. *Mathematics Teaching in the Middle School, 16*(4), 208-215.

Gephart, R. P. (1988). *Ethnostatistics: Qualitative foundations for quantitative research.* Thousand Oaks, CA: Sage Publications.

Hildreth, L. A., & Green, J. L. (2016). Using pig die and simulation to explore probability and expected values. *Teaching Statistics, 38*(2), 67-71.

Huberty, C. J. (2000). Judgment in quantitative research. *The Mathematics Educator, 10*(1), 5-10.

Katsap, A., & Silverman, F. L. (2008). A case study of the role of ethnomathematics among teacher education students from highly diverse cultural backgrounds. *Journal of Mathematics and Culture, 3*(1), 66-102.

Katsap, A., & Silverman, F. L. (2016). *Ethnomathematics of Negev Bedouins' existence in forms, symbols, and geometric patterns*. The Netherlands: Sense Publishers.

Kisker, E. E., Lipka, J., Adams, B. L., Rickard, A., Andrew-Ihrke, D., Yanez, E. E., & Millard, A. (2012). *Journal of Research in Mathematics Education, 43*(1), 75-113.

Krause, M. C. (2000). *Multicultural mathematics materials* (2nd ed.) Reston, VA: National Council of Teachers of Mathematics.

Laridon, P., Mosimege, M., & Mogari, D. (2005). Ethnomathematics research in South Africa. In R. Vithal, J. Adler & C. Keitel (Eds.), *Researching mathematics education in South Africa: Perspectives, practices and possibilities* (pp. 133-160). Cape Town, South Africa: HSRC Press.

Lesser, L. (2006). Book of numbers: Exploring Jewish mathematics and culture at a Jewish high school. *Journal of Mathematics and Culture, 1*(1), 8-31.

Lesser, L. (2007). Critical values and transforming data: Teaching statistics with social justice. *Journal of Statistics Education, 15*(1), 1-21.

Lesser, L. (2008a). Reunion of broken parts: Experiencing diversity in algebra. In P. C. Elliott & C. M. Elliott Garnett (Eds.), *Getting into the mathematics conversation: Valuing communication in mathematics classrooms* (pp. 99-107). Reston, VA: National Council of Teachers of Mathematics.

Lesser, L. (2008b). Slippery slopes. *Mathematics Teacher, 102*(2), 116-119.

Lesser, L. (2009a). Social justice, gender equity, and service learning in statistics education: Lessons learned from the DOE-funded Project ACE (ACtion for Equity). *Proceedings of the 2008 Joint Statistical Meetings, Section on Statistical Education*, pp. 424-431. Alexandria, VA: American Statistical Association. https://www.amstat.org/membersonly/proceedings/papers/300471.pdf

Lesser, L. (2009b). Equity, social justice, and the mission of TODOS: Connections and motivations. *Teaching for Excellence and Equity in Mathematics, 1*(1), 22-27.

Lesser, L. (2010a). An ethnomathematics spin on statistics class. *Notices of the North American Study Group in Ethnomathematics, 3*(2), 5-6.

Lesser, L. (2010b). Sizing up class size: A deeper classroom investigation of central tendency. *Mathematics Teacher, 103*(5), 376-380.

Lesser, L. (2010c). Equity and the increasingly diverse tertiary student population: challenges and opportunities in statistics education. In C. Reading (Ed.), *Data and context in statistics education: Towards an evidence-based society. Proceedings of the Eighth International Conference on Teaching Statistics*, Ljubljana, Slovenia. Voorburg, The Netherlands: International Statistical Institute. http://iase-web.org/documents/papers/icots8/ICOTS8_3G3_LESSER.pdf

Lesser, L. (2010d). Opening intentions. *The Teaching Professor, 24*(9), 4.

NOTES

Lesser, L. (2011). Supporting learners of varying levels of English proficiency. *Statistics Teacher Network*, 77, 2-5.

Lesser, L. (2013). Poetic reactions. *Journal of Humanistic Mathematics, 3*(1), 156-161.

Lesser, L. (2013a). Letter to the Editor: The odds of academic usage of statistics terms in everyday contexts such as lotteries. *Journal of Statistics Education, 21*(1), 1-5.

Lesser, L. (2013b). Jewish pi day: Making a secular subject more Jewish…and more engaging. *The Jewish Educator: NewCAJE's Journal of Jewish Education*, 43-47. http://thejewisheducator.files.wordpress.com/2013/03/lesser1.pdf

Lesser, L. (2013c). Cultivating empathy for English language learners in mathematics: A reflection. *Notices of the North American Study Group on Ethnomathematics, 6*(1), 3-9.

Lesser, L. (2013d). Integrating ethics with the teaching of introductory statistics: Rationale and resources. *Proceedings of the 2013 Joint Statistical Meetings, Section on Statistical Education* (pp. 91-95). Alexandria, VA: American Statistical Association.

Lesser, L. (2013e). Equity and engagement with culturally relevant mathematics: Reflections and resources, *invited* breakout workshop, National Council of Teachers of Mathematics Interactive Institute for Grades 9-12, Washington, DC, August 2013.

Lesser, L. (2013f). Dare to share: Tell us how you've used *TEEM* articles. *Teaching for Excellence and Equity in Mathematics, 5*(1), 35.

Lesser, L. (2013g). Review of [Mary Q. Foote (Ed.). (2010). Mathematics teaching and learning in K-12: *Equity and professional development*. New York: Palgrave Macmillan] J*ournal of Mathematics and Culture, 7*(1), 140-141.

Lesser, L. (2014a) Staring down stereotypes. *Mathematics Teacher, 107*(8), 568-571.

Lesser, L. (2014b). Mathematical lyrics: Noteworthy endeavours in education. *Journal of Mathematics and the Arts, 8*(1-2), 46-53.

Lesser, L. (2014c, June 27). Opening intentions for the first day of class. *Faculty Focus*. http://www.facultyfocus.com/articles/teaching-and-learning/opening-intentions-first-day-class/

Lesser, L. (2015a). Learning language: A mathematics educator's reflection on empathy and privilege. *Teaching for Excellence and Equity in Mathematics, 6*(1), 25-32.

Lesser, L. (2015b). 'American Pi': The story of a song about pi. *Journal of Mathematics Education, 8*(2), 158-168. http://educationforatoz.com/images/2015_Larry_Lesser.pdf

Lesser, L., An, S., & Tillman, D. A. (2016). The use of song to open an educational development workshop: Exploratory analysis and reflection. *To Improve the Academy: A Journal of Educational Development, 35*(2), 284-302. http://onlinelibrary.wiley.com/doi/10.1002/tia2.20047/full

Lesser, L. (2016a). Sounding off on stereotypes. *Noticias de TODOS: News from TODOS Mathematics for ALL, 11*(2), 1-3. http://www.todos-math.org/assets/documents/noticias/noticiasv11n2winter2015.pdf

Lesser, L. (2016b). Very ocean variation: Statistics learned on vacation. *Teaching Statistics, 38*(2), 73.

Lesser, L. (2016c). Engendering understanding in mathematics. *Mathematics Teacher, 110*(1), 8.

Lesser, L., Wagler, A., & Salazar, B. (2016). Flipping between languages? An exploratory analysis of the usage by Spanish-speaking English language learner tertiary students of a bilingual probability applet. *Statistics Education Research Journal, 15*(2), 145-168. http://iase-web.org/documents/SERJ/SERJ15(2)_Lesser.pdf

Lesser, L. (2017). Moving between inner and outer worlds. *Journal of Humanistic Mathematics, 7*(1), 275-284. http://scholarship.claremont.edu/jhm/vol7/iss1/21/

Lesser, L., & Marsan, L. (in press). Class size matters: Lessons learned from average class size. In Linda Fulmore & Anita Bright (Eds.), *Answering the call: Social justice in the secondary mathematics classroom*. Reston, VA: National Council of Teachers of Mathematics.

Lesser, L. & Winsor, M. (2009). English Language Learners in introductory statistics: Lessons learned from an exploratory case study of two pre-service teachers. *Statistics Education Research Journal, 8*(2), 5-32. http://iase-web.org/documents/SERJ/SERJ8(2)_Lesser_Winsor.pdf

Levin, D. Z. (2002). Which way is Jerusalem? Which way is Mecca? The direction-facing problem in religion and geography. *Journal of Geography, 101*(1), 27-37.

McCoy, L. P., Buckner, S., & Munley, J. (2007). Probability games from diverse cultures. *Mathematics Teaching in the Middle School, 12*(7), 394-402.

Moore, D. S. (1988). Should mathematicians teach statistics?, *The College Mathematics Journal, 19*(1), 3-7.

Moschkovich, J. (2013). Equitable practices in mathematics classrooms: Research-based recommendations. *Teaching for Excellence and Equity in Mathematics, 5*(1), 26-34.

Nechauser, D., & Diaz, M. (2004). Daniel: Using the Bible to teach quality improvement methods. *Quality and Safety in Health Care, 13*(2), 153-155.

Ogunkunle, R. A., & George, N. R. (2015). Integrating ethnomathematics into secondary school mathematics curriculum for effective artisan creative skill development. *European Scientific Journal, 11*(3), 386-397.

Perkowski, D. A., & Perkowski, M. (2007). *Data and probability connections: Mathematics for middle school teachers*. Upper Saddle River, NJ: Pearson.

Pfannkuch, M. (2011). The role of context in developing informal statistical inferential reasoning: A classroom study. *Mathematical Thinking and Learning, 13*(1/2), 27-46.

Rabinovitch, N. L. (1973). *Probability and statistical inference in ancient and medieval Jewish literature*. Toronto: University of Toronto Press.

Robertson, W., & Lesser, L. (2013). Scientific skateboarding and mathematical music: Edutainment that actively engages middle school students. *European Journal of Science and Mathematics Education, 1*(2), 60-68. http://www.scimath.net/articles/12/123.pdf

Romagnano, L. (2001). The myth of objectivity in mathematics assessment. *Mathematics Teacher, 94*(1), 31-37.

NOTES

Sankaran, S., Sampath, H., & Sivaswamy, J. (2009). Learning fractions by making patterns: An ethnomathematics based approach. In S. C. Kong et al. (Eds.), *Proceedings of the 17th International Conference on Computers in Education* (pp. 341-345). Hong Kong: Asia-Pacific Society for Computers in Education.

Schechter, M. (2007). Which way is Jerusalem? Navigating on a spheroid. *College Mathematics Journal, 38*(2), 96-105.

Shargel, R. (2012). Genesis encounters Darwin: A case study of educators' understandings of curricular integration in a Jewish high school. *Journal of Jewish Education, 78*(1), 34-57.

Smith, S. (1996). *Agnesi to zero: Over 100 vignettes from the history of math.* Berkeley, CA: Key Curriculum Press.

Zaslavsky, C. (1973). *Africa counts.* Boston: Prindle, Weber & Schmidt.

Zivotofsky, A. (2008). What's the truth about… pomegranate seeds? *Jewish Action: The Magazine of the Orthodox Union, 69*(1), 64-67. https://www.ou.org/jewish_action/09/2008/whats_the_truth_about_pomegranate_seeds/

APPENDIX A ("Culture-Free Version" of Worksheet)

Here are the rules of a game played with a "die" or "number cube", whose six sides are numbered 1, 2, 3, 4, 5, and 6.

In the game, players sit in a circle and each start with an equal number of (let's say, ten) items (e.g., we will say colored plastic **chips**). Each player places one chip in a center pile called the "**pot**". Before play begins, players decide on whether to say the winner will be the person with the most chips after a fixed number of rounds (a **round** consists of each player getting one turn to roll the die) or to say the winner will be the last player remaining after all other players have run out of chips (this can take a large number of rounds!).

The six-sided die or number cube is available in physical and online stores and can also be handmade by folding the diagram on the right into a cube:

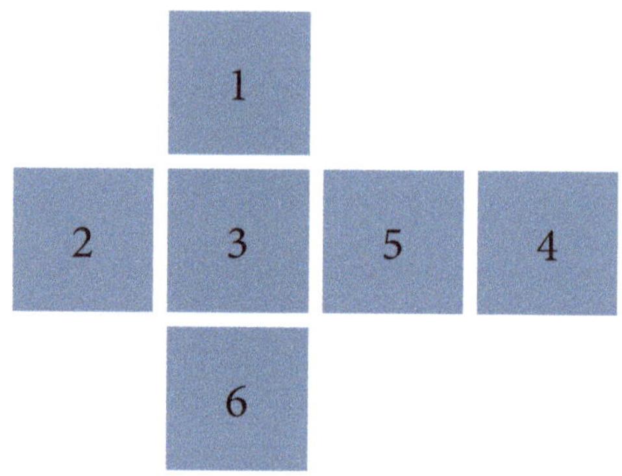

This table lists what each roll of the cube means in terms of the game:

What the face-up side of the cube is	1	2	3	4	5	6
What player does who just rolled	Take 1 chip from pot	Take 2 chips from pot	Take all chips from pot	Put 1 chip in pot	Put 2 chips in pot	Every player puts 2 chips in pot

Players agree on what order they will take turns rolling. The first player rolls the number cube and follows the instructions (see above table). If the pot gets too low or empty (e.g., after a player rolls a 3), it gets renewed by each player putting in 2 chips before the next player's turn. A player who runs out of chips is eliminated from the game.

Assume there are 5 players playing. Let's say that each player starts off with 12 chips and that the pot starts off with 10 chips in it. Suppose the first player is Elena, and the second player is Alonzo.

1. What is the **expected value** (i.e., mean) of the number of chips that Elena will win on her first turn? (To get credit, you must show your work and reasoning. Hint: see formula in textbook)

2. What is the probability that Elena will have an increase in the number of chips she has after she takes her first turn? (Remember to show your work and reasoning.)

3. Considering the six possible outcomes that can happen with Elena's first turn, what is the **expected value** (i.e., mean) of the number of chips that will be in the pot when it's time for Alonzo to take his first turn? (Remember to show your work and reasoning.)

4. Is the **expected value** of the number of chips that Alonzo wins with his first turn less than, greater than, or equal to the correct answer to Question #1? (You don't have to calculate the exact number to be able to answer the question, but remember to give your reasoning.)

5. Based on your answer to question #4, is it accurate to say that Elena (the first player) has an advantage over Alonzo (the second player), or that Alonzo has an advantage over Elena, or does neither of them have an advantage over the other person? Explain the thinking behind your answer.

6. For each row in this table, circle YES or NO in the third column:

Event 1	Event 2	Are Events 1 and 2 **independent**?
Elena rolls a '6' on her first turn	Elena rolls a '6' on her second turn	YES NO
Elena rolls a '6' on her first turn	Elena rolls a '5' on her second turn	YES NO
Elena rolls a '6' on her first turn	Elena rolls a '5' on her first turn	YES NO
Elena rolls a '6' on her first turn	Alonzo rolls a '6' on his second turn	YES NO
Elena rolls a '6' on her first turn	Alonzo rolls a '5' on his second turn	YES NO
Elena rolls a '6' on her first turn	Alonzo rolls a '5' on his first turn	YES NO
Elena rolls a '6' on her first turn	Alonzo rolls a '6' on his first turn	YES NO

For each of the following statements, circle your level of agreement or disagreement, using this scale:

1 = strongly disagree
2 = disagree
3 = lightly disagree
4 = lightly agree
5 = agree
6 = strongly agree

7. Before today, I was familiar with this game (or a slight variation of it).

   1 2 3 4 5 6

8. Thinking about this game increased my interest in the math/statistics content.

   1 2 3 4 5 6

9. Thinking about this game helped my understanding of the math/statistics content.

   1 2 3 4 5 6

10. If I end up teaching math/statistics to children in this region, I would be interested in using this game as a way to engage the children with the content.

    1 2 3 4 5 6

11. Any other comments or questions?

APPENDIX B ("Mexican Culture Version" of Worksheet)

Many cultures have games involving a "spinning top". **Toma Todo** (sometimes called "la pirinola", which is the name of the top used in the game) is a game common in México, and some other parts of Latin America. There are several versions of the rules existing, but the following is a composite of the most standard versions of the children's game (i.e., we are not considering variations some adults may have later modified for gambling or drinking):

In **Toma Todo**, players sit in a circle and each start with an equal number of (let's say, ten) items (e.g., we will say **chips,** but instead of using colored plastic chips, players could instead use uncooked beans, raisins, nuts, stones, buttons, wrapped hard candies, etc.). Each player places one chip in a center pile called the "**pot**". Before play begins, players decide on whether to say the winner will be the person with the most chips after a fixed number of rounds (a **round** consists of each player getting one turn to spin the pirinola top) or to say the winner will be the last player remaining after all other players have run out of chips (this can take a large number of rounds!).

The **pirinola** is a six-sided wood or plastic top or "topa" available in physical and online Mexican gift shops and can also be handmade by sticking a toothpick halfway through a hexagonal piece of cardboard divided into 6 triangles as shown on the right:

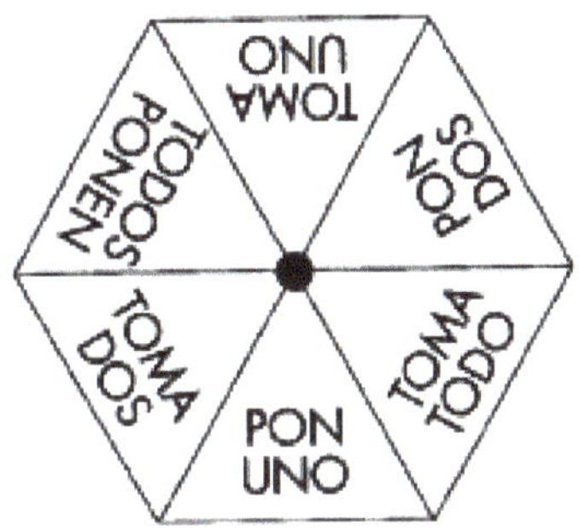

This table lists what Spanish words are printed on each of the six sides of the pirinola, along with what it means in terms of the game:

What the face-up side of the pirinola is	Toma uno (take 1)	Toma dos (take 2)	Toma todo (take all)	Pon uno (put 1)	Pon dos (put 2)	Todos ponen (all put)
What player does who just spun	Take 1 chip from pot	Take 2 chips from pot	Take all chips from pot	Put 1 chip in pot	Put 2 chips in pot	Every player puts 2 chips in pot

Players agree on what order they will take turns spinning. The first player spins the pirinola and follows the instructions (see above table). If the pot gets too low or empty (e.g., after a player spins "toma todo"), it gets renewed by each player putting in 2 chips before the next player's turn. A player who runs out of chips is eliminated from the game.

Assume there are 5 players playing. Let's say that each player starts off with 12 chips and that the pot starts off with 10 chips in it. Suppose the first player is Elena, and the second player is Alonzo.

1. What is the **expected value** (i.e., mean) of the number of chips that Elena will win on her first turn? (To get credit, you must show your work and reasoning. Hint: see formula in textbook)
2. What is the probability that Elena will have an increase in the number of chips she has after she takes her first turn? (Remember to show your work and reasoning.)
3. Considering the six possible outcomes that can happen with Elena's first turn, what is the **expected value** (i.e., mean) of the number of chips that will be in the pot when it's time for Alonzo to take his first turn? (Remember to show your work and reasoning.)
4. Is the **expected value** of the number of chips that Alonzo wins with his first turn less than, greater than, or equal to the correct answer to Question #1? (You don't have to calculate the exact number to be able to answer the question, but remember to give your reasoning.)
5. Based on your answer to question #4, is it accurate to say that Elena (the first player) has an advantage over Alonzo (the second player), or that Alonzo has an advantage over Elena, or does neither of them have an advantage over the other person? Explain the thinking behind your answer.
6. For each row in this table, circle YES or NO in the third column:

Event 1	Event 2	Are Events 1 and 2 **independent**?
Elena spins 'toma dos' on her first turn	Elena spins 'toma dos' on her second turn	YES NO
Elena spins 'toma dos' on her first turn	Elena spins 'toma uno' on her second turn	YES NO
Elena spins 'toma dos' on her first turn	Elena spins 'toma uno' on her first turn	YES NO
Elena spins 'toma dos' on her first turn	Alonzo spins 'toma dos' on his second turn	YES NO
Elena spins 'toma dos' on her first turn	Alonzo spins 'toma uno' on his second turn	YES NO
Elena spins 'toma dos' on her first turn	Alonzo spins 'toma uno' on his first turn	YES NO
Elena spins 'toma dos' on her first turn	Alonzo spins 'toma dos' on his first turn	YES NO

Volume 3 Teaching

For each of the following statements, circle your level of agreement or disagreement, using this scale:

1 = strongly disagree
2 = disagree
3 = lightly disagree
4 = lightly agree
5 = agree
6 = strongly agree

7. Before today, I was familiar with this game (or a slight variation of it).

   1 2 3 4 5 6

8. Thinking about this game increased my interest in the math/statistics content.

   1 2 3 4 5 6

9. Thinking about this game helped my understanding of the math/statistics content.

   1 2 3 4 5 6

10. If I end up teaching math/statistics to children in this region, I would be interested in using this game as a way to engage the children with the content.

   1 2 3 4 5 6

11. Any other comments or questions?

APPENDIX C ("Jewish Culture Version" of Worksheet)

Many cultures have games involving a "spinning top." **Dreidel** (which is the name of the top used in the game) is a popular game in Jewish culture. There are several versions of the rules existing, but the following is a composite of the most standard versions of the children's game (i.e., we are not considering variations some adults may have later modified for gambling or drinking):

In **Dreidel**, players sit in a circle and each start with an equal number of (let's say, ten) items (e.g., we will say **chips**, but instead of using colored plastic chips, players could instead use uncooked beans, raisins, nuts, stones, buttons, foil-wrapped chocolate coins, etc.). Each player places one chip in a center pile called the "**pot**". Before play begins, players decide on whether to say the winner will be the person with the most chips after a fixed number of rounds (a **round** consists of each player getting one turn to spin the dreidel top) or to say the winner will be the last player remaining after all other players have run out of chips (this can take a large number of rounds!).

The **dreidel** is a four-sided wood or plastic top available in physical and online Jewish gift shops and can also be handmade by sticking a toothpick halfway through a square cardboard shape divided into 4 triangles as shown on the right:

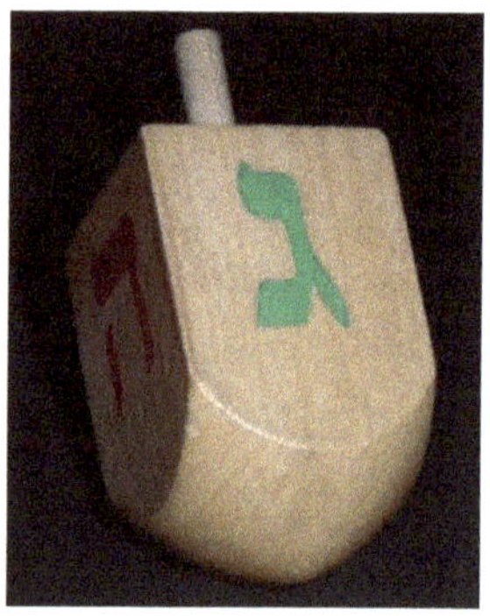

This table lists what Hebrew letters are printed on each of the four sides of the dreidel, along with what each means in terms of the game:

What the face-up side of the dreidel is	נ (the letter "nun")	ג (the letter "gimel")	ה (the letter "hay")	ש (the letter "shin")
What player does who just spun	Take no chips from the pot	Take all chips from the pot	Take half of the chips from the pot	Put 1 chip in the pot

Players agree on what order they will take turns spinning. The first player spins the dreidel and follows the instructions (see above table). If the pot gets too low or empty (e.g., after a player spins "gimel"), it gets renewed by each player putting in 2 chips before the next player's turn. A player who runs out of chips is eliminated from the game.

Assume there are 5 players playing. Let's say that each player starts off with 12 chips and that the pot starts off with 10 chips in it. Suppose the first player is Elena, and the second player is Alonzo.

1. What is the **expected value** (i.e., mean) of the number of chips that Elena will win on her first turn? (To get credit, you must show your work and reasoning. Hint: see formula in textbook)

2. What is the probability that Elena will have an increase in the number of chips she has after she takes her first turn? (Remember to show your work and reasoning.)

3. Considering the four possible outcomes that can happen with Elena's first turn, what is the **expected value** (i.e., mean) of the number of chips that will be in the pot when it's time for Alonzo to take his first turn? (Remember to show your work and reasoning.)

4. Is the **expected value** of the number of chips that Alonzo wins with his first turn less than, greater than, or equal to the correct answer to Question #1? (You don't have to calculate the exact number to be able to answer the question, but remember to give your reasoning.)

5. Based on your answer to question #4, is it accurate to say that Elena (the first player) has an advantage over Alonzo (the second player), or that Alonzo has an advantage over Elena, or does neither of them have an advantage over the other person? Explain the thinking behind your answer.

6. For each row in this table, circle YES or NO in the third column:

Event 1	Event 2	Are Events 1 and 2 **independent**?
Elena spins 'nun' on her first turn	Elena spins 'nun' on her second turn	YES NO
Elena spins 'nun' on her first turn	Elena spins 'hay' on her second turn	YES NO
Elena spins 'nun' on her first turn	Elena spins 'hay' on her first turn	YES NO
Elena spins 'nun' on her first turn	Alonzo spins 'nun' on his second turn	YES NO
Elena spins 'nun' on her first turn	Alonzo spins 'hay' on his second turn	YES NO
Elena spins 'nun' on her first turn	Alonzo spins 'hay' on his first turn	YES NO
Elena spins 'nun' on her first turn	Alonzo spins 'nun' on his first turn	YES NO

For each of the following statements, circle your level of agreement or disagreement, using this scale:

1 = strongly disagree
2 = disagree
3 = lightly disagree
4 = lightly agree
5 = agree
6 = strongly agree

7. Before today, I was familiar with this game (or a slight variation of it).
8. Thinking about this game increased my interest in the math/statistics content.
9. Thinking about this game helped my understanding of the math/statistics content.
10. If I end up teaching math/statistics to children in this region, I would be interested in using this game as a way to engage the children with the content.
11. Any other comments or questions?

(Ethno)mathematical Tasks in the Context of Proportional Reasoning

Filipe Sousa
CIEC, Institute of Education
University of Minho
filipe.fcm@gmail.com

Pedro Palhares
CIEC, Institute of Education
University of Minho
palhares@ie.uminho.pt

Abstract

Investigating mathematical knowledge used in everyday contexts of fishing communities has been our area of interest in mathematics education. Initially we investigated the fishing community of Câmara de Lobos (Madeira - Portugal) and later we also investigated the largest Portuguese fishing community (Caxinas - Vila do Conde in the North of Portugal). The initial purpose was to collect everyday mathematical knowledge of these fishing communities and then contextualize it within the curriculum documents that guide mathematics teaching. A second purpose was the development of mathematical tasks based on the information previously gathered from these fishing communities, for future use in the classroom.

Keywords:

- Everyday Mathematics

Introduction

This research was developed in order to address some specific aspects which broadly fall into mathematics education, making it relevant to investigate the existing mathematical knowledge in specific cultural contexts, in this case, Portuguese fishing communities; their use in the classroom, seeking to improve the teaching and learning of mathematics in such specific contexts, and improving the educational success of children from these contexts.

The theoretical foundation of this research is centered in ethnomathematics, in its various aspects, but especially in the investigative and educational aspects. Investigative because it seeks to collect / recover mathematical knowledge of cultural daily life of the members of the fishing communities of Câmara de Lobos and Caxinas; educational because it seeks to recover and carry this mathematical knowledge to the classroom, trying to help teachers and improve the quality of mathematics teaching and learning by the consideration of these communities students mathematical reasoning in classroom.

As first step we gathered information in the fishing settings of Câmara de Lobos and Caxinas, with caulkers (in boats building) and with fishermen and their families in their daily lives. The information collected fits the mathematical topic related to proportionality and proportional reasoning as being the one fitting more significantly with the mathematics used in the daily life of members of these communities. In this context, the methodology fits the qualitative benchmarks focusing on ethnography and multiple case studies. Two cases were considered, constituted by two sets of students: one is a set of students of a fishing community in Caxinas, and the other a set of students of a more urban location away from the shore. Data collection tools included: participant observation, document analysis (boats projects, tasks implemented in the classroom), video recordings, field notes and interviews. The use of these instruments has enabled the collection of information from the fishing communities of Câmara de Lobos and Caxinas. After transcription of caulkers interviews and of information taken from field notes, we selected data which was analyzed and organized descriptively and with a strong interpretive component, as there was an interest on particularizing and understanding the particular, instead of generalizing.

Based on data obtained after long periods of time in the fishing communities settings, a few tasks were developed (in the mathematical umbrella of proportional reasoning), framed in problem solving, validated both by the subjects themselves, but also by a panel of experts in mathematics education and ethnomathematics.

But the main motivation for this research is related to the possibility of passing the everyday context of fishing communities to the classroom teaching of mathematics also exploring the possible influence of school mathematics in students' performance. In a first phase we collected information in the ethno(mathematics) used within the fishing community. In a second phase, and based on the previous information, we constructed mathematical tasks validated both by the elements of the fishing community and by a panel of specialist in mathematics education.

As the data collection was more productive in the techniques used in the construction of boats, a very interesting topic from both a

NOTES

mathematics education and ethnomathematics point of view arose which was proportional reasoning and in special direct proportionality.

Another relevant aspect to the development of this research was the scarcity of ethnomathematics research on fishing communities in the context of proportional reasoning (direct proportionality).

The start point for the present research occurred in 2006, when we finished a comprehensive collection in the fishing community of Câmara de Lobos of mathematical knowledge used in day-to-day, verifying that due to the low education level of this community members, sometimes this everyday knowledge was clearly beyond the knowledge they had acquired in schools (Author, 2015; Author, 2006; Author, 2008; Author, 2010; Pires, 2008).

Ethnomathematics

In this study we considered appropriate D'Ambrosio's definition of an etymological nature of the word "Ethnomathematics", as the "modes, styles, and techniques (*tics*) of explanation, of understanding, and of coping with the natural and cultural environment (*mathema*) in distinct cultural systems (*ethnos*)" (D'Ambrosio. (1999, p. 146).

Ethnomathematics supports a cultural approach to mathematics, because each culture develops and uses its mathematics. A mathematics education based on a cultural approach will necessarily imply curricular reorientation that includes different mathematical contributions of cultural groups. We are advocating a curriculum that works as a link between educational theory and teaching practice, necessarily considering cultural values, recognizing, valuing and integrating all existing mathematics.

Proportional Reasoning

Proportionality is a mathematical theme used by people for several millennia. In nature we can find situations or artifacts involving proportionality, particularly measures and geometric shapes related to the golden ratio.

Proportionality and proportional reasoning are typically fundamental concepts of school mathematics. They allow students to establish connections with other mathematical topics, with other disciplines and with students' daily life (Ponte, 2006), presenting itself as the pinnacle of elementary mathematics and a foundation of advanced mathematics (Lesh, Post & Behr, 1988; Silvestre & Ponte, 2005).

The concept of proportionality seems an easy concept, but in reality proves to be quite difficult and it is estimated that over 90% of adults do not reason proportionally (Lamon, 2007). Still, we must distinguish between proportional reasoning and proportionality. Proportionality involves relationships between magnitudes, may refer to a situation in which there is an invariable relationship between two covariate quantities. Proportional reasoning is used to understand contexts and applications based on proportionality (Lamon, 2005). It implies both an understanding of a constant relationship between two quantities as well as the perception that these same quantities are related and vary jointly.

In proportionality it is important to understand the concepts of ratio, proportion, as well as their meanings, applications and properties.

The type of tasks used is also an important issue because it defines the approach in the context of proportional reasoning, but also the strategies to be used. Vergnaud (1983, 2014) proposed the theory of conceptual fields based on multiplication and division. He proposes three multiplicative classes: measures isomorphism, product of measures and multiple proportion. Other types of tasks can be offered to students such as: problems of missing value, comparison problems, transformation problems, average value problems, problems involving the conversion between ratio, rate and fractions, problems involving measurement units, problems of conversion between representation systems and pseudo proportional problems.

In this work we have investigated the mathematics used in the daily life of the members of fishing communities. We have verified that they regularly use mathematical knowledge involving proportional reasoning. Given this information, it is pertinent to understand how children of one of these communities would react in a classroom situation, to proportional reasoning problems.

Methodology

Research has been of qualitative type, as we tried to investigate the problem without isolating it from its cultural context, seeking to understand using impartial, inductive and descriptive processes (Amado, 2013).

Our view is based on the assumptions of the interpretative paradigm, defending the subjective study of human experience and in this study trying to approach it from the standpoint of the members of the fishing community (Sarmento, 2003; Cohen & Manion, 2002; Ramos & Naranjo, 2014), in which the interest is to particularize instead of generalizing (Vilela, 2009).

The research designs used in this study are ethnography and case study. Ethnography is used in the first phase of data collection where the researcher remained within the fishing communities of Câmara de Lobos and Caxinas for considerable periods of time, collecting faithfully information of the everyday life of these communities, especially in the context of building boats. In the second phase we used multiple case studies where the researcher, in the context of the classroom implemented mathematical tasks, intentionally and carefully prepared as to being similar to real experiences in a fishing community. The set of tasks was introduced in classrooms of two schools in different cultural settings. In one school all the tasks were introduced to a group of students all from a fishing community, while in the other the students were from a more urban and far from the shore.

In the fishing community of Câmara de Lobos the entrance on the ground has proved difficult, as it is a community somewhat closed in itself. In the fishing community of Caxinas entry into the community was easier and faster. Two important rules were followed: a) the consent of the people being researched; b) their protection from any damages (Bodgan & Biklen, 1994).

The fishing community of Câmara de Lobos and its territory is a landmark of the county, for its picturesque aspect, people's strong personality and the beautiful bay where fishermen repair their boats and prepare fishing material. They live and interact "isolated" from the surrounding society. They have acquired over time a way of thinking and acting, a way of life and a culture that characterizes as unique. Although disparaged by the remaining society, the fishermen neighborhood dominates the village and fishermen are the majority. They are characterized by their shyness, they are not very expansive but sure are very supportive one to the other.

Caxinas is a major population center to the north of Vila do Conde, which concentrates the largest fishing community in Portugal.

NOTES

The place where now stands Caxinas began to be occupied by some fishermen in 1840. Until the 70s they lived in very precarious situation, and like other fishing communities, were for decades socially discriminated, despised and stigmatized. They have developed a cultural identity, marked by a particular slang and behavior patterns, showing however some approximation to the other folks of the town, the result of schooling.

This research emphasizes the socio-cultural component of the fishing communities which meets the essence of ethnography. Based on Ethnomathematics, we tried (in the first phase) to collect accurate information to describe in detail the mathematical knowledge of the participants.

Data collection involves a close contact between researcher and researched and the use of various techniques that enable obtaining relevant information. We used participant observation, unstructured interviews, document analysis and video recording. It was participant observation in the strict sense that in the classroom the researcher was having contact with the groups working, questioning their reasoning and ways of thinking.

Regarding data analysis, it was carried out to compare information provided by various sources, presented descriptively and with a strong interpretive component.

We have constructed some tasks, two of them will be presented below.

Task 1:

Task Type: Problem of inverse proportionality.

> *The "God provides" travels the distance between the fishing port of Sesimbra and the fishing port of Matosinhos in 8 hours at 25 miles per hour. How long will it take to make the same trip if the average speed was 16 miles per hour?*

Description and Development of the Task

The task was presented to students through a short statement, and all necessary information is presented in the statement itself. There is no image, scheme or tip that could "help" students to interpret and induce a resolution strategy. Portrays a fishermen's common situation, which in addition to requiring mathematical calculations, also imposes a proportional reasoning somewhat different in its resolution.

The task is part of the theme proportionality, however involves a proportional reasoning that requires from the students a careful analysis of the information provided, also involves knowledge of inverse proportionality. Proportionality is worked on in schools, but the focus is completely in direct proportionality. Still and since proportional reasoning is not synonymous with proportionality (Lamon, 2005), we think that students may be able to develop strategies to answer this problem. It may also be a way to ascertain students' informal knowledge especially of those coming from the fishing communities.

Students, organized in groups of three to four elements, enabling discussion of ideas and reasoning, would have to make a careful reading so that they could correctly interpret the information.

This task does not require only the use of relations and reasoning. In other tasks there was a direct proportionality relation, in that as a magnitude increases, the other magnitude increases in proportion. In this case also there is a proportional relation, but as the speed increases, the time decreases. Apparently it seems a problem that goes beyond the mathematical knowledge of students and therefore they will not have enough tools to solve it. However, if students develop proportional reasoning in order to seek an intermediate magnitude (distance between the two sea ports), the problem becomes a problem of direct proportionality. Students must reason as follows based on the information provided in the statement: the boat travels in one hour 25 miles. How many miles will travel in 8 hours? Using the strategy that they deem most appropriate, they can conclude that in eight hours it will travel 200 miles, corresponding to the distance between the two ports. Once this intermediate magnitude is found, this is a direct proportion problem with which students are more familiar. Applying the cross product strategy or the "rule of three", they can calculate easily and quickly the time needed to travel the same distance at an average speed of 16 miles per hour. Of course, students may choose other strategies with which they identify better or in alternative they think to be more effective. Still, it is expected that students show their proportional reasoning to establish a proportional relationship between the magnitudes involved mainly multiplicative relationships within and between variables. The estimated time for the resolution of this task was 50 minutes.

Task 2:

In everyday life, with some frequency, come up situations that seem to be one thing, but, well analyzed, are not what at first seemed to us. In the context of direct proportionality, there are real-life situations that appear to involve direct proportionality but in reality this is not the case.

The task presented below is one such case, and may be suitable for students distinguish real situations involving relations of direct proportionality from other situations that are not of direct proportionality, and are hereby designated as pseudo proportional problems.

Task type: pseudo proportional problem.

When a sardine is born, its length is 1.3 cm. With 12 months of age its length is 14 cm. What will be the length of a sardine with 48 months?

Description and Development of the Task

The scenario presented in this task is a usual situation of everyday fishing, in which the results, denounce the situation and brings immediately the perception that it is not a situation of direct proportionality. Especially for the students of the fishing community of Caxinas it is expected that this happens, as it is not reasonable to have a sardine 56 cm long when 4 years old. Besides being rather strange to see such a large sardine, students of the fishing community should be aware that a sardine has an age limit not exceeding four years and a measure of length not exceeding 21 centimeters (Coelho, 2009; Castello, 2015).

The task is presented in a simple way, with a small statement, little information and numbers which are easy to operate with. In fact, they are small numbers and with all the conditions to be operated through mental arithmetic.

The most important part of the task is supposed to be the justification of the answer.

This task has been purposely prepared to determine if students are able to distinguish situations of direct proportionality from situations in which there is no such relationship.

Students in small groups of three or four elements, should resolve this problem and justify the answer they give in an approximate time of 40 minutes. Each group will have

NOTES

to analyze the problem, select strategies and discuss ideas related to the answer and respective justification.

It is expected that students foresee that the situation proposed does not fit in a direct proportionality situation.

It is also expected that they are capable of presenting valid justifications with strong arguments. It will be a good sign that the knowledge they have on direct proportionality, are properly mobilized and adjusted to the degree of difficulty of the task.

As referred before, these tasks, focusing on proportional reasoning used in the daily activities of the fishing community, display some of the mathematics the fishermen and the remaining members of the community use (chores, boat construction, daily fishing on land, etc.). As the students of one of the schools are from a fishing community we expected that perhaps their performance in these tasks matched their adult family members. If true, then before being taught proportionality, one could reasonably expect that these students would perform better than the students of the other school away from shore.

Conclusion

Throughout the research it was important to collect information relating to some fishing communities' cultural contexts, trying to realize what mathematical skills are used in the everyday life of these communities. The comprehensive collection of information on the ground, gave us ground to realize that proportional reasoning is used in both fishing communities and is very important in their day-to-day.

The primary interest of this research was focused on mathematics education, particularly in classroom work (There are no strong evidences of the connections of ethnomathematics to classroom work). Thus, research was directed accordingly, producing significant mathematical tasks for students of the fishing communities, but significant also from a mathematical point of view. We consider them relevant tasks because: (a) their construction took into account the cultural background of students; (b) they fit into curriculum documents, involving a mathematical topic considered of immense importance in the mathematical development of students; (c) these are tasks that by their nature (problem solving) promote the development of reasoning and the ability to develop mathematical skills necessary for students' daily life; d) these are tasks validated by the scientific community and by those investigated and therefore can be considered good tasks for teachers to use in classroom considering they were constructed in accordance with the common mathematical knowledge of the members of the fishing community, that they use on their daily activity.

References

Amado, J. (2013). A investigação em educação e seus paradigmas. In: *João Amado (Coord.) Manual de Investigação qualitativa em educação*. Coimbra: Imprensa da Universidade de Coimbra.

Bogdan, R. & Biklen S. K. (1994). *Investigação Qualitativa em Educação: Uma introdução à teoria e aos métodos*. Porto: Porto Editora.

Castello, J. P. *Síntese sobre distribuição, abundância, Potencial pesqueiro e biologia da sardinha verdadeira (sardinella brasiliensis)*. Departamento de Oceanografia. Fundação Universidade do Rio Grande. Accessed in February 2015. Available in https://www.google.pt/#q=JORGE+PABLO+CASTELLO+sardinha

Coelho, P. C. T. (2009). *Modelação do crescimento de Sardina pilchardus (Walbaum, 1792) em duas zonas (oceanograficamente distintas) da costa Portuguesa.* Tese de mestrado. Faculdade de Ciências do mar e do ambiente. Universidade do Algarve.

Cohen, L. & Manion, L. (2002). *Métodos de investigación educativa.* Madrid: La Muralla S.A.

D'Ambrosio. (1999). Literacy, Matheracy, and Technoracy: A Trivium for Today. *Mathematical Thinking and Learning 1*(2), 131-153.

D'Ambrosio. (2001). *Ethnomathematics: Link Between traditions and modernity.* Rotterdam: Sense Publishers.

Lamon, S. (2005). *Teaching fractions and ratios for understanding: Essential content knowledge and instructional strategies for teachers* (2ªed.). Mahwah, NJ: Lawrence Erlbaum.

Lamon, S. (2007). Rational numbers and proportional reasoning. In: F. Lester (Ed.), *Second handbook of mathematics teaching and learning* (pp. 629-667). Greenwich, CT: Information Age Publishing.

Lesh, R.; Post, T., & Behr, M. (1988). Proportinal reasoning. In J. Hiebert & M. Behr (Eds.) *Number Concepts and Operations in the Middle Grades* (pp. 93-118). Reston, VA: Lawrence Erlbaum &National Council of Teachers of Mathematics.

Ponte, J. P. (2006). Números e álgebra no currículo escolar. In: I. Vale, T. Pimentel, A. Barbosa, L. Fonseca, L. Santos & P. Canavarro (Eds.), *Números e álgebra na aprendizagem da matemática e na formação de professores* (pp. 5-27). Lisboa SEM-SPCE.

Ramos, S. T. C. & Naranjo, E. S. (2014). *Metodologia da Investigação científica.* Lobito: Escolar Editora.

Sarmento, M. J. (2003). O Estudo de Caso Etnográfico em Educação. In: N. Zago; M. P. Carvalho; R. A. T. Vilela (Org.). *Itinerários de Pesquisa: Perspectivas qualitativas em Sociologia da Educação* (pp. 137-179). Rio de Janeiro: DP&A Editora.

Silvestre, A. & Ponte, J. P. (2005). Uma experiência de ensino da proporcionalidade no 2.° ciclo do ensino básico. In: Atas do Encontro 2005. *Números e Álgebra na Aprendizagem da Matemática e na Formação de Professores*. Sociedade Portuguesa de Investigação em Educação Matemática: Caminha.

Vergnaud, G. (1983). Multiplicative structures. In: R. Lesh & M. Landau (Org.), *Acquisition of mathematics concepts and processes* (pp. 127-174). New York, NY: Academic Press.

Vergnaud, G. (2014). *A criança, a matemática e a realidade.* Curitiba, UFPR. (Maria Lucia Faria Moro).

Vilela, J. (2009). *Investigação. Processo de Construção do Conhecimento.* 1ª Edição. Lisboa: Edições Sílabo.

Culturally-Relevant Algebra Teaching: The Case of African Drumming

Janet Sharp
sharp@oakland.edu
248.370.2613
Associate Professor, Mathematics Education
Oakland University
Teacher Development and Educational Studies
485D Pawley Hall
Rochester, Michigan48309-4494

Anthony Stevens
Chair, Psychology Department
Des Moines Area Community College
PhD student
Iowa State University
Percussionist, The Bone People
Ames, Iowa

Abstract

Aside from being its own field of study, algebra has long been a servant to upper division courses and arguably to productive citizenship. Algebra plays a key role in courses like chemistry and trigonometry, which are the kinds of upper level courses required of high-tech, usually high-paying, occupations such as engineering. Keeping learners out of upper division courses also means limiting their potential for active citizenship in society and economic success (Povey, 2003). Many (see Burton, 2003) argue more philosophically that access to and knowledge of mathematics leads to an empowered populace, making it a key to social justice. Unfortunately, algebra can appear to be a tool of injustice when it protects upper division mathematics courses from potential students. Intentions are understandable since success in those courses requires algebraic thinking. Under-prepared students often fail. But, the net result has been to preclude students from those courses rather than to ensure students are prepared for those courses. Robert Moses (Moses & Cobb, 2001) suggests that algebra must move beyond its role as a pre-requisite hurdle. He refers to algebra as a *civil right*, suggesting that all learners have the *right* to know algebra. Viewing algebra in this manner elevates its justice status, allowing all students greater opportunity to economic success.

If such equity is a desirable goal, the mathematics education community should develop teaching strategies that resonate with all learners. For algebra teachers to provide teaching and learning opportunities that promote access for more students, we need insightful pedagogical knowledge and deep algebra knowledge. Practicing culturally relevant pedagogy is an approach that may require teachers to develop these kinds of knowledge. By posing problems that are relevant (Hart, 2003) and that allow students to respect, appreciate and/or celebrate other cultures (Boyer, 1990) teachers' pedagogical knowledge may become stronger. Driscoll (1999) suggests that a useful strategy to support efforts to deepen algebra knowledge is to provide teachers with experiences learning algebra in new ways and in unexpected settings. Together these kinds of knowledge impact teachers practice. In this paper, we describe an atypical context, African/Afro- Cuban drumming that has the potential to connect teachers' knowledge about pedagogy and algebra by stimulating algebra learning.

Teacher knowledge

Deep content knowledge is significant because it frees teachers from having to think about their own mathematics knowledge and instead "to attend to the mathematics in what children say and do" (Schifter, 2001, p. 71). Teachers who know mathematics deeply have forged many conceptual connections, both within and outside of mathematics. Ma and Kessel (2001) put forward the notion that teachers' mathematical knowledge is linked to pedagogical knowledge in ways that could not and should not be separated. They note that *profound* mathematical knowledge manifests itself in a teacher's ability to "reveal and represent ideas and connections in terms of mathematics teaching and learning" (p. 12). When teachers develop profound knowledge, they understand something about the nature of learning that allows them to engage with students, recognizing unique ways of thinking about a problem. That is, teachers more acutely understand thinking and the learning process (Ma, 1999).

Much of the weight of learning about mathematical connections should be carried by mathematics teacher educators first, so it can be transferred to teachers. Driscoll (1999) provides insights about content for professional development in the specific case of algebraic thinking and learning. When teachers discover algebraic connections in what he calls "unexpected" places, their pedagogy is invigorated. When students experience algebra lessons in unusual settings, they are encouraged to develop algebraic habits of mind. Creating innovative lessons requires teachers to rethink their understanding of algebra and embrace goals and methods that differ from traditional presentations. This is a somewhat recursive experience. As teachers explore algebraic ideas in atypical settings, they gain new insights about how to provide this sort of instruction. And, as Ma & Kessel (2001) noted, pedagogy and content become/are intertwined.

Culturally-Relevant Algebra Teaching and Content

The premise here is that by adapting existing curricula to tap into the different knowledge bases of all students, we provide arenas for studying algebra in unexpected contexts, thus providing more opportunity for success in the critical algebra pre-requisite. This suggestion is a tall order. But, it is in line with theory about how students learn mathematics. Pirie & Kieren (1994) suggest that mathematical learning must begin with personal knowledge in order to invest the student in the process of learning. Most algebra-based textbooks include a variety of word problems with a goal of providing learners with real-world situations. Many contexts are restricted to the expected, scientific situations like paddling up and down rivers, two trains meeting on railroad tracks, finding areas of

NOTES

rectangular gardens, or mixing chemical solutions. Although these are situations in which algebra should be employed, there are additional contexts that require rigorous algebraic thinking and learning. Cultural context have the added benefit of drawing from a more diverse collection of learners' personal knowledge bases.

As teachers, we tend to teach like we were taught (Lortie, 1975; Russell, 1997; Schifter, 1997). If we learned in classrooms where teachers literally followed traditional textbooks, it would be no surprise that we teach algebra using the same kinds of *expected* contexts –such as paddling up and down a river. To create new, innovative lessons from culturally-relevant contexts, teachers will need to expand their experiences and knowledge about algebra or teacher educators will have to provide such experiences. That is, we all must learn to "be alert to finding algebraic thinking in ostensibly nonalgebraic settings" (Driscoll, 1999, p. 21).

Algebraic thinking in these new contexts must retain rigorous mathematics. The tenor of the lessons must be coherent and purposeful. A short activity on weaving or pottery, is not enough to promote algebra to a tool for social justice (Weist, 2001). To develop and present a robust collection of lessons that honors a culture, teachers must know about cultural connections to mathematical thinking and related content. Addressing culturally-relevant connections in a classroom requires teachers to study and understand a culture before they can respect and celebrate it mathematically. Once mathematics teachers gain new connections to mathematical knowledge, because of their study of another culture, passing a celebratory attitude to their students may become more natural. As mathematics knowledge becomes visibly intertwined with cultural knowledge, rather than isolated from it, teachers forge many cultural connections to the mathematics and they will be more aware of the potential for expanded knowledge among their students (Gerdes, 1998). Ma (1999) refers to this kind of knowledge as *thoroughness* of mathematical knowledge. Hence, we describe the potential of studying some of the history of African drumming.

The cultural component: African/Afro-Cuban Drumming

The creation of a mathematically robust lesson with a meaningful cultural background requires a thick understanding of the culture as well as the mathematics. To begin to understand a culture means qualitatively understanding the roles of both typical members of the society and of the mores and codes established by the leaders of the society. This kind of knowledge can provide some measure of understanding that moves beyond superficial knowledge of a culture and to a more connected, respectful understanding. Therefore, before explaining the mathematics component, we delve slightly further into the culture of African drumming.

Many things help to define a culture. One of the most obvious things is a community- wide event. Things like county fairs, art festivals, or school plays can give a community some identity and provide opportunities to interact. Such defining events are found in all cultures and can be a way for an outsider to gain some understanding of a culture. The events of interest here are the African drumming and dancing ceremonies that trace their roots to many ancient African cultures. While we run the risk of incorrectly implying that it is appropriate to group all African ceremonies together, there are common threads among celebrations from many African cultures: the importance of the drums, drummers, dancers, and village leaders/elders.

The Burundi

In Burundi, drums are sacred and represent, along with the king, the powers of fertility and regeneration. Along with the large ingoma drums made of hollow tree trunks covered with skin, the amashako drums provide a continuous beat, and ibishikiso drums follow the rhythm of the central inkiranya drum. One of the yearly ceremonies of the Burundi people was the Mwami ceremony, during which special drums were used. The drums had their own special hut for storage and took the entire year leading up to the ceremony to be properly prepared (Grund, 1985). The purpose of the Mwami ceremony was to honor the king and his ancestors. These ancient traditions have been preserved and are on display worldwide as the Royal Drummers of Burundi, the former royal percussionists, perform worldwide, demonstrating rhythms passed down from father to son through "the oral tradition" of African drumming (indeed, African culture).

Drummers (members of the culture)

Only master drummers (Batimbo) were allowed to play the drums because only Batimbo had the proper life-long training to play the drums for the month-long ceremony. For a ceremony to be successful, it had to include dramatic dancing and intense fluctuations in heart rates. So, the master drummers had to learn not only to play the music for the ceremony, but also to learn when and how to introduce unplanned rhythms into the music. By reading their audience, ritual drummers, like the Batimbo, learned to improvise the rhythms of the traditional drum music to sway the crowd and the dancers (Charreaux, 1992; Locke, 1987). Professional drummers believe that the human heart is one of many rhythms at work in the human body (Anderson, 1994; Cohen & McFall 1994). The Batimbo were trained to recognize the cumulative pulse of their audience, lock into it, and adjust the music in ways that naturally swayed them (Steed, personal communication, 1984).

Due to the importance of this skill and the importance of the drums, drummers held a noble ranking in the culture. However, this noble ranking could be short-lived for some drummers because the Mwami ceremony was a month-long ceremony and drummers have been known to die from exhaustion at the conclusion of the ceremony. The noble ranking also afforded the drummers the honor of playing the drums for a few days following the ceremony before the drums underwent the year-long preparation for the next year's ceremony (Grund, 1985).

The Mathematical Knowledge of the Drummers

Locking into the crowd's cumulative heartbeat requires mathematical knowledge, but traditionally, drummers weren't taught to view what they were doing as being connected to math. The drummers must catch the brief moment in time when everyone's heartbeat is in synch and then find a way to control that moment and move quickly away from that point. So, how does a drummer use mathematics to do that? Mr. Stevens responds below. [Editors Note: We chose to italicize Mr. Stevens words for emphasis.]

> *I will have to forego the scientific tradition of source referencing information to explain what happens as I drum and teach math; this is because drumming isn't a science in the African/Afro-Cuban tradition, and, much of what is known about African and Afro- Cuban drumming*

NOTES

is from word of mouth histories. This is our culture. This is how I learned it and that is part of the beauty of the African drumming tradition; in its purest form it requires connectedness to the source. It cannot be fully understood from writings alone. It is earthy, breathing, and it is alive. When you drum, when you dance to the drum, you must be connected or you will miss the most important experiences.

Thus, what I've learned, what I know, has come from the source, from elders within the tradition and from those who have studied with and lived among those elders.

I feel as if I must defend the tradition, concerned that it may sound like a bit much to take in for the inexperienced. But, this is how we operate: It isn't science in our hearts it is spirit in our hearts. We feel. Asking us how we do what we do is akin to asking the Shaman where "knowing" comes from. It's like asking an empathic where feeling comes from. It's just there and it's always been there. I was supposed to do this. I'm born to do this. My elders did this and passed it down to me. It's always been there with me. It's like a 6th sense.

The "it" I speak of is an innate knowing of rhythm. And, because rhythm is central to all life and living things we can come to know the state of a person through it. The master drummer understands that intuitively. We are taught to find unity in the crowd we play to. That unity will reflect the contents of the listener's heart. The heart that is clouded will not respond to rhythm in the way that a clear heart responds. The clouded heart resists the unity called for by rhythm and meter. If one wants to see an example of this, watch children when they are in the presence of rhythm. They move instantly and (typically) vigorously. They unite with the rhythm, the message. Their hearts are clear. Watch any concert; sooner or later a foot will tap, a head will nod; that person at that time has responded as evidenced by their movement. Their heart wants to unify with the rhythm and their body tells us so through their movements. This is very natural.

When we play to you we will eventually see your true intentions. You will reveal what is in your innermost being. The doubtful heart moves awkwardly, almost shamefully, with reservation, but it will move (especially when others are moving with it). Anger and rage, hurt and loss, joy and peace are all made visible by our rhythm. We know you well shortly after we begin playing, and we will, if we are attempting to move you, adjust our cadence until we do exactly that.

The above point illustrates the core of the African drumming tradition; we ultimately are concerned more with the heart of a person than with a person's doings. Transform the heart and the rest will take care of itself. The drum can do that. The drum can speak. The drum can heal.

The marriage of drumming and math is to me most special because the elders of the tradition, I would argue, would never have been concerned with this connection. Drumming for them was likely far more serious an endeavor than education or mathematics. They played for birth, for regeneration, for the hearts of their villages, kings, and nations. They played for their gods. As we fast forward from say 13th century Mali to the present, we have become inundated with the products of the scientific movement, the technological movement, the depersonalization movement. Product is now king and students suffer because they lack connectedness to their subjects, their teachers, to each other. The drum does some of its best work here. When I enter a classroom and begin to play I take the students to the most elemental common denominator, movement via rhythm. It's ok to move again, to feel again; and they're feeling math - ratios, fractions, polyrhythms. That's math. The subdivisions are apparent as their hearts, feet, and my drums communicate. You can hear it, you can see it, and you can feel it. It's like the ancient Nigerian tradition of playing a rhythm directly on the body of the student. The student doesn't just watch the hands of the teacher and try to mimic their movements, they feel the pattern. It is internalized. So it is with math and students once they've felt a ratio or a fraction via my drum. It's a whole body type of learning. It's not cerebral it's guttural; natural; it's easy. That's the "system".

Once the student has experienced rhythm they've experienced math. They just don't know it yet. I simply explain what they have experienced in a language we call mathematics…I could have explained it in Swahili instead…math is just another language to me. And languages are important to our oral tradition. In fact, we are trained to vocalize the rhythms we play, to sing them, as we play them. There are entire systems of vocalizations for African drumming. Those systems preserved the rhythms when Africans were enslaved and their drums were taken from them as they were sent throughout the world. So, once a child has felt math there's not much left but the vocalization. This is our culture. And it lends itself perfectly to the teaching of mathematics.

African drumming is a collection of rhythms that are played simultaneously. Just as "polygon" is a term for a geometric shape that is many-sided, "polyrhythm" is a term used to describe music that is made from simultaneously playing many rhythms. These repetitive rhythms are of different lengths and bring different vibrations into play, based on how the drummer strikes the drumhead or the kind of drum used. The drummers may have at their disposal rhythms composed of several different counts, for example, 8 counts, 6 counts, or even 12 counts. They decide which rhythms to play and when to play them based on their read of the audience. In essence, the moment at which the participants' heartbeats are in synch occurs when these different rhythms are aligned. Then, the drummer incorporates the rhythms of different lengths to bring the audience into the music. By analyzing this process algebraically, we can gain new insights into the mathematical thinking of the drummers.

Of course, many rhythms make up African drumming and many drummers would be playing many rhythms at any given moment. For simplicity, in this paper, we analyze only two rhythms: the tumbao (toom-BAH-oh) and the bembe ostinato (BEM-bay aus-tin-AH-toe). The tumbao is an 8 count rhythm and the bembe ostinato is a 6 count rhythm.

NOTES

Analyzing the result of playing these two rhythms simultaneously allows us to understand how the drummer's awareness of and use of mathematics caused the king, dancers, and on-lookers to interact with the songs. The tumbao is a melodic rhythm that incorporates several sounds and vibrations. It is typically played on the tumbadora, a large conga drum (Cohen, Cohen, & Fedele, 1992; Bossierie & Panicka, 1992). The bembe ostinato is a short repetitive rhythm, much less diverse than the tumbao. It incorporates one kind of vibration and tends to be a constant underscore for more diverse rhythms.

Assuming both individual rhythms are played simultaneously, there are moments when they both begin on the same count of "one". Ritual drummers are trained to recognize when the aligned count of "one" matches the dancers' heartbeats. At this moment in the music is the count of the polyrhythm is 1, but the difference between the counts of the individual rhythms is zero. The drummers recognize this moment, and move through the rhythms, which are of different length, in order to fall away from the syncopated heartbeats. For instance, when the tumbao rhythm is on count #7, the bembe ostinato has returned to count #1. So, at that moment, the difference between the individual rhythms (7 – 1) jumps to 6. As dancers react to this drastic change, they experience a kind of rhythmic dissonance that needs to be physically resolved by their bodies. See Figure 1 for some examples of the counts of the polyrhythm and the two individual rhythms being analyzed. Resolving the dissonance results in dramatic swaying and keenly-accentuated dance movements, which are deeply appreciated by the king. The drummers move cyclically through their rhythms, sometimes with the rhythms aligned, when the dissonance is resolved, and sometimes with the rhythms misaligned, when the dissonance is at its height. Each time the rhythms are aligned, the dancers and listeners are brought back into the music. In this manner, the drummer controls the ceremony, which explains their importance in the culture. However, experienced dancers can also silently communicate a desire to maintain the dissonance. In those cases, the drummers introduce flairs and other off-cadence beats that appear to "trick" the audience into thinking that the dissonance is about to resolved but that send the rhythms off into different directions. Thus, maintain the heightened adrenaline and exaggerated dancing moves.

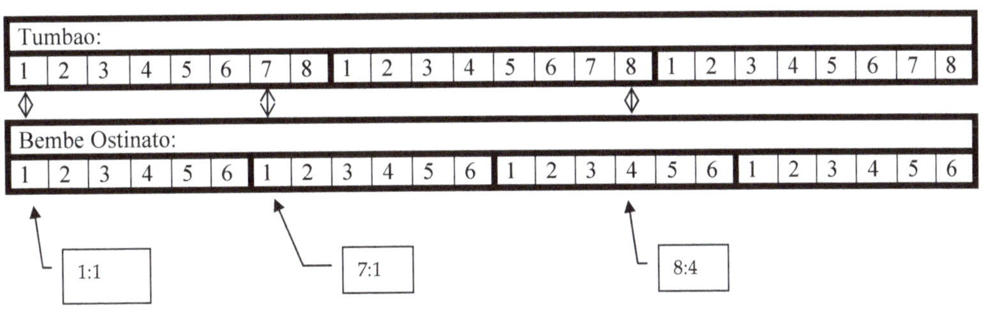

Figure 1 – Counting the Beats of the Two Rhythms

The algebraic component: Functions

Studying the rhythmic dissonance in the different African drumming rhythms within a polyrhythm may be accomplished with algebra. The dissonance may be literally described as the difference between the numerical values of the individual rhythms. By comparing the "counts" of competing rhythms making up the polyrhythm, interesting patterns are formed. Patterns often provide a background for algebraic thinking.

Beginning in Kindergarten, children develop algebraic thinking when they copy, extend, or create patterns (National Council of Teachers of Mathematics, 2000). The patterns in the rhythmic dissonance of African drumming can become a natural place to connect algebra to prior patterning knowledge (Stevens, Sharp, & Nelson, 2001). When the differences of the competing African Drum rhythms are studied, a mathematical structure may be imposed upon the polyrhythm.

Lesson Set - up

The events described here occurred in a summer session, teacher education graduate course in algebra. The algebra teachers hailed from different locations across the United State (e.g., Alaska, Montana, Virginia, Texas and Arizona.) Many of these teachers were pursuing a master's degree. All had taught algebra at least once. As the teachers flowed into the classroom, they were exposed to an on-going video clip of a conga player playing the Tumbao and Bembe Ostinato rhythms simultaneously (Center for Technology in Learning and Teaching, Curry School of Education, & Center for Learning and Teaching in the West, 2006). Naturally the conga player introduced his own flairs into the polyrhythm. Discussion of the mathematical ramifications of his flairs goes well beyond the scope of an initial lesson and furthermore discussion of the flairs is unnecessary to experience the algebra of the polyrhythms. After a few minutes of listening to the rhythms, the teachers then viewed a corresponding video clip of a dancer interacting with the same music.

Discussion questions

The following questions were offered for discussion:

1. Do you notice any consistent patterns in the dancer's movements?
2. What do you feel while listening to the music?
3. What sorts of emotions do you think the dancer is experiencing as she dances to the music?
4. Why do you think the drummer has his eyes closed as he plays the drums?
5. Watch his hands. Do you notice any differences in the way his hands strike the drumhead?

Their answers were used to weave the cultural information about the Burundi (described previously) into the teachers' knowledge bases. They noticed how the drummer's eyes were closed. Our drummer (and co-author, Stevens) says he sees and hears rhythms in his head, as though they exist in three-dimensional space. He imagines vectors projecting along various planes. He feels as though he's in this space and is communicating for the vectors. "The vectors are of differing intensities and lengths. When they are aligned in their distances the rhythm has unified at "the one", the point where the polyrhythm resolves and all players are at the beginning of their patterns together,." says Stevens. This provided the impetus for a brief lecture about the solemn nature of the drummer's role in the African culture to which the drummer in the video felt connected. The teachers noticed the varying intensity of the dancer's movements. This allowed for

NOTES

discussion of the importance of dancing and a brief description of the celebration dance from which the dancer was drawing many of the free-form movements she exhibited. It also allowed us to segue to the need to analyze the numerical differences involved in the music, for it is those differences within the polyrhythm that cause the dancer's movements to vary.

Differences & Patterning

A good way to analyze the mathematics of these rhythms is to first symbolically represent the rhythms. The teachers were introduced to the following symbols developed by Dworksy & Sansby (1994) as a way to represent hand strokes. Four hand strokes were used by the drummer: Open tone (○), Slap (▲), Heel & Toe (◒ & ●) and a rest (~). (note: The Heel & Toe generally appear together as a back and forth movement and so are represented here together. However, they are considered separate hand strokes.) The 8-count tumbao can be represented as ◒ ● ▲ ● ◒ ● ○ ○. The 6-count bembe ostinato can be represented as ○○~○○~. Although some might argue that the rhythm is a 3-count rhythm, it is important to think of it as a 6-count rhythm since that is the manner in which drummers and dancers feel the rhythm. Together the two rhythms look something like Figure 2.

1	2	3	4	5	6	7	8	1	2	3	4	5	6	7	8	1	2	3	4	5	6	7	8
◒	●	▲	●	◒	●	○	○	◒	●	▲	●	◒	●	○	○	◒	●	▲	●	◒	●	○	○
1	2	3	4	5	6	1	2	3	4	5	6	1	2	3	4	5	6	1	2	3	4	5	6
○	○	~	○	○	~	○	○	~	○	○	~	○	○	~	○	○	~	○	○	~	○	○	~
1	2	3	4	5	6	7	8	9	10	11	12	13	14	15	16	17	18	19	20	21	22	23	24

Figure 2 – The Tumbao Rhythm and the Bembe Ostinato Rhythm

Creating this diagram allowed a discussion about polyrhythms. The rhythms move through phases of being in sync (as at the counts of 1, 2, 3, 4, 5, 6 where the rhythms all result in a zero difference) and out of sync (as at counts like 12 where the rhythms creates a non-zero difference of -2.)

Here, the teachers had their first realization of familiar mathematics in the patterns. The rhythms re-aligned every 24 beats, which is the least common multiple of 8 and 6. The dancer's movements were the most calm at counts around the least common multiple and she appeared the most animated at counts further from 24. This happens because immediately following the least common multiple number of beats the rhythms are re-aligned at a count of "one", which is the point at which the dancer is (perhaps unknowingly) brought into the music and to some extent brought under the control of the drummer. Simply viewing the table of handstrokes in the polyrhythm caused a few of the teachers to ponder their prior knowledge. "Wow! I always figured there was a connection between music and math, but I never thought about it this way." And "I've watched some videos of African Dancing before and I it always seemed to me to be so random. But the math here makes me think it's more organized than I thought."

Finding the algebra

After the teachers analyzed the rhythms with basic arithmetic ideas like least common multiple, they looked more deeply at the two rhythms. Assuming one could judge the intensity of the dancer by simply watching, we noticed that her intensity seemed to remain constant for a few beats, then change to a different level (sometimes more intense and sometimes more calm.) It appeared that there were beats of the polyrhythm when the dancer's moves might be clustered into short bursts of energy.

Ordered pairs

The teachers decided to analyze the rhythms algebraically. They wanted to develop a general rule that might describe the rhythm. A good first step for them was to try and record the information into an *x-y* coordinate grid. To do that, they needed ordered pairs. The teachers created a list of ordered pairs where the counts of the polyrhythm: 1, 2, 3, …, 23, 24 served as the *x*-coordinates (the domain of the relation). The *difference* between the counts of the individual Tumbao and Bembe Ostinato rhythms served as the *y*-coordinate (the range). For instance, the 12th count of the polyrhythm corresponds to the Tumabo at count #4 and the Bembe Ostinato at count #6. The difference between these two counts is -2. So this ordered pair is (12, -2). The 24th count of the polyrhythm occurs at Tumbao #8 and Bembe Ostinato # 6. In this way, the ordered pair: (24, 2) is created. Using this strategy, the teachers created an *x-y* table (Figure 3) to record the ordered pairs to describe the polyrhythm.

x	T – BO	Diff (y)
1	1 – 1	0
2	2 – 2	0
3	3 – 3	0
4	4 – 4	0
5	5 – 5	0
6	6 – 6	0
7	7 – 1	6
8	8 – 2	6
9	1 – 3	-2
10	2 – 4	-2
11	3 – 5	-2
12	4 – 6	-2
13	5 – 1	4
14	6 – 2	4
15	7 – 3	4
16	8 – 4	4
17	1 – 5	-4
18	2 – 6	-4
19	3 – 1	2
20	4 – 2	2
21	5 – 3	2
22	6 – 4	2
23	7 – 5	2
24	8 – 6	2

Figure 3

Written as ordered pairs, we have the relation: {(1,0), (2, 0), (3,0), (4,0), (5,0), (6,0), (7,6), (8, 6), (9, -2), (10, -2), (11, -2), (12, -2), (13, 4), (14, 4), (15, 4), (16, 4), (17, -4), (18, -4), (19, 2), (20, 2), (21, 2), (22, 2), (23,2), (24,2)}.

NOTES

Domain & Range

Creating these ordered pairs led to discussions about domain and range, and discrete versus continuous graphs. Since the rhythms repeat continuously the domain for the relation is the set of all natural numbers, not just the numbers 1 – 24. However, the *y*-coordinates cycle through the same values every 24 counts. Therefore, the range consisted only of {0, 6, -2, 4, -4, 2}. A side discussion occurred and the teachers determined that the ordered pairs represent a discrete graph and the points should not be connected. Their graph is shown in Figure 4.

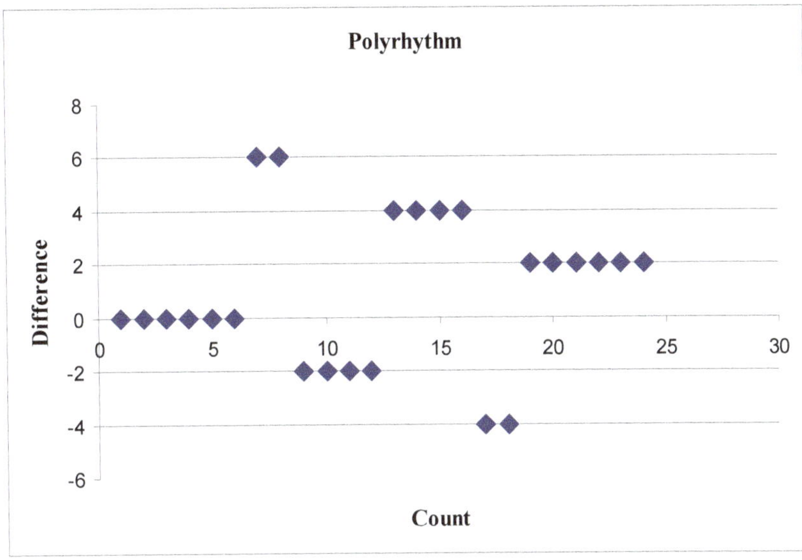

Figure 4 – A graph of the Polyrhythm

Function notation

Then, the teachers became curios to know whether or not the graph represented a relation that was a function. The vertical-line-test satisfied their inquiry. The relation is a function. They were then eager to represent the function with familiar and traditional algebra symbolism: *f(x)*. In order to determine an equation for the polyrhythm, they had to analyze the manner in which the coordinates were created. In particular, developing a symbolic representation for the repetitious nature of the *y*-coordinate proved to be quite a challenge. The first task was to determine which *x*-coordinate always matched up with any given *y*-coordinate. That is, which *y*-coordinate generated from the same counts in the polyrhythm. For instance, at *x*-coordinates 2, 26, 50, 74, … of the polyrhythm both individual rhythms were at counts of 2, so the *y*-coordinate would always be 0. At *x*-coordinates 10, 34, 58, 82, … ,the Tumbao is at count 2 and the Bembe Ostinato is at count 4. The *y*-coordinate would be – 2. (See Table 2). The repetitious nature of the *y*-coordinate was interesting to the teachers. After noting that the *x*-coordinates 1, 25, 49, 73, … always yielded a *y*-coordinate of 0, one teacher said, "The same thing is happening every 24 counts. That must be related to the least common multiple." Another teacher looked quite perplexed and said, "It seems like we have two functions here, somehow."

Transferring this information to more traditional and familiar functional notation ended up to be quite difficult. The teachers had to find a way to describe the seemingly-structured activity of the *x*-coordinate and to then match the correct cycle of the *y*-coordinates with those *x*-values. After involved discussions among the groups of teachers, they finally agreed to use the following notation: $x = 24a + c$, where a ranges through the values: $\{0, 1, 2, 3, \ldots\}$ and c ranges through the values: $\{1, 2, 3, \ldots, 23, 24\}$. In this way, x generates the specific beats in the polyrhythm. For example, $x = 24(0) + 1$ is the 1st beat of the polyrhythm and $x = 24(1) + 1$ is the 25th beat of the polyrhythm (also marking the return of the two individual rhythms to the counts of 1 and 1). In general, then, if $x = 24a + 1$, where $a \in \{0, 1, 2, 3, 4, \ldots\}$, then $y = 0$. Further generalization of this approach allowed the teachers to determine the following representation for the function *f(x)* of the polyrhythm.

$$f(x) = \begin{cases} 0 \text{ if } x = 24a + c, \text{ where } a \in \{0,1,2,3,4,\ldots\} \text{ and } c \in \{1,2,3,4,5,6\} \\ 6 \text{ if } x = 24a + c, \text{ where } a \in \{0,1,2,3,4,\ldots\} \text{ and } c \in \{7,8\} \\ -2 \text{ if } x = 24a + c, \text{ where } a \in \{0,1,2,3,4,\ldots\} \text{ and } c \in \{9,10,11,12\} \\ 4 \text{ if } x = 24a + c, \text{ where } a \in \{0,1,2,3,4,\ldots\} \text{ and } c \in \{13,14,15,16\} \\ -4 \text{ if } x = 24a + c, \text{ where } a \in \{0,1,2,3,4,\ldots\} \text{ and } c \in \{17,18\} \\ 2 \text{ if } x = 24a + c, \text{ where } a \in \{0,1,2,3,4,\ldots\} \text{ and } c \in \{19,20,21,22,23,24\} \end{cases}$$

Representing a function in this way is known as *piecewise*. One teacher said, "I have studied piecewise functions, but never really found them interesting or thought-provoking before." Another said, "I don't remember ever studying piecewise functions." The teachers became interested in a mathematical understanding of what was happening to the dancer's emotions for each piece of the function. They analyzed the function, determining that the x – coordinates for which the dancer seemed to exhibit the most dramatic movements occurred at x – coordinates $\{7, 8, 9, \ldots 22, 23, 24\}$ and the slower movements occurred at x – coordinates $\{1, 2, 3, 4, 5, 6\}$. They recognized that she seemed to prepare to slow down at the x – coordinate of 24, the LCM.

Evaluation component

The main point of this lesson was to provide the teachers with a learning experience where they encountered algebra in an unexpected location. However, of equal importance was to demonstrate how the mathematics could be much more sophisticated than simple patterning. Too often, culturally-based mathematics lessons are vague and generally over simplified. After experiencing this lesson, the teachers were to create a set of culturally-based lessons for their algebra students. They selected a culture and studied it, looking for meaningful and rigorous algebraic connections. One teacher studied the artistry of Celtic knots, another studied the importance of phases of the moon in Norway, a third studied the importance of salmon to the Native American Lummi, and a fourth studied the role of conch shells in ancient Aztec cultures of South Central Mexico. All of these teachers built algebra units from their studies.

NOTES

Summary

Readers of this paper may feel overwhelmed as they think about the effort, time and energy we put into learning about African Drumming and developing it as an alternative context for algebra lessons. When asked to develop their lessons, the teachers often found themselves short on time. One teacher noted, "I cannot believe how much time it has taken me to really learn about the culture, let alone figure out the math." We all had to carefully manage other pressing responsibilities to develop these lessons from scratch. But if culturally-relevant mathematics lessons is our goal, then we have to make time to pursue true, deep knowledge. Finding a colleague who may serve as an expert for the cultural and/or mathematical components was a good place for us to start. We encourage the readership to look for unexpected contexts that may provide an avenue for teacher educators to provide some measure of social justice. We offer not only the African Drumming information that we learned, but also suggest that our teachers' units also provide some evidence of a successful effort.

References

Anderson, J.R. (1994). *The Big Bang: In the Beginning was the Drum* [CD and book]. Roslyn, NY:Ellipsis Arts.

Boisserie, G. (Producer) & Panica, K. (Director), (1992). *Conga Drumming and Afro-Caribbean Rhythms* [film]. (Available from Alchemy Pictures, PO Box 4128, Parkside Queens, NY 11375)

Boyer, J.B. (1990). *Curriculum Materials for Ethnic Diversity*. Manhattan, KS: Center for Black Leadership Development and Research.

Burton L. (Edr.) (2003). *Which Way Social Justice in Mathematics Education? International Perspectives on Mathematics Education*. Westport, CT: Praeger Publishers.

Center for Technology in Learning and Teaching, Iowa State University, Curry School of Education, University of Virginia & Center for Learning and Teaching in the West, Montana State University (producers) Sharp, J., Stevens, A., & Carter, L. (writers and directors) (2006) *Rhythmatics Connections between African/Afro-Cuban drumming & dancing and mathematical ratios*. [DVD] (Available from authors)

Charreaux, Jacques. (1992). *Percussions Mandinques, by Adama Drame* [CD and book]. Playa Sound Collection.

Cohen, M., and Cohen, W. (producers) and Fedele, J. (director). (1992) *LP Adventures in Rhythm: Close-Up on Congas, Vol. 1 with Richie Gajate-Garcia*. [film] Garfield, N.J.: LP Music Group.

Cohen, W. (Producer) & McFall, M. (Director), (1994). *Community Drumming for Health , Happiness with Jim Greiner* [Film]. Garfield, NJ: LP Music Group.

Driscoll, M. (1999). *Fostering Algebraic Thinking: A guide for Teachers Grades 6 -10*. Portsmouth, NH: Heinemann:

Dworsky, A. & Sansby, B.. (1994) *Conga Drumming: A beginner's guide to playing with time.* Dancing Hands Music: Minneapolis, MN

Gerdes, P (1998). On culture and mathematics teacher education. J*ournal of Mathematics Teacher Education, 1*, 33-53

Grund, F. (1985). *Les Maitres-Tambours Du Burundi* [CD and book]. Arion: Paris.

Hart L. E. (2003). Some Directions for Research on Equity and Justice in Mathematics Education. In L. Burton (Edr.) *Which Way Social Justice in Mathematics Education? International Perspectives on Mathematics Education* (pp. 27 – 50). Westport, CT: Praeger Publishers.

Locke, D.(1987) *Drum Gahu The Rhythms of West African Drumming*. Crown Point, IN.: White Cliffs Media.

Lortie, D.C. (1975) *Schoolteacher*. University of Chicago Press: Chicago.

Ma, L. (1999) *Knowing and teaching elementary mathematics: Teachers' Understating of fundamental mathematics in China and the United States*. Mahwah, NJ: Lawrence Erlbuam

Ma, L. & Kessel, C. (2001). Teachers' Understanding of Fundamental mathematics. In National Academy of Sciences *Knowing and Learning Mathematics for Teaching* (pp. 12 – 16), Washington DC: National Academy Press.

Middleton, J A. & Spanias, P. A. (1999). Motivation for Achievement in Mathematics: Findings, Generalizations, and Criticisms of the Research. *Journal for Research in Mathematics Education, 30*(1), 65-88.

Moses, R. & Cobb, R. (2001). *Radical Equations*. Boston, MA: Beacon Press.

National Council of Teachers of Mathematics. (2000). *Principles and Standards for School Mathematics.* Reston VA: National Council of Teachers of Mathematics.

Pirie, S. & Kieren, T. (1994). Growth in mathematical understanding: How can we characterize it and how can we represent it? *Educational Studies in Mathematics, 26*(2-3), 165-190.

Povey, H. (2003). Teaching and Learning mathematics: Can the Concept of Citizenship Be Reclaimed for Social Justice. In L. Burton (edr.) *Which Way Social Justice in Mathematics Education? International Perspectives on Mathematics Education* (pp. 51 – 64). Westport, CT: Praeger Publishers.

Russell, T. (1997). Teaching teachers: How I teach IS the message. In J. Loughran and T. Russell (Eds.) *Teaching about teaching: Purpose, passion and pedagogy in teacher education.* (pp. 32-47) London: Falmer Press.

Schifter, D. (1997). *Learning mathematics for teaching: Lessons in/from the domain of fractions.* Newton MA: Education Development Center, Inc. (ERIC Document Reproduction Service No. ED 412 122)

Schifter, D. (2001). Perspectives from a mathematics educator. In National Academy of Sciences *Knowing and Learning Mathematics for Teaching* (pp. 69-71), Washington DC: National Academy Press.

Stevens, T., Sharp, J. & Nelson, B. (2001). The intersection of two unlikely worlds: Ratios & Drums. *Teaching Children Mathematics, 7*(6), 376-383.

Weist, L. R. (2001). Teaching mathematics from a multicultural perspective. *Equity & Excellence in Education, 34*(1), 16-25.

Williams, Steven R. & Ivey, Kathy M.C. (2001). Affective Assessment and Mathematics Classroom Engagement: A Case Study. *Educational Studies in Mathematics, 47*, 75-100.

Book of Numbers: Exploring Jewish Mathematics and Culture at a Jewish High School

Lawrence Mark Lesser
Lesser@utep.edu
postal address:
Department of Mathematical Sciences
The University of Texas at El Paso
500 W. University Avenue
El Paso, Texas 79968-0514

Abstract:

At a pluralistic Jewish community high school in the southern US, the author sought, adapted, and integrated into his teaching examples of culturally relevant mathematics (in ways adaptable for other grades or cultures). Topics/techniques explored included: quotations, mathematical "firsts," gematria, counting, infinity, pi, and connections to Jewish text/customs/holidays/games. In addition to their intrinsic interest and value, these enhancements connected to school culture/activities and appeared to help motivate additional students towards a broader view of and deeper engagement with mathematics, and possibly with Judaism as well. This article offers both scholarly background as well as a collection of diverse classroom-tested examples.

Background notes:

The word Torah refers narrowly to the books of Genesis, Exodus, Leviticus, Numbers and Deuteronomy, but can also include the entire Hebrew bible (the body of scripture non-Jews call the Old Testament and Jews call the Tanakh or Written Torah). In this article, the reader will see the usual Talmudic citation convention where Ketubot 93a refers to a particular side (in this case, front) of a particular two-sided leaf (93) from a particular tractate (Ketubot). Talmud refers to Jewish rabbinical oral tradition, which was written down over 1300 years ago. The designations BCE (Before the Common Era) and CE have been in use for several decades as more neutral substitutes for BC and AD, respectively, to keep the same dating system while being more respectful of non-Christian faiths. Hebrew words are written right to left.

Introduction

The domain of culturally relevant mathematics includes the recognition that mathematics has been present in every culture, the mathematical achievements of cultures, the effect of mathematics on any culture, and the right for all people to acquire mathematical power (Hatfield, Edwards, Bitter, et al. 2000). While aware of some literature on culturally relevant mathematics (Gutstein, Lipman, Hernandez, et al., 1997) and supportive organizations (Benjamin Banneker Mathematics Association, www.bannekermath.org; TODOS: Mathematics for ALL, www.todos-math.org), I had not had the encouragement and opportunity to integrate culturally relevant mathematics into my teaching in a comprehensive way. The opportunity came recently when I took a leave of absence from a university mathematics education position to upgrade my experiential base with full-time precollegiate teaching experience. For the next two years, I worked as mathematics department chair and teacher at a pluralistic Jewish community high school in a large city in the southern United States. Most students were affiliated with Reform or Conservative congregations, though a few students had Modern Orthodox affiliation and a few were non-Jews. In the spirit of NCTM (2005) that "highly qualified teachers of mathematics not only understand – but also invest in – the particular culture of their students and school", I sought, compiled, created, and implemented connections to Jewish culture with a range of students (their standardized test scores went from the 40th to 99th percentile) in a range of courses (algebra, geometry, precalculus, calculus) as well as in school assembly presentations. To support this endeavor, I supplemented my own Judaics knowledge and got more acquainted with the school's approach to Judaics by attending daily assembly presentations and sitting in on an 11th grade Judaics course during my planning period.

Significant material on Jewish mathematics appears in a few mathematics history textbooks (Katz 1993) and in moderately advanced Judaics books (Littman 1989, Gabai 2002), but there does not appear to be a single book or bibliography specifically and comprehensively on Jewish mathematics from which a classroom teacher could readily teach. This article shares and reflects upon a broad cultural overview of examples from many sources I shared with my high school students (and, occasionally, with general adult audiences at lectures). Clearly, I used only a subset of these examples in any one course I taught, and a few examples I did not get around to using in any, but am including them to offer a more comprehensive collection for other instructors.

The examples are intended to be of interest to Jew and non-Jew because: (1) most examples have (or may point to) counterparts in other cultures; (2) it is possible and profound to help students experience their culture as something dynamic and interdisciplinary with a nurturing egalitarian worldview that places "their history within a universal context where being part of an ethnic group is a reflection – not a separation – of their humanity" (Aceves 2004, p. 275); (3) in many countries, a large fraction of students belong to a faith tradition (e.g., Christianity, Islam) whose origin includes significant Jewish context; also, there has been increasing mainstream interest in various aspects of Jewish culture, due to media coverage of the Middle East, involvements in Kabbalah (Jewish mysticism) by Madonna, the vice-

NOTES

presidential bid of Joseph Lieberman, the chart-topping reggae of Mattisyahu, the edgy comedy of Sacha Baron Cohen ('Borat'), etc.

While Judaism has been viewed as many things (e.g., culture, race, religion, civilization, nationality), this article's focus is on culture for several reasons. As Lemish (1981) states: "All attempts to categorize or identify Jews as an ethnic, religious, or national group are simply inadequate and incomplete... Perhaps the closest any identification can come is to view the Jews holistically as a culture" (p. 28). Indeed, students at the Jewish high school generally viewed and described their own identity that way (e.g., "Jewishly-identified, but not real religious") and the school acknowledged this by emphasizing culture more than religion at assemblies, working in small doses of religion gradually in a participatory, non-coercive manner. So when we reference Jewish religious ideas, it should be kept in mind that many students (and adults) refer to or follow certain Jewish religious practices more as a matter of cultural solidarity and familiarity than as an active explicit theological belief.

However, there has been much exclusion (by Jews as well as by non-Jews) of Jews from multiculturalism, the extent of which and various explanations for are reviewed by Langman (1999). One of my goals is to help articulate and explore broader, more visible roles for Jewish culture in multicultural mathematics. While Jews are not underrepresented in careers requiring significant mathematics, it is a loss for all that Jewish culture is underrepresented in the area of multicultural mathematics. A popular commercially available set of 16 multicultural classroom posters includes posters such as "Math of Egypt", "Math of Babylon", and "Math of Arabia", but not "Mathematics of Israel". The NSF-funded Ethnomathematics Digital Library (2006) has entries for 132 cultural group categories, but "Jewish" is not one of the categories (though "Hindu", "Islamic" and "Muslim" are, for example).

Teaching Jewish mathematics as history and culture also clearly follows the separation of church and state that would support implementation in public schools as well. Thus, we are not "teaching religion", even if doing some teaching *about* religion-- in the balanced spirit of the broadly-endorsed school textbook by Schippe and Stetson (2005).

Using Culture for General Motivation: Quotations and 'Firsts'

One of the simplest and most immediate ways I found to begin integrating culturally relevant mathematics was to add to the physical environment of the classroom. To motivate students who personally took their Judaism much more seriously than their mathematics, I prominently posted on the walls of my classroom some of the explicit support within Jewish tradition for learning mathematics. Here are two examples:

> *"[One] who wishes to attain human perfection should study Logic first, next Mathematics, then Physics, and lastly Metaphysics." -- 12th-century rabbi, physician, philosopher Moses Maimonides, in his* ***Guide for the Perplexed***

> *"The statement of Galileo that 'the great book which ever lies before our eyes—I mean the Universe—is written in mathematical language and the characters are triangles, circles, and other geometrical figures' applies as well to the Halakhah [Jewish Law]. And not for naught did the Gaon of Vilna [a very influential 18th century rabbi who also authored mathematics books] tell the translator of Euclid's geometry into Hebrew [R. Barukh of Shklov], that 'To the degree that a man is lacking in the wisdom of mathematics he will lack one hundredfold in the wisdom of the Torah.'" – Rabbi*

Joseph Soloveitchik, one of the 20th century's most influential scholars in politically-conscious Modern Orthodox Judaism, in his ***Halakhic Man***

Mathematical "firsts" have come from many cultures, and can be a source of inspiration for students from that culture. Because the Jewish written record is unusually old, and because it has received a great deal of scholarly attention, it has been credited with a large variety of mathematical firsts. For example, Wainer (1996) shows a representation of possibly the first statistical graphic (made around 1400 BCE and uncovered from the Qumran Caves) -- a bar chart of 2 population counts 38 years apart for each of the 13 tribes.

The Talmud (Ketubot 93a) contains the first "fair division problem" in recorded history: how to divide an estate between three creditors owed 100, 200, and 300 units. An estate of 100 is divided using equal division so that each creditor gets 33 $^1/_3$ units. An estate of 300 is divided using proportional division, so that each creditor gets half of what he is owed: 50, 100, and 150. For an estate of 200, however, the rationale for the Talmud's allocation (50, 75, and 75) seems mysterious. These 3 cases were not reconciled mathematically until 1984 (Aumann & Maschler 1985; Hill 2000) as the "nucleolus solution" minimizing the largest dissatisfaction among all subsets of creditors.

Jewish mathematician Levi ben Gerson's 1431 work *Maasei Hoshev* introduces the technique of mathematical induction "somewhat more explicitly than his Islamic predecessors" (Katz 1993, p. 279). Using this technique, he proved that the number of permutations of *n* elements is the product of the first *n* natural numbers. (More about this mathematician appears in Simonson (2000).) Interestingly, this particular result was illustrated over 1200 years ago in the *Book of Creation (Sefer Yetsirah)*, written 2nd-8th century in which the act of creation was related to forming possible sequences:

> *"Two stones [Hebrew letters] build two houses [words], three build six houses, four build twenty-four houses, five build one hundred and twenty houses, six build seven hundred and twenty houses, seven build five thousand and forty houses (4:12)."*

Lumpkin (1997) offers a related classroom reproducible activity that extends to the astrology calculations of the 12th-century Spanish scholar Rabbi Abraham ibn Ezra.

Instructors, however, need to help ensure that this does not encourage statements of cultural (or national or religious) superiority. There are sometimes genuine uncertainties about which culture was truly the first to discover something, especially if a more extensive written record survived for some cultures than for others. One should aim to cultivate and model a humble awareness that such claims are always subject to being updated or changed as new scholarship or archaeological findings emerge. And the culture that first explored a mathematical idea may not be the culture that explored it the deepest. In any case, these pitfalls were natural to avoid at this Jewish high school, perhaps because of that school's explicit commitment to pluralism and the presence of non-Jews among the student body, faculty, and staff. Avoiding the pitfalls of insularity (that a more religious day school might be susceptible to), the school supported much intergroup interaction such as an extended (post-9/11)

NOTES

visit by the school to a local mosque. Such particular events supported the general goal of ethnomathematics in bridging cultural gaps as well as added rich context to later mathematics classroom explorations (e.g., the Islamic use of tessellations). Kraft (2005) reports the even more inspiring example of Israeli and Palestinian high school students learning and working together at a computer summer camp in the MIT-supported Middle East Education through Technology program (meet.mit.edu).

Cultural Counting

There is cultural significance even in the way a Jewish society marks time. Days of the week are not so much named as numbered (by how they head towards that week's Sabbath day): Sunday is *Yom Rishon* ("first day"), Monday is *Yom Sheni* ("second day"), etc. Like the Muslim lunar calendar, the Jewish calendar has lunar months of 29 or 30 days and the moon is prominent in symbolism and holidays (e.g., *Rosh Chodesh*). Students were further intrigued by how the Jewish calendar is actually luni-solar, having the occasional addition of an extra month (in years 3, 6, 8, 11, 14, 17, and 19 of each 19-year cycle) to retain alignment with the solar year, since a strictly lunar calendar falls 11 days behind a solar calendar each year. This adjustment ensures, for example, that the holiday of Passover always happens during the spring as specified in the Torah (Deuteronomy 16:1).

Also, in Jewish law, an hour is not simply "an hour", but is $^1/_{12}$ the amount of that day's daylight. Students can use proportional reasoning to verify that at a place and time when the difference between sunrise and sunset times is 13 hours, a Jewish halachic (legal) hour would actually be 65 minutes long. A specific application of this still practiced by many Jews today is determining the proper time or range of permissible times for certain rituals or prayers. For example, the *Shema* prayer affirming God's oneness must be said by the end of the third halachic hour in the morning. A complete mathematical analysis of the Hebrew calendar assuming no technical knowledge beyond high school algebra is in Gabai (2002), with a more detailed account in Feldman (1978).

It appears that in ancient times, Jews used other numeral systems (such as the Egyptian hieratic and then the Babylonian base 60), before deciding (by the second century) to use the letters of the Hebrew alphabet to represent numerical values assigned to the letters, as the Greeks had already been doing with the Greek alphabet. This numerology is called gematria. (Some examples of Hebrew gematria appeared in the 1998 movie *Pi*, which won the Director's Award at the Sundance Film Festival. Teachers should be advised, however, that the film is rated R for language and disturbing images.) The 22 letters of the Hebrew alphabet (first letter is *aleph*: א)

א ב ג ד ה ו ז ח ט י כ ל מ נ ס ע פ צ ק ר ש ת

are typically assigned the respective values 1-9, the multiples of 10 from 10-90, and then 100, 200, 300, and 400. An Internet search for "Hebrew numbers" or "gematria" yields more detailed charts, see Appendix. While today's Israeli/Hebrew culture uses the modern Hindu-Arabic base-ten system for most purposes, Hebrew letters are still used for numbers in calendars (days of the month, years within a millennium), and religious books (page numbers; chapters and verses of scripture). Jewish celebrations such as *Lag*(גל) *B'Omer, Tu*(וט) *B'Shevat*, and *Tu B'Av* have information about when they occur numerically embedded in their names (like "Cinco de Mayo" or "Fourth of July"). Note that the two letters used in *Tu*(וט) *B'Shevat* correspond to 9+6, instead of the expected 10+5, so as to avoid unnecessarily spelling one of the names of God.

Unlike today's modern Hindu-Arabic decimal system, the ancient Hebrew number system is nonpositional and additive. Therefore, anagrams (words formed by permuting letters) have the same numerical value, thanks to the commutative property of addition!

Gabai (2002) gives the example of the Torah's second word, ארב (*Bara*, "He created"), and connects it to one of its (3! = 6) permutations, ראב (*Be'er*, "He explained/elucidated"). A gematria-generating mathematical tool that can be used with (or instead of) permutations is partitions: all the ways of writing an integer as a sum of positive integers, not counting the order of addends. For example, there are 5 partitions of 4: 4, 3+1, 2+2, 2+1+1, and 1+1+1+1. Thus, there is a numerical equivalence between the Hebrew words *Gey* (proud; אג; 1+3) and *Abbah* (father; אבא ; 1+2+1). An Internet search for "gematria software" turns up several software packages that can find words which are numerically equivalent to each other.

The Hebrew word for life, יח (*chai*), is spelled with the two letters *chet* and *yud*, whose numerical equivalents 8 and 10 sum to 18, which explains the modern Jewish cultural practice of giving (and soliciting) charity in multiples of 18. Speaking of charity, Gabai (2002) notes that the Hebrew word for charity (*tzedakah*; הקדצ) is one of the few words whose gematria is equal to its "reverse gematria" (from assigning numbers to the Hebrew alphabet in reverse order), suggesting that this moral value of giving charity "will remain irreversible and unchangeable throughout the person's life" (p. 76). It is interesting to let students deduce the mathematical condition for this – if the nth letter of the Hebrew alphabet is in the word, so must the kth letter, where $k = 23 - n$.

There are variations on how to assign numerical values to the letters, such as using the ordinal numbers 1 – 22 or such as assigning each word the one digit you get when you keep adding up the sum of the digits of its letters' usual numerical values until you have a 1-digit number (e.g., a word with the value 618 would reduce to , which in turn would yield). In this latter option, students can discover that two words are "linked" if the sums of their letters' numerical equivalents are congruent modulo 9. Such "links" are interpreted in a great variety of ways, ranging from a simple educational mnemonic device (useful for Judaics study) to a mystical, Kabbalistic revelation. Gabai (2002) makes it a point to remind the reader that Jewish mystical and rabbinical writings warn of overuse or misuse of gematria, and it is interesting to encourage students to reflect upon statements such as Gabai (2002):

> *From a religious, spiritual, or mystical point of view, we note the repeated appearances of some whole numbers and their multiples, and we wonder about their meaning and significance. From the point of view of logic and mathematics, the repetition of numbers is not unexpected because whole numbers are mentioned so often in the Bible. (pp. 18-19)*

The Infinite: How We Count

The Talmudic statement (Mishna Sanhedrin 4:5) that someone who saves a life is viewed "as if he saved an entire world" may be familiar as the tagline of the 1993 Academy Award-winning movie *Schindler's List*. Telushkin (1991) notes that one consequence of this idea of the infinite value of human life is that "saving many lives at the expense of one innocent life is not permitted, since by definition, many 'infinities' cannot be worth more than one 'infinity'" (p. 530). This, of course, is perfectly consistent with arithmetic

NOTES

of transfinite cardinal numbers (e.g., the sum of any countable number of aleph nulls equals aleph null, where aleph null is the cardinality of natural numbers; *aleph*, א, is the first letter of the Hebrew alphabet)!

In Jewish tradition, counting people is permitted only in an indirect manner, such as a population census via 1-1 correspondence with the proxy of a coin donation (e.g., Exodus 30:13). A current example of this is a traditional way to see if there is a *minyan* ("prayer quorum") of at least ten eligible people by reciting a 10-word verse of Torah such as Psalms 28:9, assigning a word to each eligible person present. Interestingly, Zaslavsky (1973) also relates an African taboo on counting people. (As an aside, it is interesting to discuss the extent to which individual dignity is similarly respected by the statistical technique of randomized response (Warner 1965). This survey method asks about personal behaviors or beliefs in a way that gives privacy to potentially embarrassing answers but still allows estimation of the overall answer for the group of people.)

In addition to the idea of infinite value of human life, there may also be a way to apply transfinite arithmetic to *mitzvot* (plural of *mitzvah*). A *mitzvah* is viewed culturally as a "good deed" and religiously as a "commandment". It is generally accepted that the Torah contains 613 *mitzvot*, including both matters of ritual and matters of ethical behavior. It is tempting to assume that some *mitzvot* are "worth" more than others, and indeed there are seven instances (e.g., *tzedakah*) in which Talmud rabbis refer to a particular mitzvah as equal in value to all the others combined (Donin 1980). While this may seem to be mystifying hyperbole, we can see from the resulting system of mathematical equations that this is perfectly consistent with each *mitzvah* having an infinite value such as $\aleph_0$ (or, technically, a value of 0).

Some Jewish writing (Kaplan 1990; Saks 2002; Saks 1990; Schochet 1979) refers to different levels of infinite spiritual worlds, an idea similar to Georg Cantor's idea that there is an infinite sequence of (mathematical) infinities: $\aleph_0 < \aleph_1 < \aleph_2 < \ldots$. Cantor's demonstration that "the infinite cube contains exactly the same number of points as in the one-inch cube" (Gabai 2002, pp. 41-42) helps us understand "the paradox of God's simultaneous transcendence and immanence" (ibid). Gabai (2002) helps us understand other aspects of the Jewish view of God's presence by referencing Edwin Abbott's classic mathematical novel *Flatland*. One of God's many names in Judaism is *Ein Sof*, which was introduced in the 13th century and has been translated as "without end" or "infinity", and is the term for infinity you would hear in a mathematics class in Israel today.

Connections to Content: Exponential and Logarithmic Functions

While I was teaching at the high school, an article from the local paper (Copans 2003) gave my precalculus students a real-life vehicle to discuss and apply the carbon dating model from their book (Larson, Hostetler, Edwards 2001): the ratio of C_{14} to C_{12} in dead organic material equals $10^{-12}e^{-t/8223}$. The newspaper story discussed how carbon dating confirmed that a stone tablet detailing repair plans for Solomon's temple contained writing from the 9th century BCE. The class verified that this would be consistent with a measurement of about 70.5% of the radioactive carbon 14 (whose half-life is 5700 years) remaining. This Biblical archaeology context appeared to engage the class far more than did a textbook carbon dating problem about a generic fossil, because of the Judaics connection, not out of any religion versus science tension. (Indeed, evolution and big bang theories were taught at this school and can be viewed in a way that is consistent with Jewish religious teachings.)

Another day, after the precalculus students had returned from a school field trip to see a *mikvah* (a ritual immersion pool), I had them explore an excerpt from Talmud (Mikva'os

7:2) that required not only applying logarithms, but also examining modeling assumptions: "If there were exactly forty *se'ahs* [of suitable water] in a *mikvah*, and a *se'ah* [of unsuitable liquid] was put in, and then a *se'ah* [of the resulting mixture] was removed, the *mikvah* is still valid." A *se'ah* is about 8.25 liters. This excerpt from Littman (1989) provided a vehicle to explore how many times this process can be repeated until the majority of liquid is unsuitable, and to discuss what modeling assumptions might be made about how well the liquids mix. In particular, what is the answer if a *se'ah* of uniform mixture is removed each time and what is the answer under the "worst-case scenario" that a *se'ah* of suitable water is removed each time? Littman (1989) contains a detailed analysis of this and other Talmudic examples, including a maximization problem in geometry and how the Euclidean algorithm applies to calculating the occurrences of the new moon.

Value of Pi

A value of π can be obtained from I Kings 7:23:

> *"He made the 'sea' of cast [metal] ten cubits from its one lip to its [other] lip, circular all around, five cubits its height; a thirty-cubit line could encircle it all around."*

It appears the value of π implied here is simply $^{30}/_{10}$ (an error of 4.5%) until a student asks if we need to consider the tank's thickness -- given three verses later as one-handbreadth, so the inner diameter is 10 cubits minus 2 handbreadths. (Of course, this is also a chance to discuss issues of measurement!) Using the Talmudic value of $^{1}/_{6}$ cubit for one handbreadth, the inner diameter becomes $9^{2}/_{3}$ cubits and dividing 30 by $9^{2}/_{3}$ yields more accuracy (error: 1.2%). Applying a more subtle and technical approach to I Kings 7:23 (see Posamentier & Lehmann 2004 or Tsaban & Garber 1998), the ratio of gematrias for the written and spoken forms of a key Hebrew word (for "line") in that verse is 111/106, which when multiplied by 3 yields a very refined approximation for π: $^{333}/_{106}$ (error: 0.0026%). Very few words in the Torah have different oral and written forms.

Geometry Connections

Like many cultures, Jewish culture has a very prominent role for food. Bodner (2001) offers a rich variety of geometry formulas applied to the matter of assessing whether various shapes (e.g., rectangular prism, sphere, cylinder, torus) of food are of sufficiently large volume to allow and require the saying of a particular post-meal blessing. This threshold unit of volume is one *k'zayis*, which is about the size of a large olive or half an egg.

A more mainstream example is exploring the geometry of the Star of David, which has become a universally recognized symbol of Jewry and appears on the Israeli flag. My geometry students enjoyed how the Star of David appears in the geometry construction activity of Lide (2004) and how the sum of its 24 interior angles turns out to be a very nice multiple (100) of *chai*. When we discussed fractals, students were interested that "stage 2" of the particular fractal known as the Koch snowflake (the interested reader is referred to www.shodor.org/interactivate/activities/koch/) is a Star of David. And more profoundly, fractal self-similarity has been

NOTES

used as a metaphor for thematic microcosms in Jewish text. For example, Ellis (1997) states:

> *The focus in* ***parashah*** *Be-har [Leviticus 25:1-26:2] on Shabbat, on the giving of the land, on ethical behavior, on living in peace and love with our sisters and our brothers, on living in active support of our sisters and our brothers: this is the focus of the entire Torah....Line by line, breath by breath, stem by stem, blossom by blossom, each verse of our* ***parashah****, like each verse of Leviticus, mirrors fractal-like the glorious and transcendent vision of the whole. (p. 34)*

When the *sukkah* (temporary hut-like dwelling from the Biblical holiday of Sukkot that non-Jews may know by the name Feast of Tabernacles) was built on the school campus grounds, we had a "geometry treasure hunt" to find various types of angles and polygons amidst the beams and poles.

I had students explore various geometric shapes and concepts through origami constructions of various Jewish cultural symbols and holiday objects (Temko 1991, 1992, 1994), such as a ram's horn (for the Jewish New Year) or a Star of David. One test problem referenced shapes and lengths from diagrams of the ancient Temple (Wigoder 1974).

Also, examples of geometry in the Talmud include: using the circumference to calculate a circle's diameter (Eruvin 13b, 14a, 76a-b; Succah 7b), calculating the area of a square's circumscribed or inscribed circle (Eruvin 76a-b; Succah 8a-b), using a side to find a square's diagonal (Eruvin 57a, 76b; Succah 8a), and using ratios to find faraway objects' heights (Eruvin 43b).

Structural Parallels between Judaism and Mathematics/Logic

With my geometry class, I discussed the observation by Gabai (2002) that Judaism can be viewed as a deductive system with the fundamental undefined term of God, the Torah contents as axioms, and the Talmud as theorems derived from axioms. Davis and Hersh (1981) make an interesting juxtaposition with the axiom that an infinite set exists and the "axiom of God as presented by Maimonides (*Mishneh Torah*, Book 1, Chapter 1): 'The basic principle of all basic principles and the pillar of all the sciences is to realize that there is a First Being who brought every existing thing into being.'" Davis and Hersh (1981) also quote Maimonides as giving a "proof by contradiction (p. 118)" of that assertion of God's existence ("for if it be supposed that he did not exist, then nothing else could possibly exist") and quote Saddia Gaon (10th century theologian and leader of Babylonian Jewry) as supplying an indirect proof of God's uniqueness: "For if He were more than one, there would apply to him the category of number, and he would fall under the laws governing bodies."

During our chapter on deductive reasoning and logic, I supplemented problems from the book by having them critique reasoning using Jewish content such as: "If someone is a Jew, then she believes in the Torah. If someone believes in the Torah, then she observes the Sabbath. Therefore, if someone is a Jew, then she observes the Sabbath." On another assignment, students were asked to construct a logical argument to respond to a not-uncommon proselytization attempt (by a "Jew for Jesus" named Mitch, role-played by R. Tovia Singer), which had been the focus of a recent school assembly. One 9th-grade student wrote:

> *"If Jesus is not the Messiah, then Mitch is wrong. If Jesus is the Messiah, then he would have built the 3rd Temple in Jerusalem, bring all*

the Jews to Israel, bring world peace and worldwide recognition of God. There is no 3rd Temple or world peace. There are atheists and Jews living outside Israel. Therefore, Jesus is not the Messiah (Modus Tollens). Therefore, Mitch is wrong (Modus Ponens)."

Another example was to decide if the statement "If an animal chews its cud, then it is kosher" is true – and also assess the validity of its converse. (Note: a kosher animal chews its cud *and* has split hooves.) While formal data was not collected, there was anecdotal evidence that students found these examples among the most memorable for the course.

Resources such as Sion (1997) make explicit connections between logical argument structures in Jewish text (e.g., "*qal vachomer*", meaning 'lenient and stringent' and often referred to as 'all the more') and their secular Aristotelian counterparts (*a-fortiori ratione*, Latin; meaning 'with stronger reason'). Sion gives the example of the first occurrence of *qal vachomer* in the Torah ("and perhaps historically, in any extant written document") in Genesis 44:8, paraphrasing the argument as:

> *You will agree to the general principle that more honesty is required to return found money than to refrain from stealing a silver goblet.*
>
> *We were honest enough to return found money.*
>
> *Therefore, you can trust we were honest enough to not steal the silver goblet.*

Reflections

While I was teaching at the Jewish high school, I had the opportunity to reflect not only on content, but also on the process of teaching. I was struck by parallels between Jewish views of creation and the active, student-centered pedagogy supported by NCTM (2000). As Kanter (1997) explains:

> *Through an act of **tzimtzum** [contraction], the teacher allows the child to be actively engaged and to learn by doing (**na'aseh v'nishma**). Through that same act, the teacher acknowledges that children are now and will more fully become images of G-d (**tzelem elohim**), each in his/her own unique way (**ein domeh l'haverio**)." (p. 45)*

And I also gained more appreciation for Rabbi Hanina's statement in the Talmud (Taanit, 7a):

> *"Much have I learned from my teachers, more from my colleagues, but most from my students.*

Even though there are few Jews among the inservice and preservice teachers I now work with in my university position, I find many aspects that transfer or transcend. First, by sharing some of my experiences with integrating Jewish mathematics, I can serve as a model to give teachers a tangible sense for the process and power of making similar connections between mathematics and their own cultures (and the cultures of the students they will teach). Second, every culture has distinctive mathematical contributions to discuss and interesting quotations from respected figures that can be posted, etc. (For example, in my recent work with a predominantly Hispanic middle school, I

NOTES

facilitated the posting of Latino/a biographies from the SACNAS Biography Project, http://bio.sacnas.org/biography/.) Also, techniques and issues of topics such as numerology can be applied to and critically discussed with respect to any culture's writings. It is not hard to imagine that making such connections could help one teach from a more integrated self and also do a better job of teaching "the whole child". Cultural mathematics is a powerful way to build community on many levels, which can spill over to a more receptive and supportive classroom culture even when talking about mathematics that may not (yet) be connected to culture.

Students (and most faculty, for that matter) at the Jewish high school had typically viewed Judaics and secular subjects as compartmentalized experiences. While the school's literature claims that the school's Judaic studies are "integrated into the general studies curriculum", this had occurred only in isolated instances and only in humanities classes such as history and literature. Sion (1997) asserts that integration across disciplines is

> *...a fundamental need of the human mind and culture; the alternative of compartmentalization involves a sort or mental or cultural forcing, and can only be a temporary measure. Conflicts between experience and reason, on the one hand, and faith, on the other, can best be diffused by a dispassionate consideration of the underlying logical issues. Many disagreements can be harmonized, by showing at the least that the sides are simply alternatives in a disjunctive proposition, which though they may inductively be of varying probability are deductively on equal footing, for all practical purposes a matter of personal choice. (p. 216)*

By bridging the Jewish and secular worlds in the mathematics classroom, I noticed a broadening of their view of and their appreciation of mathematics. This felt important, considering that many students (and faculty) at the school had not had a favorable or robust view of mathematics beyond skills that had to be learned to succeed on standardized tests required for college admission. A possible consequence of giving students the experience of mathematics as culture is to allow them to see that mathematics, like culture, can be fallible. Sion (1997) and Littman (1989) both discuss some apparent imperfections of logic or computation in the Talmud. This supports more of a progressive educator (or even public educator) ideology, as described by Ernest (1991), and arguably encourages students to become more active and inquisitive in their engagement with mathematics.

As a secondary byproduct, it was interesting that there may have also been a broadening and appreciation of their Judaism as well. This may be enthusiastically welcomed in light of the recent study commissioned by the pluralistic Jewish youth group B'nai Brith Youth Organization in which Teenage Research Unlimited surveyed 1,153 respondents aged 10-18 (of all religions) about their attitudes towards religion (Weiss 2006). It was found that "92% of the teenagers who view religion and faith as an important part of their lives say they want a better connection and 52% of these teens are looking for new and unconventional ways to connect to their religion." (BBYO 2006). Ethnomathematics has appeared to be useful in some cultures as a component of cultural revitalization. The Jewish outreach organization Aish HaTorah (http://www.aish.com/) has been known to include in its "Discovery Seminar" presentations the equidistant letter sequence "Bible codes" à la the best-selling Drosnin (1997). It should be noted, however, that there have been Jewish and scientific critiques of these codes (Simon 1998, McKay, Bar-Natan, Bar-Hillel, et al. 1999), and these can be discussed in more advanced classes.

Some Pedagogical Considerations

As with any area of application or integration, teachers who include culturally relevant mathematics in their assessments should ensure that the presentation is sufficiently self-contained so that students with more culturally-specific knowledge do not have an undue advantage. This was usually done by providing the necessary context or by having the context closely parallel standard textbook scenarios. For example, one of my geometry assessments of this type closely followed their textbook (Schultz, Hollowell, Ellis, et al. 2001) "archaeologist" scenario of estimating the size of an ancient plate by finding the circumcenter from a surviving outer fragment of the plate. The context for my version was the Jewish engagement ritual (*tana'im*) of the couple's mothers' breaking of a plate (the fragments of which are sometimes offered by the bride to unmarried relatives for good luck). As Kaplan (1983) explains:

> *The reason for breaking the dish is to show that we mourn for Jerusalem and other martyred Jews even at the height of our joy. A china dish is broken to show that, just as a china plate can never be fully repaired, a broken engagement is an irreparable breach. Even if the bride and groom are later reconciled, the breaking of the formal Tanaim contract is considered very reprehensible. (p. 28)*

Many topics in Jewish culturally relevant mathematics are rich enough to explore at multiple points on the curriculum ladder. For example, the children's Hanukkah game of spinning a *dreidl* (4-sided top) yields an equiprobable experiment for Grades 1-6 (Krause 1983), but can also yield more sophisticated probability analyses all the way to the derivation (suitable for a first university course in probability) of the expected payoff on the nth spin with N players (that shows the usual game is biased in favor of the first player) and what rule changes would make the game "fair" to all players (Feinerman 1976; Trachtenberg 1996). Also, there are many levels to use concepts of measurement (whose importance is reflected in Leviticus 19:35-36), such as how one might estimate the volume of Noah's Ark given statements such as Genesis 6:15.

Jewish culture has been greatly influenced not only by Biblical times, but also by modern world history. Wainer (1994, 2004) shows graphs that could supplement a (Judaics or history) unit on the Holocaust, giving striking context to the three square feet of space each prisoner of Belsen Concentration Camp was allotted, for example. For that matter, students may be moved to learn how the Nazi WWII experiments on prisoners led to the creation of research ethics codes that now protect all of us (Lesser & Nordenhaug 2004). And Shulman (2002) cites two further discussion-provoking examples: how Nazi ideology infiltrated even German mathematics textbook problems at the time and subsequent controversy about whether already-collected Nazi "research data" should be used if it might save lives. Yet another kind of ethnomathematics on this topic relates to the 2004 documentary movie *Paper Clips*, in which middle school students in rural Tennessee decide to collect six million paper clips in an attempt to grasp the concept of six million Jewish Holocaust victims. The students and their community are transformed by the process, during which they meet Holocaust survivors from around the world (Schroeder & Schroeder-Hildebrand 2005).

It is hoped that this discussion of a rich variety of examples and issues associated with Jewish culturally relevant mathematics may inform and inspire teachers and curriculum writers of all backgrounds.

References

Aceves, C. (2004). The Xinachtli Project: Transforming whiteness through mythic pedagogy. In V. Lea & J. Helfand (Eds.), *Identifying race and transforming whiteness in the classroom* (pp. 257-277). New York: Peter Lang.

Aumann, R.J.. & Maschler, M. (1985). Game theoretic analysis of a bankruptcy problem from the Talmud. *Journal of Economic Theory, 36*(2), 195-213.

B'nai B'rith Youth Organization (2006). New survey reports 68% of teenagers say religion and faith are important to them, but 92% of these same teens say they want a better connection. Accessed on the World Wide Web on March 8, 2006 at: http://www.bbyo.org/index.php?c=529&kat=Press+Release

Bodner, Y. P. (2001). *Halachos of k'zayis*. New York: Feldheim.

Copans, L. (January 14, 2003). Ancient stone tablet may confirm Bible narrative. *Houston Chronicle, 102*(93), p. 10A.

Davis, P. J., & Hersh, R. (1981). *The mathematical experience*. Boston: Birkhäuser.

Donin, H. H. (1980). *To pray as a Jew: A guide to the prayer book and the synagogue service*. New York: Basic Books.

Drosnin, M. (1997). *The Bible code*. New York: Simon & Schuster.

Ellis, R. S. (1997). The book of Leviticus and the fractal geometry of Torah. *Conservative Judaism, 50*(1), 27-34. Also available at: www.math.umass.edu/~rsellis/Fractal-Geometry.html.

Ernest, P. (1991). *The philosophy of mathematics education*. Bristol, PA: Falmer Press.

Ethnomathematics Digital Library (2006). Accessed on March 9, 2006 at: http://www.ethnomath.org.

Feinerman, R. (1976). An ancient unfair game. *American Mathematical Monthly, 83*(8), 623-625.

Feldman, W. M. (1978). *Rabbinical mathematics and astronomy*. New York: Sepher-Hermon Press.

Gabai, H. (2002). *Judaism, mathematics and the Hebrew calendar*. Northvale, NJ: Jason Aronson.

Gutstein, E., Lipman, P., Hernandez, P., & de los Reyes, R. (1997). Culturally relevant mathematics teaching in a Mexican American context. *Journal for Research in Mathematics Education, 28*(6), 709-737.

Hatfield, M. M., Edwards, N. T., Bitter, G. G., & Morrow, J. (2000). *Mathematics methods for elementary and middle school teachers*. New York: Wiley.

Hill, T. P. (2000). Mathematical devices for getting a fair share. *American Scientist, 88*(4), 325-331.

Kanter, E. (1997). B'hai lishna: A Jewish philosophy of Jewish pedagogy. *Conservative Judaism, 49*(3), 43-45.

Kaplan, A. (1983). *Made in Heaven: A Jewish wedding guide*. Brooklyn, NY: Moznaim.

Kaplan, A. (1990). *Innerspace*. Brooklyn, NY: Moznaim.

Katz, V. J. (1993). *A history of mathematics: An introduction*. New York, NY: HarperCollins.

Kraft, D. (2005). Key(board) to coexistence. *Hadassah Magazine, 87*(3), 12-15.

Krause, M. C. (1983). Multicultural mathematics materials. Reston, VA: National Council of Teachers of Mathematics.

Langman, P. F. (1999). *Jewish issues in multiculturalism: A handbook for educators and clinicians*. Northvale, NJ: Jason Aronson.

Larson, R., Hostetler, R. P., & Edwards, B. H. (2001). *Precalculus with limits: A graphing approach* (3rd ed.). Boston: Houghton Mifflin.

Lemish, P. (1981). Hanukah bush: The Jewish experience in America. *Theory into Practice, 20*(1), 26-34.

Lesser, L. & Nordenhaug, E. (2004). Ethical statistics and statistical ethics: Making an interdisciplinary module. *Journal of Statistics Education, 12*(3), www.amstat.org/publications/jse/v12n3/lesser.html.

Lide, A. M. (2004). Circles of humanity. *Mathematics Teaching in the Middle School, 9*(5), 280-285.

Littman, M. (1989). *Approaching infinity*. New York: Feldheim.

Lumpkin, B. (1997). *Algebra activities from many cultures*. Portland, ME: J. Weston Walch Publishers.

McKay, B., Bar-Natan, D., Bar-Hillel, & M., Kalai, G. (1999). Solving the Bible code puzzle. *Statistical Science, 14*(2), 150-173.

National Council of Teachers of Mathematics. (2000). *Principles and standards for school mathematics*. Reston, VA: NCTM.

National Council of Teachers of Mathematics. (2005). Position statement: Highly qualified teachers. Accessed on November 18 at www.nctm.org/about/position_statements/qualified.htm.

Posamentier, A. S. & Lehmann, I. (2004). *Pi: A biography of the world's most mysterious number*. Amherst, NY: Prometheus Books.

Saks, T. V. (2002). Different levels of infinity in Torah and mathematics. *B'Or Ha'Torah*, 13E, 113-120.

Saks, T. V. (1990). Applications of mathematical infinity in Jewish philosophy. In H. Branover, A. Gotfryd, & S. Lipskar (Eds.), *Fusion: Absolute standards in a world of relativity* (pp. 123-142). Jerusalem, Israel: Feldheim.

Schippe, C., & Stetson, C. (2005). *The Bible and its influence* (student text). Fairfax, VA: Bible Literacy Project.

Schochet, J. I. (1979). *Mystical concepts in Chassidism*. Brooklyn, NY: Kehot Publication Society.

Schroeder, P. W., & Schroeder-Hildebrand, D. (2005). *Six million paper clips: The making of a children's Holocaust memorial*. Minneapolis, MN: Kar-Ben Publishing.

Schultz, J., Hollowell, K. A., Ellis, W., & Kennedy, P. (2001). *Geometry* [high school textbook]. Austin, TX: Holt, Rinehart, and Winston.

Shulman, B. (2002). Is there enough poison gas to kill the city?: The teaching of ethics in mathematics classes. *College Mathematics Journal, 33*(2), 118-125.

Simon, B. (1998). A skeptical look at the Torah codes. *Jewish Action, 58*(3), 17-24.

Simonson, S. (2000). The mathematics of Levi ben Gershon. *Mathematics Teacher, 93*(8), 659- 663.

Sion, A. (1997). *Judaic logic: A formal analysis of Biblical, Talmudic and rabbinic logic*. Geneva, Switzerland: Slatkine.

Telushkin, J. B. (1991). *Jewish literacy*. New York: William Morrow.

Temko, F. (1991). *Jewish origami*. Torrance, CA: Heian International, Inc.

Temko, F. (1992). *Jewish origami 2*. Torrance, CA: Heian International, Inc.

Temko, F. (1994). *Bible origami*. Torrance, CA: Heian International, Inc.

Trachtenberg, F. M. (1996). The game of dreidel made fair. *College Mathematics Journal, 27*(4), 278-281.

Tsaban, B., & Garber, D. (1998). On the rabbinical approximation of pi. *Historia Mathematica, 25*(1), 75-84.

Wainer, H. (1994). Visual revelations. *Chance, 7*(3), 50-53.

Wainer, H. (1996). Visual revelations. *Chance, 9*(2), 43-52.

Wainer, H. (2004). Visual revelations. *Chance, 17*(1), 44-48.

Warner, S. L. (1965). Randomized response: A survey technique for eliminating evasive answer bias. *Journal of the American Statistical Association, 60*(309), 63-69.

Weiss, J. (February 17, 2006). Group wants teens to connect with religion. *El Paso Times, 126*(48), p. D1.

Wigoder, G. (Ed.) (1974). *Encyclopedic dictionary of Judaism*. New York: Leon Amiel Publisher.

Zaslavsky, C. (1973). *Africa counts*. Boston: Prindle, Weber & Schmidt.

APPENDIX: Hebrew Alphabet and Numerical Equivalents

Source: www.omniglot.com/writing/hebrew.htm

ט	ח	ז	ו	ה	ד	ג	בּ/ב	א
טֵית	חֵית	זַיִן	וָו	הֵא (הֵי)	דָּלֶת (דָּלֶד)	גִּימֶל	בֵּית/בֵית	אָלֶף
tet	chet	zayin	vav	he(y)	dalet(d)	gimel	bet/vet	alef
t	h/ch	z	w	h	d	g	b/v	'
[t]	[x]	[z]	[v]	[h]	[d]	[g]	[b, v]	[?, Ø]
9	8	7	6	5	4	3	2	1

ס	ן	נ	ם	מ	ל	ך	כּ/כ	י
סָמֶךְ	נוּן סוֹפִית	נוּן	מֵם סוֹפִית	מֵם	לָמֶד	כָּף סוֹפִית	כָּף/כָף	יוֹד (יוּד)
samech	nun	nun	mem	mem	lamed	kaf	kaf/khaf	yod (yud)
s	sofit	n	sofit	m	l	sofit	k/kh	y/i
[s]	final nun	[n]	final mem	[m]	[l]	final kaf	[k, x]	[j]
60		50		40	30		20	10

ת	שׂ/שׁ	ר	ק	ץ	צ	ף	פּ/פ	ע
תָּו/תָף	שִׁין/שִׂין	רֵישׁ (רֵיִשׁ)	קוֹף (קוּף)	צָדִי סוֹפִית	צָדִי (צָדִיק)	פֵא סוֹפִית	פֵּא	עַיִן
tav(f)	sin/shin	resh (reyish)	kof (kuf)	tzadi	tzadi(k)	pe	pe(y)/fe	ayin
t	sh/s	r	k	sofit	tz/ts/z	sofit	p/f	'
[t]	[ʃ, s]	[ʁ]	[k]	final tzadi	[t͡s]	final pe	[p, f]	[?, Ø]
400	300	200	100		90		80	70

"Math is Everywhere": Connecting Mathematics to Students' Lives

Javier Díez-Palomar, Ph.D.
CEMELA Associate – University of Arizona
Fulbright Post - Doc
jadiez@math.arizona.edu

Ksenija Simic, Ph.D.
CEMELA – University of Arizona
Post – Doc k
simic@math.arizona.edu

Maura Varley, EdM
CEMELA – University of Arizona
maurav@math.arizona.edu
520.626.8559

CEMELA Central Office
Department of Mathematics University of Arizona
617 N. Santa Rita
Tucson, AZ 85721

Abstract

Este artículo describe una experiencia de enseñanza y aprendizaje de las matemáticas en un programa extraescolar ubicado en una ciudad de suroeste norteamericano, donde los/as estudiantes trabajaron en proyectos mediante los que establecían conexiones entre las matemáticas y sus respectivas vidas. En el proyecto que describimos aquí, los/as estudiantes discuten y resuelven problemas sobre su escuela, su barrio y la inmigración, desde un punto de vista crítico. El enfoque pedagógico de nuestro "Club de Matemáticas" se basa en una perspectiva sociocultural, con una orientación crítica. Esta experiencia destaca las conexiones entre cultura, lenguaje, y diálogo entre los/as estudiantes de matemáticas. Se presentan implicaciones interesantes basadas en el desarrollo de la agencia de los estudiantes a través del aprendizaje de las matemáticas.

This article describes an experience of teaching and learning mathematics in an after- school program based in a southwestern city, in which the students worked on projects that connected mathematics to their lives. In the project that we describe, the students discussed and solved problems about their school, their neighborhood, and immigration from a critical perspective. The pedagogical approach of our "Math Club" is grounded in socio-cultural approaches, from a critical point of view. This experience highlights the connections between mathematics and everyday life that are brought out when emphasizing culture, language, and dialogue among mathematics learners. Implications are made for drawing on student agency through mathematics learning.

Introduction

"Math is everywhere," claimed Dolores, a ten year old Latina and participant in an after- school Math Club. One of the undergraduate facilitators of the Math Club reflected, "I was so proud of her; yes, she understands." What she understands is an explicit intent of our Math Club: to highlight and draw on the connection that mathematics has to the knowledge and life experiences of the participants. This is often not the case for Latino/a students, as their languages, cultures, and experiences are not always represented in traditional classrooms. Dolores's quote demonstrates one result of placing the knowledge and experience of Latino/a students at the center of mathematics learning, where our goal is to transform it from the scary subject many students fear and subsequently remain disconnected from, to one in which they are engaged and empowered mathematics learners. So, what does this look like? We intend to outline a community-based project conducted in this bilingual learning community in order to illustrate how learning mathematics in a learning community inclusive of students' language, culture, and everyday experiences might look.

Over the course of the past year, a team of undergraduate, graduate, and post-doctoral researchers facilitated an after-school mathematics club, in which we attempted to create a transformative, after-school learning space where student experiences and lives outside of school are incorporated into their mathematics learning. Our project is a part of the research of the Center for the Mathematics Education of Latinos/as, which seeks to better understand "the interactions among mathematics learning, language, and culture (especially community knowledge)" (www.math.arizona.edu/~cemela). In this unique learning environment, where we employ strategies of culturally relevant mathematics instruction, we seek to connect mathematics to students' everyday lives and experiences in order to understand the impact on Latino/a students' mathematics learning.

The Math Club takes place at Agave Elementary, a small (~ 300 students) urban school, located in a predominantly Latino/a neighborhood. At this school, 95% of the students qualify for free or reduced lunch and 26% of the students are classified as English Language Learners (ELL) (District statistics). The participants in the Math Club reflect the 90% Latino/a student body of the school, all considering themselves Latino/a, and most of them speaking either primarily Spanish, or both Spanish and English. We encourage a bilingual environment with students often switching fluidly from one language to another depending on whom they are speaking to or who is near them. This is particularly salient as the state of Arizona has recently passed legislation to virtually eliminate bilingual education (Proposition 203). The few bilingual education classrooms that are left are severely limited due to increased testing pressures for ELLs. The result of this is that the students that attend the Math Club who are talented bilinguals or predominant Spanish speakers are limited in their ability to draw on this knowledge in their classrooms. One intention of the Math Club is to provide a bilingual environment that encourages students to draw on their everyday and academic knowledge and skills, including language, and share them collectively.

NOTES

Practice Based on Research

Research demonstrates that when teachers incorporate the historically accumulated knowledge and skill bases of their students and the communities from which they come (their "funds of knowledge"), learning is enhanced (González, Moll, Floyd-Tenery, Rivera, Rendon, Gonzaeles, 1995; González, Moll & Amanti, 2005). Civil & Andrade (2002) and Civil (2002) further expand on the theory of funds of knowledge to focus on household *mathematical* practices while stressing the importance of creating a connection between the academic and the everyday. Because students often feel that mathematics is the subject least relevant to their daily lives, it is an important arena in which teachers can attempt to incorporate students' funds of knowledge in the context of culturally relevant mathematics instruction (Gutstein, Lipman, Hernandez, & de los Reyes, 1997). In this situation, the funds of knowledge of each student would become an integral aspect of the mathematics classroom. Inherent in this approach to pedagogy is the decentering of the source of knowledge from only the teacher or textbook to include student knowledge and skills. The challenge is to then ensure that the mathematics tasks are rich and allow students to wrestle with difficult concepts, while bridging school and out of school mathematics practices (M. Civil, personal communication, June 24, 2006).

Research calling for multicultural learning communities also highlights the importance of creating inclusive, transformative, classrooms in order to engage students in learning (Nieto, 1999; Banks, 2001). Rather than superficially incorporating an activity that serves to represent *the* "Latino/a" culture, which can lead to a misrepresentation of an internally diverse ethnic group such as Latino/as, multicultural education must focus on creating learning environments in which student knowledge is central and equally valued by all participants (Gutiérrez, 2002; Elboj, et.al., 2002). An important aspect of this type of learning community is dialogic learning, based on the notion that students' engaging in dialogue about their thinking and ideas would allow them to include their own knowledge and value that of others (Flecha, 2000). In our Math Club and through the project we describe below, student knowledge is central to the mathematics activities and is brought out in dialogue among all members of the Club.

The Math Club included fourth and fifth grade students who attended two days a week for two after-school hours during the school year. The project we describe below began in February, and lasted for about three months, until the end of the academic year for the after- school programs at Agave. We attempt to describe how a project based in students' own lives and experiences bridges their life experiences with mathematics, with the potential to develop their conception of mathematics as a tool with which it is possible to transform the world around them.

A Community Project

Our project consisted of three parts, set in three different contexts, starting with the one most immediate to students, moving out to those increasingly more global. We began with the students looking at their own classrooms, moved to an exploration of their neighborhood through photographs and discussion, and ended with their creation of mathematical problems that pertain to the entire Latino community in the United States. For the most part, we executed the project as a continuous unit, but with occasional "breaks." These breaks occurred when we were not sure which direction the project should take next, when the students were resistant or uninterested in the activities (as we will explain later), or when other activities or tasks arose based on student interest, holidays, etc... On these occasions, we would engage in other types of activities, such as collective reading of children's mathematics books or problem-solving. This accounts for the three months it took to complete the project.

Measuring the school

The first part of the project lasted three weeks. The students first computed the areas of their respective classrooms in order to determine whose classroom had the greatest amount of space per student. They all concluded that each student had a sufficient amount of space and that there is no overcrowding at the school. As a part of this activity they also made scale drawings of their classrooms and computed the area and perimeter of the entire school with the use of a satellite picture and scale. The last part of the school measurement activity was more open- ended: each group created a mathematical question about the school, made a poster with the solution and presented it to the other members of the Club. For example, Andres' group explored how many students (holding hands) it would take to (in their words) "go around the school." In other words, their question was how many students it would take to circumscribe the school (see right).

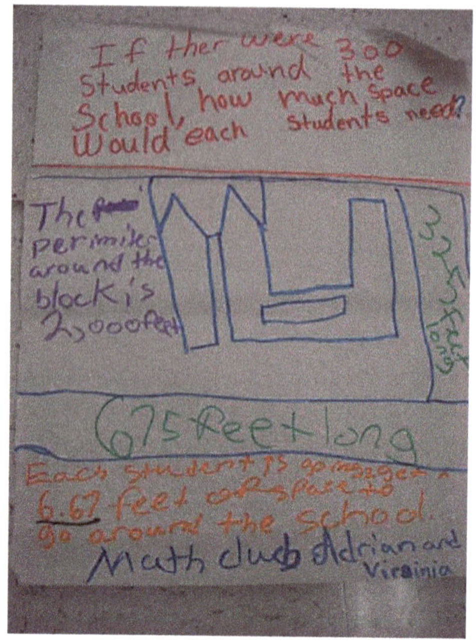

"Go Around the School"

The students probed Andres during his presentation to explain what the group meant by "around the school," how they found the measurements and how they computed the number of students, until they were satisfied with the answer. They even helped him when he got stuck with showing how he calculated the perimeter because he was only adding up two sides. Another group decided to measure the Peace Patio of the school, a favorite location of most students as it is an inner courtyard decorated with colorful mosaics and murals. This is also the location of school assemblies and so they decided to figure out how much space each student would have if the entire student body was evenly spaced out in the Peace Patio.

The students were able to pose their own problems and worked collaboratively to find a solution that made sense based on the context. Each group worked with an adult facilitator

NOTES

serving as a resource rather than a judge of "correctness" on all parts of the activity. As these were all real-life problems, they did not have "clean" solutions and the students had to grapple with a number of issues in order to solve them. Since none of the classrooms were rectangular in shape, all the groups discussed how to measure irregular shapes and find their areas. They also had to decide how to convert units (among yards, feet and inches or between meters and centimeters, depending on which unit system they chose to work in) and how to choose a reasonable scale for the classroom drawings. They often used original, creative strategies rather than previously memorized formulas. The solutions were the result of mathematical discussions and negotiations of solution methods among the group members, in which every student's voice was heard and valued.

As we share the approach taken in funds of knowledge research, which regards culture as people's daily practices, "what it is that people do, and what they say about what they do" (Gonzalez, Moll & Amanti, 2005, p. 16), we consider the school measurement activity culturally relevant. Students utilized multiple methods and knowledge bases to document space distribution in the classrooms where they spend considerable amounts of time, and engaged in dialogue about the meaning of space per student in this context. Being able to communicate in their home language facilitated their engagement in rich mathematical discussions. Furthermore, as this was the first project in the Math Club that had a considerable social component, it was important for the students to begin by looking at equity and community issues, even if on a minor scale (classrooms having sufficient space) in a place most familiar to them, the school, before turning their attention to the "outside world."

Community walk

The second part of the project expanded our focus from the school to the community in which our students live. Equipped with digital cameras, on two occasions (on two consecutive weeks) we took walks around the neighborhood in which the school is located. The students were asked to document what they are proud of as well as what they would like to change. Based on student interests, we walked to the library, past numerous murals, to a local church, to the neighborhood convenience store, and paid a visit to the fire station. The students were able to bring in their expertise about the neighborhood to determine the route we would take. Among other things, we took pictures of library books in Spanish, fire trucks, murals, graffiti, trash, and run-down houses. The following photograph was taken by a Math Club student:

Neighborhood Pride

The firefighters gave us a full tour of the station and answered all of the students' questions. While answering these questions, they occasionally used mathematics. For example, one firefighter stated that the fire station gets 1800 calls per year, on average, but that the actual number of calls per day varies greatly, which tied in with the conversations we had been having with the students about the meaning of average. We later made a "thank you" card for the firefighters in which the students calculated how many calls the fire station gets, on average, in one month and in one day.

After each walk we discussed the photos with the students. This gave them the opportunity to talk about their expert cultural and experiential knowledge, which, according to Gutstein, Lipman, Hernandez, and de los Reyes (1997), is a component of a model for culturally relevant mathematics teaching. Among other things, they talked about their families, the stories behind the murals, local businesses, and the history of the neighborhood. They shared considerable information about the life of the community. Patri, who had in the past reprimanded her peers for using Spanish in discussions instead of English, pointed out a mural that portrayed a woman making tortillas and commented that this was how her Nana made tortillas, too. Andres was talking about the significance to him, of the Deer Dancer[1], and most of the

[1] A dancer performing the traditional Yaqui deer dance. There is a significant Yaqui population in the area where the school is located.

NOTES

students referred to La Virgen[2] as an essential part of their lives. Some were familiar with the location of the WIC (Women, Infants and Children)[3] office and of the local boxing gym. The photographs they had taken also showed their concern with trash and (mostly gang) graffiti that can be found in the neighborhood. When we asked them how all of this was related to mathematics, Marisol responded, "They are not related at all!" But then Briona said that you could organize pictures by topic, which steered the conversation in the mathematical direction. A lively discussion ensued. The facilitators ensured that the conversation didn't stray from the topic and were constantly probing the students with the question: How could we connect these issues to mathematics? The students responded by generating ideas for possible projects. Due to the content of our two walks, most of the ideas were related to data collection. In particular, the students suggested two such projects: looking at the number of thefts, murders, and drug trade in gangs, and following the daily transactions of the local convenience store. Although the data for the former project would be difficult to obtain, we (the facilitators) suggested we could find information about local crime statistics from the nearby police station and track the convenience store transactions ourselves. The students were also interested in doing a project about the fire station, saying that we could look at the number of received calls, the number of fires put out, or the amount of water used to put out a fire, which is especially relevant during the summer heat and drought in our area.

Although the community walk part of the project was less mathematical than the others, we feel it was equally important. One reason is that the students were encouraged to think about applications of mathematics outside of their classroom. In some previous conversations with students that age, we found that probably due to the nature of their mathematics instruction, students often think of mathematics mostly in terms of numbers or shapes. School mathematics seldom offers connections between mathematics and the students' daily lives, something that we were able to provide in this phase of the project. Another important aspect of the community walk is that we (the facilitators) learned more about the students and their community, which is a critical component of culturally relevant mathematics teaching (Gutstein, et al., 1997). We have found that to establish a mathematics learning community with students, it is equally important to provide rich mathematical contexts and to constantly make conscious efforts towards community building. Finally, it is an important aspect of incorporating students' funds of knowledge into a culturally relevant mathematics curriculum for teachers to learn more about their students' lives and experiences.

Immigration

Despite the students' active participation in the mathematical discussion following the community walk, they were not convinced there was a good reason for addressing a social issue such as, for example, littering or gang violence with a mathematical project. A belief most of them held was that things are the way they are supposed to be, that they cannot change, and that "nobody would listen to children anyway." Basically, they felt they had no "agency" to change things in the world around them. Gutstein (2006) ascribes this sense of powerlessness that students often have when faced with injustice both to the educational system that discourages children from being critical thinkers and to the paralyzing effect that injustice can have on those experiencing it. To further develop agency in students, he recommends that students learn their own history, that they be presented with examples of positive social change, and that they be given the opportunity to take action.

Following these recommendations, we intended our next step in the project to be picking one or more of the students' ideas

[2] Virgin of Guadalupe
[3] A Federal mother and child wellness program

from the discussion following the community walk and formulating specific mathematical questions to address. We began to share examples of youth activism enacted by students their age in the hopes of convincing the students that young people like themselves can affect change. However, around this time (April 2006), events happened that especially reinforced the recommendations of Gutstein (2006), changed the course of our project, and affected most of our students. Namely, high school and middle school student walk-outs were occurring at local schools, based on the community's concern with the national immigration debate. The students began to reveal their own connections to, knowledge of and confusion over this debate. Many felt the need to get involved and were no longer convinced that they were powerless. We decided that the only way to proceed with the project would be to put aside the prior topics and to develop mathematical problems in the context of immigration. Current events provided the students with the opportunity to truly see how mathematics can be used to describe and explore a problem based in their daily lives.

Because this was a sensitive topic, it was especially important for dialogue to take place among all participants. We had a few discussions on the rug (where most of our discussions take place), when the students shared personal or news stories. We encouraged them to think about how they could contribute their stories and information to a mathematics project related to immigration. Veronica, for example, decided to conduct a survey in her neighborhood about the proposed legislation (see figure below).

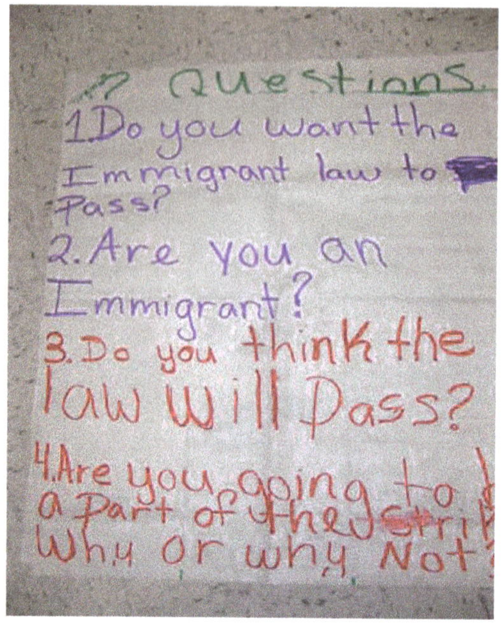

Veronica's Survey

Although initially one of our quietest students, she had undergone a profound change during the course of the year, and took the initiative to do the survey entirely on her own. The children collaborated to quantify the survey data, creating a table on the whiteboard and calculating fractions and percentages for each response. They also conducted the survey in the Math Club and compared the results, being able to generate data to show the overwhelming community support for the protests, marches and walk-outs. Jacinta found the time it takes a person to walk to the city in which we live from a nearby Mexican border town, while Mónica computed the amount of water and food such a person would need. As Mónica presented the material, she explained: "I figure(d) out how long it takes by walking to [city we live in] by writing one hour and three miles to

NOTES

see how many hours to walk sixty miles[4]. It was twenty hours. How much water would a person need? They would need one hundred and twenty eight ounces and it would be eight pounds. Most people will need eight cups or ten cups. I used eight cups for my chart. How much food would a person need? A medium guy would need two thousand calories with him while walking. Two thousand calories is one pound." When probed by the facilitator to explain what this made her think about, Mónica responded, "Like to see what a person would need while they're walking. If they don't have the same amount they might die." Because it would take 23 hours to cross the border, Jacinta also had to think about the real-life meaning of this number: she had to decide how many hours a day a person could really walk, and how many days the trip would take in order to understand the reality of crossing the border. Dolores and Marisol made a graph of the immigrant population in the United States based on country of origin, in order to understand the complexity of this national issue. In discussing the high percentage of undocumented immigrants who are Mexican, Andres remarked, "They want to have more money and keep working and have a lot of food for their children." The issue was real and salient to the students as they interpreted the meaning of the statistics presented.

Because of the unforeseen early end of the after-school program in the school the Math Club is based, we had to finish the project earlier than expected. For this reason we were unable to try to present our findings to some of the local officials, our intended audience for the data. We think this is unfortunate, because we feel that it would have had a positive influence on our students' sense of agency in that they could engage in community activism related to the mathematics problems they addressed.

[4] By "writing one hour and three miles" she means that she used the information that an average person walked three miles per hour.

Final Remarks

In this paper we have focused our attention on several examples in which we illustrate how children carried out mathematics in the Math Club environment. The Math Club is a friendly environment, where children are encouraged to make their own decisions and conjectures, and to discover the application of mathematics to real life. This project allowed children to draw on their multiple abilities as well as their prior funds of knowledge in collaboratively exploring a variety of approaches to finding meaningful answers to mathematical problems. These projects were also important opportunities for the children to make meaningful connections between mathematics and their everyday lives, such as through discussions about immigration. This illustrates how the Math Club is not only a place oriented to the study and reflection of mathematics, but also a space where children are free to express and explore their own real life connections to mathematics.

One of the most relevant contributions of the Math Club is that it combines rich mathematics projects and a multicultural learning community that encourages dialogic learning (Flecha, 2000). Through dialogue, children reflected on each other's solutions and connected the solutions to their real life contexts. Dialogue among children about their peers' answers was a useful way to spark that interest in analyzing a solution's relevancy, as reflected in Jacinta's example in the immigration project. Jacinta found that it would take a person approximately 23 hours to walk from a Mexican border town to the city in which we live. When she presented this answer, someone asked her if it was possible for a person to walk for 23 hours without breaks. She reflected on this, and after a brief discussion with other students and facilitators, Jacinta modified her answer, concluding that, realistically, it would take more than a whole day for a person to cover the distance between the two cities. Jacinta's solution to the problem, presented as simply the number 23, though correct, would not have been an accurate reflection of the real context in which the problem was posed.

An important aspect of multicultural learning communities is that they are not only inclusive of multiple views, but that they are also transformative locations for learning. The project we have described allowed children in the Math Club to find a bridge between their position in their community and mathematics; so mathematics, far from being knowledge separated from the real world, became a tool to understand the role that each person plays in his/her community and the possibilities for making changes in that reality. This project provides a suggestion for a way to think about contextualizing a mathematics curriculum, enabling children to connect their own life experiences with mathematics. Furthermore, we suggest that mathematics can be a powerful way to foster a sense of agency in order to raise critical mathematics thinkers who are prepared to change their world, because "mathematics is everywhere."

Next Steps

As we reflect on the experience of this project, we see areas where we would make changes when done again to enhance the mathematics and draw out the students' sense of agency. During certain parts, we could have made the mathematics content more explicit, such as during the community walk. We could have incorporated more mathematics into learning how to take good pictures, such as with proportion and geometry. In terms of choosing a topic to focus on, we would offer more structure in terms of identifying a problem, working on the mathematics and providing a forum and more meaningful purpose for presenting. With this, students would see the purpose from the beginning and potentially remain more engaged throughout. Also, in the end we were unable to bring the students' findings to a wider audience, which would have further developed the students' sense of mathematical agency. As we enter the new school year, our main challenges are to explore children's thinking in more depth when bridging home and school mathematics in the context of rich mathematical tasks, as well as to then utilize this knowledge in a meaningful forum for expression of student agency.

Acknowledgements

We would like to thank CEMELA for allowing us to conduct this study as a part of their overall research agenda. CEMELA (Center for Mathematics Education of Latinos/as) is funded by the National Science Foundation under grant ESI-0424983. We would also like to thank the Fulbright Commission for the support given to one of the researchers and authors of this study, through the Fulbright Visitor Program. The views expressed here are those of the authors and do not necessarily reflect the views of the funding agencies.

NOTES

References

Banks, J. A., (2001). Citizenship education and diversity: Implications for teacher education. *Journal of Teacher Education, 52* (1), 5-16.

Civil, M. & Andrade, R. (2002). Transitions between home and school mathematics: Rays of hope amidst the passing clouds. In G. de Abreu, A.J. Bishop, and N.C. Presmeg (Eds.), *Transitions between contexts of mathematical practices* (pp. 149-169). Dordrecht: Kluwer.

Civil, M. (2002) Culture and mathematics: A community approach. *Journal of Intercultural Studies, 23*(2), 133-148.

Elboj, C., Puigdellívol, I., Soler, M., & Valls, R. (2002). *Comunidades de aprendizaje. Transformar la educación.* Barcelona: Graó.

Flecha, R. (2000). *Sharing words: Theory and practice of dialogic learning.* Lanham, MD: Rowman & Littlefield.

González, N., Moll, L., Floyd-Tenery, M., Rivera, A., Rendon, A., & Gonzales, R. (1995). Funds of knowledge for teaching in Latino households. *Urban Education, 29* (4), 444-471.

González, N., Moll, L., & Amanti, C. (2005). *Funds of knowledge: Theorizing practices in households and communities.* New Jersey: Lawrence Erlbaum.

Gutstein, E., Lipman, P., Hernandez, P., & de los Reyes, R. (1997). Culturally relevant mathematics teaching in a Mexican American context. *Journal for Research in Mathematics Education, 28* (6), 709-737.

Gutstein, E. (2006). *Reading and writing the world with mathematics: Toward a pedagogy for social justice.* New York: Routledge.

Gutierrez, R. (2002). Beyond essentialism: The complexity of language in teaching mathematics to Latina/o students. *American Educational Research Journal, 39* (4), 1047-1088.

Nieto, S. (1999). *The light in their eyes: Creating multicultural learning communities.* NY: Teachers College Press.

Proposition 203, English for the Children, Arizona voter initiative (2000).

Using Vedic Mathematics to Make Sense of the Finger Algorithm

Dr. Bethany Noblitt
noblittb@nku.edu
Northern Kentucky University

Blaire Richter
richterb1@nku.edu
Northern Kentucky University

Abstract

Vedic mathematics, based on sutras from ancient Hindu texts, was rediscovered in the early twentieth century by Bharati Krishna Tirthaji (Stemn & Collins, 2001). Vedic mathematics deals mainly with mentally carrying out tedious arithmetical operations. While studying a particular Vedic algorithm and its connections with standard arithmetic operations, the authors discovered connections to a nonstandard multiplication algorithm, the finger algorithm, found in a mathematics textbook for preservice elementary teachers. These connections provided justification for why the finger algorithm works. This paper discusses a justification of the Vedic algorithm and how it led to justification of the finger algorithm.

Using Vedic mathematics to make sense of the finger algorithm

"Why in the world does that work?" That was the question asked in my mathematics for elementary and middle grades teachers course in response to seeing a finger algorithm for multiplication for the first time. My answer was, "I have no idea." While this answer was accepted by students in the class, I could not let it go. When I was unable to answer my students' question about why the finger algorithm works, I challenged myself to figure it out. In this article, we will explain the finger algorithm and why it works through an exploration of Vedic mathematics.

Vedic mathematics was rediscovered in the early twentieth century by Swami Bharati Krishna Tirthaji and is based on sutras from ancient Hindu texts known as the Vedas (Stemn & Collins, 2001; Tirthaji, 1965). Vedic mathematics and our use of it may be viewed as ethnomathematics. D'Ambrosio (1985), described ethnomathematics as the field which bridges anthropologists and historians of mathematics. According to D'Ambrosio (1985), building this bridge is "an important step towards recognizing that different modes of thoughts may lead to different forms of mathematics" (p. 195). In the early twentieth century, Tirthaji used his knowledge of the Vedas, which included areas of thought presumably unrelated to mathematics, to develop Vedic mathematics (Tirthaji, 1965). Vedic mathematics deals mainly with mentally carrying out tedious arithmetical operations (Tirthaji, 1965). Tirthaji's work created an example of the bridge described by D'Ambrosio. Our explanation of the relationship between the finger algorithm and Vedic mathematics also describes how a mode of thinking very different from ours led us to deeper mathematical understanding.

My exploration of Vedic mathematics was not a solitary one. At the same time that I was grappling with understanding why the finger algorithm works mathematically, I was studying Vedic mathematics and its connections with standard arithmetic operations with one of my students (who was not in the mathematics for elementary and middle grades teachers course), Emily Book (a pseudonym). The synchronicity of discussing the finger algorithm in class and the study of Vedic mathematics was imperative to our progress in understanding why the finger algorithm works.

The Finger Algorithm

The finger algorithm is introduced in a problem in the exercises in the text *Mathematics for elementary teachers: A contemporary approach* by Gary L. Musser, William F. Burger, and Blake E. Peterson (2004). The problem is:

> *The use of finger numbers and systems of finger computations has been widespread through the years. One such system for multiplication uses the finger positions shown for computing the products of numbers from 6 through 10.*

NOTES

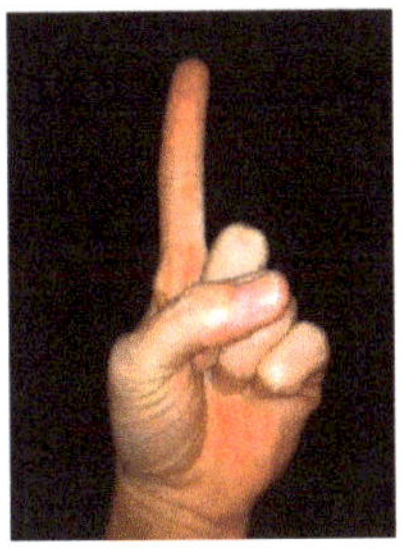

6

7

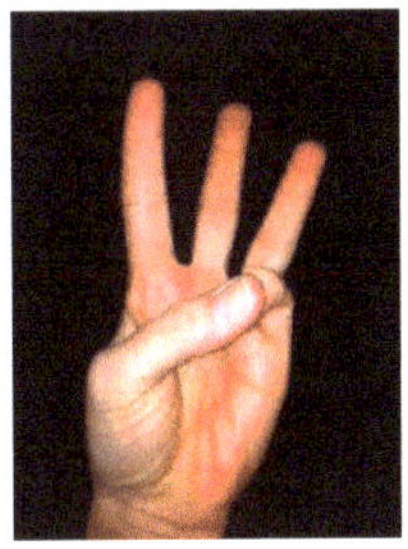

8

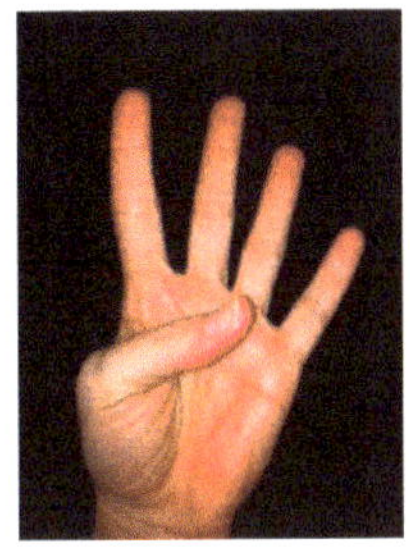

9

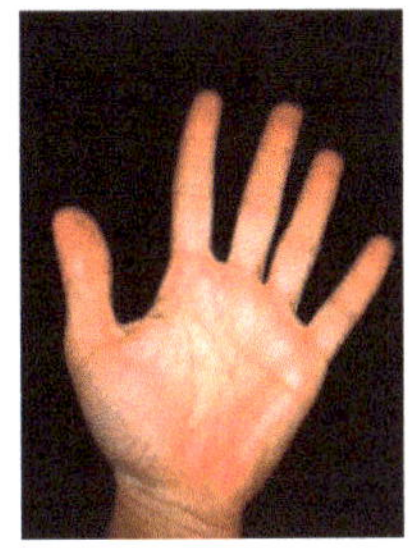

10

The two numbers to be multiplied are each represented on a different hand. The sum of the raised fingers is the number of tens, and the product of the closed fingers is the number of ones. For the example below, 1+3=4 fingers raised and 4X2=8 fingers down yields 6X8=48 (Musser, Burger, and Peterson, 2004, p. 180-181).

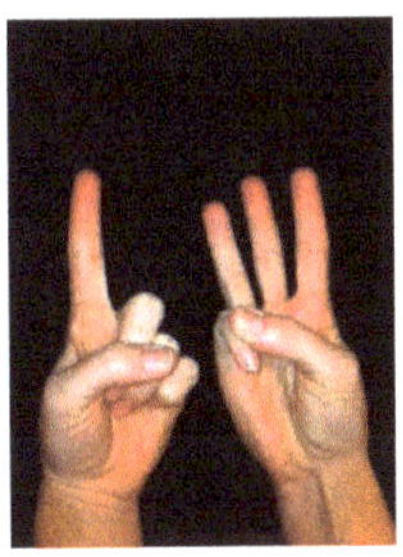

6x8

When we discussed this problem in class, we spent a few minutes just talking about the representation of the numbers. For this algorithm, first think of representing a number between 6 and 10 using both hands. For the numbers 6 through 10, one hand always has all five fingers raised. The finger algorithm representation of a number is obtained by taking away the hand that has all five fingers raised, leaving a hand with between 1 and 5 fingers raised. Because the finger algorithm is used to determine the product of two numbers between 6 and 10, two hands will be used in the algorithm, each hand representing a factor in the product.

To reiterate and clarify the algorithm, let's look at another example. To determine the product of 8 and 7, one hand will have three fingers raised (representing 8), and one hand will have two fingers raised (representing 7).

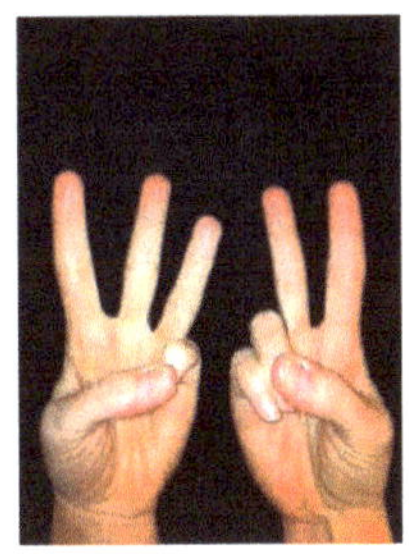

8x7

The sum of the raised fingers is the number of tens in the product. In the 8x7 example, there are a total of five fingers raised, representing 5 tens, or 50. The number of ones in the product is given by the product of the closed fingers on each hand. In our example, there are two closed fingers on the hand representing 8 and three closed fingers on the hand representing 7. Since 2 times 3 is 6, there are 6 ones in the product of 8 times 7. Adding the tens to the ones, 50+6, we get the product of 56.

Figuring it out

One day, as Emily and I studied a particular Vedic algorithm for the multiplication of integers based on the Nikhilam sutra, I mentioned the finger algorithm and the fact that I did not have an answer to my students' inquiries about why it works. While our conversation ended there, Emily's thinking did not. At our next meeting, Emily shared some insights into the finger algorithm based on her understanding of the Nikhilam approach. As she thought about the finger algorithm, she kept coming back to two aspects of the algorithm that she recognized from the Nikhilam approach. First, in the finger algorithm, numbers are represented on one hand by taking five away. Also, in the algorithm, the "down" fingers are complements from ten. These are both also aspects of the Nikhilam approach. In the rest of this paper, we describe the Nikhilam approach and its relation to the finger algorithm.

The Nikhilam algorithm

Let's consider the above example 8x7 using Vedic mathematics. To use the Nikhilam algorithm, express the multiplication vertically.

$$\begin{array}{r} 8 \\ \times\ 7 \\ \hline \end{array}$$

Next, take the complement of each number from ten and write it as shown below.

$$\begin{array}{rcl} 8 & \searrow & 2 \leftarrow (10-8) \\ \times\ 7 & & 3 \leftarrow (10-7) \\ \hline \end{array}$$

At this point there are two columns with which we work. We begin with the column on the right. Multiplying these two numbers together yields the number of ones in the product. Since 2x3=6, there are 6 ones in the product of 8x7. Place that product directly under the 2 and 3.

$$\begin{array}{rcl} 8 & \searrow & 2 \leftarrow (10-8) \\ \times\ 7 & & 3 \leftarrow (10-7) \\ \hline 5 & & 6 \end{array}$$

To get the number of tens in the product, we turn our attention to the numbers in the column on the left. Subtract diagonally 8-3 (or 7-2; both differences will always be the same. This difference is the number of tens in the product. Since 8-3=5, there are 5 tens in the product of 8x7. Place that difference directly under the 8 and 7.

$$\begin{array}{rcl} 8 & \searrow & 2 \leftarrow (10-8) \\ \times\ 7 & & 3 \leftarrow (10-7) \\ \hline 5 & & 6 \end{array}$$

Since 5 tens and 6 ones gives 56, we get 8x7=56.

Regrouping

We typically do not think about regrouping when multiplying two one-digit numbers. However, regrouping can be seen using the finger algorithm, as well as the Nikhilam algorithm, with certain products. Let's consider an example using the finger algorithm where regrouping occurs.

Consider 6x7.

NOTES

6x7

One hand has 1 finger raised (representing 6) and the other hand has 2 fingers raised (representing 7), so 1 + 2 gives the number of tens in the product of 6 times 7. So, there are 3 tens, or 30, so far. There are 4 fingers closed on one hand (representing 6) and 3 fingers closed on the other hand (representing 7). Multiplying 4 by 3 gives the number of ones in the product 6 X 7. So, there are 12 ones. Adding the number of tens to the number of ones, 30 + 12, gives a final product of 42. Regrouping occurs because there are more than 9 ones obtained, thus creating the need to regroup 10 ones as a ten.

We can also see regrouping in the Nikhilam algorithm with the product 6x7. Again, begin by writing the multiplication vertically.

$$\begin{array}{r} 6 \\ \times 7 \\ \hline \end{array}$$

Then, taking complements from ten we get the following.

$$\begin{array}{rr} 6 & 4 \\ \times 7 & 3 \\ \hline \end{array}$$

Multiplying 4x3 we get 12 ones.

$$\begin{array}{rr} 6 & 4 \\ \times 7 & 3 \\ \hline 3 & 12 \end{array}$$

Subtracting diagonally 6-3 (or 7-4) we get 3 tens.

$$\begin{array}{rr} 6 & 4 \\ \times 7 & 3 \\ \hline 3 & 12 \end{array}$$

The product of 6x7 is 3 tens and 12 ones, or 42.

$$\begin{array}{rr} 6 & 4 \\ \times 7 & 3 \\ \hline 3 & 12 \end{array}$$

tens / ones

4 ← 2

The 10 ones in the 12 are regrouped to form a ten, leaving 2 ones. The 1 ten from the regrouping is added to the 3 tens. This gives 4 tens and 2 ones for a total of 42.

Justifying the Nikhilam algorithm

During our exploration of the Nikhilam algorithm, Emily sensed a connection to the finger algorithm. In order to explore this connection further, we answered the question of why the Nikhilam algorithm works. First, we considered a specific example. Let's look again at the example 8x7.

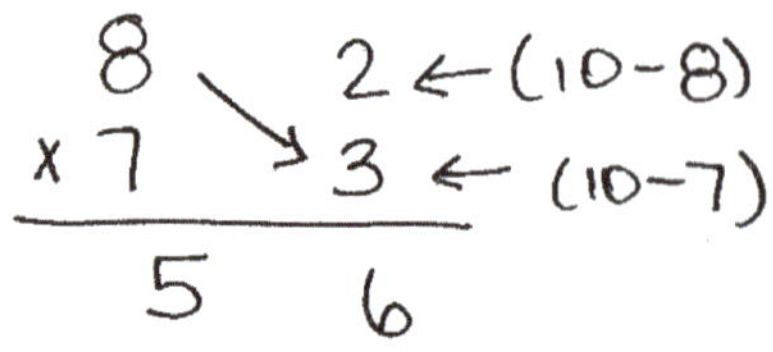

Remember, in the multiplication of 8x7, the number of tens in the product came from 8-3 (illustrated by the diagonal arrow) and the number of ones in the product came from (2)(3) (the product of the complements from 10). Considering this, we can write the algorithm as a single expression, 10[8-(10-7)]+(10-8)(10-7), where [8-(10-7)] is the number of tens and (10-8)(10-7) is the number of ones.

To generalize, we substitute x and y into this expression and simplify. From this, we see how the Vedic algorithm does indeed yield the product xy.

$10[x-(10-y)]+(10-x)(10-y)$

$= 10[x-10+y]+(10-x)(10-y)$

$= 10x-100+10y+100-10x-10y+xy$

$= xy$

Volume 3 Teaching

Relating the Nikhilam algorithm to the finger algorithm

Once we justified the Nikhilam algorithm, we were able to uncover the connections between the Nikhilam algorithm and the finger algorithm. We did this by manipulating the expression 10[8-(10-7)]+(10-8)(10-7) in order to make sense of the finger algorithm as a result of the Vedic algorithm. We also tried to relate our manipulations to the earlier connections Emily had made between the finger algorithm and the Nikhilam approach. Emily had focused her connection-making on two important aspects of the finger algorithm: the "up" fingers being a representation of a number by taking five away and the "down" fingers being a representation of a number's complement from ten. Consider the following manipulations with these two aspects in mind.

10[8-(10-7)]+(10-8)(10-7) [the Nikhilam algorithm written as a single expression]

= 10(8+7-10)+(10-8)(10-7) [distributing the negative sign]

= 10(8+7-5-5)+(10-8)(10-7) [rewriting -10 as -5-5]

= 10((8-5)+(7-5))+(10-8)(10-7) [associativity]

At this point, I could see what Emily sensed all along. The Nikhilam algorithm really illustrates the finger algorithm. First, we address the meaning of the expression ((8-5)+(7-5)). Traditionally, you use both hands to represent the number 8, with 8 of the ten fingers raised.

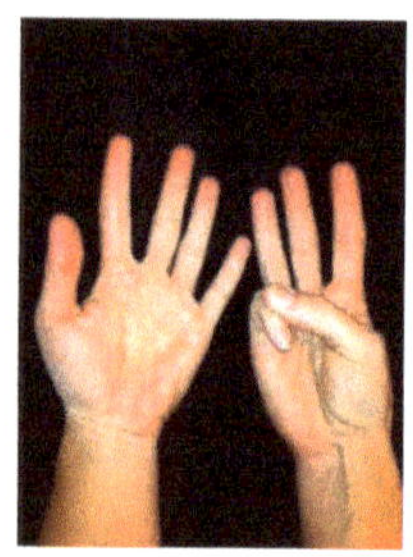

If you take away the hand with all five fingers raised, you are left with a hand with three fingers raised and two fingers down.

NOTES

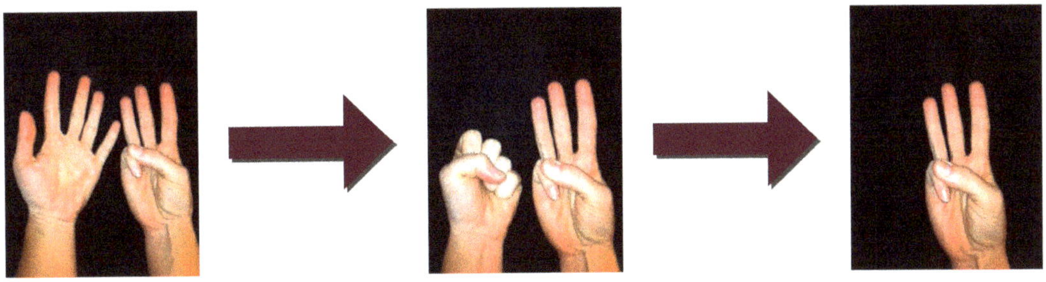

This one hand with three fingers raised is the finger algorithm representation of the number 8.

So, one way to count the "up" fingers in the finger algorithm representation of the number 8 is 8-5 (all 8 fingers take away the full hand of 5 fingers). So, the (8-5) represents the 3 raised fingers on one hand of the finger algorithm. The (8-5) can be seen clearly in the expression 10((8-5)+(7-5))+(10-8)(10-7). Similarly, (7-5) represents the 2 raised fingers on the other hand of the finger algorithm. So, ((8-5)+(7-5)) is the total number of raised fingers in the finger algorithm, which, as previously described, gives the number of tens in the product. So, it makes sense to multiply ((8-5)+(7-5)) by 10: 10((8-5)+(7-5))+(10-8)(10-7).

Again, the traditional way of representing the number 8 using fingers, is to raise 8 of the ten fingers. This leaves 10-8 or 2 (complement of 10) fingers down on the hand with three fingers up. Similarly, the representation of 7 will leave 10-7 or 3 (complement of 10) fingers down on the hand with two fingers up. As previously described, the product of the down fingers gives the number of ones in the product. This can be seen clearly in the expression 10((8-5)+(7-5))+(10-8)(10-7). Finally simplifying the expression 10((8-5)+(7-5))+(10-8)(10-7), we see the final product of 56.

10((8-5)+(7-5))+(10-8)(10-7)

= 10(3+2)+(2)(3) [sum of up fingers multiplied by ten plus the product of the down fingers]

= 10(5)+(2)(3)

= 50+6 = 56

My answer to "why in the world does that work?"

When showing the finger algorithm as an example of a nonstandard algorithm for multiplication, the question of why it works has unfortunately gone unanswered in my class. But, after exploring the finger algorithm in the context of its connections to the Nikhilam algorithm, I now can give my students the opportunity to explore and justify two nonstandard multiplication algorithms, while making mathematical connections between them. Being able to provide students with a way for them to answer the question, "why in the world does that work?" is much more satisfying not only for them, but for me as their teacher. My students and I were not the only people to benefit from this mathematical exploration. Emily's contribution to the investigation of the finger algorithm through her understanding of the Vedic algorithm was invaluable. She made necessary connections that brought clarity not only to my understanding of both algorithms, but her own understanding was deepened as well. Being able to apply her knowledge of one algorithm to justify another solidified her understanding even more.

Why study nonstandard algorithms?

One of the emphases in my Mathematics for Elementary and Middle Grades Teachers course is the understanding of different nonstandard algorithms for arithmetic operations. Prior thinking in mathematics education supported the idea that children memorize basic math facts (Adding it up,

2001). More recent research has shown that children actually progress through a series of more and more advanced and abstract methods for determining the answers to arithmetic problems (Adding it up, 2001). While not all children will progress through the same series of methods, all children will use intermediate and temporary procedures (Adding it up, 2001). Nonstandard algorithms may be the intermediate and temporary procedures that serve as a bridge between first exposure to arithmetic calculations and quick recall of basic math facts. It is this power of nonstandard algorithms that motivates me to share algorithms such as the finger and Nikhilam algorithms with my preservice elementary and middle grades teachers. The fact that one of these nonstandard algorithms is also a non-Western algorithm is also beneficial to my students as preservice teachers. Culture manifests itself through, among other things, jargons, codes, symbols, and ways of reasoning. (D'Ambrosio, 1985). Certainly Vedic mathematics has jargon, symbols and ways of reasoning that reflect a culture different from the culture most of my students have experienced, especially in a mathematics classroom. Providing them with exposure to a culturally different way of viewing mathematics, while deepening their mathematical understanding, is an exciting opportunity for me. My hope is that they take with them the mathematics that they have learned as well as a cultural awareness prompted by connecting an algorithm introduced in a traditional Western textbook to a non-Western, nontraditional multiplication algorithm.

References

D'Ambrosio, U.D. (1985). Ethnomathematics and its place in the history and pedagogy of mathematics. *For the Learning of Mathematics, 5*, 44-48.

Musser, G.L., Burger, W.F., Peterson, B.E. (2004). *Mathematics for Elementary Teachers: A contemporary approach*. 7 ed. John Wiley & Sons, Inc.

National Research Council. (2001). *Adding it up: Helping children learn mathematics*. J. Kilpatrick, J. Swafford, and B. Findell (Eds.). Mathematics Learning Study Committee, Center for Education, Division of Behavioral and Social Sciences and Education. Washington, DC: National Academy Press.

Stemn, B.S. & Collins, J.E. (2001). Do numbers have shapes? Connecting number patterns and shapes through the Vedic matrix. *Teaching Children Mathematics, 7*, 542-547.

Tirthaji, B.K. (1965). *Vedic Mathematics*. Delhi, India: Motilal Barnarsidass Publishers.

Reflections on: Relationships between the Hopi calendar and measurement concepts

Janet Sharp
Washburn University
Topeka, Kansas USA
janet.sharp@washburn.edu

Tracie Lutz
Washburn University
Topeka, Kansas USA
tracie.lutz@washburn.edu

Abstract

This paper provides reflections after co-teaching the lesson: *Using Moon Phases to Measure Time* to a group of 6th grade learners. The co-authors are a 6th grade math teacher and a university math educator, both from the United States. As a result of reflecting on this lesson, both authors enhanced their knowledge of the relationship between mathematics and culture. The 6th grade teacher described a new awareness of the impact of culture in a realistically-situated mathematics problem whereas the university professor developed knowledge about the multi-disciplinary nature of ethnomathematics. Collaborative follow-up reflection revealed pedagogical insights, such as the need to explicitly encourage students to reframe discussion outside of conventional and comfortable viewpoints and of the importance of solidifying cross-curricular topics. In the following sections, the authors described individual responses about the teaching episode. Concentration is given to instruction, curriculum, addressing standards and resistance to unfamiliar content.

Keywords:

- Time Measurement
- Pedagogy

Lesson Summary

As Tracie's students gathered, we were both excited to share what we know about how the Hopi Celestial Body Watchers used lunar cycles as a critical component of their calendar of festivals, which accompany agricultural and other societal needs (Sharp, 2015). We greeted the students and set up the first lesson, which was a science lesson about lunar cycles and the various moon phases. This information seemed familiar enough to the students, who had completed similar lessons in their elementary grades. From there, we smoothly transitioned to the mathematics lesson about the important conceptual elements of measurement (standard unit, continuity, infinite divisibility, tiling, same-size and zero-indicator). After that lesson, we were satisfied our students were ready to move to an explanation of the importance of the various moons to the Hopi. This history lesson seemed to go very well and the students seemed eager to learn about the moon names and to match moons to corresponding months. Their willingness to match moons to months should have alerted us to what would become an issue, but for the time being, it slipped our immediate notice. After that lesson concluded, we moved to the literature portion of the lesson, where students read "Broken Pattern" (Carlson, & Qoyawayma, 1985) a Hopi story which included many passages about measuring time with seasons, moons, or other natural phenomena. Finally, we presented students with the Hopi's problem of how to incorporate the lunar cycle into measurement of an Earth year.

Opening comments

6th Grade Teacher:

In order to find time to incorporate this lesson, it was important to make connections to other curricular areas, as well as to mathematical problem solving. For me, this lesson was a remarkable experience because I changed my thought patterns to incorporate a more rounded understanding of what it means to teach mathematics, in general, and measurement, in particular. Prior to this experience, I admittedly posed measuring problems that were slightly isolated from other content. After this experience, I see how a mathematics problem can (and should) permeate all of the curricular areas of the classroom. Moreover, I became more aware of deficiencies in the curricular materials adopted by her district, which paled in comparison to this experience.

> *"Equally frustrating to me, is the fact that not only have I not been prepared to teach using cultural problems, I am upset that the students apparently had no previous opportunities to experience this kind of lesson."*

University Professor:

In order to transform the previously-written work, which described the Hopi calendar and their successful use of the lunar cycle to help them measure time, the essential elements of the original work had to be revisited, with an eye toward implementation in a 6th grade classroom. First, I looked for the important content that would also appeal to 11-year olds. Then, restructured those facts into a format that would be useful for a 6th grade classroom. This effort required building and

NOTES

selecting content around scientific, literary, cultural, historical, and mathematical ideas. This process helped me develop a deeper appreciation for the connections between ethnomathematics and what might be considered ethno-science and ethno-literature. It also ignited my passion to continue to develop not only informative written work for fellow scholars and classroom teachers, but also to learn to present the information within the confines of classroom walls.

> *"The frustrating aspect of the lesson for me was despite a seeming awareness of the mathematics to be learned, I was not at all sure that the students fully grasped the strategies used by the Hopi for measuring an Earth year with a lunar cycle."*

Substantive Changes/Validations of beliefs:

Tracie (6th grade teacher):

The topic of learning about alternative methods to measure time was interesting to me and the planned approach was a unique departure from my normal teaching. I often use discussion lessons and I often integrate content, but somehow this lesson felt different. The thought of encouraging students to discuss, explore, and discover how the moon can be used, and has been used, to measure time by the Hopi, was exhilarating. I believe part of my anticipation rested on the fact that much of the information was new to me, as well. So, I was excited to share it with my students.

Partway through the lesson where we connected the moon phases to Hopi societal events and festivals, I noticed a few students looking apprehensive. Because of the manner with which Hopi astronomers demonstrated mathematical understandings, by effectively marking the passage of time with lunar cycles, many students' reactions bordered on skeptical or suspicious. A lunar cycle is 29.5 days and an Earth year is 365.25 days. Since there are not a whole number of lunar cycles in an Earth year, students appeared unsure about how to understand the Hopi's use of the lunar cycles to measure time. Watching students tilt their heads and look at one another through squinted eyes was disturbing to me as their teacher.

Why were students' reactions so unlike mine – so unlike what I expected (hoped) for them? Upon reflection, it became clear that the lesson encouraged a different or novel way of measuring time when compared with my students' experiences. The Hopi's method of measuring time was distinct and circular and incongruous with what students knew to be true or accepted as true, concerning how time is traditionally measured and conceptualized as linear. Their reaction was rooted around the term . . . 'traditionally'.

Instruction

In the United States, it has long been established that the majority of math concepts are taught and modeled in a traditional manner and using supposedly culture-free contexts. This is what students have grown accustomed to and in-turn expect. So, when a 'non-traditional' method of time measurement (linked directly to an unfamiliar frame of reference) was discussed during the lesson, students' reaction seemed uncertain, which affected how I continued with and thought about the lesson at hand. Instead of expanding on how the Hopi scientifically used the moon to measure time/seasons, I focused on encouraging the students to accept and embrace the fact that the lunar cycle can actually measure time. During the lesson, I concentrated on why the students reacted in this initial manner and what role my instruction had played or could play in addressing their reactions.

Since all culturally identifiable groups have jargons, codes, symbols, myths, and even specific ways of reasoning and inferring (D'Ambrosio 1997), I needed to work harder to deepen my students' thought processes to open them to alternative or flexible methods of reasoning and problem solving. Teaching and modeling this sort of problem solving is quite powerful. As a result of this lesson, I began to clearly see that, as teachers, we lean-on and utilize methods of measurement that are comfortable, common place, linear, and expected, just as our students do. Students often are constrained by their perceptions of mathematics and school. As the students' teacher, it became evident that I had fallen into the trap of teaching math in a traditional manner and had most often used only familiar contexts.

Curricular issues

Curricula should be organized in such a way as to encourage students to develop an understanding of what numbers are and provide them with the ability to perform mathematical operations numbers in an efficient, effective manner. A variety of current math curricula showcase various traditional and non-traditional methods students may use to figure or calculate math problems, but has the curricula embedded cultural methods of mathematic computation and problem solving as well? After decades of teaching, I would have to say no.

The National Mathematics Advisory Panel (2008) stated that, "Students should understand key concepts, achieve automaticity as appropriate (e.g. with addition and related subtraction facts), develop flexible, accurate, and automatic execution of the standard algorithms, and use these competencies to solve problems." (p.xvii). As teachers, we can encourage students' development of flexibility in executing algorithms and solving problems by modeling alternative mathematic methods. These alternative methods or techniques can be derived from studying a wide range of thinking about and solving math problems. After teaching this lesson, I personally want to develop my own awareness about how to solve unfamiliar societal problems. This kind of thinking moves me laterally from the required curriculum to a more expanded curriculum. For this to occur, I will take the initiative to learn about mathematical ideas as they are used in other cultures, which will help me pose meaningful problems that encourage a wide range of solution strategies.

Resistance to unfamiliar content

The students' dismissive reaction was parallel to the reaction students portray each year when we discuss the metric system of measurement. While teaching measurement, I would often hear students' resonating complaints that the customary way of measuring is the only form of measurement they needed to understand. Even though I would explain and demonstrate the relevance and importance of various systems of measurement, students' buy-in was limited, just as it was while teaching the moon phase lesson. Knowing that the vast majority of the world has adopted the metric system, and that we live in a global society, it is obvious that the metric system

NOTES

should be embraced, rather than debated, in the classroom. This point only reinforces the idea that diverse methods of computation and problem solving should be common place in our teaching, rather than a novel idea or suggestion.

Tracie's summary

The lesson: *Using Moon Phases to Measure Time*, taught me much more than it did the students. Their initial reactions to using a non-traditional method to measure time made me rethink, not necessarily what I taught, but how I taught. Maybe how I approach teaching math derives from how I was taught math in grades K-12, as well as during my undergraduate and graduate studies, which were all taught in a traditional manner. Traditions can be comfortable, but comfort does not necessarily grow one's thinking. I believe that encouraging and modeling unfamiliar mathematical problem solving methods and ideas gives students the opportunity to develop deeper understanding and a higher level of mastery concerning the wide range of content.

As I reflected on this experience, I consulted many resources. None touched me more than Dr. Linda Furuto's comments in an article by Cheryl Ernst (2010). When Furuto said, "I firmly believe that there exists a powerful light that burns within each and every one of us. We have unique and special heritages, languages, cultures, and traditions. Ethnomathematics allows us to tap into these treasures and find a connection between wisdom grounded in the past and hope for a bright and beautiful future." After studying this comment, I could finally organize my thoughts about this experience. She helped me see that it is important for us, as teachers, to seek out and teach about alternative methods for our students to become successful mathematical problem solvers. If diversity was included or woven throughout a substantial number of classroom lessons, maybe my students' reactions concerning using the moon to measure time would not have been so dismissive.

Janet (university professor):

The opportunity to teach a group of 6th graders about this information was both overwhelming and exciting. I culled relevant information in ways to meaningfully complement Tracie's instructional style and to address her existing curricular requirements. To put together the sequenced lessons required ongoing reciprocal communication as well as rather lengthy discussions that introduced Tracie to the information related to the Hopi's use of the moons in their society and that introduced me to her curricular requirements.

In an effort to ensure connections to all required curricular areas, we decided to pace through all appropriate background information, before we presented the culminating problem. As part of the problem, it was essential for students to understand the importance of the lunar cycle to the Hopi. All of the preceding topics would play a role in understanding the mathematics and in understanding the cultural importance of the moons. And, in the end, students would have to strike a balance between the fact that the lunar cycle ultimately fails as a standard unit, but was, nonetheless, brilliantly used by the Hopi.

Instruction

The lessons proceeded well until students attempted to directly match the lunar cycles and moon phases to the months on the currently-accepted calendar. We anticipated students would use the calendar as a means to visually count off sets of 29.5 days (lunar cycles.) We did not anticipate the implicit issues this task would evoke. Because students matched moons to months, rather than to sets of 29.5 days, they did not slip out of typically Western (linear) thinking and into a circular, overlapping, way of thinking. Students seemed to be less bothered by the half day of the lunar cycle, something that we anticipated would be a problem, than by the fact that most months are longer than 29.5 days. In the long run, our students managed the information solely within the confines of their own ways of knowing about calendars, which we should have expected. But we did not anticipate the extent to which the students would grapple with the way moons did not reset each solar year. In retrospect, we believe this struggle resulted from students' resistance to deviate from linear thought. They did not visualize time as circling around to itself and the circle beginning with a different phase in the lunar cycle.

To further study this phenomenon, I presented a similar lesson to my fourth-year college students majoring in mathematics. Among those students, I found similar conceptions of time. So, it is not an issue of strong enough mathematics knowledge, rather it is an issue of finding a hook to pull the students out of the traditions of thought that have them bound to a linear conception of time. In my own studies, to represent this unfamiliar view of time passage, and to better understand when to insert a thirteenth moon into the calendar, I created a circular moon ruler. (See figure 1).

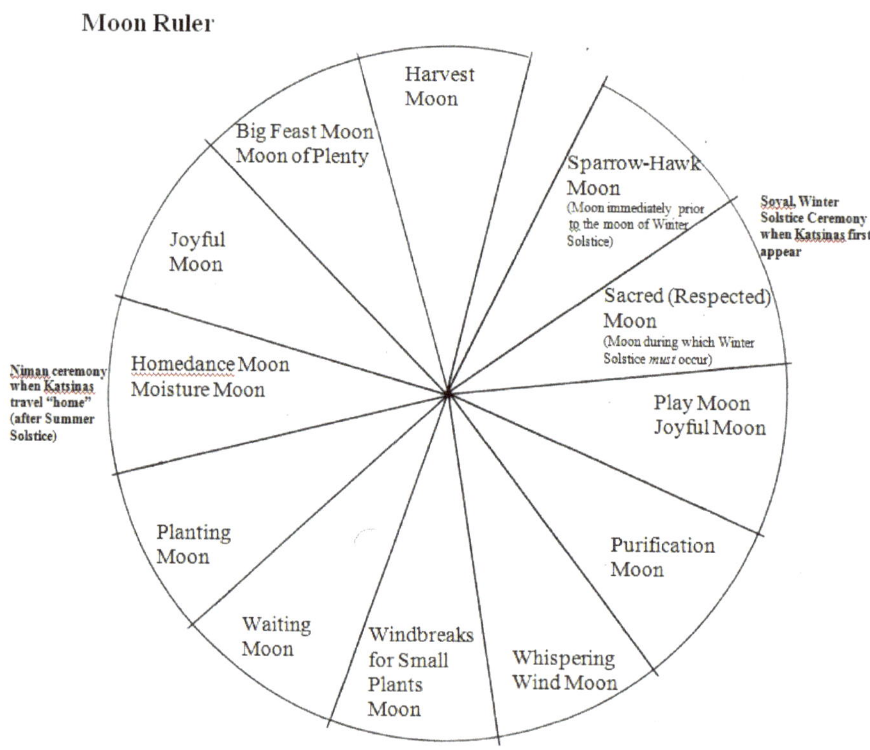

Figure 1 – Circular Moon Ruler

When I shared this ruler with the students, many were confused by it. Whereas I saw it as perfectly illuminating the unfamiliar pieces of the problem, students (6th graders and 16th graders alike) did not readily comprehend its meaning or relevance. What I learned is that

NOTES

despite the clarity this model provided for my reflection, it was presented to the students before they were genuinely aware of the circular conception of time. They remained determined to map the lunar cycles onto the traditional human-invention of the 12-month calendar, rather than understand how lunar cycles occur during the naturally-occurring solar year.

The Hopi Sky-Watchers successfully measured an Earth year with a lunar cycle (together with the solar year), because of their circular conception of time. A solid classroom experience should have negotiated the lesson toward the Hopi truth that time is a circle and the lesson content should have addressed this in a straightforward manner. Then, perhaps, the circular moon ruler would have been more meaningful. Instead, we spent an overwhelming amount of time focused on the moon names and the arithmetic calculations comparing sets of 29.5 days of the lunar cycle to the 365.25 days in an Earth year. In retrospect, we should have spoken directly to the fact that to the Hopi, time may be thought of as a turning wheel.

Curricular issues

I was pleased that the collection of lessons was able to address a long list from Tracie's required curriculum. Solving a problem deeply related to a culturally-important event, required mathematical expertise, as well as information from other disciplines. That we were able to relate this lesson to a wide range of topics, provides further motivation for me to continue to study ethnomathematics. I remain firm in the belief that there is a secure place for such problem posing, where solutions incorporate a vast array of content from across the curriculum.

Resistance to unfamiliar content

Despite our pointed questions, "Are all months the same length?" (No) "Then how can a month be a candidate for a standard unit?" (Silence), I remain unconvinced that the students truly dealt with the intended focus of a standard unit. They seemed inclined and comfortable with the idea that a "month" is what currently works, so it must be an acceptable way to measure an Earth year, rather than as a social and cultural construct whose sole purpose is to chunk the year into smaller (unequal-sized) units.

Janet's summary

Like Tracie, I learned more than the students. Often, it is our own resistance to abandoning a particular way of conceptualizing an idea that competes with understanding mathematical ideas from another point of view. Ten years ago, after reading *Thirteen Moons on Turtle's Back* (Bruchac, 1997) to a 4th grade class, one child asked, "Why are there 13 moons?" This question, which I thought I might answer in a timely fashion, actually served as a catalyst for several years of research. It took approximately 10 years for me to form a knowledge base that might have enabled me to answer her question.

Condensing the information I had learned into a one-day lesson sequence appropriate for 6th graders required that I isolate the multi-disciplinary background knowledge students needed. The next time I teach this lesson, I will include pointed discussions about a circular conception of time. I remain committed to studying mathematical concepts and ideas through a variety of lenses, precisely because it broadens my mathematical understanding, as well as strengthens my teaching.

Collaborative Summary

To consistently, rather than occasionally, acknowledge, promote and celebrate culturally-relevant mathematics teaching and learning, great care must be taken to (1) delineate the cognitively-accessible pieces of the ethnomathematics story (2) work carefully to respect the classroom teacher's tight schedule & already-packed curriculum and (3) identify unfamiliar frames of reference that may interfere with the mathematics of the lesson. For this lesson, it was critical for the professor and the teacher to collaboratively study about the Hopi calendar prior to working with the students. In the classroom, it was equally important to move slowly through the content so that students had time to absorb the information since they were being asked to make connections outside of mathematics, as well as within mathematics. Many student comments indicated that at least some of our goals had been met. Student comments like, *"12 or 13 cycles is not a good representation for a year because it unbalances when people plant crops or have celebrations. It would off-balance the way of life, ... Just the difference between 13 full moon cycles (383 days) and the normal 365 days alters each year"* give us the incentive to continue to teach mathematics lessons like this. In addition, in lessons like this, many collateral objectives were met because a problem was presented within a larger story, which included science and literature, as well as mathematics and history.

Endnote:

And, we will remember that understanding the Hopi conception of time will forever elude us, as neither of us are Hopi. Our knowledge has been primarily gleaned from secondary sources. The authors continue to ask for their patience and grace in our efforts to understand time from their point of view, given our inescapable Western viewpoint as we endeavor to put our own backgrounds on hold. We know that we can never fully experience time passage as they do, and that probably makes it difficult for us to teach this lesson well. However, it does not diminish our desire to understand and share as much as we can.

NOTES

References

Bruchac, J. (1997). *Thirteen moons on turtle's back: A Native American year of moons*. Puffin.

Carlson, V. F., & Qoyawayma, P. (1985). *Broken Pattern: Sunlight & Shadows* of Hopi History. Naturegraph Publishers.

D'Ambrosio, U. (1997). Ethnomathematics and its place in the history and pedagogy of mathematics. In A. B. Powell & M. Frankenstein (Eds.), *Ethnomathematics: Challenging Eurocentrism in mathematics education* (pp. 13–24). Albany: State University of New York Press. (Original work published in 1985)

D'Ambrosio, U. (1997). Foreword. In A. B. Powell & M. Frankenstein (Eds.), *Ethnomathematics: Challenging Eurocentrism in mathematics education* (pp. xv–xxi). Albany: State University of New York Press.

Ernst, C. (2010). Ethnomathematics makes difficult subject relevant. *University of Hawai'i Malamalama Magazine (see http://www. hawaii. edu/malamalama/)*

National Mathematics Advisory Panel. (2008). *Foundations for success: The final report of the National Mathematics Advisory Panel*. US Department of Education.

Sharp, J. (2015) Relationships between the Hopi calendar and measurement concepts. *Journal of Mathematics and Culture, 9*(1), 30-51.

Culture *That* Counts

Volume 4 ◆ Teacher Prep

Volume 4
Teacher Prep

Reflection Within Mathematics Methods: Prospective Teachers Develop A Classroom Diversity Teaching Assignment

Jenni L. Harding
University of Northern Colorado USA
Jenni.harding@unco.edu

Abstract

Prospective teachers can learn about and practice ethnomathematics teaching to students in the elementary classroom through multicultural children's mathematics literature. Forty-seven prospective elementary teachers taking a mathematics methods course at a Doctoral Intensive University in the Rocky Mountain Region participated in a Classroom Diversity Teaching Assignment which includes planning, teaching the lesson to elementary students, reflection using transcriptions with focus questions, and oral reflection with the professor. Prospective teachers found this assignment beneficial.

Keywords:

- Ethnomathematics
- Diversity Assignment
- Prospective Elementary Teachers
- Case Study
- Transcription

Introduction

Ethnomathematics provides a natural means for students to access a framework for conceptual understanding in mathematics. Each student's cultural reference, worldview, and history are unique to the experiences this child has encountered. Furthermore, these experiences may be incorporated into the classroom to help students learn authentic mathematics. This structure of culture is a mind-set integrating the prior knowledge of students in a way that builds upon what they already know leading to mathematical understanding. Nuri Robins, Lindsey, Lindsey, and Terrell (2006) posit to be a culturally proficient teacher one needs to teach "in a manner that builds understanding of the teacher's and learner's world that engenders a value of diversity" (p. IX). The National Council of Teachers of Mathematics (2005) take this one step further by explaining "...all students need the opportunity to learn challenging mathematics from a well qualified teacher who will make connections to the background, needs, and cultures of all learners".

Ethnomathematics provides teachers the building blocks connecting students' out of school mathematics knowledge with their classroom mathematics knowledge. By making the school to home mathematical connection explicit teachers can create connections with the mathematics they know already found in their worlds. Doing so also allows teachers to enhance and restore cultural dignity to the children in their classrooms with meaningful mathematics understanding (D'Ambrosio, 2002).

Prospective elementary teachers may use children's literature to incorporate cultural ideas and build upon student experiences connecting real life to classroom mathematics learning. These books can cover a wide variety of mathematics understanding based on the characters and concepts in the story. An example is *Gabriela's Beautiful Carpet/ La bella alfombra de Gabriela* by Thompson and Thompson (2003). This story is set in Antigua Guatemala where the village is creating "carpets" in the road out of flowers, colored sawdust, and stencils (cardboard with wooden frames) for the religious Lenten procession that takes place in their community. These carpets use the mathematics concepts of patterns, symmetry, proportions, and geometry. By reading this book together as a whole class, (1) the foundational knowledge of "carpets" can help the students who have participated in these processions to connect real world mathematics knowledge to more abstract mathematics principles and (2) the other students in the classroom, who have never participated in a "carpet" procession, will have the knowledge and information to think about these mathematics principles in new ways to extend their knowledge to the real world. Multicultural children's mathematics literature is a means to create mathematics connections by opening the world for the students through this shared experience.

Literature Review

Many authors have defined "ethnomathematics." D'Ambrosio (2006), known as the father of ethnomathematics, defines it with the "roots *tics*, *mathema*, and *etho* to signify that there are various ways, techniques, abilities (*tics*) to

NOTES

explain, understand, deal with, and live with (*mathema*) distinct natural and socio-cultural contexts of reality (*ethos*)" (p. 48, *italics original*). Borba (1990) defines ethnomathematics as the mathematical knowledge articulated in the language code of a specified socio-cultural group. Similarly, Eglash, Bennett, O'Donnell, Jennings, and Cintrorino (2006) define ethnomathematics as the study of mathematical ideas and practices positioned in people's cultural context. Moreover, Powell and Frankestein (1997) assert, "Ethnomathematics includes the mathematical ideas of peoples, manifested in written or non-written, oral or non-oral forms" (p. 9).

In addition, Barton (1996) states ethnomathematics is an instrument by which people may make sense of their world, as they see it, and as others see it. Furthermore, Powell (2002) claims ethnomathematics departs from a universal notion of mathematical knowledge benefiting male, European, heterosexual, racist, and capitalistic interests and values. Izmirli (2011) necessitates "exposing students to the mathematics of different cultures so as to increase their social awareness, to reinforce cultural self-respect, and to offer a cohesive view of cultures" (p. 39). However, Ascher and Ascher (1997) contend "ethnomathematics is the study of mathematical ideas of nonliterate peoples" (p. 26).

The above definitions of ethnomathematics can be brought to life in the kindergarten through fifth grade classroom through multicultural children's mathematics literature. Prospective teachers need the guidance in being able to teach mathematics in culturally relevant ways while equipping them with various approaches to teach students from different cultures and backgrounds (Blakeney-Williams & Daly, 2013; Harding-DeKam, 2014). Multicultural children's mathematics literature represents a unique art form in children's literature because they encompass representations of the variety of ethnic, racial, and cultural groups around the world. Young children can develop awareness and understanding of cultural groups outside of their own through multicultural children's mathematics literature (Loyd, Harding-DeKam, & Hamilton, 2015). For children from non-dominant cultures, the books also can affirm their own cultural identities (Mendoza & Reese, 2001) as well as serve the dual purpose of a mirror to reflect one's own culture and of a window for learning about other cultures (Sims Bishop, 1997).

Multicultural children's mathematics literature can be a powerful way to facilitate culturally responsive math instruction as they extend students' literary experiences beyond the lived experiences of the various members of the learning community. "The characters and situations in books introduce children to what the world may look like through others' eyes and offer a change to further construct their own views of self and the world" (Mendoza & Reese, 2001). With a rich resource of multicultural picturebooks, all the members of the learning community can expand their cultural references, worldviews, and color their cognitive histories in pluralistic ways.

Methodology

Research Questions

The goal of this research is to evaluate the impact of ethnomathematics teaching through multicultural children's mathematics literature in a mathematics methods course from the perception of prospective elementary teachers. The following research questions were addressed in this study:

1. How effective is the ethnomathematics instruction through implementing WIDA Standards and student learning objectives into lesson plans, and the Classroom Diversity Teaching Assignment?

2. Is transcription effective for prospective teachers as a reflection tool?

Participants

These participants are forty-seven prospective elementary teachers taking a mathematics methods course while completing a math practicum in an elementary classroom (grades kindergarten through fifth grade) using multicultural children's mathematics literature. This is a purposeful sample of pre-service elementary teachers who are in the professor's mathematics methods courses (early childhood, undergraduate, and graduate). These prospective elementary teachers are female (90%), twenty to twenty-five years of age (59%), and of white ethnicity (94%). The participants are at a Doctoral Intensive University in the Rocky Mountain Region.

Data Collection Procedures

Data was collected during a mathematics methods course from August to December 2015 as part of the regular course assignments focusing on quality math instruction using multicultural children's mathematics literature. The following data were collected (1) classroom diversity math lesson plan with professor feedback, (2) audio recordings of prospective teachers teaching with multicultural children's mathematics literature to elementary children, (3) reflection paper of teaching experience by prospective teachers after transcribing and analyzing their own teaching, and (4) audio recording of interviews with prospective teachers after teaching cycle: lesson plan, teaching, and reflection (semi structured questions).

Data Analysis

This is a case study design providing a holistic portrayal of perceived prospective teachers' integration of multicultural children's mathematics literature into mathematics instruction (Creswell, 2009; Yin, 2009). The data above was collected, transcribed, and organized. These data were coded and analyzed for themes and patterns (Marriam, 1998; Saldaña, 2009) to answer the before mentioned research questions. Credibility issues were addressed by using a second coder, expert/peer checks, triangulation, and member checks (Merriam, 1998).

Mathematics Methods Course Structure

The timing of the mathematics methods course, where how to teach mathematics to children is the focus, has changed from being the same semester as student teaching to the semester prior to student teaching. This change has positioned the mathematics methods course with a mathematics and science practicum. With access to this practicum, my prospective teachers have the opportunity to apply what they are learning to students in the kindergarten through fifth grade classroom. This application of teaching mathematics allows prospective teachers to put their philosophy and math content knowledge into action.

Mathematics Methods

My mathematics methods course over the past decade has evolved to include more ethnomathematics and culturally diverse classroom experiences than the Farsi video, human graph, and a specific section in the lesson plan for children learning second languages and diverse cultures (Harding-DeKam, 2007). These additions include: adding WIDA standards and learning objectives to the lesson plan and the

Classroom Diversity Teaching Assignment enriching my prospective elementary teachers understanding of teaching mathematics to the diverse children in their classroom.

Modeling Mathematics Instruction with Language Supports

Modeling in my mathematics methods course now focuses on specific language supports allowing students to access the content. Specific math vocabulary instruction, representation, and differentiation have a more purposefully focus of teaching. Mathematics vocabulary is taught with visuals (photographs, images, symbols) and written in a language other than English to support academic language. This mathematic vocabulary is then posted around the classroom as a resource. Mathematics concepts are presented with multiple modes of representation such as nonverbal, oral, graphic, written, objects, manipulatives, and technology. These concrete examples support conceptual understanding for all students, but give English Learners (ELs) an understanding even if there is a language barrier. Differentiation is made explicit by modeling how the mathematics concepts can be taught to a diverse group of students at the same time such as ELs, students who understand the concept, students who do not understand the concept, Gifted Learners, Special Needs learners, students who have negative attitudes about math, etc. These specific language supports are modeled in my mathematics methods classroom giving prospective teachers strategies to take with them into their practicum teaching experience.

WIDA Standards and Learning Objectives

WIDA advances academic language development and academic achievement for kindergarten through twelfth grade linguistically diverse students through high quality standards, assessments, research, and professional development for educators (WIDA, 2016). Previously, prospective teachers in my mathematics methods course would create lesson plans including specific content standards from the Colorado Academic Standards with appropriate learning objectives. Now, my prospective teachers add the mathematics WIDA standard with a specific learning objective to their lesson plans as you can see in the example below.

> Standards-Based Instruction
>
> WIDA: English Language Development:
>
> Standard 3 – English Language Learners communicate information, ideas, and concepts necessary for academic success in the content area of mathematics. To meet this English Language Development standard, students will:
>
> - Use mathematic vocabulary to describe to the class how they began their math process: I started solving the problem by ___________.

By adding the WIDA standard and learning objective to mathematics lesson plans it guides my prospective teachers in thinking about how they will differentiate their instruction for English learners in their classroom. This opportunity allows prospective teachers to specifically plan for English leaners in their classroom supporting them to be successful mathematics learners.

Classroom Diversity Teaching Assignment

The Classroom Diversity Teaching Assignment using multicultural children's mathematics literature gives prospective teachers the opportunity to bring culture and diversity into mathematics teaching allowing all students to see themselves as mathematicians. The impact of this teaching happens when the prospective teachers have students from diverse cultures participate in the mathematics learning; furthermore, these students see their culture, group, or identity represented in the children's book being successful with mathematics.

This is a four-part assignment for my mathematics methods prospective teachers: (1) complete a Classroom Diversity Math Lesson Plan for children in your partner school. The math content and date for teaching this lesson plan needs to be discussed and approved by your elementary classroom teacher prior to creating this lesson plan. You should plan for teaching at least half of an hour to the whole class. Lesson must use multicultural children's mathematics literature book to teach math like modeled in class. You may use the list of multicultural children's mathematics literature book list to find a book to use in your classroom. (2) After being approved to teach your Diversity Math Lesson Plan by the professor, then you may teach children in your classroom while audio recording your lesson. Make sure you and the children can be heard on your audio recording. You will submit the raw audio recording and transcription focusing on classroom dialogue and learning conversations to observe how effective your mathematics teaching is through conversation. (3) Written Teaching Reflection discussing what went well with the teaching of your math lesson and what you would do differently next time. Reflection questions are below to help guide in this reflection. (4) Oral Reflection for fifteen minutes with the professor to give feedback on classroom teaching, audio transcription, and reflection. You will have time to discuss you teaching and ask any questions to help improve your practice through the planning, implementation, or reflecting stage.

The Classroom Diversity Math Lesson Plan Rubric is how the professor evaluates the mathematics lesson plan prior to the prospective teachers using the lesson with children. There are four areas of evaluation in this rubric: (1) Lesson Plan Format: all sections of the lesson plan are detailed including big understanding, materials, standards with learning objectives, engagement, investigation, explain/clarify, conclusion, meeting the needs of all students, and assessment. (2) Math Content: covers mathematics concept(s) in depth by acquiring, analyzing, organizing and processing information. Math content is differentiated to meet the diverse children in your classroom. (3) Multicultural Children's Mathematics Literature Book: enhances mathematics and cultural math concepts while reading aloud a diverse mathematics picturebook to your students focusing on active student participation and genuine learning experiences (Loyd, Harding-DeKam, & Hamilton, 2015). (4) Questions asked during the lesson: are high-level questions with rigor allowing students to think mathematically and culturally. Specific questions are in lesson plan to guide math content understanding and integration with multicultural children's mathematics literature.

After the teaching and transcription of the Classroom Diversity Math Lesson Plan, prospective teachers were asked to reflect on their teaching by analyzing their teaching transcriptions to focus on student and teacher mathematics conversations in general and then with specific guided written reflection questions to answer. The following question were reflected upon:

NOTES

1. How do you feel the lesson went (what are your perceptions of your strengths of this lesson and what are your perceptions for areas of growth within this lesson)?

2. How did your lesson planning strengthen your teaching?

3. How did you further the students' mathematical knowledge, skills, and engage them intellectually in understanding math subject matter?

4. How did integrating mathematics and diverse children's picturebook work in your lesson?

5. What strategies did you use to monitor student learning during this teaching? How many students were successful in understanding this math content because of your teaching? Which students were not successful in understanding this math content because of your teaching?

6. How did you scaffold instruction (sheltered instruction, differentiation, modeling, vocabulary, etc.) for all children in your classroom (English language learners, students who understand the math concepts, students who do not understand the math concepts, special needs students, students who have negative attitudes about math, and gifted students)?

7. What did you learn about your teaching from your transcription? How can you understand what children understand or know about content from listening to what they say?

8. Overall, how would you rate the quality of your mathematics teaching to children on a 1 to 5 scale (1=no student learning, 2=a few students learned, 3=half of the class learned, 4=most of the class learned, and 5=every student in the class learned)? Why do you rate yourself this way?

Findings

Resoundingly the prospective teachers found the ethnomathematics focus within their mathematics methods course as beneficial through adding WIDA standards and learning objectives to the lesson plan as well as with the Classroom Diversity Teaching Assignment. Prospective teachers specifically found the teaching cycle as a positive experience of planning their lesson, receiving feedback from the professor on their lesson plan, teaching the lesson while audio recording it, answering focused reflection questions, and then meeting with the professor to discuss it.

Prospective teachers taught their lessons in kindergarten through fifth grade classrooms. For most of the prospective teachers this was their first mathematics lesson planned and taught in the elementary classroom. The mathematics content taught with the integration of ethnomathematics was varied including fractions, doubling, decimals, multiplication, integers, factorials, addition, area, perimeter, subtraction, patterns, measurement, counting, data analysis, graphing, money, mental math, geometry, estimation, even/odd, and symmetry. Examples of multicultural children's mathematics literature taught from include: *One Grain of Rice: A Mathematical Folktale* by Demi, *If America Were a Village: A Book About the People of the United States* by D.J. Smith, *The Patchwork Path: A Quilt Map to Freedom* by B. Stroud, *Islamic Art: recognizing Geometric Ideas in Art* by J. Levy, *The Tortoise Who Bragged: A Chinese Tale with Trigrams* by B. Franco, *Lucky Beans* by B. Birtha, *Emeka's Gift* by I. Onyefulu, and *Can You Count Ten Toes?: Count to 10 in 10 Different Languages* by L. Evans and D. Roche. The data was analyzed from the written reflection, interviews, and course evaluation.

Classroom Diversity Teaching Assignment

Prospective teachers found the lesson planning process valuable allowing them to be prepared for teaching in the classroom. One hundred percent of the prospective teachers thought the lesson planning process were beneficial. There is a preponderance of planning benefits where it helps "imagine the whole picture of your lesson", "create questions that lead students to think about content", "deliver concepts clearly and correctly", "boosted my confidence", "reach content standards and WIDA objectives", "pace the lesson", "establish expectations", "prepare all materials needed", "decide what type of manipulative to use", "connect to grade level expectations", "prepare for modeling", "identify vocab that needs to be explained", "estimating time", "integrating culture", and "assessing students".

The multicultural children's mathematics literature book was effectively used to make the connection between math concepts and culture through discussion allowing classroom conversations about "families", "food", "real life applications in terms of culture: people, families, clothes, and beliefs", and "realia". Prospective teachers gave examples of students from other cultures who were engaged in the mathematics lesson because the characters, context, or activities in the stories were relatable to them: "cultural celebrations", "nesting dolls", "Native American blankets", "home language", "currency", "measuring systems", "type of food", "their ethnicity", and "their culture". One prospective teacher mentioned about a stereotype misunderstanding one student in her classroom had and how "through discussion we were able to clarify that those things were not true".

There were some logistical issues with incorporating multicultural children's mathematics literature into the classroom because it was not part of the normal mathematics curriculum. Several prospective teachers stated, "It was difficult planning a book into math because that isn't what normally happens in the classroom" and "in my school they believe students should not be read to in math". Furthermore, with some mathematics content areas it was difficult to find a multicultural children's book at the appropriate developmental level for students, especially "a diverse one—to fit the lesson genuinely". Despite these challenges, a majority of the prospective teachers believed the multicultural mathematics children's book enhanced the lesson through student engagement, integration of content areas (social studies, language arts, and mathematics), and providing a visual or concrete understanding for the mathematics.

Transcriptions

Prospective teachers were required to listen to their audio-recorded teaching experience and to transcribe it, in order to reflect upon this experience. There were specific reflection questions given to guide the focus of the transcription on conversations taking place during their teaching (see above). Ninety-seven percent of the prospective teachers thought the transcription process were beneficial allowing them to evaluate their teaching. The following quotes reinforce using transcription to reflect upon what mathematics children are learning through

NOTES

classroom conversations: "allowed me to think about what students said", "showed me where students made connections", "listening to students lets me hear what they know", and "tells me what I need to teach next".

Prospective teachers felt the transcription process was useful as a tool to analyze their teaching through the themes of focusing on student thinking, identifying where students made math and culture connections, and focusing on student understanding. One prospective teacher encompasses the overall sediment about transcribing with "it was helpful to know what I have done and what I need to change in my future [teaching]". On the other hand, a small group of prospective teachers (3%) thought transcribing was tedious and a "lot of work", "not helpful", and "didn't seem to be worth the time".

Prospective teachers identified many areas of strength as well as areas for improvement within their teaching through transcription. The areas for improvement included, (1) verbal communication: "talking too fast" or "saying filler words", (2) instruction: "not much student conversation", "should of asked more questions", or "need to go deeper into concepts", and (3) management: "heard students misbehaving that I missed while teaching" or "more wait time needed".

Course Evaluation

Prospective teachers in the course evaluation found the Classroom Diversity Assignment as beneficial with their qualitative comments: "the diversity lesson provided a meaningful experience" and "the multicultural lesson was really helpful and it gave me the initiative to plan a lesson independent from my practicum teacher". On the course evaluation, one hundred percent of the prospective teachers agreed or strongly agreed assignments and tests in the course were related to course objectives and the instructor created a classroom environment that as inclusive and respectful of diversity.

Discussion

Prospective elementary teachers benefit from the experience of planning, teaching, and reflecting upon their instruction. These prospective teachers perceive planning the lesson as a valuable process in helping to think through and anticipate what was going to occur. Feedback and grading of their lesson plan prior to teaching it with children was appreciated allowing them to be more prepared and think through everything within the lesson. The act of teaching is a complex and multifaceted endeavor; furthermore, prospective teachers cannot catch everything happening in the moment because they are focused on classroom management, content being taught, pedagogy of teaching, classroom conversations, connections, and differentiation to name a few. Transcription provides the opportunity to slow down the teaching and learning process where prospective teachers can concentrate on what happened during their instruction through classroom conversations. Specific reflection questions guide the focus of what to focus on during the transcription. Additional oral reflection with the professor contributes to another level of reflection by getting feedback from an experienced teacher on what went well and ideas to improve upon in the future. The data suggests the Classroom Diversity Teaching Assignment is worth the time and effort for prospective teachers to complete in their mathematics methods course.

Conclusions

The modeling of mathematics instruction with language supports, addition of WIDA standards and learning objectives in lesson plans, the Classroom Diversity Teaching Assignment, and audio transcription experiences were deemed meaningful by the prospective elementary teachers in my mathematics methods course. Multicultural children's mathematics literature seemed to make an impact on the children in the classroom by allowing them to relate to the stories, put mathematics into a concrete context, and see how everyone is a mathematician. However, more authentic multicultural children's mathematics literature books need to be written for a variety of mathematics content areas at different developmental levels in order to use them as a mathematics-teaching tool in the elementary classroom. Current mathematics curriculum in many schools is a narrow way of teaching mathematics where there is not any integration of subjects. This mathematics curriculum needs to be re-conceptualized in order to bring multicultural children's mathematics literature to enlarge the conceptual understanding of mathematics and its connection to all people around the world.

Prospective teachers need to have high mathematical expectations for all students in the classroom. These expectations can be achieved by bringing ethnomathematics learning into the classroom where teachers know their students enough to individualize mathematics in a way that builds upon their prior knowledge of their personal history and home life. This can be accomplished through multicultural children's mathematics literature. It is important to leave our students' difference intact and their multiple identities recognized in order for them to make personal connections to mathematics.

References

Ascher, M., & Ascher, R. (1997). Ethnomathematics. In Powell & Frankenstein (Eds). *Ethnomathematics: Challenging eurocentrism in mathematics education*. (pp. 25-50). Albany: State University of New York.

Barton, B. (1996). Making sense of ethnomathematics: Ethnomathematics is making sense. *Educational Studies in Mathematics, 31*(1/2), 201-233.

Blakeney-Williams, M. & Daly, N. (2013). How do teachers use picturebooks to draw on the cultural and linguistic diversity in their classroom?. Set, 2, 44-50.Borba, M. C. (1990). Ethnomathematics and education. *For the Learning of Mathematics, 10*(1), 39-43.

Creswell, J. W. (2009). *Research design: Qualitative, quantitative, and mixed methods approaches* (3rd ed.). Los Angeles: Sage.

D'Ambrosio, U. (2002). Ethnomathematics: an overview. Proceedings of the II Congresso Internacional de Etnomatematica, Ouro Preto, Brazil.

D'Ambrosio, U. (2006). *Ethnomathematics: Link between traditions and modernity*. AW Rotterdam, The Netherlands: Sense Publishers.

Eglash, R., Bennett, A., O'Donnell, C, Jennings, S., & Cintrorino, M. (2006). Culturally situated design tools: Ethnocomputing from field site to classroom. *American Anthropologist, 108*(2), 347–362.

Harding-DeKam, J. L. (January 2007). *Foundations in ethnomathematics for prospective elementary teachers*. Journal of Mathematics and Culture,2(1), 1-19.

Harding-DeKam, J. L. (2014). Defining culturally responsive teaching: The case of mathematics. *Cogent Education Journal 1*(1), doi: 10.1080/2331186X.2014.972676.

Izmirli, I. M. (2011). Pedagogy on the ethnomathmatics-epistemology nexus: A manifesto. *Journal of Humanistic Mathematics, 1*(2), 27-48.

Loyd, S., Harding-DeKam, J. L., & Hamilton, B. (2015). The power of multicultural mathematics picturebooks. Colorado Reading Journal, 25, 5-10. Retrived from: http://www.ccira.org/colorado-reading-journal/WINTER-2014-15.html

Mendoza, J., & Reese, D. (2001). Examining Multicultural Picture Books for the Early Childhood Classroom: Possibilities and Pitfalls. *Early Childhood Research & Practice, 3*(2), 1-33. Retrieved from: http://files.eric.ed.gov/fulltext/ED458040.pdf

Merriam, S. B. (1998). *Qualitative research and case study applications in education.* San Francisco: Jossey-Bass.

National Council of Teachers of Mathematics. (2000). *Principles and standards for school mathematics*. Reston, VA: Author.

Nuri Robins, K., Lindsey, R.B., Lindsey, D. B., & Terrell, R.D. (2006). *Culturally Proficient Instruction: A guide for people who teach* (2nd ed.). Thousand Oaks, CA: Corwin Press.

Powell, A. B. (2002). Ethnomathematics and the challenges of racism in mathematics education. In *Proceedings of the Third International MES Conference*. Copenhagen: Centre for Research in Learning Mathematics.

Powell, A. B., & Frankenstein, M. (1997). *Ethnomathematics: Challenging eurocentrism in mathematics education*. Albany: State University of New York.

Saldana, J. (2009). *The coding manual for qualitative researchers*. Los Angeles, CA: Sage Publication.

Sims Bishop, Rudine. (1997). Selecting literature for a multicultural curriculum. In Violet J. Harris (Ed.), *Using multiethnic literature in the K-8 classroom* (pp. 1-20). Norwood, MA: Christopher-Gordon.

Thompson, K, & Thompson, S. (2003). *Gabriela's beautiful carpet/La bella alfombra de Gabriela*. Guatemala City, Guatemala: Vista Publications.

WIDA. (2016). Wisconsin Center for Education Research. Online available from https://www.wida.us/aboutus/mission.aspx.

Yin, R. K. (2009). *Case study research: Design and methods* (4th ed. Vol. 5). Thousand Oaks, CA: Sage.

Investigating Students' Perceptions of Indigenous Mathematical Knowledge Systems through Immersion in Transcultural Settings

Iman Chahine
Georgia State University
ichahine@gsu.edu

Abstract

A handful of research has shown the importance of integrating cultural practices that resonate with students' ethnic and background experiences in everyday instruction. Particularly in teacher education courses, and in light of the diverse demographics that thrive in schools today there has been an unprecedented focus on valorizing the use of community based, culturally oriented learning experiences. Notwithstanding the extensive literature that urged the utilization of cultural immersion experiences as means of increasing cultural sensitivity for students across disciplines, little effort is documented regarding the usefulness of immersion for teaching culturally responsive mathematics. The purpose of this research is to document initial, firsthand field experiences of graduate mathematics education students while engaged in exploring indigenous knowledge systems by immersion in the daily experiences of indigenous cultures in South Africa.

Keywords:

- Indigenous knowledge systems
- Ethnomathematics
- Immersion
- Study abroad

Introduction

Indigenous Knowledge Systems (IKS) have been recently established by the international world organizations as top global priority for empowering traditional and local communities in their striving efforts towards improved and sustainable development, particularly in developing countries (Shizha & Kapoor, 2010). Generally, the term Indigenous Knowledge (IK) refers to traditional and local knowledge that was coined to differentiate the knowledge developed by and within distinctive indigenous communities from formal knowledge systems generated by institutions and universities (Semali & Kincheloe, 1999), government research centers, and private industry, which is sometimes called the 'Western Knowledge System' (Maurial, 1999). However, despite its proclaimed importance and sound pedigree of recognition for strengthening native communities' preservation of social and traditional capitals towards more independence (Chahine, 2011a), no clear effort has been cited to uncover the contributions of indigenous cultures to the development of mainstream knowledge and epistemologies. Immersion through studying abroad as a college program is an area where some meaningful, but limited, research has been published. While the particular focus of studying abroad as a program that facilitates language acquisition and exposure to international contexts has been researched, limited investigations have been attempted to explore the effect of immersion on teaching and learning mathematics.

This study capitalizes on the role that IKS play in exploring successful and meaningful practices to teach mathematics in nonconventional contexts (McQueen, Goings, Shaheed, & Chahine, 2010). In this study, we argue that immersing students in indigenous cultures will acquaint them with a broad spectrum of knowledge systems and structures, which covers a plethora of contents and contexts that incorporate mathematical artifices. These structures include indigenous music, architecture, mural decorations, indigenous games, beadwork, weaving, cultural artifacts, and other social systems and practices.

This study is part of a newly designed course uniquely offered at an urban, highly diverse university in the southeast. The course particularly focuses on teaching an emergent field called *Indigenous Mathematical Knowledge Systems* (IMKS) through cultural immersion in indigenous contexts (Chahine, 2010, 2011b, 2012). The primary purpose of the course is to examine how different cultural groups interpret mathematical concepts in ways that are quite different from what we might expect from typical mathematical texts. This instructional experience dovetails two components:

Component 1: Class work. In this component, students and in-service teachers investigated mathematical ideas inherent in cultural activities such as creating calendars, art and decoration, divination, and counting schemes. Students were also introduced to IMKS as a program of study and they further explored the implications of integrating such a program in the school curriculum. Connections between IMKS and mathematics education in schools are also emphasized. Students also examined the epistemology of IKS and explored various paradigms that support the cultural basis of mathematical knowledge.

The class met on campus once per week for two and half hours and the in- class activities were conducted over a period of 12 weeks.

Component 2: Cultural immersion. This component afforded students firsthand field experiences to explore IMKS by immersing students in the authentic daily experiences of master crafts people and knowledge holders of vital indigenous technologies. The primary purpose of the immersion component was to provide an alternative lens to view mathematics that transpires in the workplace with greater precision. Students traveled to several destinations in South where they visited artisan markets, cultural villages, museums and many historic sites. During the site visit, students explored the integration of hand and mind tools that indigenous cultures continually employ to plan, conceptualize, visualize and execute myriad activities as part of their daily practices. They also investigated the problem solving behavior of craft masters and apprentices in their traditional work environment and examined various heuristics that emerge in the process. The goal of this experience was to help students develop awareness and appreciation of the complex dynamics that instigate the emergence of mathematical tools. The ultimate aim of the immersion intervention is to encourage students to think more critically and humanely about what it means to do mathematics in contexts characterized by high tolerance to risks and uncertainties. Students conducted their research activities off-campus as they visited several locations in South Africa for the duration of ten days. To motivate critical thinking, we employed various educational modalities in the on-campus portion of the course including small group inquiries, whole-group discussion, problem-based investigations and tiered assignments. For the immersion component, students engaged in participative forms of inquiry investigating ideas that transpired in out-of-school settings. Throughout the course, students were encouraged to experiment with innovative ideas, to make their experimental thinking public, and to develop new epistemologies that guide the teaching and learning of nonconventional mathematics. Students' effort was evaluated based on the intellectual merit of their research projects, the investigative procedures they developed during fieldwork, validity of their claims, and communicability of their research accounts.

Purpose of the Study

This study examined graduate students and mathematic teachers' reflections on an immersion experience that facilitated first-hand observation of the mathematical ideas that emerge across indigenous cultures as creative expressions of human thought. The primary goal of this immersion was to offer students a transcultural experience as they explored the realities of knowing and doing mathematics amidst contextual challenges. We argue that a critical implication of immersion in indigenous practices is to sensitize students to the delicacies of what it means to be part of a different cultural fabric, leading to their appreciation of various modes of thinking across cultural boundaries.

The role of Indigenous Mathematical Knowledge Systems (IMKS) is paramount for the survival of many South African citizens particularly in the countryside. This study introduced students to an assortment of tacit, implicit and explicit mathematical knowledge, skills, which people in a particular geographic area possess, and which enable them to get the most out of their environment.

Theoretical Framework

The study is principled by two paradigms: *Embodied, Situated, and Distributed Cognition* (ESDC) and Activity Theory. The basic tenet underlying ESDC as an epistemological paradigm is the view that cognition is physiologically embodied, socioculturally situated, and ostensibly distributed among individuals (Anderson, 2008). Within the ESDC paradigm, we argue that student cognition can be motivated and embodied by their abilities to use multiple modalities and translations within and between such modalities. This assumption coalesced with an overarching conjecture that, during immersion, students will be able to employ multiple modalities as "instruments of thought" to assimilate and reflect on knowledge systems inherent in indigenous contexts. Moreover, through collaborative group work activities, we argue that students can distribute and externalize their thinking by deliberating and sharing their understanding among their peers and other community members participating in immersion.

The second theoretical framework is Activity Theory (AT), which has at its core the *subject-tool-object* triad built off Lev Vygotsky's (1978) expansion of the basic stimulus-response relation fundamental to behaviorist models of psychology. However, Leont'ev (1978, 1979) developed first-generation Activity Theory that refined the meanings and relationships of the components of the triad shown in Figure 1. For Leont'ev, the meaning of subject was broadened to give more attention to a *collective subject* – a group of individuals acting upon a common object. This concern for the collective subject means that Leont'ev explored complex *activities*, where different actors need to engage in separate *actions* in order to transform the object into the desired outcome of the activity. The efforts of all individuals are directed towards such transformations, but in order to address the differences between the individual and the group forms of such efforts, Leont'ev distinguished three levels of efforts. At the highest level, there is the group activity in which the transformation of the object into the outcome is driven by the over-arching group *motive*. At the intermediate level, this transformation is directed towards the attainment of an individual's *goal* (which may or may not contribute directly to achieving the motive); and at the lowest level, there are subconscious *operations* which are influenced by the *conditions* under which the activity takes place and can contribute to realizing goals and motives.

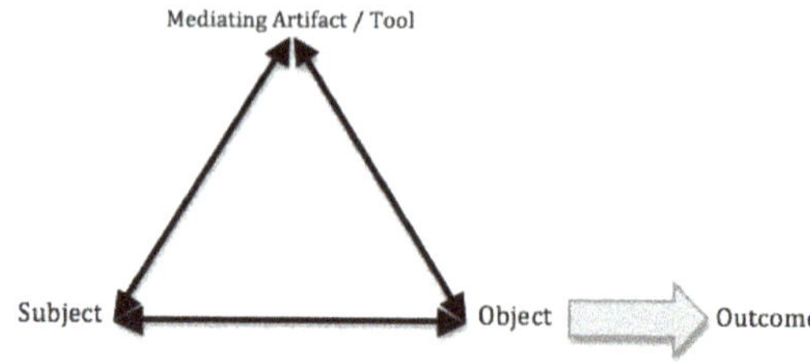

Figure 1: The Triad of 1st-generation Activity Theory

AT framework is guided by the belief that thinking and doing are inseparable –a central assertion of AT, which assumes that our knowledge of the world is mediated by our interaction with it, and thus, human behavior and thinking occur within meaningful contexts as people conduct purposeful goal-directed activities (Nardi, 1997). Using AT framework, we focused on the premise that subjects, i.e., IK holders and teachers acting in different contexts, i.e. community and schools, both create and represent their intentions and desires as objects, while the tools (i.e., concepts and artifacts) used in this process mediate goals between groups' (Kaptelinin, 2006). Activity theorists suggest that knowledge is socially situated or context bound because social interactions are fundamental and inseparable from the development of ways to think, understand, or mediate mental activity (Stetsenko, 2005). This perspective aligns with IKS epistemology and calls for a focused consideration of the teaching context in which meaningful learning emerges in situ.

Within an AT framework, we studied the activity comprising those actions undertaken by IK holders and teachers while interacting with different tools which they normally use in their respective work environments, i.e. community and schools. Specifically, we investigate how immersion in cooperative activities could arguably motivate IK holders and teachers to engage in knowledge production that ultimately serves their community and schools. The AT framework afforded a rich description of activity systems that emerge as IK holders and teachers employed different tools to mediate their activities and reach their objectives. Since AT incorporates socially organized human activity as the major unit of analysis, we believe that AT methodology is well positioned to capture the totality of activities undertaken by IK holders and teachers while interacting during professional development.

During the study, we drew on the relationship between these elements (tools, subject, and object) as our participants engage in a set of activities, actions and operations conducted during immersion.

Methodology

Research methods and data collection

Participants. The participants in this research study included 14 graduate students and high school mathematics teachers who travelled to South Africa on a study abroad program. The 10-day immersion program is part of a 3-credit graduate course on Indigenous Mathematical Knowledge Systems; a repeatable course required of graduate students in the mathematics education program and is offered twice per year.

Data Collection and Procedure. We conducted a 4-month qualitative study during a course on IMKS that included a highly integrated immersion experience (i.e., students living with host families) in South Africa. We employed two data collection techniques to examine graduate students' reflections on the immersion experience: collection of artefacts including photos and reflection journals, and informal conversations.

At the beginning of class, students were informed of the purpose of the trans-cultural experience and the duration of the program. Students were asked to voluntarily participate in a research study that will examine their internal reactions and personal reflections vis-á- vis the study abroad program. Prior to immersion, students attended a week long orientation on campus during which students were informed of the logistics related to travel, research, and program implementation.

During immersion in South Africa, students spent 4-6 hrs per day observing the artisans practices in the market place and reflecting on the mathematical ideas that emerged in situ. Accompanied by the researcher as the immersion director and course facilitator, students were continuously encouraged to observe, explore, and acculturate into the authentic host culture to understand and make meaning of the local community daily practices. Students were assigned to different sites and were specifically encouraged to particularly focus during their observations on capturing those nuances in which indigenous communities are involved in a process of mathematizing. Students were also asked to keep a daily journal to document and reflect on their lived experiences during immersion.

The focus of immersion was to engage students in an intense action research experience to explore mathematical ideas prevalent in numerous practices in indigenous contexts. During immersion, students were presented with several opportunities to immerse themselves within several indigenous societies including the Zulu and Ndebelle communities and to explore their heritages by visiting local museums (Phansi, Campbell, Vukani, etc.), and other historic sites. Students also explored myriad social networking settings such as campfire storytelling and other practices that facilitated knowledge transfer to younger generations in an effort to preserve strategies fundamental for survival in native ecological niches. Students also engaged in open dialogues with the knowledge holders, who are elderly key figures in the community who learned and mastered the elements of those knowledge systems despite the ever-changing societal, political, and environmental circumstances that attempted to disintegrate traditional indigenous structures. Students visited a number of Zulu cultural villages in South Africa that preserved the traditions of authentic south Zulu civilization, such as the Lesedi Village, and investigated a number of various Zulu artwork forms prevalent throughout the village. For example, students examined the significance of numerous geometric designs and strip patterns that thrive in the traditional Zulu craftwork by studying the mathematical ideas inherent in a wide variety of authentic crafts including clay pots, basket ware, utensils and clothing. During the immersion experience, students were asked to document their encounters as they consciously view, actively participate, and critically reflect on how other cultures use and develop their own mathematics to respond and make sense of their world. Students recorded their reactions and reflections in their daily journals. Upon returning from the trip, students' journals including their reflections were collected for inspection and analysis. After immersion, students completed a critical reflection paper describing their immersion experiences and comparing their newly gained knowledge and experience as a result of immersion to preconceived notions and dispositions before the immersion. The culmination in written and oral presentation of students' work to faculty and peers took place during the final week of class. Some of the projects were presented at national and international conferences. (Bass, 2010; Byrd, 2010; Frazier, 2011; Johnson, 2011; Porter, 2011; Smalls, 2011).

Analysis and Results

We used the open coding procedure using grounded theory research design (Strauss & Gorbin, 1998) to analyse students' reflective narratives and journal writings. Using this approach, the qualitative data was subjected to several stages of analysis. First, an initial in depth reading of the written scripts was conducted to allow familiarity with the qualitative data. Second, a coding system was identified to facilitate the organization and clustering of specific terms in the narratives. Third, potentially important variables that capture students' internal reactions to the immersion experience were extracted and further clustered into categories and subcategories. Since research on the effects of cultural immersion in the field of mathematics education is relatively new, identifying the latent variables that describe students' experiences during immersion is invaluable.

It is worth noting that the analysis of the findings of this study does not focus on outcomes such as improved test scores, but rather on the lessons, conflicts and stories that were created along the way and that affected students' conceptualization of the nature of mathematical knowledge from an IKS perspective. We employed the AT subject-tool- object triad for coding the data in preparation for analysis. We argue that learning emerged from processes, interface, content, and attitude shifts that occur at the junction of the AT triad.

NOTES

We conducted the data analysis on several levels:

1. Participants' initial reactions to the immersion experiences and their appreciation of transcultural encounters.

2. Participants' learning as a result of engagement in indigenous practices and the consequent impact on their cognitive, affective and socio-cultural norms.

3. Students' willingness to transfer what they learned during immersion to their mathematics classrooms.

Results of the qualitative analysis revealed several interconnected themes that emerged and viewed through AT lens.

Theme 1: Two Communities, Two Activity Systems

As members of the formal, school community, participants who are mathematics teachers, perceived immersion as a way to explore new means to enact some changes in their teaching practices and in their activity systems, i.e., schools, to better support their understanding of mathematics. As such, they viewed their students as the object of their activity and student learning the outcome sought. The primary artifacts that mediated teachers' student-oriented activity included curricular resources such as unit and lesson plans, instructional media, and technologies (Suthers, Yukawa, & Harada, 2007). In such activity systems, we argue that mathematical knowledge is perceived in isolation, disengaged from the environment and detached from authentic real life contexts.

On the other hand, IK holders and practitioners are community members and elders that are holders of traditional knowledge and resources that are transmitted orally (stories and explanations) (Chahine, 2013). As members of local communities, practitioners and IK holders employ their indigenous practices in everyday activities to facilitate the resolution of day-to-day issues. The object of their activity systems is survival and adaptation to continuously evolving situations. As such, indigenous communities are in constant learning modes appropriating variety of artifacts including hand and mind tools to navigate the world around them. Such tools are typically the results of generations of observation, trial, and experimentation.

During immersion, IK holders and graduate students interacted with each other exchanging ideas to create and represent their goal directed activities using numerous tools. We argue that, as a result of immersion, a collaborative partnership between IKS holders and teachers emerged, which helped motivate, inform and empower both parties to address priorities related to their activities (See Figure 2).

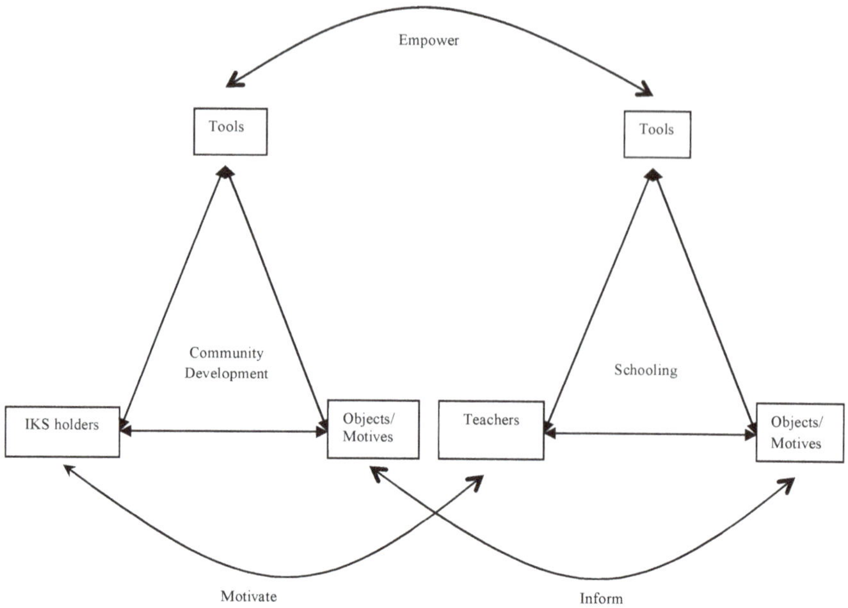

Figure 2. A depiction of a collaborative partnership between IK Holders and students as a result of immersion.

Theme 2: Self-reflexivity and pedagogies built through global connections

Overall, students felt that the immersion experience enriched them culturally, linguistically, academically, and personally. The perspectives students gained on the nature of mathematical thinking were considered useful in informing their future instructional decisions and the range of experiences they offered their students. For example, one student noted: "we also participated in a workshop where we had to create beaded items, which was a challenge for me. I reluctantly went into the workshop because I was never good at arts and crafts growing up and I just didn't like it. I was ready to give up before I even started. This experience reminded me of how many students must feel about mathematics, but many times the fact is that it's not that the content is too hard for them to learn, it's just that their attitude and their perception of their teacher's attitude towards them as learners impacts and impedes their growth and understanding. So, as educators, we really need to be mindful of how we approach and act towards our students when they struggle to grasp a concept." Also, another student explained: "The next day, we walked around the village and participated in a carpet-weaving workshop where we each had the opportunity to weave the carpet with the help of the master weaver. Thankfully, I had great teachers, some of whom didn't even speak English, but we were able to communicate and understand each other well. I felt more comfortable this time around and I enjoyed the experience much more. This further highlights the fact that the teacher's attitude towards the student greatly impacts their success."

Theme 3: Appreciation of mathematics as a creative expression of worldviews

Engaging in the humanity of indigenous communities and participating in their daily practices promoted students' respect of cultural heritages and strengthened a belief that all people are capable of doing mathematics in their own unique and personal perspectives. Students experienced the generosity of indigenous people, especially Elders, who welcomed them, appreciated their diversity and shared their teachings. To illustrate, one students argued: "When I was conducting my observations [of beadwork] I began to realize how mathematics played a large part in the work that we were doing, but there were also so many other important aspects at play here. Some of these aspects included teaching, cultural practices and navigating language barriers." And another student explained: "Impressively enough a simple form of mathematics was used in the counting of the beads and the pattern of the sequence of beads collected yet the real skill came in the knots and how those beads were strung together to form the bracelet. It took great insight to be able to visualize what the design would look like. And then to be able to translate that into easily followed instructions in a second language was noteworthy. They made sure to keep the pattern somewhat simple and used counts that were under five beads to construct the bracelet, which is a teaching strategy found in most education programs. Keep it simple and break the information into small chucks so that the students can understand. I believe that is a great example for teachers in general and a good example of ethnomathematics at work."

Discussion

Investigating the impact of using indigenous techniques and practices to investigate relevant concepts in the mathematics classroom is an emerging research interest. Ample research has shown the importance of integrating cultural practices that resonate with students' ethnic and background experiences in everyday instruction (Jurdak & Shahin, 1999; 2002; 2011; 2012; Ascher, 2002; Presmeg, 2007; Noss, Hoyles, & Pozzi, 2000). Furthermore, extensive research has urged the utilization of cultural immersion experiences as means of increasing cultural sensitivity for students across disciplines. Particularly in teacher education, the value of using community-based learning has been heightened in light of the diverse demographics that thrive in schools today. In mathematics education, the yet unseen value of immersion experiences transpires inadvertently when participants consciously view, actively participate, and reflect upon how other cultures use and develop their own mathematics to respond and make sense of their world.

Building on AT paradigm, we sought to understand how much the immersion experience upset students' existing equilibria, requiring fresh adaptation toward a higher developmental level. Using the AT paradigm, our charge in this study was to provide evidence supporting the embeddedness of a thinking-in-acting model as a fundamental framework that explains and supports students' learning. Specifically, the range of research studies that the students report at the completion of the course provided evidence of the role that immersion played in enhancing students' procurement and understanding of the mathematical concepts and problem solving strategies that indigenous people employ in their daily practice.

The study produced interesting results on the scale measuring development in ways of thinking and learning. Usually in our teacher preparation courses, we see students typically progress from a simple, more absolute understanding of the nature of knowledge and academic authority to a more complex and contextual understanding. We found that students who participated in the immersion experience showed growth toward the belief that developing one's own point of view is important, and that it is important to seek an interdisciplinary approach to knowledge. Additionally, we noticed that students returned home with a reduced tendency toward dualistic thinking and an inclination toward the related belief that there are no absolute right or wrong answers, good and bad information, as well as sources thereof.

Experiencing IMKS through cultural immersion emphasized the importance of unfolding various knowledge systems inherent in the cultural practices of indigenous communities. We argue that this immersion experience afforded an intellectual landscape for students to:

- Develop an awareness of and appreciation for IKS as a well-founded research model delineating those cross-generational endeavours aimed at investigating and understanding the physical world. As such, the immersion experience afforded mathematics.
- Examine research areas in mathematics education that integrate race, ethnicity, social class, and language issues. By observing IK holders, students experienced upfront the unfolding of those culturally embedded mathematical competences, which empower communities and grant them opportunities for survival and self-development.
- Employ multiple, malleable approaches to teaching and learning mathematics inspired by mathematical ideas that emerge in the workplace.
- Use real world problems as opportunities for reflective thinking. By exploring mathematics via tasks, which come from the workplace and everyday settings, students were able to discern the relevance of mathematical concepts and hence teach it in meaningful ways.

The position of this paper is that immersion in indigenous contexts provided an opportunity for graduate students and mathematics teachers to expand their understanding of mathematical ideas that emerge from different worldviews. Such understanding provided firsthand experiences to explore mathematical structures inherent in indigenous practices thereby valuing the diversity and complexity of indigenous knowledges. We contend that encouraging teachers to participate in study abroad experiences that focus on exploring mathematical knowledge systems of indigenous communities could broaden their perspectives regarding the nature of mathematical knowledge and how best to teach it. A direct implication of affording teachers opportunities to explore mathematics across worldviews is to be mindful of the fact that children from diverse backgrounds have efforts on tasks depending on personal criteria, which they deem useful. Such perspective will help teachers design novel models for teaching mathematics in meaningful, engaging, and self-directed ways.

References

Anderson, M. (2008). Evolution, embodiment, and the nature of the mind. In B. Hardy- Vallée & N. Payette (Eds.), *Beyond the brain: Embodied, situated and distributed cognition* (pp. 15-28). United Kingdom: Cambridge Scholars Publishing.

Ascher, M. (2002). *Mathematics Elsewhere: An Exploration of Ideas across Cultures*. London, UK: Princeton University Press.

Bass, E. (October 2011). The isicholo Zulu hat: An investigation into rotational and transformational symmetry. Paper presented at the *52nd Annual Conference for Georgia Council of Teachers of Mathematics*, Rock Eagle, GA.

Byrd, S. (October 2011). The beaded crown of the Yoruba Obas. Paper presented at the *52nd Annual Conference for Georgia Council of Teachers of Mathematics*, Rock Eagle, GA.

Chahine, I.C. (2010). Experiencing the prodigies of ethnomathematics: Cultural immersion studies in the old City of Fez, Morocco. Paper presented at the *Institute of Science and Technology Education* (ISTE), Pretoria, South Africa.

Chahine, I.C. (2011a). Beyond Eurocentrism: Situating ethnomathematics within the history of mathematics narrative. *International Journal for Studies in Mathematics Education, 4*(2), 35-48.

Chahine, I.C. (2011b). An ethnomathematical encounter: A cultural immersion of mathematics teachers in the daily practices of craftsmen in the Old City of Fez- Morocco. *International Study Group on Ethnomathematics Newsletter (ISGEm), 5*(2), 11-14.

Chahine, I.C. (July 2012). Embracing indigenous mathematical knowledge systems: Mathematics teachers' investigations of ethnomathematics through immersion in indigenous African cultures. Paper presented at the *International Conference on Mathematics Education* (ICME), Seoul, Korea.

Chahine, I.C. (2013). Ethnomathematics in the classroom: Unearthing the mathematical practices of African cultures. In D. Martin, & J. Leonard (Eds.), *The Brilliance of African American Students in Mathematics*, (pp. 195-220). Charlotte, NC: Information Age Publishing.

Dwellingham, T. (2010). The mathematical structures inherent in hair cutting. Paper presented at the *51st Annual Conference for Georgia Council of Teachers of Mathematics*, Rock Eagle, GA.

Frazier, L. (October 2011). The link in the ink: Examining curvatures in the tattooing practice of six tribes in South Africa. Paper presented at the *52nd Annual Conference for Georgia Council of Teachers of Mathematics*, Rock Eagle, GA.

Johnson, A. (October 2011). Deciphering the bar code design of the Zulu warrior shields. Paper presented at the *52nd Annual Conference for Georgia Council of Teachers of Mathematics*, Rock Eagle, GA.

Jurdak, M., Shahin, I.C. (2002). Problem solving activity in the workplace and the school: The case of constructing solids. *Educational Studies in Mathematics, 47*, pp. 297-315.

Jurdak, M., Shahin, I.C. (1999). An ethnographic study of the computational strategies of a group of young street vendors in Beirut. *Educational Studies in Mathematics Education, 40*, pp. 155-172.

Kapoor, D., & Shizha, E. (2010). *Indigenous knowledge and learning in Asia/Pacific and Africa*. NY, New York: Palgrave Macmillan.

Kaptelinin, V. (2006). The object of activity: Making sense of the sense-maker. *Mind, Culture, and Activity, 12* (1), 4 – 18.

Leont'ev, A. N. (1978). *Activity, consciousness, and personality*. Upper Saddle River, NJ: Prentice-Hall.

Leont'ev, A. N. (1979). The problem of activity. In J.V. Wertsch (Ed.), *The concept of activity in Soviet psychology* (pp. 37 – 71). Armonk, NY: M.E. Sharpe.

Maurial, M. (1999). Indigenous knowledge and schooling: A continuum between conflict and dialogue (pp. 59-77). In L.Semali, & J. Kincheloe (Eds.). *What is indigenous knowledge? Voices from the academy*. NY, New York: Falmer Press.

McQueen, M., Goings, C., Shaheed, S., & Chahine, I.C. (2010). Voices, echoes, and narratives: Multidimensional experiences of three teachers immersed in ethnomathematical encounters in Morocco. *Journal of Urban Mathematics Education, 3*(2), 47-56.

Nardi, B. (1997). Studying context: A comparison of activity theory, situated action models, and distributed cognition. In B. Nardi (Ed.) Context and consciousness: Activity theory and human-computer interaction (pp. 69 – 102). Cambridge, MA: MIT Press.

Noss, R. Hoyles, C. & Pozzi, S. (2000). Working knowledge: Mathematics in use. In A. Bessot & J. Ridgway (Eds.), *Education for mathematics in the workplace* (pp. 17-36). Dordrecht, The Netherlands: Kluwer Academic Publishers.

Porter, E. (October 2011). Using Mathematics Music Theory to examine indigenous Sub-Saharan African Music. Paper presented at the *52nd Annual Conference for Georgia Council of Teachers of Mathematics*, Rock Eagle, GA.

Presmeg, N.C. (2007). The role of culture in teaching and learning mathematics. In F. K. Lester (Ed.), *Second handbook of research on mathematics teaching and learning* (pp. 435-458). Charlotte, NC: Information Age Publishing..

Semali, L., & Kincheloe, J. L. (1999). *What is indigenous knowledge?: Voices from the academy*. NY, New York: The Falmer Press.

Smalls, N. (October 2011). Isometries in beadwork designs: A look at indigenous mathematical knowledge systems of the Zulu culture in South Africa. Paper presented at the *52nd Annual Conference for Georgia Council of Teachers of Mathematics*, Rock Eagle, GA.

Stetsenko, A. (2005). Activity as object-related: Resolving the dichotomy of individual and collective planes of activity. *Mind, Culture, and Activity, 12* (1), 70 – 88.

Suthers, D. D., Yukawa, J., & Harada, V. H. (2007). An activity system analysis of a tripartite technology-supported partnership for school reform. Research and Practice in Technology Enhanced Learning, 2(2), 1-29.

Vygotsky, L. (1978). *Mind in society: The development of higher psychological processes*. Cambridge, MA: Harvard University Press.

Aspectos Retrospectivos e Introspectivos de una Experiencia de Capacitación en Etnomatemática

Oswaldo Jesús Martínez Padrón

Universidad Pedagógica Experimental Libertador-Venezuela
ommadail@gmail.com

Abstract

This document contains some reflections inspired by another published paper by this author in the *Journal of Mathematics and Culture*, which relates an experience of training in Ethnomathematics addressed to Venezuelan indigenous teachers. It refers to the initial motivations that made possible its concretion and what has been its impact on the research work of the author. Such references pose a retrospective introspection that moves between past and present experiences that are disseminated in two key moments: the first thought from the interior insight that I direct toward myself and the second is held by the reflections raised from the spirit of the initial thought.

Key words:

- Teacher Training
- Ethnomathematics
- Venezuelan Indigenous Teachers

Resumen

Este documento contiene algunas reflexiones inspiradas en otro publicado en la *Journal of Mathematics and Culture*, donde relato una experiencia de capacitación en Etnomatemática, dirigida a docentes indígenas venezolanos. Aquí hago referencia a las motivaciones iniciales que hicieron posible su concreción y cuál ha sido su impacto en mi quehacer investigativo. Desde tales referente planteo una introspección retrospectiva que se mueve entre vivencias pasadas y presentes, diseminadas en dos momentos clave: el primero pensado desde la mirada interior que dirijo hacia mí mismo, el segundo sostenido por las reflexiones planteadas desde el espíritu del pensamiento inicial.

Palabras Clave:

- Capacitación Docente
- Etnomatemática
- Docentes Indígenas Venezolanos

Introducción

Hace poco leí un documento de Delibes (2016) donde hace alusión a un investigador a quien calificó como curioso. Señala que dicho sujeto a veces se conformaba con satisfacer sus curiosidades sobre determinados temas y cuando ya no le divertía lo que hacía, al respecto, solía suspender esas investigaciones, sin importarle si los demás quedaban convencidos o no de lo que había encontrado o reportado. Por fortuna, considero que ese no es mi caso. Cuando estudio e investigo, por ejemplo, asuntos que tienen que ver con la Etnomatemática, me comprometo y me satisface seguir hurgando sobre este tema, sin que la posibilidad de abandonar se apodere de mí accionar investigativo. No obstante, a veces surgen situaciones que invitan a abortar los procesos investigativos que se emprenden, bien por falta de recursos financieros, por ausencia de apoyo institucional o por otras razones que se convierten en adversidades. Cuando esto ocurre, lucho contra los molinos del viento y suelo mantener mi quehacer investigativo con miras de salir airoso de esas situaciones en contra.

Situaciones como las anteriores vienen a mi mente y están asociadas con lo que me correspondió experimentar previo a la concreción de la publicación de mi artículo en la *Journal of Mathematics and Culture* (JMC) que relata una experiencia de capacitación en Etnomatemática, dirigida a docentes indígenas venezolanos. Aunque no es el momento de listarlas, no resultó fácil vencer las adversidades, pero el compromiso por desarrollar mi primera investigación sobre Etnomatemática, la cual estuvo asociada con esa propuesta de capacitación, bien valió la pena ya que luego de desarrollarla pude presentarla en el 5th *International Congress on Ethnomathematics* (ICEm5) en los Estados Unidos y, posteriormente, hacerla visible a través de la JMC, en el año 2012, de ese mismo país.

Esa experiencia de capacitación surgió a raíz de mi participación en un curso sobre Multiculturalidad y Etnomatemática (Oliveras, 2009), celebrado en Venezuela en el año 2009, el cual fue guiado por la Dra. María Luisa Oliveras Contreras de la Universidad de Granada, España, y desarrollado en las aulas de la Universidad Pedagógica Experimental Libertador (UPEL). Luego de entender que lo multicultural va más allá de la suma de tradiciones culturales que convergen en un contexto geográfico, me obligué revisar materiales asociados con el tema, poniéndolos en escena para su discusión y aplicación en contextos concretos, tal como lo experimenté cuando diseñé un proyecto de aprendizaje inspirado en un instrumento musical típico venezolano.

Como complemento de ese curso, formé parte de un grupo de docentes de la UPEL que realizó una visita académica-investigativa, junto con la Dra. Oliveras, a varias comunidades indígenas del estado Amazonas, Venezuela, lo cual permitió hacer observaciones de prácticas escolares en esas comunidades. En esa visita, también presentamos opciones educacionales al cuerpo de docentes que gerencia lo que tiene que ver con la Educación Intercultural Bilingüe en ese estado y con la Coordinación de Programa de Educación Intercultural Bilingüe (PEIB) de la UPEL, a nivel regional y nacional. Todo esto fue avizorando posibilidades para poner en práctica lo aprendido en ese curso.

NOTES

De inmediato, un grupo de docentes de la UPEL, en el cual estuve incluido, comenzó a diseñar un Curso de Capacitación Pedagógica dirigido a los docentes indígenas que laboran en esas comunidades. Su diseño estuvo conformado por cursos que versaban no sólo sobre Etnomatemática sino sobre otras áreas tales como Lengua Indígena, Pedagogía Indígena y Mapeo Sociolingüístico.

El Curso de Etnomatemática se puso en práctica algunos meses después (en Enero 2010) y en la capacitación se atendieron, en primera instancia, más de 25 indígenas que laboran como docentes en la Educación Primaria del estado Amazonas que en ese momento tenían la responsabilidad de enseñar contenidos matemáticos, y otros contenidos, en escuelas cuyos estudiantes son originarios de las etnias tales como *Hiwis, Piaroas, Piapocos y Ye'kuanas*. Igualmente se atendieron otros profesionales de la zona que, a futuro, servirían como facilitadores del Curso en otras comunidades indígenas lejanas, pero del mismo estado.

La experiencia de capacitación previamente descrita fue, justamente, la que publiqué en la JMC, según detalles que aparecen en Martínez Padrón (2012a). El hecho de asumir el compromiso de conducir ese curso de Etnomatemática fue una osadía ya que, para la fecha, no tenía experiencia ni había tenido la oportunidad de hacer algún trabajo de campo donde aplicara este tipo de conocimientos. Sin embargo, se hizo posible y me nutrí de las diferentes informaciones que surgieron desde los participantes, sobre todo cuando se discutía sobre la Matemática que subyace en sus prácticas habituales.

Tal experiencia permitió apasionarme por la investigación en Etnomatemática, hecho que se acrecienta a medida que pasa el tiempo. Desde allí mantengo el compromiso de seguir estudiando, investigando, escribiendo y dando a conocer los resultados de mis producciones al respecto.

Una de esas producciones es la que presento en este momento y su discurso ha sido pensado mediante una retrospección e introspección conjugada, destacando dos grandes momentos. El primero sostenido por las distintas reflexiones planteadas desde el espíritu del pensamiento inicial que originó la experiencia de capacitación desarrollada en el año 2009, en el estado Amazonas, presentada en el ICEm en el año 2010, y publicada por la JMC en el año 2012. Todo eso se sustancia en un proceso retrospectivo que sigue una línea de tiempo perceptible mediante las producciones que fueron emergiendo en fechas posteriores. Mostrar la trayectoria de dichas producciones, desarrolladas y en proceso, gestadas desde el espíritu del pensamiento inicial hasta lo que se piensa hacer en el futuro inmediato, informa sobre el impacto que generó el hecho de escribir este primer artículo sobre Etnomatemática. En el segundo momento, sustentado en la idea de una introspección, dirigí la mirada hacia mí mismo, generando episodios cargados de una reflexión referenciada en observaciones y análisis de mis propios pensamientos. Desde allí develo lo que ocurre y ocurrió dentro mí en esa trayectoria que regresa al momento inicial que motivó la propuesta, posteriormente publicada, hasta avizorar lo que podría hacerse en el futuro. El proceso introspectivo permitió experimentar la comprensión de lo hecho, sin excluir o eludir aspectos prospectivos requeridos para visionar lo que se puede hacer. El Diccionario de Psicología Científica y Filosófica (2016) señala que cuando esta capacidad reflexiva se ejerce a través de los recuerdos y sobre lo pasado, este proceso se denomina introspección retrospectiva. De manera que la introspección se mueve entre vivencias pasadas y presentes que al desmenuzarlas permiten repensar, detenidamente, no sólo la esencia del documento que se publicó en la JMC en el año 2012, sino otros que emergieron consecutivamente. En tal sentido, involucra la evolución vivida por el autor.

A la luz de lo mencionado, es posible capturar las bondades de la Etnomatemática y, en un recorrido temporal, relatar semblanzas que permiten dar cuenta del creciente, aunque tenue, uso de la Etnomatemática en ese corto lapso. Igualmente, describo acciones que sirven de apoyo para desarrollar nuevas investigaciones que involucran al mismo tipo de docentes y contextos. Por tanto, se siguen abriendo senderos para abordar la Cultura Matemática de esos pueblos y comunidades indígenas del estado Amazonas.

Contextos de Referencia

En países como Brasil, Guatemala, Costa Rica, España y Colombia varios investigadores han puesto en escena las múltiples bondades de la Etnomatemática, como perspectiva sociocultural de la Educación Matemática, sobre todo por estar signada por prácticas cotidianas y por la posibilidad de permitir abordar la diversidad cultural de grupos socioculturalmente diferenciados. Para hacer referencia a las mismas basta observar lo que dicen autores como:

1. D´Ambrosio (2003; 2005), y Rosa y Orey (2010): Brasil;
2. Blanco (2008): Colombia; y
3. Oliveras (2005a; 2005b; 2009): España.

Rosa y Orey (2010) mencionan que la Etnomatemática no es sólo el estudio de ideas matemáticas, también contempla la antropología y la historia, permitiendo abordar las contribuciones matemáticas de diferentes culturas. Destacan que su foco de interés tiene que ver con procesos de generación y producción de conocimientos matemáticos y con los mecanismos sociales en su institucionalización y difusión. Con frecuencia, sustentan sus ideas en las del brasilero Ubiratán D´Ambrosio al citar que todas las culturas hacen matemáticas expresadas en los lenguajes específicos. Según D´Ambrosio (2003; 2005) una de las bondades de la Etnomatemática está en el hecho de permitir explorar la Matemática usada fuera de la escuela y analizarla desde las prácticas que acontecen en diversos grupos socioculturales: etnias, costureras, constructores de viviendas y otros.

Siguiendo con las ideas de D'Ambrosio, es conveniente resaltar lo que reporta Blanco (2008) mediante la publicación de una entrevista que le realizó, la cual resulta muy alumbradora para quienes nos iniciamos en el campo de la Etnomatemática. D´Ambrosio la piensa como una manera de hacer Educación Matemática a la luz de las necesidades ambientales, sociales y culturales de la comunidad abordada. Siendo así, la Matemática no es vista de manera aislada de la realidad de los sujetos, por lo que es propiciatoria para llevar las prácticas que acontecen en las comunidades a la escuela.

En el mismo orden de ideas, se hace necesario poner en escena los planteamientos de Oliveras (2005a; 2005b; 2009) en cuanto a las actividades que se pueden desarrollar al momento de ser pensadas desde la Etnomatemática. En la mayoría de los casos, Oliveras sugiere que se viabilicen mediante la elaboración y puesta en escena de micro-proyectos pensados como materiales formativos derivados del contexto multicultural donde se lleva a cabo la acción docente. También sugiere tomar en cuenta el contexto social de la escuela, los aspectos socioculturales de la clase y las particularidades de cada estudiante; por lo que han de emerger de la cotidianidad de los grupos donde pertenecen los estudiantes, sin descuidar los variados referentes

NOTES

multiculturales que matizan las aulas de clase, evitando consideraciones monoculturales que suelen imperar en la mayoría de los casos. Cuando habla de las aulas de las comunidades indígenas, indica que deben ser pensadas para atender la diversidad cultural, dado que en ellas suelen confluir estudiantes que provienen de diversas etnias (Oliveras, 2005a). Tales micro-proyectos se perfilan como aptos para arrostrar, integralmente, aspectos culturales específicos de cada pueblo o comunidad indígena, donde se ponga en escena un trabajo didáctico centrado en los estudiantes y en los grupos que lo conforman. Al acceder estas particularidades culturales, con sentido en sí mismas, Oliveras (2005b) considera que ello provoca aprendizajes con significados globales que, para materializarlos, requieren del uso de metodologías activas y fundamentadas en una perspectiva socioconstructivista capaz de reconstruir la realidad, secuenciándola de modo análogo a cómo se produce realmente.

Este contexto inicial de referencia esboza suficientes bondades que sirven de pilares para edificar cualquier pretensión investigativa que piense poner en escena elementos potencialmente útiles y robustos que configuran objetos o prácticas seguidas por grupos particulares. Eso quiere decir que desde cualquier actividad propia de cada cultura se pueden concretar procesos educativos contextualizados con significación para cada cultura.

Todos estos nutrientes, servidos con apoyo de las actividades pautadas por Oliveras (2009) en el curso de ya mencionado, sirvieron para elaborar mi primer micro-proyecto curricular, tomando en cuenta signos relevantes de una cultura particular. Luego, surgieron tareas y se hicieron planteamientos que fueron avizorando un Programa de Capacitación para Docentes Indígenas (Martínez Padrón, 2010) que obligó a seguir hurgando nuevos referentes contentivos de actividades matemáticas cargadas de profundo significado social.

Otros referentes también marcaron nuevos accionares. Yojcom (2006) realizó un trabajo donde analizó el uso actual del sistema de numeración vigesimal, en cinco comunidades de Guatemala, y contempló la elaboración de una propuesta curricular acorde con las exigencias y necesidades de la población *maya-q'eqchi'* de ese país. Allí encontró la existencia de una "relación entre las prácticas sociales vivenciadas por los *q'eqchi'* con los procesos matemáticos de contar, calcular y medir... [sosteniendo que]... "el sistema vigesimal no sólo es importante para cuantificar los bienes materiales, resolver situaciones-problemas y conservar la cultura, sino que también para manifestarse espiritualmente" (p. 10), lo cual es una clave que hay que tener presente en investigaciones que aborden estos ámbitos. Vale acotar las aseveraciones de Gerdes (2012) al señalar que los estudios sustentados en la Etnomatemática procuran elementos culturales que pueden servir como puntos de partida para organizar actividades de enseñanza de la Matemática.

Al igual que Oliveras (2005a), Gavarrete (2012) sigue sosteniendo que la Etnomatemática es una alternativa que permite hacer Educación Matemática desde la diversidad cultural. En los entornos indígenas, también invoca a D´Ambrosio para dar cuenta del desafío que suscita el proceso de enseñanza-aprendizaje de la Matemática en dichos entornos, sobre todo por la predominancia mundial de las Etnomatemáticas denominadas occidentales que han incidido en la descontextualización curricular en estos espacios y su consecuente exclusión social derivada de la no consideración de las distintas formas de conocer sostenida por D´Ambrosio (2005). Para el caso de Costa Rica, Gavarrete sostiene que grupos étnicos de ese país exhiben formas de conocer que dependen de su manera de percibir e interpretar sus mundos, por lo que es necesario hacer las consideraciones pertinentes a este tipo de contextos.

En este paisaje de experiencias, hay suficientes evidencias para asumir a la Etnomatemática como marco de referencia para reforzar el accionar formativo del PEIB de la Universidad donde laboro: UPEL-Venezuela. Por ende, la admito como sustento de una propuesta educativa emancipadora y esperanzadora.

El Espíritu del pensamiento original

La propuesta de capacitación que fue desarrollada en los docentes indígenas del Estado Amazonas se elaboró pensando en la posibilidad de construir proyectos dirigidos a la población estudiantil asentada en las escuelas atendidas por estos docentes que enseñan contenidos matemáticos.

En otros apartados, informo que esta capacitación cubrió varias áreas del saber pero aquí sólo preciso lo que tiene que ver con el Curso de Etnomatemática, donde se solicitó, a los participantes, construir Proyectos "alumbrados por el uso de determinados objetos y prácticas como vías para recolectar, identificar y sistematizar los conocimientos y los saberes matemáticos ancestrales que le son propios" (Martínez Padrón, 2012a, p. 287) a los pueblos y comunidades indígenas a las que ellos pertenecen. Como se hizo hincapié en que exploraran la Matemática usada por su comunidad, tenían que apoyarse en sus prácticas rutinarias.

Para construir los Proyectos se tomó "en cuenta el contexto sociocultural de la escuela y su entorno, los aspectos socioculturales de la clase y las particularidades de los docentes y sus estudiantes, considerando la procedencia de los diversos pueblos y comunidades indígenas" (p. 287).

Las producciones de los participantes estuvieron relacionadas con: (a) la caza: la Cerbatana; (b) la cestería y el cernido: El Manare; (c) la ornamentación: La Cortina de Palitos; y (d) el transporte: La Curiara.

Figura 1: Objectos de referencia de los Projectos de Aprendiziaje elabrados por los Docentes

En todos los casos, los participantes reportaron que en el diseño, construcción y uso de esos objetos subyacen abundantes contenidos matemáticos que pueden servir de insumo para la organización de actividades escolares. Por ende, aflora un mundo de ideas de talante matemático que pueden resultar útiles para la organización de actividades escolares desde esos contextos. Tales consideraciones abrieron alternativas para atacar los tradicionales problemas de aprendizaje que tienen los estudiantes con determinados contenidos matemáticos. En la Figura 1 se visualizan algunos de esos objetos de referencia.

En Martínez Padrón (2012a) pueden encontrarse otros detalles que informan sobre toda la riqueza de los pensamientos matemáticos involucrados en el diseño, construcción y uso de los objetos que sirvieron de guía para la construcción de los

NOTES

Proyectos mencionados. A saber, contemplan procesos de medición y conteo, ángulos, figuras geométricas, cuerpos geométricos, segmentos paralelos, altura, ancho, largo, diámetros, adiciones y otros elementos aritmético-geométricos.

Impacto del Pensamiento Original en las Actuaciones Posteriores.

Aunque en el contexto educativo venezolano quedan muchas cosas por hacer, en torno a la Etnomatemática, sigue estando pendiente la popularización de esta opción investigativa como alternativa para enseñar y aprender matemática, pues, a pesar de estar explícita la necesidad de desarrollar actividades contextualizadas en todos los espacios escolares, aún se precisa aprovechar sus bondades para convertirla en una alternativa propiciatoria que ponga en escena prácticas contextualizadas en las aulas de clase. A pesar de esta realidad, han venido acaeciendo actuaciones posteriores que informan que dicha opción sigue siendo deseable y propiciatoria para el aprendizaje de contenidos matemáticos en grupos socioculturales diferenciados.

De manera específica, se mencionan algunos avances derivados de las motivaciones asociadas con el estudio original:

1. Se concretó la primera etapa del desarrollo de un Proyecto de Investigación sobre los Sistemas Numéricos de las comunidades *Hiwi, Ųwǫttują y Puinave*, del Estado Amazonas. Se pretende elaborar un material informativo sobre sus sistemas numéricos propios, si los hubieren. Los ejes rectores del estudio lo conformaron sus saberes cotidianos y sus conocimientos ancestrales manifiestos en sus procesos de cuantificación presentes en las prácticas cotidianas que tienen que ver con contar y medir. El trabajo se corresponde con una investigación de campo complementada con un estudio lingüístico-matemático donde se analizan los contenidos numéricos que emergen de la observación y de las entrevistas sobre las prácticas. Las últimas se han venido utilizando para concretar vocablos, expresiones y algoritmos que pudieran utilizar estos indígenas en sus procesos de cuantificación. Se interrogó, entre otros, sobre los procesos que seguían para realizar determinados tipos de conteo de la cantidad de productos cultivados o comercializados, así como de las dimensiones del terreno utilizado en estos cultivos. Este proceso también indagó sobre cantidades, buscando la escritura de vocablos, simbologías y patrones. La escritura de los vocablos se apoyó en dos instrumentos: uno donde deben escribir varios números considerados como clave para detectar regularidades o irregularidades lingüísticas, prefijos aditivos o multiplicativos o combinaciones de palabras que sigan o den luz a algún proceso operatorio, y otro para detectar los procesos que siguen para cuantificar después de conocer conteos previos. Entre otros aspectos, se aspira concretar: (a) la cultura aritmética presente en las actividades de los pueblos mencionados, a la luz de las actividades de contar y medir; (b) la determinación, escritura y uso de números propios y distintos al sistema de numeración decimal occidental; (c) la manera de hacer conteos en sus posibles sistemas numéricos; (d) la regularidad lingüística usada en los vocablos numéricos; y (e) la determinación de simbologías, bases y sistemas numéricos, si los hubiere, por cada población abordada (Martínez Padrón, Magro, González, Berríos, Lara, Martínez Madail, Bello & Morillo, 2014).

   En un avance de esa investigación, Martínez Padrón, González y Berríos (2015) encontraron que tales comunidades no poseen simbologías propias para expresar sus números, así como tampoco para dar cuenta de las operaciones, lo que restringe la posibilidad de hablar de bases y, por ende, de sistemas numéricos autóctonos en esas comunidades. Llamó la atención que en comunidades equivalentes, tanto nacionales como internacionales, se leyeron documentos relacionados con el tema donde viene hablando de sistemas numéricos que tienen determinadas bases, por ejemplo el de base vigesimal maya. Igual se habla de otros sistemas que ni siguiera gozan de representaciones simbólicas. Pareciera que tales determinaciones son hechas aludiendo a determinadas bases y consideramos que ello merece nuevas discusiones mientras no se declare cuál es su acepción. Tampoco se establecen, con claridad, los basamentos matemáticos que siguen algunos investigadores para determinar que algo es un sistema numérico. Mientras no queden claras las acepciones quedarán pendiente varias de esas aseveraciones. Nos atrevemos a conjeturar que en las comunidades que estudiamos en nuestro proyecto sólo podría hablarse de sistemas de conteo, sin mayores pretensiones y compromisos teóricos. Retomando lo de los mayas, vale preguntarse: ¿cuál es la base de este sistema numérico denominado vigesimal, a sabiendas que está sustentado en el uso de sólo tres símbolos?

2. Diseño otro Proyecto que aborda procesos de tejidos elaborados por varias comunidades de la misma región: Amazonas. Adelanto varias investigaciones documentales sobre estos procesos para luego complementarlas con nuevas observaciones *in situ*. Presumo profundizar en el tejido de cestas donde no sólo interese el pensamiento matemático alineado con su diseño y su correspondiente carga aritmética-geométrica, sino con la cosmovisión soterrada en las figuras dibujadas en los objetos que elaboran.

3. Existen varios trabajos ya presentados en eventos, tal es el caso de: **(a)** Martínez Padrón (2012c; 2015) donde hace alusión, respectivamente, a los elementos matemáticos y a las urdimbres y tramas de la cestería indígena del estado Amazonas. Ambos profundizan en dos tipos de cestas particulares: el Manare y la Wapa, con el propósito de seguir guiando las actividades escolares subsumidas en el tejido de cestas y sus formas; **(b)** Martínez Padrón, Oliveras y González (2014), y Martínez Padrón y Oliveras (2015) donde hace referencia a la producción científica encumbrada desde el Curso de Multiculturalidad y Etnomatemática; **(c)** Martínez Padrón, Martínez, González y Oliveras (2013; 2014) que también emergió del curso de Multiculturalidad y Etnomatemática, pero en este caso con referencia a un instrumento musical venezolano; y **(d)** Martínez Padrón (2013; 2014) que repercuten en el enriquecimiento de la Etnomatemática como teoría sustantiva. El escrito en el 2013 contempla acepciones, dimensiones y corrientes de la Etnomatemática, incluyendo la posibilidad de hablar de varias matemáticas, según las diferentes culturas.

4. Propuse la creación de un Curso de Etnomatemática (Martínez Padrón, 2012b) con el propósito de incorporarlo al PEIB de la UPEL. Prevé interconectar la Matemática escolar oficializada con la creada por las comunidades indígenas que atiende este programa, sin pretender imponer los conocimientos matemáticos de una cultura en otra. Se presume generar espacios para la construcción de conocimientos matemáticos a partir de la propia cultura de los estudiantes. No obstante, pueden derivarse validaciones de contenidos matemáticos escolares propios de la cultura occidental, desde las prácticas y abordajes de cada grupo sociocultural. En todo caso, contempla la organización de actividades escolares donde los estudiantes reflexionen y discutan sobre situaciones que incorporen los conocimientos matemáticos propios de su cultura.

5. Estoy emprendiendo dos investigaciones para analizar la producción científica publicada en la Agenda Latinoamericana de Matemática Educativa y en la Revista Latinoamericana de Etnomatemática. Su propósito es conocer cuáles áreas temáticas se vienen publicando en Etnomatemática, quiénes lo hacen, desde dónde, cómo, cuántos, para qué, por qué y para quien lo hacen.

A manera de Cierre: Prospectiva

Desde el momento en que viví la experiencia de diseñar y desarrollar el Curso de Capacitación en Etnomatemática, he experimentado cambios que considero a favor de mi configuración como investigador en esa área. No temo en reconocer que ese inicio fue exigente, pues, aunque ya tenía algunas informaciones generales sobre Etnomatemática, aprendidas en el Curso, manejaba pocos aspectos. Me hacía falta la práctica en grupos socioculturalmente diferenciados. Particularmente, precisaba abordar y observar situaciones y prácticas en una Cultura Matemática como la indígena, pues, hacia ellos está volcado mi interés indagatorio.

Declaro que estaba convencido de la necesidad de desplegar cambios en las tradicionales maneras de organizar actividades ya que el fracaso en el aprendizaje de la Matemática se había puesto de manifiesto en esos contextos. Por tanto, era necesario apoyar a la educación indígena con una capacitación propiciatoria de mejores experiencias de aula, tal como la publicada en mi artículo en la JMC, año 2012. Ello me invitó a atender lo que tiene que ver con el conocimiento profesional de los docentes que tienen el compromiso de enseñar contenidos matemáticos en las escuelas, sobre todo en el PEIB de la UPEL, en Venezuela. Aunque tengo claro que dicho conocimiento sustenta las decisiones que se toman en el aula e impacta en las experiencias desarrolladas por los protagonistas de la clase, no resulta fácil capacitarlos o formarlos para que mitiguen o anulen la aversión que sus estudiantes, e incluso ellos, sienten por la Matemática.

Cuando la educación está dirigida a los pueblos y comunidades indígenas es importante no sólo reconocer su diversidad cultural sino valorar la realidad donde se desarrollan sus acciones. Esto conduce a desarrollar una Enseñanza de la Matemática desde diversas cosmovisiones, valores, saberes y prácticas sociales que impregnan la cotidianidad. Hay que establecer relaciones entre la Matemática y los grupos donde se desarrolla, tomando en cuenta la cultura a la cual pertenecen los aprendices y sin imponer los conocimientos de una en la otra. Por tanto, se debe propender por un programa de acción que busque no sólo elementos matemáticos sino lenguajes para su concreción y difusión entre sus miembros. En todo caso, hay que conjugar los saberes matemáticos propios de cada grupo sociocultural con la enseñanza de los mismos en diferentes ambientes, incluyendo ideas, perspectivas y prácticas que le son propias a esos grupos diferenciados.

REFERENCIAS

Blanco, H. (2008). Entrevista al profesor Ubiratán D'Ambrosio. *Revista Latinoamericana de Etnomatemática*, [en línea], 1(1): 21-25, Recuperada de http:// www.etno matematica.org /v1-n1-febrero2008/blanco.pdf.

D´Ambrosio, U. (1993). Etnociencias. *Enseñanza de la Matemática, 3*(1), pp. 4-15.

D´Ambrosio, U. (2005). *Etnomatemática. Elo entre as tradições e a modernidades*. Coleção Tendências em Educação Matemática. Brasil: Autêntica Editora.

Delibes, M. (2016). *Ciencia y Literatura. Revista IDesqbre*. Documento en línea recuperado el 10 de Abril de 2016 de https://idescubre.fundaciondescubre.es/2016/04/21/ ciencia-y-literatura/

Echegoyen, J. & Blanco, I. (2016). *Diccionario de Psicología científica y filosófica,* Portal de Filosofía, Psicología y Humanidades en Internet. Ediciones: Torre de Babel. Documento en línea recuperado el 11 de abril de 2016 de http://www.e-torredebabel.com/Psicologia /Vocabulario/ Introspeccion.htm

Gavarrete, M. E. (2012). *Modelo de aplicación de Etnomatemáticas en la Formación de Profesores para Contextos Indígenas de Costa Rica*. Tesis Doctoral no publicada, Universidad de Granada, España.

Gerdes, P. (2012). *Etnomatemática, cultura, matemática, educação*. Colectânea de Textos 1979-1991. Instituto Superior de Tecnologias e Gestão, Belo Horizonte, Boane, Moçambique.

Martínez Padrón, O. J. (2010). Programa de Curso: Etnomatemática. *Programa de Capacitación en el marco del convenio con la Gobernación del Estado Amazonas*, Documento no publicado. Universidad Pedagógica Experimental Libertador: Autor

Martínez Padrón, O. J. (2012a). Una Experiencia de Capacitación en Etnomatemática, en Docentes Indígenas Venezolanos. *Journal of Mathematics and Culture, 6* (1), pp. 286-295.

Martínez Padrón, O. J. (2012b). *Propuesta de curso de Etnomatemática para el programa de educación intercultural bilingüe*. Ponencia presentada en VIII Encuentro de Educación Matemática e Informática de las Regiones Nor-Oriental, Insular y Guayana, Maturín, Venezuela.

Martínez Padrón, O. J. (2012c). *Elementos matemáticos que subyacen en la producción de cestería indígena del estado Amazonas*. Ponencia presentada en VIII Encuentro de Educación Matemática e Informática de las Regiones Nor-Oriental, Insular y Guayana, Maturín, Venezuela.

Martínez Padrón, O. J. (2013). Etnomatemática: una reseña crítica de sus acepciones. *Revista Científica*, Edición Especial, pp. 427-431.

Martínez Padrón, O. J. (2014). La Etnomatemática como Perspectiva sociocultural de la Educación Matemática. Revista Observatorio de la Ciencia, Ministerio del Poder Popular de Ciencia, Tecnología e Innovación, 2 (3), pp. 99-108.

NOTES

Martínez Padrón, O. J. (2015). Urdimbres y tramas de la cestería indígena del estado Amazonas. En *La educación en el cambio de época en América Latina y el Caribe desde las perspectivas pedagógicas*. Tomo II, (Comp. E. Martínez), Centro de Investigaciones Educativas, Universidad Central de Venezuela, pp. 531- 531.

Martínez Padrón, O. J., & Oliveras, M. L. (2015). Surcando caminos de interculturalidad sustentados en la Etnomatemática. *Revista Latinoamericana de Etnomatemática: Perspectivas Socioculturales de la Educación Matemática*; Una mirada a la Etnomatemática a través del ICEm-5, Vol. 8, No 2. pp. 341-363.

Martínez Padrón, O. J., Martínez, A., González, A. & Oliveras, M. L. (2013). *Etnomatemáticas en el diseño y construcción del instrumento musical "cuatro": propuesta didáctica para el desarrollo del currículo escolar*. Ponencia presentada en RELME 27, Argentina.

Martínez Padrón, O. J., Martínez, A., González, A. & Oliveras, M. (2014). Una experiencia en Etnomatemática sustentada en el diseño y construcción del instrumento musical cuatro. *Actas del VII Coloquio Internacional Enseñanza de las Matemáticas*, pp. 537-546.

Martínez Padrón, O. J., Oliveras, M. & González, F. (2014). Multiculturalidad y Etnomatemática: se hace camino al andar. *Actas del VII Coloquio Internacional Enseñanza de las Matemáticas*, pp. 1027-1044.

Martínez Padrón, O. J.; González, A. & Berríos, J. (2015). *Cultura aritmética de algunos pueblos indígenas venezolanos*. Ponencia aprobada, no presentada, en la XIV CIAEM, México.

Martínez Padrón, O., Magro, M., González, A., Berríos, F., Lara, J., Martínez Madail, A. A., Bello, I. & Morillo, R. (2014). *Sistemas numéricos de las etnias venezolanas. Casos: pueblos Hiwi, Ųwǫttują y Puinave*. Proyecto de investigación no publicado. Universidad Pedagógica Experimental Libertador, Instituto Pedagógico Rural El Mácaro, Turmero, Venezuela.

Oliveras, M. L. (2005a). Microproyectos para la educación intercultural en Europa. *Uno: Revista de Didáctica de las Matemáticas*, N°. 38, 2005, p.p. 70-81.

Oliveras, M. L. (2005b). Microproyectos [Presentación en Multimedia]. España: Autor.

Oliveras, M. L. (2009). *Programa de curso interculturalidad y Etnomatemáticas*. Universidad Pedagógica Experimental Libertador: Maracay, Venezuela.

Rosa, M. & Orey, D. (2010). Etnomodeling as a Pedagogical Tool for the Ethnomathematics Program. *Revista Latinoamericana de Etnomatemática, 3*(2). 14- 23

Yojcom, D. (2006). *Análisis del uso actual del sistema de numeración vigesimal en cinco comunidades q'eqchi' de Guatemala*. Trabajo de Maestría, no publicado. Pontificia Universidad Católica de São Paulo, Brasil.

Volume 4
Teacher Prep

Foundations in Ethnomathematics for Prospective Elementary Teachers

Jenni L. Harding-DeKam
Jennifer.HardingDekam@unco.edu
970.351.1029
University of Northern Colorado
Assistant Professor of Elementary Education
School of Teacher Education
McKee 216
Campus Box 107
Greeley, CO 80639

Abstract

Ethnomathematics is a term that has been coined to elucidate that everyone uses and can learn mathematics. This manuscript demonstrates how a college Mathematics Methods Course at a Doctoral Intensive University in the Rocky Mountain Region can implement the idea of using the culturally diverse background of students as a foundation to teach children learning second languages and children of diverse cultures. This includes one year of quantitative and qualitative research about prospective elementary teachers who were instructed on how to teach children learning second languages during their Mathematics Methods Course. These prospective teachers then applied these methods while completing their student teaching.

Introduction

Ethnomathematics is a "broader view of how mathematics relates to the real world" (D'Ambrosio, 2002, p. 1). Ethnomathematics in the elementary classroom is where the teacher and the students value cultures, and cultures are linked to curriculum. Ethnomathematics "represents a complex and multifaceted dynamic, which illustrates the culturally influenced uses of mathematics through its applications" (Barta & Shockey, 2006, p. 79). Teachers must emphasize the importance of building upon the students' foundational knowledge in a way that brings in their culture and their history in order to promote value in these cultures. This Mathematics Methods Course takes the first step towards ethnomathematics where the qualities of culture are explored to open up this previously unknown topic to prospective teachers. Staats (2006) explains that "ethnomathematics at the undergraduate level must be transformative, not only for how students understand their abilities in mathematics but also in how they understand their relationship to others in the world" (p. 41). These prospective teachers must learn how to enhance and restore cultural dignity to the children in their classrooms (D'Ambrosio, 2002).

This research study was two-semesters in length with ninety-two prospective elementary teachers participating. These prospective teachers were taught explicit strategies for working with children learning second languages in their Mathematics Methods Course through the specific course activities and projects of the Enhancing English Language Learning in Elementary Classrooms (Center for Applied Linguistics, 2003) often referred to as the "Farsi Video,", human graph, and lesson plans. The readings from Darder (1991); Echevarria, Vogt, and Short (2004); and Howard (2004) provided context for theoretical knowledge about the importance of focusing on children learning second languages in the elementary classroom. The theme of ethnomathematics philosophical underpinnings is grounded in having respect and some knowledge of how a students' native culture impacts the understanding of mathematics (Reyhner & Davison, 1992), understanding other cultures through *Material World: A Global Family Portrait* (Mendel, 1994), creating mathematics homework based on children's culture and surroundings is one way to bring in the success and family life of the child's community (Zimmerman, 2006), understanding the Yupik Eskimo's of Alaska including the elders, parents, and children in this community and their tribal use of mathematics (Lipka, Wildfeuer, Wahlberg, George, & Ezran, 2001), and the Northern Ute culture of using mathematics as you live (Barta & Shockey, 2006). The results from the study were two-fold. First, quantitative results on a pre- assessment and post-assessment measuring the prospective teacher's comfort level of teaching children learning second languages and diverse cultures were measured. Secondly, qualitative results from the prospective teacher's student teaching assignment demonstrated themes in meeting the needs of children learning second languages and diverse cultures in the elementary classroom.

Participant Information

The majority of the prospective elementary teachers who participated in this study were female (90%), twenty-five to thirty-five years of age (48%), and of white ethnicity (94.2%) as documented in Table 1.

Table 1: Prospective Teacher Demographics

Gender		Age		Ethnicity	
Female	90%	Under 25 years	15%	White	94.2%
Male	10%	25-35 years	48%	Hispanic	3.8%
		36 and over years	37%	Asian	1.0%
				Other	1.0%

These prospective teachers will be teaching in national public schools where "nearly 40 percent of children are minorities, but just 11 percent of the teachers are" (Kochuk, 2005, p. 17). As represented by the prospective teacher demographics above, one can see that even fewer of these participants (about 6%) are minorities. It is important to bring this foundational knowledge of ethnomathematics to these prospective elementary teachers so that they will obtain the knowledge required to effectively teach children learning second languages and of diverse cultures. This knowledge of ethnomathematics is at the beginning level where these prospective elementary teachers have to have their worldview opened to understand there are ways of learning mathematics beyond their white, female perspective. These prospective elementary teachers need to self-introspect so they can understand the importance of ethnomathematics in the classroom.

Mathematics Methods Course

The Elementary Education Program where this research takes place graduates about 175 students per school year with their Elementary Education Teaching License for grades kindergarten through sixth. The mission of this Elementary Education Program is to help prospective teachers become professionals who are able to perform proficiently in elementary school classrooms situated in increasingly complex, diverse, and technologically advanced communities. The four themes of reflection, modeling, collaboration, and integration flow as the philosophical underpinnings of this program. Both the National Council of Accreditation of Teacher Education (NCATE) and the Association for Childhood Education (ACEI) accredit this Elementary Education Program.

The Interdisciplinary Studies Liberal Arts Major had been designed to prepare students to become elementary teachers. This foundational coursework comes from the core courses of (a) mathematics (b) science, (c) history and social studies (d) language and literature. The courses taught from the mathematics department examine ways of thinking within and across broadly related disciplines. Concentration areas are provided to students for more in-depth study of academic disciplines while allowing them to be sufficiently well rounded to teach all subject areas at the elementary level. There are three required mathematics courses (nine credit hours) taken before beginning the Mathematics Methods Course (three credit hours). This Mathematics Methods Course is an intensive 45 contact hour course that prospective teachers take in five weeks after the prospective teachers have completed their nine credit hours of mathematics for their Interdisciplinary Studies Liberal Arts Major. These prospective teachers are then placed at a school to complete their student teaching assignment for the eleven weeks left in the semester.

Prospective Teachers' Learning

There are many ways to prepare prospective elementary teachers to educate children who are learning second languages and diverse cultures within the realm of ethnomathematics. The most effective way to teach children utilizes the National Council of Teachers of Mathematics' (NCTM) *Principles and Standards for School Mathematics* (2000). Specifically, the prospective teachers in my course learned how to educate children through the content standards of Number Sense, Geometry, Measurement, and Data Analysis and Probability through the Process Standards of Problem Solving, Reasoning and Proof, Communication, Connections, and Representation.

Problem solving is taught and modeled to provide meaningful problems, solving strategies, and opportunities for students to design their own problems, as well as to provide cooperative learning opportunities, encourage multiple ways of solving problems, and encourage the use of visuals in drawings, models, props, or acting out the problem. Reasoning and Proof is taught and modeled to use both examples and non-examples, provide graphic organizers, encourage students to use multiple representations, and provide Venn Diagrams. Communication is taught and modeled to build community and rapport with students, provide vocabulary in context to develop meaning, allow teachers to seek an understanding of what students are trying to communicate, provide visual scaffolding, provide visual opportunities to review vocabulary with clear labels and word walls, identify specific objectives in lesson planning, increase peer interactions to provide language acquisition opportunities, and provide modeling. Connections are taught and modeled to start with prior knowledge, move the concept from concrete to abstract, relate to the real world, relate to the culture and history of the student, provide graphic and advanced organizers, and integrate to connect to other content areas building on knowledge already acquired. Representation is taught and modeled by using graphs to vary from concrete to more abstract, have students demonstrate reasoning with manipulatives, act the problem out, provide graphic organizers for student recording, and show how to classify objects based on cultural knowledge.

To prepare prospective teachers to teach mathematics to all students, including children learning second languages, best teaching practices are modeled to my prospective teachers. Just as I expect the teacher candidates to use the above strategies, so must I while teaching this Mathematics Methods Course. First, I have to access my prospective teachers' prior knowledge, culture, and histories to get to know them. This allows instruction to be differentiated, meaningful to each individual, and taught at the appropriate level of maturity. Second, I build a safe community where all are ready to learn and share through the Responsive Classroom approach (Charney et al. 1997, 1999). This safe community allows students

NOTES

to take risks in mathematics in a way that challenges the prospective teachers cognitively and places them in cognitive dissonance allowing misconceptions to be challenged. Third, I model how to plan a lesson that considers teaching children who are learning second languages and diverse cultures. This lesson thinking is done out loud for modeling the concept to the prospective teachers. The lesson is then taught to the group allowing them to see how a plan progresses from the planning stages to action in the classroom. Fourth, I demonstrate through the Process Standards of teaching mathematical content with active learning rather than lecturing (NCTM, 2000). This encompasses giving the prospective teachers time to understand the mathematical content through constructivist learning (Crotty, 1998).

Specific Course Activities and Projects

These prospective elementary teachers participated in many course activities and projects to develop their understanding of children learning second languages and of diverse cultures in addition to learning about the national standards (NCTM, 2000) and our state standards. The prospective teachers were asked to think about the following considerations when teaching to children learning second languages and diverse cultures in this case of Latino students: "(1) honor the diversity of Latino students' experiences, (2) know the students and their experiences, (3) avoid deficit models, and (4) provide opportunities for mathematical discussions" (Moschkovich, 1999, p. 9). The instructional elements of the Farsi video, human graph, and a specific session in the lesson plans for children learning second languages and of diverse cultures were the most meaningful activities taught in the Mathematics Methods Course to these prospective teachers based on their Course Evaluation Survey.

Farsi Video

My Mathematics Education Course uses the *Enhancing English Language learning in Elementary Classrooms: Video* as an introduction to children who are learning second languages and of diverse cultures. The first lesson in this video presents a three minute and twenty second health lesson taught in the Farsi Language. The first health lesson is lectured to the audience with a quick rate and pace of speech. When this section of the video is watched in my course, the prospective elementary teachers are given the task of determining what Colorado Model Content Standards are being met as well as evaluating the teaching style of the presenter. These students are then given five minutes to discuss their findings in small groups (five to six students). The groups then report their discussions back to the rest of the class. The class responses generally encompass a wide range of content areas being guessed for the video lesson from mathematics (as this content covered in the course) to social studies, as well as science and language arts. The actual health content has never yet been chosen as the content being taught in the video from the lectured teaching presentation.

The students then watch the next part of the video wherein the same health lesson is taught to the audience in Farsi, but includes props, gestures, emphasis on key content vocabulary, and a slower rate and pace of speech. The students participating in this experience can now understand that the lesson is a health lesson being taught as well as comprehend the actual components of the lesson. Students then go back to their groups and are able to complete the task of determining what Colorado Model Content Standards are used.

The following questions are posed to the small groups of prospective teachers. Group answers are then reported back to the whole class. The typical class responses follow the question below.

1. How many of you knew Farsi when we started the video? The answer is that no one yet speaks the Farsi language.

2. How did you understand the health lesson being taught the second time if you do not speak the Farsi language? The answer here is that the teacher is doing more than just talking as she is engaging students into the lesson being taught. The moment is taken here to teach the prospective teachers that instruction can be made comprehensible for children who are learning second languages and of diverse cultures through building on the student's prior knowledge (culturally and historically); moving from concrete to abstract; focusing on making meaning of new vocabulary comprehensible; incorporating the use of pictures, props, demonstrations and hands-on activities to show meaning; talking about the content being shown; and simplifying the speech (Franco, 2003).

3. What can you do to succeed in teaching all students in your future elementary classroom regardless of if they speak English or not? This answer focuses on the observable teaching strategies used in the video including props, gestures, emphasis on key content vocabulary, and a slower rate and pace of speech. The answer also includes strategies for mathematic content such as (1) application or expansion when mathematics vocabulary is introduced based on their cultural experiences, opportunities for student interaction with the vocabulary, and a concrete example of the concept is created, expanded, or included into correct schema, (2) discussing vocabulary by using it in context to progress concept(s) from concrete to abstract (3) exploring questions and pictures from their community to help define vocabulary and (4) differentiating instruction to meet all learning needs (Franco, 2003).

It becomes clear to the prospective elementary teachers in my course how much easier it is to understand the lesson through these teaching techniques even though they do not speak the language in the video. This fifteen-minute experience makes a vivid, lasting impression on those who participate to help create an understanding of the challenges of learning through a language barrier; furthermore, it demonstrates teaching strategies and practices that help overcome that language barrier and models good teaching for all learners regardless of their culture.

Human Graph

The human graph deals with how the prospective elementary teachers perceive empathy towards children who are learning second languages and other cultures. The human graph is a physical number graph on one wall in the room. This left side of this wall represents the quantity of zero, the middle of this wall represents the quantity of five, and the farthest right hand side of this wall represents the quantity of ten with all the numbers in between zero through five and six through ten being represented. The students then physically go stand next to the number on the wall that represents the quantity they wish to express. If there is more than one student who chooses the same quantity than

NOTES

the students stand one in front of another creating a visual human graph for comparison and discussion.

The definition of empathy is discussed with the whole class giving the foundational knowledge for the human graph activity. The Oxford American College Dictionary (2002) defines empathy as the ability to understand and share the feelings of another (Lindberg). Students are then given the task to physically move to the section on the human graph that represents the quantity they believe to be true based on the one-to-ten scale for the following questions:

1. During your first student teaching experience, what empathy do you believe your host teacher had toward children learning second languages? Why?

2. During your first student teaching experience, what empathy do you believe your host teacher had toward children with diverse cultures? Why?

3. During your first student teaching experience, what empathy do you believe your elementary school had toward children learning second languages? Why?

4. During your first student teaching experience, what empathy do you believe your elementary school had toward children with diverse cultures? Why?

5. What do you think the empathy level is for children learning second languages in the city of Greeley? Why?

6. What do you think the empathy level is for children with diverse cultures in the city of Greeley? Why?

7. What do you think the empathy level is for children learning second languages in the state of Colorado? Why?

8. What do you think the empathy level is for children with diverse cultures in the state of Colorado? Why?

9. What do you think the empathy level is for children learning second languages across this nation? Why?

10. What do you think the empathy level is for children with diverse cultures across this nation? Why?

11. What is your personal empathy for children learning second languages? Why?

12. What is your personal empathy for children with diverse cultures? Why?

The "why' portion of each question is discussed while the group of prospective teachers are part of the human graph. This human graph activity helps prospective elementary teachers look at the issue of children learning second languages and children with diverse cultures in the classroom in contexts ranging from social to personal based on their observations and experiences at the community, school, state, and national levels. This activity also engages students to think outside of their own culture and empathize with cultures that are different from their own based on the population of these prospective elementary teachers who are female (90%), twenty-five to thirty-five years of age (48%), and of white ethnicity (94.2%).

Lesson Plans

A section specifically addressing children who are learning second languages and of diverse cultures has been added to the lesson plan requirement in my Mathematics Methods Course. This helps prospective elementary teachers consider students of varying backgrounds who will be in their classroom while continuing to practice teaching methods and strategies that will help each student learn. This requirement has been added to multiple assignments including the

Mathematics Concept Lesson Plan, Integrated Mini Unit Lesson Plan, and the Thematic Unit Lesson Plan. The first two individual student lessons are taught to classroom peers in order to develop mathematical concepts and enhance the integration of language arts, social studies, and other content areas with the mathematics subject matter. The final Thematic Unit of lessons (five hours worth of teaching) is taught to elementary children while the prospective teachers are completing their student teaching. This unit is developed, planned for, taught, and evaluated by the prospective elementary teacher and integrates many content areas. This section, which specifically addresses children learning second languages and children with diverse cultures, has been added to all three assignments encouraging prospective teachers to consider these questions: (1) how will you ensure all students will learn from this activity or lesson? and (2) how can you access your student's prior and cultural knowledge? This practice of thinking about all children will greatly enhance the instruction provided in the elementary classroom for children of varied backgrounds and experiences.

Quantitative Mathematics Course Results

A colleague, Carol Picard, and I examined the effect of children who are learning second languages through the prospective elementary teachers in our Mathematics Methods Courses. In Fall semester of 2003 and Spring semester 2004, a pre-assessment on the first day of the course and a post-assessment on the last day of the five-week course were conducted to examine how comfortable our prospective teachers were in teaching children learning second languages. Ninety-two prospective teachers participated in both assessments. The instrument used to measure their comfort level with children learning second languages had a Cronbach's alpha reliability of 0.96 for the Likert-type instrument of 24 questions. The factor analysis completed was a Varimax rotation with the Principle Components Method. The factor analysis through the Scree Plot and Component Matrix confirm that we are measuring one trait: confidence in teaching children learning second languages and of diverse cultures at the elementary level.

Out of the 24 questions on the survey, there were 13 that were significantly statistic ($p<0.001$) showing the growth of the prospective elementary teachers in the area of preparing instruction and teaching children learning second languages and of diverse cultures. The specific areas of growth include creating grouping configurations that will support language and content objectives (67.4% increase), providing hands-on material and/or manipulatives for students to practice using new content knowledge (91.3% increase), clearly supporting language objectives through lesson delivery (82.6% increase), understanding how to emphasize and apply key vocabulary (78.2% increase), and understanding how to use a variety of techniques to scaffold content concepts for students who are learning second languages (e.g. modeling, visuals, hands- on activities, demonstrations, gestures, body language, and culture) (71.7% increase). The Mathematics Methods Course demonstrates the confirmed comfort level the prospective teachers felt about teaching children learning second languages when they left the course (Picard & Harding-DeKam, 2004).

Qualitative Research Results

The thematic unit assignment for prospective elementary teachers is a holistic look at the teaching and learning cycle in the classroom requiring prospective teachers to plan for at least five hours of instruction that is an integration of mathematics, social studies, and language arts or an integration of mathematics, science, and language arts. The thematic unit is then taught to elementary students where the prospective teachers are conducting their student teaching. After the unit has been taught, the prospective teacher reflects upon what learning took place for their students in both quantitative and qualitative evaluations, evaluates the level to which the Thematic Unit helped prepare students to meet the state standards, discusses changes they would make the next time they taught this unit, and demonstrates how they meet the Performance-Based Standards for Colorado Teachers while teaching this unit. Units on average are about fifty pages in length and show the specific planning, teaching, learning, and reflection that take place while teaching in the elementary classroom.

Table 2: Grade Levels Taught During Student Teaching

Grade Level	Frequency	Percent
Kindergarten	8	8.7%
First Grade	18	19.6%
Second Grade	16	17.4%
Third Grade	15	16.3%
Fourth Grade	11	12.0%
Fifth Grade	19	20.6%
Sixth Grade	5	5.4%

The qualitative diversity research takes place in the natural setting of the elementary classroom through the prospective teachers reflections on how they met the Performance-Based Standards for Colorado Teachers. The grade levels taught represent kindergarten through sixth grade as seen in Table 2. This qualitative inquiry (Schwandt, 2001) is aimed at understanding how prospective teachers applied the course learnings about teaching children learning second languages and diverse cultures through ethnomathematics in the elementary classroom during student teaching. The research methodology implements emergent and systematic ways of finding themes (Creswell, 2003; Merriam, 1998). The three Performance-Based Standards for Colorado Teachers focused on for this research of children learning second languages are as follows:

1. Standard 5 – Knowledge of Classroom and Instructional Management: The prospective teacher is knowledgeable about classroom practices in order to successfully manage time, communications, and record keeping procedures that will support and enhance student learning.

2. Standard 6 – Knowledge of Individualization of Instruction: The prospective teacher is responsive to the needs and experiences children bring to the classroom, including those based on culture, community, ethnicity, economics, linguistics, and innate learning abilities. The prospective teacher is knowledgeable about learning exceptionalities and conditions that affect the rate and extent of student learning, and is able to adapt instruction for all learners.

3. Standard 9—Knowledge of Diversity: The prospective teacher creates an inclusive learning environment for all students by drawing upon representations from students' experiences, knowledge, and beliefs.

Results

The qualitative data was collected from the Thematic Units and the themes from the data emerged. These themes were taken from the reflective essay section where the prospective elementary teachers demonstrated how they meet the Performance-Based Standards for Colorado Teachers in the areas of diversity and ethnomathematics. These prospective elementary teachers demonstrated how they met Standards 5, 6, and 9 throughout the teaching of their thematic unit and the following major themes emerged from the data:

1. Prospective teachers used the Thematic Unit lesson plans to develop ideas to help meet the needs of children learning second languages and of diverse cultures (98% of the prospective teachers).

2. Prospective teachers successfully applied differentiated instruction to their students in the classroom (64.2% of the prospective teachers).

3. Prospective teachers implemented different teaching techniques to build upon student's prior content knowledge and cultural knowledge (32.4%).

4. Prospective teachers demonstrated the respect for their students' culture when developing lesson plans, teaching the lessons, and assessing the students (23.1% of the prospective teachers).

5. Prospective teachers created a safe learning environment for all students in their classroom (16% of the prospective teachers).

6. Prospective teachers held different students accountable for content in different ways (8.7% of the prospective teachers).

7. Prospective teachers reflected upon how they would use student's cultural and foundational content knowledge if they were to teach this thematic unit again (7.2% of the prospective teachers).

This qualitative data demonstrates how these prospective elementary teachers support children learning second languages and of diverse cultures in the classroom during their student teaching experience through the use of ethnomathematics.

Conclusions

These prospective elementary teachers are building their own foundation of learning when it comes to teaching children learning second languages and diverse cultures. This is a beginning understanding of how the culture and history of a child play into the elementary classroom in the realm of ethnomathematics. This groundwork of foundational knowledge in ethnomathematics for these prospective teachers was established through the Mathematics Methods Course with the most effective and highest impact activities recorded by the

prospective teachers as being the Farsi video, human graph, and adding children learning second languages and diverse cultures into lesson plans. This growth in teaching children learning second languages and of diverse cultures is supported by the quantitative data of the prospective teachers' comfort level and the qualitative data of themes based on the reflection of the prospective teachers.

NOTES

These experiences allow prospective elementary teachers to see the need to teach all students through ethnomathematics. This need for ethnomathematics must be established in prospective elementary teachers before specifically looking at different ways of teaching mathematics culturally in the classroom to give the awareness of the need to know ethnomathematics and why it is important in the classroom. After these prospective teachers have this philosophical underpinning, then we move into learning the specifics of mathematics from other cultures.

References

Barta, J, & Shockey, T. (May 2006). The mathematical ways of an Aboriginal People: The Northern Ute. *Journal of Mathematics and Culture 1*(1), 79-89.

Center for Applied Linguistics (Producer). (2003). *Enhancing English language learning in elementary classrooms: Video*. [Motion picture]. (Available from Delta Systems Co., Inc., www.delta-systems.com).

Charney, R. S., Clayton, M. K., & Wood, C. (1997). *Guidelines for the responsive classroom*. Greenfield, MA: Northeast Foundation for Children.

Charney, R. S., Clayton, M. K. & Wood, C. (1999). *The responsive classroom advanced guidelines*. Greenfield, MA: Northeast Foundation for Children.

Colorado Department of Education: Performance-Based Standards for Colorado Teachers, adopted 1/13/00, http://www.cde.state.co.us/

Creswell, J. W. (2003). *Research design qualitative quantitative, and mixed methods approaches* (2nd ed.). Thousand Oaks, CA: Sage Publications.

Crotty, M. (1998). Constructionism: Making of meaning. In M. Crotty (Ed.), The foundations of social research (pp. 42-65). London: Sage Publications.

D'Ambrosio, U. (2002). Ethnomathematics: an overview. Proceedings of the II Congresso Internacional de Etnomatematica, Ouro Preto, Brazil.

Darder, A. (1991). *Culture and power in the classroom: A critical foundation for bicultural education.* Westport, CN: Bergin & Garvey an imprint of Greenwood Publishing Group, Inc.

Echevarria, J., Vogt, M. E., & Short, D. J. (2004). *Making content comprehensible for English learners: The SIOP model* (2nd ed.). Boston, MA: Pearson Education, Inc.

Franco, L. (2003). *Enhancing English language learning in elementary classrooms*. Greeley, CO: Grant funded by Colorado Teacher Preparation Enhancement Project through the Colorado Department of Education, Title II.

Howard, G. (2004). How we are white. *Teaching Tolerance, 26*, issue number, 50-52. Kochuk, N. (Feb. 2005). Clearing a path for teacher diversity. NEA Today, 17.

Lindberg, C. A (Ed.). (2002). *Oxford American college dictionary* New York: G. P. Putnam's Sons Publishers.

Lipka, J., Wildfeuer, S., Wahlberg, N., George, M., & Ezran, D. R. (2001). Elastic geometry and storyknifing: A Yup'ik Eskimo example. *Teaching Children Mathematics, 7*(6), 337-343.

Mendel, P. (1994). *Material World: A Global Family Portrait*. San Francisco, CA: Sierra Club Books.

Merriam, S.B. (1998). *Qualitative research and case study applications in education: Revised and expanded from case study research*. San Francisco, CA: Jossey-Bass Publishers.

Moschkovich, J. N. (1999). Understanding the needs of Latino students in reform- oriented mathematics classrooms. In W. G. Secada, *Changing the faces of mathematics perspectives on Latinos* (p. 5-11). Reston, VA: National Council of Teachers of Mathematics.

National Council of Teachers of Mathematics (NCTM). (2000). *Principle and standards for school mathematics*. Reston, VA: National Council of Teachers of Mathematics.

Picard, C. & Harding-DeKam, J. L. (2004, April). *How do you get preservice teachers to reach second-language math learners?* Paper presented at the meeting of the National Council of Teachers of Mathematics, Philadelphia, PA.

Reyhner, J., & Davison, D. M. (1992). Improving mathematics and science instruction for LEP middle school and high school students through language activities. *Third National Research Symposium on Limited English Proficient Student Issues: Focus on Middle and High School Issues*. Retrieved May 30, 2003, from source.

Schwandt, T. A. (2001). *Dictionary of qualitative inquiry* (2nd ed.). Thousand Oaks, CA: Sage Publications.

Staats, S. (May 2006). The case for rich contexts in ethnomathematics lessons. *The Journal of Mathematics and Culture*, V1(1), 39-56.

Zimmerman, G. (August 2006). It all adds up: Connecting math homework to the community math problems rotted in neighborhood life ease the homework burden for students, parents, and teachers. *The Responsive Classroom*, 18(3).

A Case Study of the Role of Ethnomathematics among Teacher Education Students from Highly Diverse Cultural Backgrounds

Ada Katsap
Mathematics Education
Department Kaye Academic College of Education
Beer-Sheva, Negev Region Capital of Israel 84536 Israel
08-640-2777 adak@013.net

Fredrick "Rick" Silverman
School of Teacher Education
College of Education
University of Northern Colorado
Greeley, Colorado 80639 USA
970-351-2760 flsilver@aol.com

Abstract

Mathematics prospective teachers (MPT), Jewish and Bedouin alike, who came together in the Ethnomathematics Program (EP) in the Kaye College of Education (Israel), explored, identified, and learned naturally occurring mathematics in the cultures of their own people. They developed projects based upon their explorations and shared them in class. The MPT participated in collaborative discussions and formed intercultural understanding within a purposefully conducive classroom atmosphere. In particular, they shared with one another classroom activities for their classmates, plans for transferring their accomplishments to school children, ethnomathematics activities derived directly from cultural experiences, as well as mathematics content and pedagogy. This case study investigated and documented the intentional integration of two people's cultures, traditions, and their ethnomathematics, on the one hand, with mathematics content, history, and pedagogy, on the other. The MPT learned ethnomathematics in the cultures of their own people and came to perceive mathematical concepts as a mix of two significances: one meaning stemming from their abstract courses in academic mathematics, a focus preeminent in western culture, and the second, stemming from comprehension of practical math concepts rooted for both Jews and Arabs in their cultures of the East. The following perception arose during class debates, discussions, and sharing of projects: that there is some common religious knowledge (e. g., that of the number seven in passages in the Torah and in the Koran) and common ritual (e. g., Jewish and Muslim religious women must cover their heads) in both lines of the descendants of Abraham. However, the big take-away point is that just an exposure to mathematics rooted in socio-cultural settings of their own people was not sufficient to have altered the standpoints of the class members toward the way in which mathematical concepts should be seen. Inquiry into situations where mathematical practice actually occurs was necessary.

Introduction

Globalization has brought about demographic changes and a plurality of cultures in various countries. The different aspects relating to multi-cultural society are becoming increasingly significant to education in general and mathematics education in particular. Discussing the face of mathematics from this point of view has the potential of understanding mathematics as an area whose face is diverse and can be interpreted in both formal and local aspects. The first interpretation as recognized in its formal definition, which can be conceptualized by the mathematician and is sufficiently absolute, perceptually speaking, is to be understood by learners thousands of miles apart – one, perhaps, in Australia and the other in Brazil or Africa. The second interpretation stems from expanding the world of the mathematical concepts by local aspects that are influenced by various explanations – culture, era and human history.

This realization requires us to hold a purposeful discussion about the added values in mathematics education. It is possible that Ethnomathematics, which is an expression of a Multicultural View of Mathematics Education and was launched as a new area at the 1985 NCTM by D'Ambrosio, helps us understand the role of mathematics in a multi-cultural society. The meaning of the term, Ethnomathematics, expressed by its socio-cultural aspect, includes within it the language, verbal lexicon, behavioral norms and symbols of a certain group, and in its mathematical aspect it includes such actions as "the corpus of knowledge derived from quantitative and qualitative practices, such as counting, weighing and measuring, sorting and classifying" (D'Ambrosio, 1985), encoding, sequencing, drawing a conclusion, exemplification, formation, location, explanation and reconstruction (Bishop, 1988). Regarding the actions that are identified with the mathematical aspect, there are a wide variety of existing implementations, enabling explanation of the processes that characterize practical mathematics in traditional work. The reference is to the accumulated store of mathematical knowledge that over time becomes "the oral law that is passed from father to son and from mother to daughter" (Katsap, 2004). For instance, a Bedouin woman observing her rituals will instruct her daughters how to embroider, and put into practice compositions requiring basic comprehensive knowledge in the field of Geometry, such as squares, triangles, and symmetry. However, often the people involved in math education have been unaware of the actual fact that "mathematical procedures, patterns, and structures are developed by human beings, living and working in societies," or ignoring them altogether (Shirley, 2006).

Ethnomathematics and Teachers Education: Foreword

Professional literature, including the research literature, on the subject of multiculturalism in mathematics, responds at two levels: pure mathematics and mathematics education. On the first level, the following aspects are discussed: the nature of multiculturalism in mathematics, identification of cultural components in algorithms and in mathematical models, and the disclosure of differences in the presentation of mathematical objects in different cultures. On the second level, multiculturalism in

NOTES

mathematics is examined through the prism of mathematics education in all its components. The subjects discussed on this level include such topics as explanation by researchers of the points of view of those engaged in mathematics education and of the students in different cultures, regarding their concepts of algorithms or mathematical models, besides the training of mathematics teachers.

The researchers concluded that "there is a great necessity for ethnomathematics-based programs to identify and seek teaching-learning practices directed towards pedagogical action"(Orey & Rosa, 2005). Hence, the direction is to search for a way to design, through collaboration with the teachers, a curriculum linked to the cultural needs of the learners (Masingila, 1996), to identify the preferable strategy of applying ethnomathematics in teaching (Bisher, 2001), and to use ethnomathematics "as an educational tool to help students understand what mathematics is about, and to help them make it part of their own knowledge" (Adam, 2004).

Shirley (2001), who conducted a survey of published projects, researches and essays, concludes that today the field of Ethnomathematics is central to the process of teaching and teaching methods, and that there is no reason to avoid its inclusion it in the curricula of mathematics teacher training. However, there is a gap between the ideal and the reality, and the current state of affairs is that in most cases mathematics departments of educational colleges are preparing teachers for mathematics discipline without any cultural context. This situation induces the future teachers to think of mathematics as universal and "acultural" (D'Ambrosio, 2001). Since this is the way things look at present, educators must think how to change the situation and help the teachers to establish cultural models in the classroom. This solution seems to be potential pedagogical guidance that takes into account the knowledge the students learn outside of the classroom. Likewise, educators need to broaden the understanding of mathematics prospective teachers (MPT) regarding the influence socio-cultural values have on the grasping of mathematical concepts on the one hand, and how mathematics is involved in the communal life- experience of different cultural groups, on the other.

In this article, we will focus on some characteristics of the second level.

The background of course population and EP conditions

The History of Mathematics seemed an appropriate course at Israel's Kaye College of Education for incorporating the Ethnomathematical Program (EP). It provided opportunities for aspiring teachers to examine the cultural contexts of mathematics concepts, to further their understanding regarding multicultural mathematics education, and to explore the presence of naturally occurring mathematics in their national, ethnic, and family activities and heritage. What tipped the scales in this case was the heterogeneity of the cultural background of the MPT population enrolled in the course. The demographic situation in the Negev region in the area of Beer-Sheva where Kaye College is located drives it to accept students in approximately equal numbers of Negev Arab-Bedouin (hereinafter: "Bedouins") (48%) and Jewish (52%) of the total of about 1,400 students, which was the number of students matriculating in all of the various academic tracks of the College during the year our research took place. Let us explain about the college population.

First, let us focus on the Negev Arab-Bedouins. This is a unique Palestinian community of Arabic-speaking desert nomads of the Middle East (Encyclopædia Britannica, Inc., 2007) that has lived in the Negev for centuries. In the 19th century the Bedouins began a transition to permanent settlements, which continued throughout the years. The process intensified from 1948, when the State of Israel was established.

According to Boteach (2005), approximately 160,000 Arab Bedouins have settled in the Negev region of Israel. Arabs in Israel make up about 22 percent of the total population, which was about 6,990,700 at the end of 2005. Arabs living in the Negev constitute about two percent of the total Arab population of the State of Israel (Israel Central Bureau of Statistics, 2006).

The Bedouin population can be divided into two groups: roughly half of them lives in approximately 45 localities unrecognized by the State of Israel, follow the traditional Arab- Bedouin way of life, and in most cases lives in tents and shacks. The other half of the Bedouin population is concentrated in seven government-planned towns: Rahat, Tel Sheva, Kseife, etc, planned as urban centers. The majority of Bedouins work in agriculture, construction, and traditional lifestyle works, among other fields. While formal education has been a part of Jewish tradition for centuries, the nomadic Bedouins began sending their children to educational settings only in 1948 under the Compulsory Education Law. Since then, the Bedouin of Israel have succeeded in reducing illiteracy from 95% to 25% (Ben-David, 2001). Most of Bedouins are religious Muslims that follow the Islamic calendar and culture. The cultural gap between Bedouin minority versus Jewish majority created an identity crisis for the Bedouins, when "56% of them regard themselves as Israelis and 61% as Palestinians" (Abu-Saad, Yonah and Kaplan, 2000). During the last thirty years a recognizable "process of integrating the Bedouin into Israeli society", has taken "place on two levels – the formal, by government policy; and the informal, by changing relationships with Israeli society in general and Jewish society in particular" (Ben- David, 2001). Moreover, Negev Bedouins have been changing/moving "from controllers of the desert region" to "fringe dwellers of a growing, modernizing Beer-Sheva city region (Yiftachel 2004). This can be seen in the range of Bedouins employed as clerks, lawyers, doctors, and teachers (at the time in small numbers).

Kaye College is the favorite Bedouin choice for higher education. The main reason for this is that the B.Ed. degree received with their graduation ensures them a job in the Negev area, close to their places of residence. Bedouins attend separate school systems than Jews, schools where the spoken language is Arabic. These school systems need competent Bedouin teachers to address the continual shortage. The Bedouin population attending the College from the Negev settlements still follows the traditional ways of life and preserves the ancient customs and traditions, as well as follows the religious imperatives of the Islam. In the college, these Bedouin students come in contact with their Jewish peers to perform mutual tasks during courses, but on other occasions they prefer their own company, where the language they use is Arabic, which is also their communication language with Arab lecturers. Otherwise, with the Jewish students, Jewish lecturers and the management of the college, the language they speak is Hebrew.

Let us focus now on religious and secular Jewish population. The words 'secular' and ‘religious’ for Jews refer to the type of culture or lifestyle. From this point of view, secular and religious Jews are two separate groups. The secular Jews accept the Bible in general as the central source of "Jewish collective memory", and celebrate the Jewish Festivals and traditional ceremonies such as Brit Milah (circumcision) and Bar-Mitzvah

NOTES

(ceremony of a Jewish boy who has reached the age of 13). Despite the distinctions between different Jewish religious groups, the faithful observing the classical rules of Judaism (mostly Orthodox) and those like Conservative and Reform Jews who are not obliged to adhere to the religious traditions as the Orthodox do, nevertheless share a common traditional knowledge of religious rituals, habits, and behaviors. These variations explain the differences in the traditional practical mathematical knowledge (calendar calculations, mathematical calculations required to observe the religious commandments, etc) possessed by the religious MPT and their secular Jewish peers when they enter the college.

In some tracks the Bedouin and Jewish sectors in Kaye College learn separately; for example: MPT of Early Childhood Education - Jewish sector learn in Hebrew, and for the Bedouin sector in Arabic. However, the language used in the Math Education Department, including for the The History of Mathematics, is Hebrew for all. Kaye College contributes to educational activities in the Negev region and conducts many projects in both the Jewish and the Bedouin sectors, such as the project in which several Bedouin teachers are teaching Arabic in the Jewish school. In several tracks and departments, computer forums and social activities, co- existence discussions take place, discussions which are in keeping with College policy. The History of Mathematics course is the only course of the Math Department in which intercultural conversation has been one of the pillars of the course syllabus since 1998. One of the course's goals is to direct the students to think cross-culturally in accordance with the integrated tone of Kaye College (Katsap, 2002).

Research in the course

The research that has accompanied the implementation of EP and was two-semesters in length, dealt with examining possible effects likely to accrue for mathematics teachers' education from learning and experiencing Ethnomathematics. As value added, the research attempted to discover the nature of the social-cultural-mathematical discussion that takes place in the college classroom during EP. 19 MPT (10 Jewish and 9 Bedouins) took active part in the EP by preparing and presenting study units combining ethnomathematical contents in school math. These students constituted the research population.

The research was carried out by qualitative study, collection and analysis of data following the method of Grounded Theory (Glaser & Strauss, 1967).

The objectives of the research were:

1. To observe the educational process undergone by MPT
2. To examine the effects of EP on MPT by documenting:
    - a. MPTs' attitudes toward interlacing of ethnomathematical content in school mathematics;
    - b. MPT assessment of the EP;
    - c. The feelings of MPT affected by exposure to ethnomathematics in the culture of the their own people and the surroundings in which they live;
    - d. The influence of learning environment during EP on the mathematical- social-cultural dialogue conducted in the college classroom;
    - e. The values emerging in the educational process of the MPT during EP.

The Grounded Theory (GT) research process was guided by the framework formed by two research models. The first model (proposed by Dick, 2005) is characterized by an algorithm built from six stages executed in the course of the research: 1) data collection; 2) note-taking; 3) coding; 4) memoing; 5) sorting; and 6) writing the report. The second model (proposed by Glasser & Strauss, 1967; Strauss & Corbin, 1998) supplements the first model through its focus and the great detail this model has with respect to the process of encoding. The combination of the two models enables optimum implementation of the GT method for construction of a theory regarding the phenomenon under investigation. Every sub-group of raw data passed through the first four stages according to Dick's model, with the primary coding being in the form of an open coding, close to the time of the gathering of the data. The method made it possible to receive memoranda that gradually built the theory on the basis of "the categories" and "the characteristics" that were coded, as well as the connection between them. It should be indicated also that in most cases, the names given to these categories were taken from the expressions of the teacher candidates themselves. The next phase of the data analysis was axial coding, described by Brown et al (2002) by means of four analytical processes: 1) continuing attribution of categories; 2) comparison between categories and data; 3) reducing the number of categories by determining their characteristics and fields; and 4) examination of the differences in the phenomena. With the continuation of this phase, the fifth and sixth stages were carried out, according to Dick (2005).

Integrating Ethnomathematics in school mathematics: contents and process

Two strand frameworks are identified with the field of Ethnomathematics: Geometry Patterns and Time Calculations were added to the syllabus of the course and served as mathematical content for the execution of the program. The four topics included in the first strand framework, "geometry patterns," were as follows:

- Learning Transformations: Researching Bedouin Dress Embroidery
- The Hemisphere Bowl Shape, The Circle, and The Jewish Kippah (small skullcap)
- Identifying Geometry Motifs in Bedouin Carpets
- Learning the Characteristics of the Equilateral Triangle and Investigating the Magen David (Star of David) Symbol of Jewish Culture

And in the second strand framework, "time calculations", such as the sequence of festivals, days, months and years, the following three topics were presented:

- Who Knows Eight? In the Jewish Festivals
- Mathematics in the Hebrew Calendar
- Mathematics in the Islamic Calendar

Diverse activities were utilized in order for the MPT to understand the complexity and application of these themes in each culture, with the goals being:

- Presenting the developmental, human, and cultural aspects of mathematics, and developing a comprehensive perception of mathematics through interlacing academic mathematics with cultural perspectives of mathematics;
- Understanding mathematical concepts and ideas through the prism of each other's culture - Jewish and Bedouin -

NOTES

and learning and solving mathematical problems as a part of mathematical knowledge developed over generations in both cultures;

- Adapting the ethnomathematical content of mathematics to the scholastic content in teaching mathematics in elementary through high school.

Readers might wonder what the meaning of the topics titles were, and how they can connect academic mathematics and ethnomathematics. The topic titles were intended to set academic mathematics, on the one hand, and practical mathematics identified by each culture, Jewish and Bedouins, on the other, in minimum words, in language understood both by school pupils and by aspiring teachers from each respective culture. For the MPT in the course the above topics titles were very clear.

Henceforth, I will focus on three of the four topics that were presented on the theme of geometry patterns: a) Learning Transformations: Researching Bedouin Dress Embroidery; b) The Hemisphere Bowl Shape, The Circle, and The Jewish *Kippah*; and c) Geometric Motifs in Bedouin Carpets.

Learning Transformations: Researching Bedouin Dress Embroidery

The young Bedouin woman has learned fixed geometric patterns of embroidery from her mother. She has learned from her mother the requisite knowledge and skills for embroidery and has cultivated an attitude that demands she carry out the process with perfection. Moreover, she invests much thought in rearranging the patterns she has learned in order to make an original new creation. The color in the embroidery symbolizes the woman's status in society: for example, red embroidery symbolizes a fertile woman; in contrast, blue embroidery means a widow.

The group that presented this topic in the class showed that geometric patterns serve as a basis. These patterns appear in the Arabic Bedouin traditional dress embroidery identified with the composition that characterizes the dress in the Negev. At the time of the topic presentation class members are asked to learn about the tradition of embroiders, and to provide details about how they managed to construct the composition requiring basic knowledge of squares, triangles, rhombs, and the most common forms in the Muslim world – "the symmetry: of half and quarter, of repeated repetition, diamond symmetry and symmetry of color" (Tal, 1990). One of the more interesting shapes appearing on the embroidery is the octagonal star, which, according to the aspiring teachers' explanations, depicts a Bedouin settlement with the Sheikh's house in the middle (the point at the hub of the shape, see Figure 1).

Figure 1: Octagonal star in the Bedouin embroideries

To summarize, it can be said that the background and the activities the aspiring teachers have learned worked well as a conclusion of the geometry topic quadrilaterals, and they were helpful for developing the skills needed for understanding different types of isometric transformations: reflection, rotation, translation, or glide reflection that appear in the mathematics curriculum of the school children. Following is an example of one of the activities implemented in the college class (includes pictures of embroideries to be used, Activity Sheet 1):

Activity Sheet 1

Learning Transformations: Researching Bedouin Dress Embroidery

Bedouin woman

Embroideries

Dress

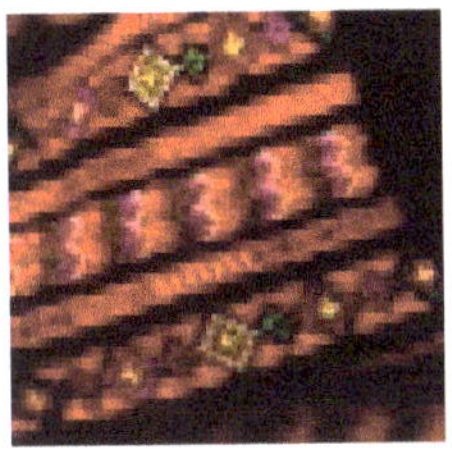

From website: http://www.snunit.k12.il/beduin/menu.html

Activity Instructions

1. Imagine and describe in your own words the feeling of the embroidering Bedouin woman.
2. Observe the designs appearing above, write a list of the symmetry types identifiable in the embroideries and sketch figures having the translational symmetry.
3. Find basic geometric patterns in the embroideries and copy them to the page. Form a new composition for the embroidery by using these figures.
4. Describe your feeling during the creation of embroidery "of your own".

The Hemisphere Bowl Shape, The Circle, and The Jewish Kippah

From 1500 CE onwards the tradition of covering the head became an obligatory practice for Jewish religious men (Raskin, 1990). Over the past three centuries, in some European communities, the hat has evolved into the smaller skullcap – **kippah** in Hebrew. In Israel, wearing a kippah also has a social significance. Religious men don't appear in public without a kippah or some other headdress, in contrast to Secular men who wear a kippah while attending religious ceremonies. Moreover, the color of the kippah indicates affiliation with a particular religious group or stream. Thus, a black cloth kippah indicates that the wearer belongs to the orthodox stream of Judaism; other believers wear colored kippah. The kippah differs from any other hat, but it is a common convention that it must cover the crown of head, it is a three- dimensional, and it ends in a size that equal to a circle circumference of the head.

A group of MPT presented the subject of the kippah in the college class. They discussed the kippah wearer's identification with socio-cultural and religious affiliations and also the symmetry in the kippah's decoration, size, and shape. The class members were required to identify types of symmetry and to examine the circle that is the basis of the kippah. Note that the circle has fascinated the Jews since biblical times, with verses depicting the calculation of π appearing in the Bible (see more about math in Jewish writings in Lesser (2006)). One of the activities required students to gather information a kippah maker would need in order to make a kippah for a class member. To determine the circumference of the kippah, they measured around the student's head as far down his head as he would wear the kippah. Suppose the kippah maker is using a knitting process similar to coiling a clay rope in making a pot, but from the lip to the bottom. To begin, the kippah maker needs to know the radius of the kippah at its base. Having the necessary circumference, the students calculated the radius by manipulating the formula $C = 2\pi r$. Actual kippah makers start knitting with a flat circular form, perhaps a cloth disc of the required circumference. As the process moves along, the kippah maker uses fewer and fewer "eyes" for each successive level of the kippah until he finally arrives at the pinnacle, having created the kippah in hemispheric shape as shown in Activity Sheet 2. Jennifer Tocker (2005) presents details of the procedure for knitting a kippah (See http://www.geocities.com/jbtocker/patterns/basicknitkippah.pdf).

Another activity examined different shapes of the kippah, inter alia, the Bukharan kippah that was influenced by the dominant Muslim culture where they lived in Uzbekistan. This kippah looks like a cylinder open at the base, flat on top like a round cake pan, not very deep, and sized with a circumference to fit one's head, a shape that contrasts with the more common hemisphere, like a round bowl. An example of these kippah shapes and the corresponding activities implemented in the college class appears on Activity Sheet 2.

NOTES

Activity Sheet 2

The Hemisphere Bowl Shape, Circle, and The Jewish Kippah

Bucharian Kippah

Crochet Kippah

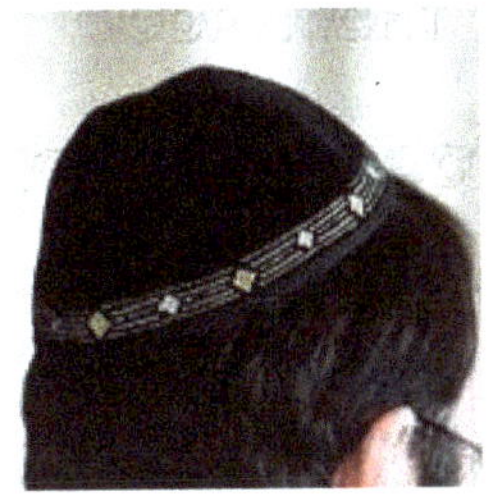

A man wear a Kippah

Activity Instructions

1. Think about the Kippah-wearing custom in Jewish Culture and the events in which the males in the family wore Kippahs.
2. Observe the three Kippah's above and describe the shape, size and the decoration of the Kippah. Write a list of the symmetry types identifiable in each Kippah.
3. Select the basic geometric shapes from one or more Kippah's, copy them to the page and use them to make new decoration of a Kippah.
4. Describe your feeling while draw the decoration of a Kippah of "your own".

Identifying Geometric Motifs in Bedouin Carpets

Carpets serve as a supplement of Bedouins way of life, especially for those observing the traditional Arab-Bedouin community lifestyle and live in tents and shacks. A Bedouin tent is customarily divided into two sections by a woven carpets, one section reserved for the men and for the reception of most guests, and used as a gathering room for Bedouin men. The other section, in which the women cook and receive female guests, is the place of the women.

Carpets for tent

Both sections are padded with hand woven carpets to help guests stay warm during frosty seasons and to decorate their sitting places. The colors and geometric figures in the carpets are designated for special purposes; for example, geometric multi-color composition or red color background carpet is used only for exclusive persons in contrast with strips in white-leaden color background indicating that that carpet is for the family.

The group presenting the topic of "Identifying Geometric Motifs in Bedouin Carpets" made a systematic investigation of different carpets found in the area. They brought to the class stories (see below) from women elders in their families. Group members reported about instances of choosing figures for woven carpets and showed photographs of original carpets of their own people.

Among the activities the class members were asked to perform during the presentation of the topic "Identifying Geometric Motifs in Bedouin Carpets" was the completion of the left side of the carpet image below:

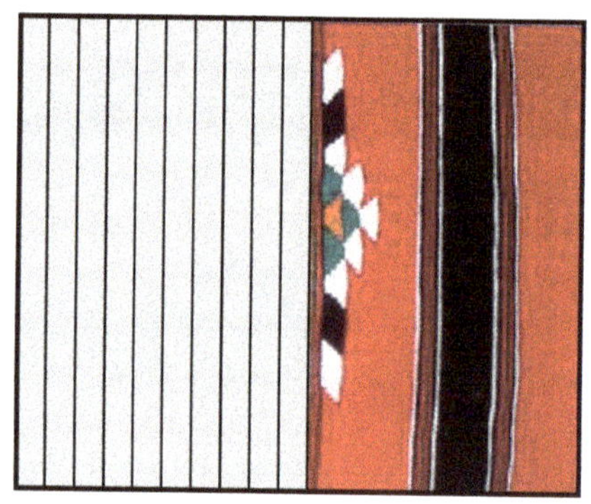

Presenters led their classmates to complete the symmetrical design in the image of the whole carpet by modeling the process, which the mother of one of the students elaborated, that Bedouin women customarily use. This woman had woven the carpet (see Activity Sheet 3, the carpet on the right), which was among the carpets that presenters had brought to class. To complete this classroom activity, class members made a drawing of vertical lines left of the midline, as shown in the image above. Next, the MPT measured the width of

NOTES

each part, starting at the right-hand edge of the carpet with the red swath and proceeding successively to the thin white strip, then to the maroon, and so on to the midline of the image. The most intricate element in the design, of course, is in the center and would seem to require careful measuring and counting of stitches, though the presenters did not state the details. In this activity, class members filled in the left half of the image, as a Bedouin woman would do in weaving a real carpet of this type, moving leftward from the midline all the way to the left edge, in essence symmetrically reflecting over the midline. And so by enacting this process on the carpet image, the MPT simulated the completion of a carpet in a way that Bedouin women use to weave actual carpets that feature reflection symmetry in method of creation and also in the design itself, as shown in Activity Sheet 3. In actual practice of weaving a carpet, of course, the Bedouin women move freely and smoothly from the right to left. The design they have in mind guides their work. Once the right-hand half of the carpet is complete, the midline of the carpet acts as a line of reflection as they proceed to weave toward the left-hand edge, matching each new section in size and color to the corresponding one they have already woven to the right of the midline.

Activity Sheet 3

Identifying Geometric Motifs in Bedouin Carpets

Two original carpets from the house of one member of the group that prepared the study unit

Activity Instructions

1. Observe the designs appearing above and write a list of the symmetry types identifiable in the carpets.
2. Find basic geometric figures on the carpets, copy them to the page and explore their features.
3. What kinds of triangles have been woven on the carpets?
4. Describe your feeling toward the Bedouins carpets.

The conversation conducted inside the college class about the geometric figures and the kinds of geometric transformations appeared in the carpets built a bridge between academic mathematics and outside classroom life, and it permitted mathematical-socio-cultural dialogue, thus facilitating better understanding of several transformations. Photos of two carpets shown in the class and the questions the class members were asked during the activity appear in Activity Sheet 3.

After all presentations, the MPT used ethnomathematical ideas from dress embroidery, knitting the kippah, woven carpets, etc, to design new activities for school math in the fields of Geometry, Transformations, and Measurement.

Intercultural and multiplex educational dialogue in the course

One of the salient characteristics of the educational process that took place in the course was an intercultural dialogue. In any other study framework in Israel, it would have been unusual to emphasize national and religious differences in the class. Let us describe a case.

In a lesson on the concept of time presented by a Jewish group, the discussion ranged from the philosophical understanding of the concept to mathematical insight. When a Bedouin student presented his own position on the matter without referring to his culture's attitude, one of the presenters immediately said, "Why are you presenting this example? It would be better if you told us how you view time in the desert. What does your sheikh (literally, "elder") think about the essence of time?" (See below regarding mathematics and Islamic calendar). In any other class, a political argument could have easily ensued. However, in this case, the Bedouin prospective teacher began to relate how Bedouin religious leaders saw the concept of time. Everyone listened with interest, as moments before the class had discussed the Western attitude of "time is money." Two opinions from the two respective cultures represented their emotional perceptions of time and their cognitive mathematical perceptions of time.

Dialogue opened up the possibility of a renewed perception or deepening of understanding of the mathematical concepts – not only from the purely rational side but also from the affective side, with the latter being based on feelings rooted in the traditions and cultures of the MPTs. The clash of perspectives common outside the classroom was introduced into the classroom during the study and discussion of the mathematical and scientific notion of time. Incorporating an ethnomathematical perspective made this interchange possible, indeed encouraged it, and humanized and augmented the study academic mathematics.

How the learning and teaching environment affects the ethnomathematics experience of the MPT

The educational process in the course was comprised from three phases: the first phase was characterized by groups each comprising two or three MPT choosing a topic of mutual interest, collecting relevant material, and studying and learning collaboratively through investigation of the topic. The second phase was typified by students' preparations for their presentations. So in their group work they attended to pedagogical considerations for activities their class mates would do, activities school students would do, and for subsequent class discussion and reflection with their peers. And the third phase, classroom presentation of the topic, was, according to all MPT, an "exciting learning experience."

The first stage was each group's search for sources of information for their topic . From what the MPT indicated, in this search they turned to the following resources: the

Internet Web literature, encyclopedias, books that focused on religious or cultural source materials, such as the Jewish Bible (Tanakh in Hebrew; Old Testament), the Talmud and other books of interpretation, among the Jews, and the Koran and the Hadith among the Bedouin. In these sources of information the MPT found descriptions regarding handcrafts, such as knitting, weaving, embroidery and creative works, containing integrated patterns of certain cultural character that could be clearly identified. Regarding mathematics concepts and processes, they formed explanations regarding counting and ordering, such as calculations for calendar arrangement and fixing the dates on which the religious festivals fall. The MPT claimed that the information sources they prized most highly in the authentic material were in fact the human sources: 1) an authority in the particular field; 2) a religious personality; 3) an expert or professional; or 4) a family member (mother, father or family elders). All Jewish and Bedouin sectors participants emphasized that most of the people to whom they turned did not know at all how to express the connection between mathematics and their work, but they had mastered the authentic field itself. These reports are reminiscent of a marvelous story that Fasheh (2002), who holds a Ph. D. in mathematics, told about his "illiterate mother" who was a seamstress, and how she encountered math in her daily work. He was amazed to discover that his "mother was illiterate in relation to his type of knowledge [i. e., academic mathematics], but he was illiterate in terms of her type of understanding and knowledge [i. e., ethnomathematics]."

Some of the statements made by the MPT regarding the sources of information are cited below. The first story relates to an information source, "Mathematics in the Islamic Calendar", by a Bedouin group of MPT. Members of this group became interested in the hour-by-hour schedule of the daily prayers that exists in the Bedouin tradition, and for this reason they went to the Bedouin settlement of Kseife (located in Beer Sheva area) in order to meet with an "information source." The story is presented below:

"I spoke with one of the elders of Kseife. I asked him to tell me how they know to set the hours of prayer without using a clock. This was his answer:"

> *We pray five prayers a day: morning, noontime, afternoon, evening and night, and the time for prayer is determined by the sun and the moon. Morning prayer begins when the dawn rises and a red horizon appears in the sky. Noontime prayer arrives when a man's shadow is equal to his height. The time for afternoon prayer comes when a man's shadow is longer than his height. Evening prayer arrives when both the sky and the sun look red in the west. And finally, the time for nighttime prayer comes when the red color in the sky can no longer be seen.*

Here is another story told by a member of the group that studied Identifying Geometry Motifs in Bedouin Carpets:

> *My group chose to present a study unit on 'Geometric Motifs in Bedouin Carpets.' I remembered that my home has carpets with different geometric forms. Then I went to my mother and to the elderly women in the family, and I asked them about the forms in the carpets. They responded*

with surprise bordering on refusal: What? You're interested in women's handicrafts? I explained to them that this was an assignment I took upon myself as part of the preparation of my paper for college, and that the subject of forms is connected to geometry, an area that I am studying in the framework of my degree studies in mathematics teaching. Their immediate reaction was that I had come to mock their work, and only when I reassured them and told about the geometric forms appearing in the carpets that they had made, such as triangles and squares, that they complied with my request, and began to tell me how the work of carpet-weaving is done. In the end, they added that they had never thought that what they were doing was connected to mathematics.

Both stories share being collected from an "in vivo" source. An ethnomathematical perspective enabled the MPT to identify and study the mathematics intrinsic in each of the two accounts, just as Fasheh's ethnomathematical perspective, acquired after his formal doctoral studies in mathematics, enabled him to discern the active geometric knowledge his mother applied in making clothing from scraps of cloth. Tribe elders told the teachers that most stories about Bedouin existence, tradition, and culture are passed orally from generation to generation by father to son and from mother to daughter.

Contrary to their Bedouin colleagues, the Jewish MPT, secular and religious alike, encountered an abundance of written material in the Jewish sources, providing very precise interpretations of how to conduct mathematical calculations to ensure that those coming to fulfill the mitzvahs (Jewish religious laws) shall indeed meet the theoretical and practical demands of the task. The texts described, explained, and directed the actions by using components linked to mathematical knowledge in the culture, religion, and tradition of the Jewish people.

The information was revealed, as mentioned above, in books which provided a lot of information on the way days are counted to calculate the date of a certain holiday, how calendars are constructed, measurements taken, or Jewish symbols are built: Hanukkah lamp, Star of David, etc. They too, just like their Bedouin colleagues, did not miss the opportunity to obtain the information from a religious authority, which usually was, quite naturally, the local rabbi:

> *I began to prepare the work on the subject of "Mathematics in the Hebrew Calendar" by collecting information from books, and then asked my family members to check and add more information on the subject, and finally visited a rabbi (religious teacher or leader), asking him to explain and help me search for verses pertaining to our subject in the Torah.*

Let us pass to the third stage because the second stage, which was preparation of the lesson array, was not characterized by any special or distinguishing components among the topics or the domains of mathematics. Implementing their projects in class was, indeed, an "exciting experience" in the opinion of the MPT; they described the process of class activation as the most significant for them, thanks to the fertile mathematical-social-cultural discussion that developed during the lesson. They have also mentioned that during the first stages of investigating their chosen subject, they anticipated that the 'simplicity' characterizing the practical mathematics in the studied topics would fail to elicit much respect from their class mates for the mathematics inherent in the tasks associated with their topics. However, the questions asked during the following class activities and discussion indicated that ethnomathematics content did, in fact, arouse curiosity and create enrichment.

One Bedouin MPT expressed appreciation and was impressed by the positive feedback her group received from Jewish MPT in the college class after the presentation:

> *I was very happy during the presentation... because the material was linked to our culture, and everyone is proud when other people learn and become acquainted with their own culture... I've noticed that during the presentation of the study unit that the participating teachers showed considerable interest and desire to learn about our culture... Our group prepared many examples of embroidery and in one activity Jewish teachers were asked to describe orally the transformation types revealed by this example. At the end of the lesson the teachers described their sensations during this task. The satisfaction they've felt was evident.*

The words in "our culture" bore two meanings. One, creation of linkage between the MPT's culture and mathematical concepts belonging to academic mathematics, and second, a sense of satisfaction with teaching math by using examples relaying the story of the MPT's own people. It can be said that the ethnomathematical experience greatly contributed to the sensation that ethnomathematics can serve the teachers for modeling, which cannot be provided by books written in the language and symbolism of formal mathematics and usually associated with mathematics content courses.

The profile of educational process undergone by the MPT

Essential components of the MPT's ethnomathematical education were examined through 5 categories:

1. The attitudes concerning humanist-social and cultural aspects of ethnomathematics;
2. The potential utilizations of ethnomathematical examples;
3. A feeling aroused by exposure to mathematics in the culture of their own people and surroundings in which they live;
4. The contribution of EP experience to intentions to promote mathematical ideas in cultural perspective;
5. The values ethnomathematics teachers' education calls forth.

Concerning the first two categories, about 90% of the MPT agreed on the importance of linking the math teacher's preparation with cultural perspectives of mathematics and with the benefit in integrating "ethnomathematical" subjects in Math teaching. A typical statement supporting this inference is the following:

> *Participation in the program helped me to understand that mathematics is in my roots and culture, that it served and still serves my people in preservation of a time-old tradition... Mathematics has an extra value... Ethnomathematics answers exactly the questions asked by so many students: "why should*

we learn mathematics?" This area offers a ladder we can use to climb from mathematics in our environment to the universal math... I see this as the main importance of ethnomathematical education.

The citation above and similar expressions point toward MPT's appreciation of ethnomathematical education and to their comprehension that mathematics rooted in the culture and environment in which they live and teach need not be detached from the academic mathematics curriculum they are expected to implement with children.

Ethnomathematical education emphasizes the connection between mathematics and other areas. The MPT acquire knowledge in interdisciplinary teaching when they appeal to history, sacred texts, and holy books; geography, and other disciplines; and so on. These connections create an "uninterrupted continuum" in their general knowledge. Answers supplied by the MPT were used to compile a list of useful characteristics provided by integrating ethnomathematical contents in the school mathematics intended for teaching. The most popular characteristics, those mentioned in various forms by more than 80% of the MPT, were these four: a) rich learning of the lesson's topic, extending the pupils' "world of knowledge"; b) changing the attitude towards mathematics, which is not only equations and laws, but also a part of national, ethnic, and family culture; c) creation of inter-disciplinary integration; and d) breaking the routine frames of the Math lesson. These data show that the MPT assigned most of the benefits to the fact that integration of ethnomathematics content in school mathematics can enhance the Math lessons and quality of the pupils' learning.

It should be noted that the explicit use of the term "ethnomathematics" in the reactions of the MPT was scarce. The most common alternative for this term was "mathematics in people's culture" or "mathematics in daily life". Likewise, the importance of ethnomathematics was always appraised for its contribution or potential to help cope with the difficulties in learning academic mathematics.

Concerning category three, the feeling aroused by learning mathematics in the context of their own cultures, the Jewish and Muslim MPT were delighted with the experience in mathematics in their own cultures, which utilized practical mathematics present for many generations. The level of marveling expressed by the MPT stemmed from the fact that the acquaintance with their own culture occurred at the "junction" named mathematics. The application of the word "our" was motivated by pride, and the appreciation showed by one culture in the class towards the practical mathematics of the other. The MPT reactions reflected three major emotions:

- **Surprise** - from the ability to arouse interest through Ethnomathematics;
- **Enjoyment** - learning and teaching ethnomathematic in the college classroom;
- **Pride** - representing their own people and culture.

Representing the fourth category, the contribution of experience with ethnomathematics during the EP to the MPT's intentions to promote mathematical ideas in cultural perspective was expressed in the course of the self-learning, and by observing the presentation of the other study units in the class. Thus, for example, the MPT discovered the mathematical knowledge required to achieve a sophisticated use of symmetry in carpet weaving.

Through their experiences with ethnomathematics and academic mathematics, the MPT showed that they refined and extended their understanding of mathematics content and pedagogy. For example, before the EP, the MPT understood symmetry as an abstract math concept, and they were much less likely to make connections between the abstract embodiments and everyday appearances of symmetry, but after their experiences with mathematics embedded in cultural context, they refined and extended their comprehension of this concept through actions of their families and elders: symmetry in evidence in Bedouin women's dress embroidery, their carpet weaving, and also in the designing and making of Jewish kippahs. Every case presented sophisticated application of principles of symmetry, not in the abstract, but in the precise enactments of the processes to create the intended products. These processes are ways of knowing and understanding in action, not in the abstract. Anticipating their own future students, the MPT also thought that children in schools could encounter these concepts through their own actions and their use of their own language to describe the processes in which they would be engaged.

Another example is knowledge of the rules for compilation of annual calendars in both cultures, Jewish and Bedouin. The extant facts pertaining to mathematical knowledge hidden, so to speak, in the life experiences of the people, spurred the desire of the MPT to transfer to their future pupils the message that Math "is here", all around them, part of their culture. The MPT expressed their hope that the same feeling of enthusiasm that they, themselves, experienced would also be experienced by their future pupils when exposed to mathematics in their culture and in the culture of others. Cultural discourse complemented the mathematical discourse and filled the classroom with a lot of friendliness, mutual respect, and marveling at the ethnomathematics inherent in both cultures.

Table 1: Distribution of values concerning Teachers' Enthnomathematical Education

Mathematical-Studied Values	Mathematical-Pedagogic Values	Social-Cultural Values
a) Discovering mathematical concepts in symbols characterizing the teacher's own culture;	a) The teacher as indicator of ethnomathematics' ability to generate pupils' Interest;	a) Contribution to the teacher's self-confidence and openness;
b) Refining and extending prospective teachers' understanding of mathematical concepts and their own teaching;	b) Improving the teacher's ability to conduct discussion with the class due to the didactis provided by the field of ethnomathematics;	b) Strengthening the bond between the teacher and the legacy of his people;
c) Holistic viewing of the mathematical concepts and mathematics around us;	c) Fostering teacher's creativity by developing new teaching materials;	c) Extending the knowledge about tradition and culture of the teacher's own people;
d) Acquaintance with popular culture traditions and customs through studying of mathematical concepts.	d) Acquiring know-how for adjustment and integration of ethnomathematical subjects in school mathematical contents.	d) Creation of a social climate of mutual respect originating in exposure to mathematical knowledge existing in both cultures: Jewish and Arabic.

NOTES

Original quotations of the MPT served to determine values formed during the educational process that they experienced (fifth category). The model proposed by Bishop (1996), which divides values into three groups - mathematical, mathematical-pedagogical, and general educational - was adjusted to the array of values extracted from the findings and grouped accordingly. MPT expressed these values (see Table 1 above) with respect to their experiences in the process of education for integration of ethnomathematical content and school mathematics.

In general we can say that the findings of the research point toward two trends: the first one is characterized by 'systemic betterment' by: a) improvement in teaching techniques; b) enriching the program of MPT education; c) creation of a college classroom environment conducive to integrating ethnomathematical and multicultural features with mathematics content and pedagogy, as one MPT puts it:

> *For the first time the mathematical concepts were taught using symbols and attributes characteristic of our culture, holidays and calendar - this was an opportunity to feel proud of mathematics present in our culture." (See Table 1: Mathematical-Studied Values, a).*

The second trend is revealed in improvement of qualities of the MPT by the achievement of the following goals: a) improvement of learning and teaching skills, boosting their interest as learners of mathematics, and deepening their own mathematical content knowledge; b) broadening of the MPT general knowledge, c) creating a linkage between the MPT with his people and their roots. Quoting one of the MPT:

> *A subject that I did not previously like, such as the theory of different symmetries, I saw suddenly in a new way in this course, after it was connected to the culture of my people. It was easy to understand and I now like it, and therefore I definitely think that the process of exposing the teacher to the cultural aspects of the mathematical ideas is one that contributes to the training of the teacher." (See Table 1: Social-Cultural Values, d).*

Conclusion

The status of the teacher in the intersection between society and pedagogy is an issue for discussion with respect to mathematics teacher education in general and with respect to Ethnomathematics as a constituent strand in mathematics teachers teacher education, in particular. In an effort to embrace and celebrate multicultural phenomena in the mathematics curriculum process, the Ethnomathematical Program was designed as a MPT professional development project integrating etnomathematics, academic mathematics for aspiring teachers, and explorations into MPT's own cultures. Orey & Rosa (2006) encouraged the initiative and claimed that "ethnomathematics may come to exert some influence on mathematics education if the researchers would develop rigorous activities that are linked to standards and curriculum that make mathematics a living subject for both teachers and students." Focusing on ethnomathematics as a part of the History of Mathematics course permitted the MPT in the course to grasp the history of practical mathematics in the culture of their own people and to conceive the perception of mathematical concepts as a mix of two significances: one meaning stemming from their abstract interpretation in academic mathematics that is preeminent in western culture, and the second, stemming from comprehension of practical math concepts in a culture, rooted for both Jews and Arabs in the East. This perception came into sight during class debates, discussions, and sharing of projects: that there is some common religious knowledge (e.g., import of number seven in passages in the Torah and in the Koran) and common ritual (e. g., Jewish and

Muslim religious women must cover their heads) in both lines of the descendants of Abraham.

However, the big take-away point is that just an exposure to mathematics rooted in the socio-cultural setting of their own people apparently would not have altered the standpoints of the class members toward the way in which they perceived mathematical concepts. More than that was required. In particular, the personal inquiries about the concepts in situations where mathematical practice actually occurs, was necessary. It was, in fact, the MPT's construction of the study units and presentation therein of the mathematical concepts, previously encountered by MPT strictly in the abstract, that stimulated these aspiring teachers to plan to integrate ethnomathematics into their mathematics teaching in their future school assignments. The Ethnomathematics Program, based on modeling and the persuasion of practical examples, that apparently was instrumental to the outcome.

In a similar vein, Shirley (2006) noted such adaptation and attributed the outcome in his example to the class discussion of these issues. As with Shirley's (2006) work, this research indicated that discussions became important to teachers and motivated their further pursuit and understanding of mathematics content and pedagogy. Indeed, the way these MPT encountered culturally embedded mathematics created a learning environment that encouraged and nurtured interdisciplinary contexts and personal connections with one another and to their lives outside the classroom walls. This approach cultivated an acceptance that academic mathematics is only one side of the mathematics coin, the flip side being the one on which mathematics shows itself as a humanistic and socially embedded discipline. This development suggests that responsibility on the shoulders of professional educators lies in caring for creation of a learning atmosphere where the two facets of mathematics shall walk side by side. Mathematicians and mathematics teacher educators must accept, as priority, the pursuit of a civilization with dignity for all, and look into the relations between two universals: 1). mathematics as a universal mode of thought and 2). survival with dignity as the most universal problem facing mankind (D'Ambrosio, 2006 and 2007). Incorporating Ethnomathematics into mathematics teacher education would seem to advance this goal, given the personal respect and appreciation for one another and their respective heritages that developed across the cultural boundary between the two often highly contentious groups of which the MPT are members. We recommend future research to investigate transfer in the schools by MPT who have experienced the Ethnomathematics Program.

Our mission as educators is to lead the MPT's experiences so that they are likely to link classroom encounters with mathematics content and pedagogy, on the one hand, to ethnomathematical activities, on the other, that have definite affective and cognitive meaning for the learners in their national, ethnic, or family cultural realms. This research has demonstrated that Ethnomathematics in mathematics teacher education can influence the MPT's professional development so that the MPT incorporate humanistic and social aspects of mathematics as they move from the role of teacher education students to that of teachers of children. Ethnomathematics activities were apparently responsible for creating a classroom climate in which appreciation and respect for different cultures and traditions emerged and flourished. One would hope that such inclusion of Ethnomathematics in teacher education will advance the ideals of peace and a better world.

References

Abu-Saad, I., Yonah, Y., and Kaplan, A. (2000). Identity and political stability in an ethnically diverse state: a study of Bedouin Arab youth in Israel. Social Identities 6(1), 49-61.

Adam, S. (2004). Ethnomathematical ideas in the curriculum. Mathematics Education Research Journal, 16(2), 49-68.

Ben-David, Y. (2001). The Bedouin in Israel. Israeli Foreign Ministry, Jewish Virtual Library, AICE.

Bisher, H. B. (2001). The effect of Bedouin ethnomathematics on achievements in primary stage mathematics courses and behavior in daily life situations. Unpublished M.A., University Ain Shams, Cairo, Egypt.

Bishop, A. J. (1996). How should mathematics teaching in modern societies relate to cultural values. Some preliminary questions. In D. T. Nguyen, T. L. Pham, C. Comiti, D. R. Green, E. Southwell, & J. Izard (Eds.), Proceedings of the seventh South East Asian Conference on Mathematics Education (SEACME 7), (pp. 19-24). Hanoi: Vietnamese Mathematical Society.

Bishop, A.J. (1988). Mathematical enculturation: A cultural perspective on mathematics education. Dordrecht: Kluwer Academic Publishers.

Boteach, E. (2005). The indigenous bedouin of the Negev Desert in Israel. Paper submitted to the UN Working Group on Indigenous Populations. Negev Coexistence Forum, July 2005. Brown, S.C., Stevens, R.A., Troiano, P.F. & Schneider, M.K. (2002). Exploring complex phenomena: grounded theory in student affairs research. Journal of College Student Development, 43(2), 1-11.

Israel Central Bureau of Statistics (2006). The Bureau. Tel Aviv.

D'Ambrosio, U. (2007). Peace, social justice and ethnomathematics. In B. Sriraman (Ed.), International perspectives on social justice in mathematics education (pp. 25-34). Monograph of The Montana Enthusiast. Montana: The University of Montana and The Montana Council of Teachers of Mathematics.

D'Ambrosio, U. (2006). Ethnomathematics. Sense Publishers. ISBN: 9077874763

D'Ambrosio, U. (2001). What is ethnomathematics, and how can it help children in schools? Teaching Children Mathematics, 7(6), 308-310.

D'Ambrosio, U. (1985). Ethnomathematics and its place in the history and pedagogy of mathematics. For the Learning of Mathematics, 5(1), 44-48.

Dick, B. (2005). Data-driven action research. Retrieved March 15, 2008, from http://www.scu.edu.au/schools/gcm/ar/arp/datadriv.html .

Doll, W.E. (1993). A post-modern perspective on curriculum. New York: Teachers College Press.

Encyclopædia Britannica, (2007). The Online Encyclopedia. Retrieved March 14, 2008, from http://www.britannica.com .

Fasheh, M. (1983). Mathematics, culture, and authority. For the Learning of Mathematics, 3(2), 2-8.

Fasheh, M. (2002, September). How to eradicate illiteracy without eradicating illiterates? Paper presented for the UNESCO round table on "Literacy As Freedom". The International Literacy Day, Paris.

Glaser, B. & Strauss, A. (1967). The discovery of grounded theory: strategies for qualitative research. Chicago: Adline.

Hernandez, A. (2004). My Kingdom for a Crown: An Around-the-World History of the Skullcap and its Modern Socio-Political Significance. Retrieved March 14, 2008, from http://www.hatsuk.com/skullcaps.pdf.

Katsap, A. (2006). Ethnomathematics: The welcoming face of mathematics arises from various world cultures (In Hebrew). Mispar Hazak 2000. Journal of Mathematics Teaching in the Elementary School, pp. 30-35.

Katsap, A. (2004). One mathematics, two cultures, and a history of mathematics college course as a starting point for exploring ethnomathematics. In Proceedings of the HPM 2004, History and Pedagogy of Mathematics (ICME 10) Conference (pp. 262-270). Copenhagen, Denmark, satellite meeting in Uppsala, Sweden. Copenhagen: Roskilde University.

Katsap, A. (2003, September). Empowerment of teachers as learners: active learning in the college mathematics classroom. In Rogerson, Alan (Ed) Proceedings of the International Conference of the Mathematics Education into the 21st Century Project: "The Decidable and the Undecidable in Mathematics Education." Brno, Czech Republic: Brno University of Technology, pp. 140-144. Retrieved March 24, 2008, from http://math.unipa.it/~grim/21_project/21_brno_03.htm .

Katsap, A. (2002, June). Humanizing mathematics: the humanistic impression in the course for mathematics teaching, Humanistic Mathematics Network Journal, Issue # 26, (pp. 12-19). Claremont, CA. ERIC_no: EJ654612.

Lesser, L. (2006). Book of numbers: exploring Jewish mathematics and culture at a Jewish high school, Journal of Mathematics and Culture, 1 (1). (pp. 8-31).

Masingila, J.O. (1996). Report on ethnomathematics research, ISGEm Newsletter, 10(2). Originally published in the International Study Group on Ethnomathematics (ISGEm) Newsletter, 10(2), June 1995.

Orey, D., Rosa, M. (2006, November). Cultural assertions and challenges towards pedagogical action of an etnnomathematics program. For the Learning of Mathematics 26(3), FLM Publishing Association, Edmonton, Alberta, Canada, pp.

Orey, D., Rosa, M. (2005). Current Trends in Ethnomathematics as a Program: Towards Pedagogical Action (Tendências atuais da Etnomatemática como um Programa: Rumo à Ação Pedagógica). Revista ZETETIKÉ: CEMPEM – Faculdade de Educação – UNICAMP. Campinas, Brasil. English Translation: Retrieved March 14, 2008, from http://www.csus.edu/indiv/o/oreyd/papers/etnotendenciasatuais.translation.doc.

Raskin, J., 1990, A History of the Yarmulke. Retrieved March 14, 2008, from http://jef.raskincenter.org/humor/yarmulke.html.

Seach W.T. & Bishop, A.J. (2002, April). Researching for validity and value: the negotiation of cultural value conflicts by immigrant teachers in australia. Paper presented at the 83rd annual meeting of the American Educational Research Association, New Orleans, LA.

Shirley, L. (2006, February). Ethnomathematics in a global education program. Paper presented at the third International Congress of Ethnomathematics (ICEM-3), Auckland, New Zealand.

NOTES

Shirley, L. (2001). Ethnomathematics as a fundamental of instructional methodology. ZDM, International Reviews on Mathematics Education, 33(3), pp.85-87.

Strauss, A. & Corbin, J. (1998). \Basics of qualitative research: techniques and procedures for developing grounded theory. Thousand Oaks, CA: Sage.

Tal, S., (1990). "Bedouin Bridal Dress as a Social Indicator," The Negev Bedouin Woman in an Era of Change (In Hebrew). Retrieved March 31, 2008, from http://www.snunit.k12.il/beduin/arti/be5.html.

Tocker, J. (2005). Basic Knit Kippah (Yarmulka) in Reverse stocking stitch, Jennifer Tocker (2005). Retrieved March 14, 2008, from: http://www.geocities.com/jbtocker/patterns/basicknitkippah.pdf

Yiftachel, O. (2004). The making of an urban ethnocracy: jews and arabs in the beer-sheva region, Israel. (Revised version of a paper co-written with Haim Yacobi). Presented at the conference Urban Informality in the Age of Liberalization, Berkeley, April 2003.

Volume 4
Teacher Prep

Una Experiencia de Capacitación en Etnomatemática, en Docentes Indígenas Venezolanos

Oswaldo Jesús Martínez Padrón
Universidad Pedagógica Experimental Libertador-Venezuela
ommadail@gmail.com; Telf: 00584124794180

Resumen

En este documento se abordan detalles referidos a la capacitación en Etnomatemática que se concretó con la construcción de Proyectos pensados como materiales formativos y derivados del contexto donde los participantes del curso ejecutan su acción docente. En los Proyectos elaborados se exploró la Matemática usada fuera de la escuela en función de la construcción y uso de objetos ancestrales de su cultura, encontrándose que allí subyacen abundantes contenidos matemáticos que sirven de insumo para la concreción de actividades escolares sustentadas en tales contenidos. Se concluye que es importante que estos docentes no solo reconozcan la diversidad cultural correspondiente a su comunidad sino que valoren, analicen y promuevan la construcción de Proyectos que emerjan de las prácticas matemáticas desarrolladas por los diferentes grupos culturales, tomando en cuenta aspectos tales como los sistemas simbólicos involucrados, los métodos y sistemas de conteo, cálculo y medición, las formas especificas de razonamiento e inferencia y otros referentes que pueden traducirse a representaciones de la Matemática formal.

Palabras Clave:

- educación indígena
- etnomatemática
- proyectos de aprendizaje

Introducción

Este documento forma parte de una experiencia de capacitación pedagógica que se llevó a cabo en algunas poblaciones y comunidades indígenas asentadas en el estado Amazonas, Venezuela. En la mayoría de los casos, los participantes de dicha experiencia son originarios de comunidades indígenas tales como la **Hiwi**, **Piaroa Piapoco** y **Ye´kuana**, y laboran allí como docentes en las dos primeras etapas de la Educación Básica.

Dicha capacitación cubrió áreas tales como lengua, pedagogía indígena y etnomatemática, pero aquí sólo se precisan detalles en torno a la capacitación en Etnomatemática centrada en la construcción de Proyectos, los cuales fueron alumbrados por el uso de determinados objetos y prácticas como vías para recolectar, identificar y sistematizar los conocimientos y los saberes matemáticos ancestrales que le son propios a los pueblos y comunidades indígenas que fueron atendidas en esta oportunidad.

La Etnomatemática fue considerada según la concepción de D´Ambrosio (1985; 2005): un programa útil de investigación que permite explorar la Matemática usada fuera de la escuela.

Los Proyectos emergieron de la cotidianidad de los grupos donde pertenecen los docentes participantes del curso y los estudiantes que atienden. En este sentido, se tomó en cuenta el contexto sociocultural de la escuela y su entorno, los aspectos socioculturales de la clase y las particularidades de los docentes y sus estudiantes, considerando la procedencia de los diversos pueblos y comunidades indígenas. Eso obligó a tomar en cuenta alternativas para atender la diversidad cultural de las aulas y para evitar, según Oliveras (2005), las consideraciones monoculturales que siguen caracterizando a las actividades de la gran mayoría de las escuelas de todo el mundo. Por tanto, se abogó para que los Proyectos construidos bajo estas pautas sean considerados como aptos para el desarrollo del currículum en la educación intercultural. Desde esta perspectiva, se convierten en apoyo didáctico para la labor docente que se desarrolla atendiendo situaciones propias de los pueblos y comunidades indígenas.

Los docentes atendidos concretaron sus producciones en la construcción y uso de objetos propios que suelen utilizarse en prácticas relacionadas con el transporte (la Curiara), la caza (la Cerbatana), la cestería y el cernido (el Manare), y la ornamentación (La Cortina de Palitos), reportando que allí subyacen abundantes contenidos matemáticos que pueden servir de insumo para la organización de actividades escolares inspiradas en esos procesos. Sobre la base de estas especificaciones, se elaboró este documento que reporta, sucintamente, algunos elementos de sólo tres (3) Proyectos y sus consideraciones puntuales en torno a la matemática que subyace en cada una de ellos.

La Produción de los Participantes

Antes de describirse los Proyectos anunciados ha de destacarse que la selección del tema, actividad u objeto de estudio que tenga sentido en la cultura fue realizada por los participantes (docentes indígenas) en función del contexto donde se desarrollará el Proyecto. Posteriormente, se realizó una reflexión grupal y general al respecto a fin de establecer su adecuación y su pertinencia

NOTES

social. A partir de allí, se esbozaron algunas Fases, según su naturaleza, y se realizó un breve análisis didáctico del tema a fin de concretar su potencial de aprendizaje, en función de los contenidos matemáticos, y de otro tipo, que subyacen en la actividad.

Luego de las generalidades anteriores, se esbozaron los referentes que conformaron el Proyecto, sugiriendo las siguientes partes: (a) **Título del Proyecto**; (b) **Contexto de Referencia**: (b.1) Descripción del Tema; (b.2) Contexto Geográfico-Histórico; (b.3) Descripción de la Matemática Implícita en el Tema; (b.4) Glosario: incluye la traducción del término al o a los idiomas que hablan las comunidades atendidas en la clase; (c) **Propuesta Didáctica** que permita concretar las potencialidades del Proyecto. En este sentido, debe considerar los siguientes aspectos: (c.1) Duración, en horas/clase; (c.2) Audiencia; (c.3) Organización de los Participantes; (c.4) Objetivos Instruccionales; (c.5) Contenidos Escolares conectados con la Matemática Implícita en el Tema: conceptuales, procedimentales y actitudinales; (c.6) Competencias; (c.7) Ejes Transversales; (c.8) Estructuración de las Actividades escolares/extra-escolares, si es el caso, donde se incluya una secuencia de actuaciones por FASE o Componentes, si es que existen, y se consideren las conexiones con otras asignaturas; (c.9) Metodología; (c.10) Recursos; y (c.11) **Evaluación**

A continuación se hace una breve descripción de la producción de tres (3) de los grupos participantes en el curso de Etnomatemática, en función de algunos elementos de los contextos de referencia que conforman los Proyectos esbozados.

Proyecto 1: La Cerbatana

Figura 1

Contexto de Referencia:

Una cerbatana es un instrumento que suele ser usado por algunos indígenas para realizar la caza. Tradicionalmente, se fabrican a partir de una rama larga cuyo interior es vaciado cavando un orificio cilíndrico por donde deben desplazarse, con fuerza, dardos de unos 25 cm de largo (ver Figura 1). En la construcción de una cerbatana se observan procesos de medición de longitudes y se toman referents debidos al diámetro externo e interno de la rama de cerbatana. De igual manera, cuando este objeto es usado en situaciones de caza, aparecen variados contenidos matematizables debidos al ángulo de inclinación del lanzamiento, distancia recorrida, tiempo de desplazamiento, precisión y fuerza del lanzamiento para lograr el blanco, y otros aspectos físico- matemáticos debidos a la construcción y el uso del objeto.

Proyecto 2: El Manare

Figura 2

Contexto de Referencia

El Manare (cernidor) es un objeto tejido mediante un entrecruzamiento de tiras. Constituye una cesta usada para tamizar la harina de yuca amarga en el proceso de elaboración del casabe. Su forma es circular con un diámetro, aproximado, de 70 cm y con una cavidad cónica que en su centro alcanza hasta, aproximadamente, 5 cm de profundidad (ver Figura 2).

Las tiras que se entrecruzan durante el tejido del manare suelen tener el mismo ancho y la misma longitud, logrando que se configuren figuras geométricas caracterizables según se crucen en forma perpendicular u oblicua. También subyacen otros elementos matemáticos cuando se usan como decorativas, agregando en muchos casos figuras geométricas. En la conformación del aro que conforma su extremo se configura una circunferencia, y la figura cóncava de su cuerpo es vista, en el plano, como una figura circular

Proyecto 3: La Cortina de Palitos

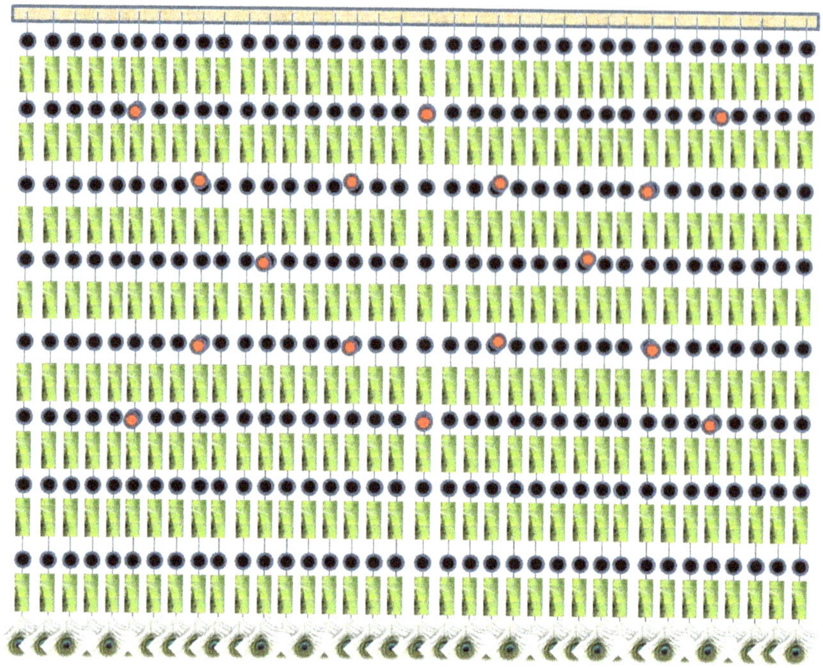

Figura 3: Diseño de una *Cortina de Palitos*

Contexto de Referencia:

Enre los grupos de trabajo, había uno conformado solamente por Hiwis que ejercen su rol de docentes en esas mismas comunidades. De acuerdo con la información reportada por ellos, en estas comunidades han aparecido nuevas artesanías con fines ornamentales y comerciales tal como lo es la **Cortina de Palitos** (Ver Figura 3), la cual está conformada por un conjunto de cuerdas colgadas de una faja de madera. En dichas cuerdas se insertan objetos decorativos que en este caso son semillas en forma de esferas y palitos en forma cilíndrica que se organizan en una secuencia "semilla-palito": (a) Las semillas son tomadas de un árbol llamado "parapara" y son de color negro; y (b) Los palitos son trozos de madera cilíndrica y su largo gira alrededor de los 10 cm cada uno.

La faja de madera es tomada de un arbusto llamado "bolla" y el grosor de este palo, que tiene forma de varilla, es aproximadamente de 10 cm de diámetro en toda su extensión, por lo que asemeja un cuerpo cilíndrico.

La longitud de dicha faja oscila entre 1 m y 1,30 m y es perforada, cada 5 cm, con orificios redondeados para colgar las cuerdas que son de nylon, insertándose en ellas las semillas y los palitos. En los extremos de cada cuerda, cuya longitude aproximada es de 1,60 m, se acostumbra colocarles plumas de aves para darle vistosidad. Tales plumas son tomadas de diferentes aves garantizando así diferentes colores.

Como se observa, en este caso se abordan abundantes contenidos matemáticos implícitos en la construcción de la cortina, los cuales se evidencian en los siguientes referentes Aritméticos-Geométricos: seriaciones (semilla- palito), medidas de longitude (altura de cada cuerda, separación entre ellas), Figuras (triángulos, rombos, círculos, circunferencias) y cuerpos geométricos (cilindros, esferas) y sus correspondientes elementos y medidas (segmentos paralelos, puntos equidistantes, altura, ancho, largo, diámetros, etc.). Además, si se mantiene la misma cantidad de semillas y palitos por cada cuerda, este objeto permite organizar procesos de conteo y adiciones, incluyendo las abreviadas (multiplicaciones).

A Manera de Cierre

La consideración de trabajar con la Etnomatemática en los espacios escolares es importante en la medida que permite integrar la Matemática con otras formas del conocimiento, utilizando así las prácticas de cada uno de los grupos culturales en los propios salones de clase. De esa manera, el trabajo se convierte en una actividad eminentemente práctica en función de las necesidades ambientales, afectivas, sociales y culturales que le son propias a los grupos. También, el aula de clase se convierte en un espacio propicio para producir conocimientos y construir saberes matemáticos no aislados de la dinámica cotidiana de los miembros de cada grupo cultural. Siendo así, el rol del docente deber estar signado por la presentación de situaciones-problemas que establezcan las condiciones necesarias para que cada uno de sus estudiantes gestione resoluciones en función del compendio de información cultural que posea, según sus experiencias de vida. En este sentido, el docente podría disponer de diferentes formas de abordar soluciones que deberán someterse a la discusión grupal, hasta concretar procedimientos más expeditos, y sus bondades, en la obtención de los resultados correspondientes.

En consecuencia, es importante que estos docentes no sólo reconozcan la diversidad cultural correspondiente a su pueblo o comunidad sino que valoren y analicen, críticamente, la realidad donde desarrollan sus acciones profesionales, lo cual conduce a incorporar una Educación Matemática relacionada con las diversas cosmovisiones, valores, saberes y prácticas de los grupos culturales. Para ello hay que propugnar la búsqueda de elementos matemáticos ancestrales y los lenguajes utilizados para la concreción y difusión entre sus miembros.

Las pretensiones anteriormente planteadas se pueden viabilizar mediante la construcción de Proyectos que emerjan de las prácticas matemáticas desarrolladas por los diferentes grupos culturales, los sistemas simbólicos involucrados, los diseños y técnicas de construcción, los métodos y sistemas de cálculo, de medición y de conteo, las figuras y cuerpos geométricos involucrados, las formas especificas de razonamiento e inferencia, las actividades cognoscitivas y metacognitivas asociadas, las acciones, los afectos y otros referentes que subyacen en las actividades y objetos que puedan permitir traducciones a representaciones de la Matemática formal.

Referencias

D'Ambrosio, U. (1985). *Ethnomathematics and its place in the history and pedagogy of mathematics.* For the Learning of Mathematics, 5(1), pp 44-48.

D'Ambrosio, U. (2005). *Etnomatemática. Elo entre as tradições e a modernidades.* Coleção Tendências em Educação Matemática. Brasil: Autêntica Editora.

Oliveras, M. L. (2005a). Microproyectos para la educación intercultural en Europa. *Uno: Revista de Didáctica de las Matemáticas*, N°. 38, 2005, p.p. 70-81.

Note: Imágenes compuestas por el autor con otras tomadas de http://www.unique-southamerica- travelexperience. com/images/yahua-blowgun.jpg y http://www. unique-southamerica-travel- experience.com /images/tucuna-darts.jpg.

Promoting Culturally Responsive Teaching through Action Research in a Mathematics Methods Course

Emily P. Bonner
The University of Texas at San Antonio Department of Interdisciplinary Learning and Teaching
San Antonio, TX
emily.bonner@utsa.edu

Abstract

Culturally responsive pedagogy is one way that teacher educators and teachers can address the unique sociocultural and cognitive needs of learners from diverse backgrounds while shifting away from traditional school practices and mathematics teaching methods. This article describes a teacher educator's inquiry project that aimed to encourage and enhance culturally responsive thoughts, discussions, understandings, and actions among pre-service mathematics teachers in a mathematics methods course for elementary school teachers. Results show that in constructing meaningful in and out of class experiences that focus explicitly on culture, pre-service teacher understandings of culturally responsive mathematics teaching are enhanced and deepened.

Introduction

As an educator, it is difficult and crystallizing to be objective about inequitable discrepancies that still exist in our schools and classrooms. It is undeniable, however, that students of color are experiencing continued struggles in instructive settings. Overall, the number of students of color in American schools is rapidly increasing (NCES, 2009) while rates of achievement among these students have reached critically low levels (Lee, 2006).

Existing research signifies that student achievement can increase when teachers utilize critical knowledge about students' backgrounds in classroom instruction (Gay, 2010; Nieto, 2010; Sheets, 2005; Howard, 2006). This foundational theory that culture plays a significant role in teaching and learning is becoming increasingly important as the student body in America continues to diversify (currently over 40% of students are non-white) while the teaching force continues to maintain homogeneity, hovering around 83% white (National Center for Education Statistics [NCES], 2008). These statistics highlight cultural incongruities, which are evidenced by staggering achievement gaps that continue to exist between students of color and their white counterparts (NCES, 2009), that often exist between students and teachers, especially in mathematics.

Ultimately, statistics show that students who perform from a European frame of reference outperform students who do not in mathematics (NCES, 2009). Traditional practices, in fact, are so engrained in the everyday complexities of mathematics teaching that such systems become "normal", keeping many teachers from recognizing the "critical connection between culture and schooling" (Sheets, 2005, p. 3) and the strong ties between culture and cognition. As mathematics teacher educators, it is imperative that we encourage educators to recognize these links while gaining new insights into their own beliefs and norms that might drive pedagogical decisions. Moreover, in allowing and encouraging our students to discuss uncomfortable realities that exist in today's society and schools (Tatum, 1997), we arm them with ideas and tools that might help them to work equitably with different types of students.

Culturally responsive teaching (CRT; Gay, 2000) provides a framework within which such action can occur, as the needs of diverse groups of learners are at the forefront. Further, in the classroom, culturally responsive teachers provide a tangible (rather than theoretical) model for pre-service teachers, explicitly embracing diversity and using it as a means for growth and learning (Nieto, 2002). In providing pre-service teachers with such frameworks and models, mathematics teacher educators can more readily illustrate how to "[use] students' culture in order to maintain it and to transcend the negative effects of the dominant culture" (Ladson-Billings, 1994, p. 17). Thus, we (two mathematics teacher educators) embarked on an inquiry of our mathematics methods course, hoping to increase our pre-service teachers' culturally responsive propensities in the context of mathematics.

Several working programs that aim to prepare teachers for diversity, such as the Center for Urban Learning/Teaching and Urban Research in Education and Schools (CULTURES; Irvine, 2003) and the Teach for Diversity Program (TFD; Ladson-Billings, 2001) have been discussed in the

NOTES

literature. These are exemplary examples of entire programs devoted to the preparation of culturally responsive teachers. We support these efforts and have readily gleaned ideas from these programs; however, similar programs do not exist at our institution. Therefore, this is a discussion of our first attempt to incorporate culturally responsive mathematics practices in our K-6 mathematics methods course (which exists within a largely traditional mathematics education program) to encourage culturally responsive thoughts, discussions, and actions among pre-service mathematics teachers. This inquiry into our own teaching methods provided great insights; we will note what worked, what did not work, and what we can do in the future to encourage cultural responsiveness in mathematics instruction. We feel that this discussion is important in learning about the beginnings of culturally based programs.

The Setting

At the time of this study, we (two white, middle class females) were beginning to explore issues of diversity and equity in our own teaching, and were becoming more critical of social structures that exist in schools and universities. We were seeing many inequitable institutional and systemic perceptions and practices among our pre-service teachers, and sought to change the discourse in our classroom to challenge such beliefs. We also hoped to capitalize on the fact that the majority of our students each semester shared our cultural background. While this homogeneity is a drawback in that perspectives in the classroom are limited, we recognized that we could not immediately change our enrollment, and committed to use our shared cultural views as an opportunity to connect with students through uncomfortable conversations about histories, perceptions, and practice. Our desire was to use these connections to push students' thinking about diversity and equity in mathematics education, ultimately impacting their future practice as elementary school mathematics teachers.

We believe that mathematics teacher education is a challenging, complex, and fascinating field. Mathematics anxiety, negative feelings about mathematics, and resistance are common self-reported issues among the pre-service elementary school teachers with whom we work. These novice educators are required to take our mathematics methods course during their fourth year of undergraduate work. Often, their discomfort with mathematics content is obvious, and they report that they are not yet comfortable teaching mathematics content in pedagogically and developmentally sound ways. Further, our course is not coupled with a field placement, making in-class experiences the sole basis of the student experience.

When students come to us, they have already had several experiences with children of color and children in high poverty situations. For example, during a student's sophomore year, he or she is required to work with and mentor one student who lives in and attends school in a high-poverty area. This interaction is largely individual, with the pre-service teacher traveling to meet with the student for various activities. Further, prior coursework for pre-service teachers included aspects of social justice and CRT in a general context, though such information is somewhat piecemeal. Moreover, most of our students had not yet had the opportunity to explore CRT in a mathematics or other classroom situation in the midst of teaching.

Over several semesters of teaching this course, we noticed that many students had general conceptions of CRT, but did not apply these understandings to lesson plans, peer teaching activities, or other written assignments. Rather, they viewed content and CRT separately from instructional planning and implementation, and would simply add culturally-based modifications onto traditional lesson plans. As such, we wanted to provide opportunities for students to integrate CRT into mathematics teaching while exploring issues of social justice. Ultimately, our inquiry was driven by this desire to change students' perspectives about equity and CRT in the context of mathematics over one semester.

Pre-service teachers at our institution report themselves to be largely female and white. This is certainly not a steadfast rule, but is the main population with whom we have worked. On average, our courses contain 25 – 30 students, about 85 - 90% of who fit this profile. While conducting this inquiry, each of us taught one section of the course. In section A, 93% of students were white females, and 7% of students were African American females. In section B, 96% of students were white females, 2% were white males, and 2% were African American females. Socioeconomically, 98% of our students reported that they identify as members of the middle class no matter what their race. With very few exceptions, our population of students is not diverse.

Integrating Culture into Mathematics Methods

We wanted to incorporate several different types of experiences into our course that would build awareness of the need for CRT in mathematics, and simultaneously provide opportunities to explore pedagogical responses to such issues in the context of the mathematics classroom. Our research was guided by the following questions:

1. Do our students come to our mathematics methods course with prior knowledge of CRT and social justice?
2. Do the culturally based experiences that we have constructed affect this knowledge base in any way?
3. What, if any, changes in student discourse about CRT occur over the course of one semester in our mathematics methods course?

In an effort to answer these questions, we collaboratively designed a pre- and post-survey, and a series of ongoing experiences that we believed would require that students focus explicitly on equity issues in the context of mathematics. These experiences included critical journaling, video analyses, attendance at cultural events, in-class discussions, selected critical readings, and modeling. The pre- and post-surveys were used to gauge the extent to which individual student perspectives changed over the course of the semester.

In order to provide a clear picture of this process, each experience will be discussed separately (except for selected readings, as these are self-explanatory and occurred on a continuous basis). It is important to note, however, that these events were integrated holistically and overlapped continuously. For example, students may have been asked to journal about a video clip or to comment in class on a reading. Thus, students' engagement in these activities was not distinct; rather, it was integrated and intense.

Pre-survey

During the first week of the semester we wanted to ensure that our culturally based activities were situated appropriately, and we hoped to get a more specific idea of who our students were as a whole. In order to gain information about students' prior knowledge and experiences, we designed several written activities for the first day of class. First, we administered a pre-survey (see Appendix I) to gather information about students' perspectives on educational strategies, the influence of culture in the classroom, and student learning in general. Further, in a separate assignment, we asked students to write about their prior educational, mathematical, and cultural experiences. This gave us a good starting point from which to implement our culturally relevant strategies.

We found that student responses on the pre-survey showed a basic understanding CRT, but with a superficial overtone. For example, a common response to question two was, "teachers should acknowledge all students' cultures." While this may be true, there is little substance to the statement alone, and no implication that social action should be an integral part of teaching diverse learners. Another student said, "I plan to do a lesson each month on a different culture." While this shows awareness of the issue of culture in the classroom, the respondent fails to realize that by giving a culture a special month, he or she is further marginalizing the group in question. Overall, we got a sense that most of our students had an 'us - them' mentality; "others" should be acknowledged, but perhaps not fully integrated into a classroom.

Students also showed a disengagement from sociopolitical issues that surround CRT. Again, there was an awareness of school and home culture, but students did not see this as a tool through which to teach. Rather, they worried, that "certain practices in our culture…may not be welcome in other cultures that just send their kids to school and expect the teachers to handle everything." Further, "some families don't value education so…the child doesn't do well in school." Our goal was to help these future teachers address these issues so that cultural diversity in the mathematics classroom is an asset instead of a burden.

Journaling

Students engaged in two overlapping levels of journaling – group and individual. Group journals were completed online on a weekly basis. Using a discussion board, each student was required to respond to a prompt posted by the teacher (see Appendix II for sample journal prompts). Since we incorporated cultural issues into our lessons throughout the semester, we mainly crafted prompts that required that students reflect on class discussions or readings that may have caused discomfort. We also encouraged students to use the discussion board collaboratively, thinking "aloud" about new ideas relating to CRT in an informal way.

In addition to posting a response each week, students were asked to log in before class and read other students' replies and responses. In requiring this, we were able to discuss responses and conversations of note in class. We wish to note that responses were short and tentative at the beginning of the semester, yet over time students began to engage in substantive online discussions, and many students often became so engaged in the discussion that they began posting numerous times each week. Other students did not fully engage, viewing the online discussions simply as another weekly assignment, but most students seemed to benefit from group journaling.

Students were also asked to journal independently throughout the semester. In order to allow students freedom to express honest thoughts without judgment, the instructors only read selected entries. Generally, students were asked to reflect on class discussions or readings during sessions that occurred at the end of the class period. Overall, the individual journals proved to be fruitful data for the researchers, though some students did not choose to reflect unless the entry was being collected.

Class Discussions

Class discussions seemed to fuel and supplement the journaling process. Over time, these two activities became continuations of one another. While most in-class discussions were deliberate and structured, students were encouraged to speak freely and openly. To encourage discourse, the instructor would periodically interject an opinion or common perception to stir up the conversation. Over time, the instructors became less and less involved, allowing students to determine the tone and direction of the discussion.

Modeling and Video Analysis

For our purposes in this project, we modeled culturally responsive mathematics instruction in two ways. First, we showed videotape of a culturally responsive mathematics teacher who was well known in our geographical area. The videos were created through a research project involving a fifth grade elementary school teacher, Ms. Foster[1], with whom we had been working for two years. During that time, our team of researchers videotaped her teaching at different points in the school year.

Ms. Foster worked with a largely African-American, high poverty population. Her students, despite "at risk" labels and previous failures on standardized tests, were extremely successful mathematicians as evidenced by state standardized tests, success rates in higher grades, and other qualitative analyses. Thus, we viewed her methods as a model of exemplary culturally responsive mathematics instruction. In specific, Ms. Foster's instruction was guided largely by student interests. She consistently used chants, rhythm, repetition, storytelling, music, and movement in her classroom. Furthermore, her involvement in the community had an impact on the children she taught. Ms. Foster truly embodied culturally responsive mathematics teaching.

As a class, we deconstructed Ms. Foster's methods, discussed ways in which her ideas could be used in elementary school classrooms, and why such instruction is important. Notably, the conversations that ensued among our students after a viewing of the video were extremely fruitful and profound. Secondly, we as the instructors of the course practiced what we preached. That is, we used the somewhat limited diversity in our own student population as a vehicle through which significant mathematics learning could take place.

Cultural Events

Another component of our course required each student to attend a cultural event of his or her choosing (see Appendix III for an outline of the assignment). The most important piece of this experience was the writing component. It was there that students were required to reflect on the event, compare it to their own cultural experiences, and discuss implications for their classroom. We found that many students gained insight into a culture other than their own through attending these events. We were impressed with the level of reflection that was evident in most student papers, and many of these reflections became topics of class discussion. The following are examples of events that students attended:

[1] A pseudonym has been used

NOTES

- Gay/Lesbian Pride Week Events
- Ramadan services at the Muslim student center
- Churches that are different from the one in which they were brought up
- A study abroad experience in South Africa
- Community picnics in high-poverty areas

Results

On the last day of class, we asked students to complete the same survey that was administered at the beginning of the semester. Overall, we saw some progressive thoughts emerge about teaching mathematics to diverse children and children living in poverty at the end of the semester. Using domain analysis, we grouped these responses into three major categories that are illustrative of students' depths of understanding about culture and mathematics: shared responsibility, a more sophisticated view of the role of culture, and continuous integration of culture into schools.

Shared Responsibility

The most common category of response related to views of community practice as an integral part of education. The "us-them" mentality we encountered at the beginning of the semester had morphed into an "us and them" attitude, with many students viewing cultural diversity and communication with the community as benefits (rather than struggles they had to overcome) in the classroom.

While some distinctive discourse was still present, most notably with respect to race and class, we considered the overall response pattern to be indicative of progress. For example, one student stated that "by having open communication teachers, students, and parents can all work together for the benefit of the child." Further, students saw their own role in a more progressive way, stating that "teachers must address their own stereotypes and expectations for certain students before they can provide them with a quality education." This ownership and sense of shared responsibility is in distinct opposition to the attitudes reported at the beginning of the semester.

Sophistication of Cultural Understandings

The discussions about culture in our classrooms grew more sophisticated as the semester came to a close. Students began to recognize that home culture and school culture are often incongruous. Moreover, we saw an increased awareness of sociopolitical issues surrounding culture and their impact on schools and classrooms. One student suggested that "[the achievement gap] exists because of the determining factors of each group. Limited English proficiency for Hispanic/Latino students, for instance, places these students at a disadvantage in school" and "students that live in poverty level neighborhoods are more likely to attend poor and under-funded-schools [sic] and thus are at a disadvantage as well." This type of response puts more responsibility on the system that has wronged the student rather than on the student himself.

Continuous Integration of Culture in Academics

Our students seemed to gain a sense of the integration that is necessary when using culturally responsive techniques. For instance, "...it is important to have a multicultural curriculum, not just celebrations and foods." Also, "it is important to use multicultural curricula and instruction practices... multicultural literature and awareness of social issues and concerns should be addressed as such." Thus, students recognize that culture should be utilized as a tool in instruction through which tremendous multicultural growth is possible. Further, students indicated that cultural contexts should infuse and alter all aspects of classroom practice, and should not be treated as supplementary material.

Discussion

We felt that significant progress was made in our classrooms with regards to culturally responsive mathematics instruction. It is important to note, however, that we treated this semester as a true inquiry and not a scientific experiment. In other words, we were not expecting to see a profound transformation that we could attribute to our course; rather, we wanted to make our pre-service teachers aware of cultural issues with which they will come in contact. The building of these experiences, then, was our first attempt at bringing these issues to the forefront of all that our students do.

There were several other limitations that should be noted. First, it was difficult for us to know if our pre-service teachers were truly thinking differently and critically about marginalized students or if they were simply saying and writing what they thought we wanted to hear. Further, it is impossible to determine if any of these experiences will transfer into the classroom and ultimately transform their teaching. This is especially relevant since we were providing these experiences in an academic setting rather than in a classroom or field experience; theory may not always translate easily into practice. It is possible that when these students begin teaching, learning to teach and survival will be so consuming, that cultural considerations will become less important. It would be helpful for our students to have a field placement as a part of our course so that some of these ideas could be tried out in a safe, supportive environment. Culturally responsive mentor teachers would also greatly benefit our students. Moreover, we believe that the transformations would be more profound if all teacher educators across our program adopted similar strategies.

Though we may not be able to solve any major social problems in one course in one semester, we did find that incorporating these experiences into our coursework was beneficial to our pre-service teachers. Further, we encourage other mathematics teacher educators to develop similar experiences with similar goals in mind. We believe that our students now have some ideas about the role that culture plays in the classroom, and have had some difficult, important discussions about sociopolitical issues in schools. In starting such conversations with our students, we can help them to be successful mathematics educators in diverse classrooms.

NOTES

References

Brown, D. (2003). Urban teachers' use of culturally responsive strategies. *Theory Into Practice, 42*(4), 277-282.

Gay, G. (2000). *Culturally responsive teaching: Theory, research, and practice*. New York: Teachers College Press.

Gay, G. (2002). Culturally responsive teaching in special education for ethnically diverse students: Setting the stage. *Qualitative Studies in Education, 15*(6), 613-629.

Gay, G. (2010). *Culturally responsive teaching: Theory, research, and practice (2nd Edition)*. New York: Teachers College Press.

Howard, G. R. (2006). *We can't teach what we don't know: White teachers, multiracial schools* (2nd Ed.). New York: Teachers College Press.

Irvine, J. J. (2003). *Educating teachers for diversity: Seeing with a cultural eye*. New York: Teachers College Press.

Ladson-Billings, G. (1994). *The dreamkeepers: Successful teachers of African American children.* San Francisco: Jossey Bass.

Ladson-Billings, G. (1997). It doesn't add up: African American students' mathematics achievement. *Journal for Research in Mathematics Education, 28*(6), 697-708.

Ladson-Billings, G. (2001). *Crossing over to Canaan: The journey of new teachers in diverse classrooms*. San Francisco: Jossey Bass.

Lee, C. D. (1995). A culturally based cognitive apprenticeship: Teaching African American high school students skills in literary interpretation. *Reading Research Quarterly, 30*(4), 608-630.

Lee, J. (2006). *Tracking achievement gaps and assessing the impact of NCLB on the gaps: An indepth look into national and state reading and math outcome trends*. Cambridge, MA: The Civil Rights Project at Harvard University.

Lortie, D. C. (1975). *Schoolteacher; a sociological study*. Chicago: University of Chicago Press.

National Center for Education Statistics (2009). *Participation in Education – Indicator 3 – Racial/ ethnic distribution of public school students*. Washington, D.C.: U.S. Department of Education.

Nieto, S. (2002). Profoundly multicultural questions. *Educational Leadership, 60*(4), 6-11.

Nieto, S. (2010). *The light in their eyes: Creating multicultural learning communities (10th anniversary edition)*. New York: Teachers College Press.

Sheets, R. H. (2005). *Diversity pedagogy: Examining the role of culture in the teaching- learning process*. Boston: Pearson.

Tatum, B. (1997). *Why are all the black kids sitting together in the cafeteria? And other conversations about race*. New York: Basic Books.

Townsend, B. (2003). "Testing while black." Standards-based school reform and African American learners. *Remedial and Special Education, 23*(4), 222-230.

Appendix I - Pre-survey

Please answer each question briefly on a separate sheet of paper.

This is for my information only. THERE ARE NO RIGHT OR WRONG ANSWERS.

1. Many student attributes contribute to student learning in the classroom. What do you feel are the most important of these things? How do they contribute to student learning? Briefly discuss your answer.

2. What role do you think that culture plays in the classroom? How might this affect student learning?

3. What types of specific practices can you use as a teacher to make sure that you are respecting and encouraging learning through student cultures in your classroom?

Appendix II - Sample Journal Prompts

1. Briefly discuss and describe the overall cultures of the schools that you attended (elementary, middle, high schools) before college. Include information pertaining to demographics, socioeconomic status, social cliques, and other influences that you feel were important in shaping your experience. What effect did the schools' cultures have on your own educational and social formation and development?

2. Using the internet and what you might already know, see what you can find out about culturally responsive teaching. What is it? How is it helpful/useful to educators and students? What ideas might you have for instruction that are culturally responsive?

3. Think about the positive and negative aspects of homogeneous grouping/tracking. Do you think that tracking is beneficial? Do you think that there is value in heterogeneous grouping? Why or why not? Defend your position.

4. Do short-term cultural celebrations (such as Black History Month) help or hinder the promotion of equity in schools? Defend your position.

Appendix III - Cultural Event Paper Assignment

Opening Your Cultural Eye

1. Attend a cultural event of your choosing. The event should be something that you would not normally experience in your own culture. For example, if you are Christian and go to a Jewish service (or vice versa), that would work. However, going to the Krishna lunch on campus might not be enough because of the very limited interaction that is required.

2. Write an essay (2-3 pages, double-spaced) in which you address the following points:

   a. Describe the event you attended.

   b. Describe a comparable event you have experienced in your own culture.

   c. Compare and contrast the two events.

   d. Discuss the educational value of this experience (e.g., as an educator, how do you feel that attending a cultural event will help your practice?)

   e. Any other ideas of relevance that you feel are important.

Culture *That* Counts

Index

Author

Keywords

Culture *That* Counts

Author Contact Information

updates as of April 2019

Thomasenia Lott Adams
Professor
Associate Dean for Research
& Faculty Development
University of Florida
tla@coe.ufl.edu

Jim Barta, PhD
Professor & Dean
Bemidji State University
College of Art, Education, and Humanities
jim.barta@bemidjistate.edu

Emily P. Bonner, PhD
Associate Professor
University of Texas at San Antonio
emily.bonner@utsa.edu

Iman Chahine, PhD
Associate Professor
University of Massachusetts, Lowell
Iman_Chahine@uml.edu

Karen François, PhD
Professor
VRIJE Universiteit Brussel
Center for Login and Philosophy of Science
Karen.Francois@vub.ac.be

Linda H.L. Furuto, PhD
Professor of Mathematics Education
University of Hawaiʻi at Mānoa
lfuruto@hawaii.edu

Thomas Gilsdorf, PhD
Professor & Chair
Central Michigan University
Department of Mathematics
gilsd1te@cmich.edu

Nicholas J. Goetzfridt, PhD
Emeritus Professor of Library
Science and Micronesian Studies
University of Guam
ngoetzfr@gmail.com

Jenni L. Harding, EdD
Professor
University of Northern Colorado
Jenni.Harding@unco.edu

Gregory K. Harrell, PhD
Professor of Mathematics
Valdosta State University
gharrell@valdosta.edu

Ada Katsap, PhD
Senior Lecturer
Kaye Academic College of Education
Beer-Sheva, Israel
adak@013.net

Brendan Larvor, DPhil
Reader in Philosophy
University of Hertfordshire
b.p.larvor@herts.ac.uk

Lawrence M. Lesser, PhD
Professor
The University of Texas at El Paso
Lesser@utep.edu

Tamsin Meaney, PhD
Professor
Western Norway University
of Applied Sciences
Tamsin.Jillian.Meaney@hvl.no

Bethany Noblitt, PhD
Associate Professor
Northern Kentucky University
noblittb@nku.edu

Daniel C. Orey, PhD
Professor
Universidade Federal de Ouro Preto
Departamento de Educação Matemática
Instituto de Ciências Exatas e Biológicas
oreydeema@gmail.com

Oswaldo Jesús Martínez-Padrón, PhD
Docente-Investigador
Universidad Técnica del Norte- Ecuador
ommadail@gmail.com

Pedro Palhares, PhD
Associate Professor with Aggregation
Doutor
CIEC, Institute of Education
University of Minho
palhares@ie.uminho.pt

Javier Díez-Palomar, PhD
Associate Professor
University of Barcelona
jdiezpalomar@ub.edu

Blaire Richter
blairerichter@cinci.rr.com

Milton Rosa, EdD
Adjunct Professor IV
Universidade Federal de Ouro Preto
milrosa@hotmail.com

Janet Sharp, PhD
Professor of Mathematics Education
Washburn University
janet.sharp@washburn.edu

Tod Shockey, PhD
Professor
University of Toledo
Tod.shockey@utoledo.edu

Fredrick Silverman, EdD
Professor Emeritus
University of Northern Colorado
flsilver@gmail.com

Anthony Stevens, MS/ABD
Group Leader, Psychology Department
Des Moines Area Community College
acstevens@dmacc.edu

Anthony Stevens
Percussionist
The Bone People
acsdrums@aol.com

Tony Trinick, EdD
Associate Professor
University of Auckland
t.trinick@auckland.ac.nz

Molly Tun, PhD
Assistant Professor
St. Olaf College
tun1@stolaf.edu

Amy E. Wagler, PhD
Associate Professor
The University of Texas at El Paso
awagler2@utep.edu

www.ingramcontent.com/pod-product-compliance
Lightning Source LLC
LaVergne TN
LVHW071342210926
848486LV00003B/210
* 9 7 8 1 9 3 9 0 4 4 5 0 1 *